W9-BVH-737

PROPERTY OF FUNK SEEDS INTERNATIONAL, INC.

NAME _____

ANNUAL REVIEW OF PHYTOPATHOLOGY

EDITORIAL COMMITTEE (1976)

K. F. BAKER

E. B. COWLING

R. G. GROGAN

R. M. LISTER

J. TAMMEN

H. D. THURSTON

P. H. WILLIAMS

G. A. ZENTMYER

Responsible for the organization of Volume 14
(Editorial Committee, 1974)

K. F. BAKER

E. B. COWLING

T. O. DIENER

J. F. FULKERSON

R. G. GROGAN

H. D. THURSTON

P. H. WILLIAMS

G. A. ZENTMYER

R. J. COOK (Guest)

M. SHAW (Guest)

Assistant Editor   T. HASKELL
Indexing Coordinator   M. A. GLASS
Subject Indexer   D. C. HILDEBRAND

ANNUAL REVIEW OF PHYTOPATHOLOGY

KENNETH F. BAKER, *Editor*
The University of California, Berkeley

GEORGE A. ZENTMYER, *Associate Editor*
The University of California, Riverside

ELLIS B. COWLING, *Associate Editor*
North Carolina State University

VOLUME 14

1976

ANNUAL REVIEWS INC.   4139 EL CAMINO WAY   PALO ALTO, CALIFORNIA 94306

ANNUAL REVIEWS INC.
Palo Alto, California, USA

COPYRIGHT © 1976 BY ANNUAL REVIEWS INC., PALO ALTO, CALIFORNIA.
ALL RIGHTS RESERVED. No part of this book may be reproduced in any form or by any means without permission in writing from the publisher.

International Standard Book Number: 0-8243-1314-3
Library of Congress Catalog Card Number: 63-8847

Annual Reviews Inc. and the Editors of its publications assume no responsibility for the statements expressed by the contributors to this Review.

REPRINTS

The conspicuous number aligned in the margin with the title of each article in this volume is a key for use in ordering reprints. Available reprints are priced at the uniform rate of $1 each postpaid. The minimum acceptable reprint order is 10 reprints and/or $10.00, prepaid. A quantity discount is available.

PRINTED AND BOUND IN THE UNITED STATES OF AMERICA

PREFACE

From its inception, the *Annual Review of Phytopathology* has sought to present critical analyses and syntheses of specialized aspects of our discipline, written by recognized authorities. As so lucidly stated by J. E. Vanderplank in the prefatory chapter of this volume, it is our role to re-view a subject rather than merely to survey the literature on it.

Plant pathologists use the Review as a means of keeping abreast of subjects outside their specialities, rather than to learn of new work in their chosen fields. Investigators learn of the research progress of colleagues through oral communication at Society meetings, conferences, symposia, congresses, by telephone, and during frequent visits to pertinent laboratories. Ideas now travel faster than publications; the international communication network is today more highly developed than ever before. However, the narrowing specialization demanded by modern science often leaves the investigator uninformed of discoveries in furrows being plowed by colleagues. It is not enough to scan abstracts of the flood of publications outside one's specialty, even if it were possible to do so; one needs the analytical syntheses of an expert to highlight significant progress and to show the remaining gaps. This is a major reason for the increase in the number of review journals since the late 1940s, and for a corresponding decline of abstracting journals.

Several points derive from these considerations: 1. Review articles should be written by authorities. In planning a volume of *Annual Review of Phytopathology* we select significant topics that are ready for analysis, and then find the best available authorities to explain them lucidly to the profession in nonspecialist language. The selection is as objective as we can make it, but necessarily is based on information available to the Editorial Committee. 2. Monographic presentations of a given disease, or the diseases of a given crop, are considered to be outside the scope of *Annual Review of Phytopathology* unless some fundamental process of wide significance is analytically discussed. 3. Because these reviews are often used by investigators and students as a point of entry into a topic, it is important that references cited be accurate and sufficiently detailed that they can be obtained on interlibrary loan if necessary. It thus requires that authors and editors carefully check references for accuracy and completeness. If a bibliography is excessively long, the core references may be presented in the review, and the additional citations are filed with the National Auxiliary Publications Service, New York City, and may be purchased by interested readers.

Within these limits, the Editorial Committee welcomes suggestions of both topics and authors for consideration at its annual spring meetings.

We welcome J. F. Tammen as a new member of the Editorial Committee replacing J. F. Fulkerson who concludes five years of valuable service; we thank R. J. Cook and Michael Shaw, the guest committeemen who helped plan this volume.

THE EDITORIAL COMMITTEE

CONTENTS

SOME RELATED ARTICLES IN OTHER ANNUAL REVIEWS

From the *Annual Review of Biochemistry,* Volume 44, 1975
Restriction Endonucleases in the Analysis and Restructuring of DNA Molecules,
D. Nathans and H. O. Smith, 273–93
Chemotaxis in Bacteria, J. Adler, 341–56
*Methods for the Study of the Conformation of Small Peptide Hormones and
Antibiotics in Solution,* L. C. Craig, D. Cowburn, and H. Bleich, 477–90
Virus Assembly, S. Casjens and J. King, 555–611

From the *Annual Review of Ecology and Systematics,* Volume 6, 1975
Experimental Studies of the Niche, R. K. Colwell and E. R. Fuentes, 281–310
Simulation Models of Ecosystems, R. G. Weigert, 311–38

From the *Annual Review of Entomology,* Volume 21, 1976
Status of the Systems Approach to Pest Management, W. G. Ruesink, 27–44
The Economics of Improving Pesticide Use, R. B. Norgaard, 45–60
Biochemical Genetics of Insecticide Resistance, F. W. Plapp, 179–97
Genetic Problems in the Production of Biological Control Agents, M. Mackauer,
369–85

From the *Annual Review of Genetics,* Volume 9, 1976
Compartmentation and Regulation of Fungal Metabolism: Genetic Approaches,
R. H. Davis, 39–65

From the *Annual Review of Microbiology,* Volume 29, 1975
Determination of Cell Shape in Bacteria, U. Henning, 45–60
Ecological Implications of Metal Metabolism by Microorganisms, A. Jernelöv
and A.-L. Martin, 61–77
Regulation and Genetics of Bacterial Nitrogen Fixation, W. J. Brill, 109–29
Bacteria as Insect Pathogens, L. A. Bulla, Jr., R. A. Rhodes, and G. St. Julian,
163–90
Patents in Relation to Microbiology, E. S. Irons and M. H. Sears, 319–32
The Genus Agrobacterium and Plant Tumorigenesis, J. A. Lippincott and
B. B. Lippincott, 377–405
Interaction of Temperature and Psychrophilic Microorganisms, W. E. Inniss,
445–65
The Genetics of Dissimilarity Pathways in Pseudomonas, M. L. Wheelis, 505–24

From the *Annual Review of Plant Physiology,* Volume 27, 1976
Hormone Binding in Plants, H. Kende and G. Gardner, 267–90
Root Hormones and Plant Growth, J. G. Torrey, 435–59
Freezing and Injury in Plants, M. J. Burke, L. V. Gusta, H. A. Quamme,
C. J. Weiser, and P. H. Li, 507–28
Phytohemagglutinins (Phytolectins), I. E. Liener, 291–319

ANNUAL REVIEWS INC. is a nonprofit corporation established to promote the advancement of the sciences. Beginning in 1932 with the *Annual Review of Biochemistry*, the Company has pursued as its principal function the publication of high quality, reasonably priced Annual Review volumes. The volumes are organized by Editors and Editorial Committees who invite qualified authors to contribute critical articles reviewing significant developments within each major discipline.

Annual Reviews Inc. is administered by a Board of Directors whose members serve without compensation.

BOARD OF DIRECTORS
1976

Dr. J. Murray Luck
Founder Emeritus, Annual Reviews Inc.
Department of Chemistry
Stanford University

Dr. Esmond E. Snell
President, Annual Reviews Inc.
Department of Biochemistry
University of California, Berkeley

Dr. Joshua Lederberg
Vice President, Annual Reviews Inc.
Department of Genetics
Stanford University Medical School

Dr. William O. Baker
President
Bell Telephone Laboratories

Dr. James E. Howell
Graduate School of Business
Stanford University

Dr. William D. McElroy
Chancellor
University of California, San Diego

Dr. Wolfgang K. H. Panofsky
Director
Stanford Linear Accelerator Center

Dr. John Pappenheimer
Department of Physiology
Harvard Medical School

Dr. Colin S. Pittendrigh
Department of Biological Sciences
Stanford University

Dr. Alvin M. Weinberg
Director, Institute for Energy Analysis
Oak Ridge Associated Universities

Dr. Harriet Zuckerman
Department of Sociology
Columbia University

Annual Reviews are published in the following sciences: Anthropology, Astronomy and Astrophysics, Biochemistry, Biophysics and Bioengineering, Earth and Planetary Sciences, Ecology and Systematics, Energy, Entomology, Fluid Mechanics, Genetics, Materials Science, Medicine, Microbiology, Nuclear Science, Pharmacology and Toxicology, Physical Chemistry, Physiology, Phytopathology, Plant Physiology, Psychology, and Sociology. In addition, two special volumes have been published by Annual Reviews Inc.: *History of Entomology* (1973) and *The Excitement and Fascination of Science* (1965).

J. E. Vanderplank

Copyright © 1976 by Annual Reviews Inc.
All rights reserved

FOUR ESSAYS   ♦3629

J. E. Vanderplank
Plant Protection Research Institute, Department of Agriculture, Pretoria, South Africa

I was pleased to be asked by the Editor to write the prefatory chapter to this volume. It is a great honor for me. Also, the *Annual Review* is my favorite reading in phytopathology, and I am sure it is many others' too. So I have written some essays which I hope might inform and divert. The first three ought to be read together. They might even start research on a new line of defense against disease. Who knows?

A SCIENCE FICTION ESSAY

A man at his bench is treating oat cell protoplasts with a concentrated solution of DNA. The DNA comes from corn from a gene for susceptibility to *Puccinia sorghi*. From these cell protoplasts he hopes to culture a flowering oat plant homozygously immune from *P. graminis* and *P. coronata,* and still with oat's immunity from *P. sorghi*. This plant will be the start of a new oat variety safe from rust.

Ledoux et al (8) treated seeds of *Arabidopsis thaliana* with foreign DNA, and corrected the cells for thiamine deficiency. *Arabidopsis* is a small flowering plant, and Ledoux et al used mutant lines with lesions in the branched pathway for thiamine synthesis. These mutant lines can be grown to maturity if they are fed thiamine. But if thiamine is withheld, plants grown on a mineral medium die within 15 to 20 days. Treating the ungerminated seed during imbibition with a solution of appropriate bacterial DNA remedied this. From the treated seeds a few plants were obtained that grew satisfactorily without added thiamine. The observed frequency of correction was far higher than the spontaneous reversion rate, and in any case the corrected plants could be distinguished phenotypically from normal plants without lesions for thiamine synthesis. The corrected plants were homozygous, and there was no segregation on selfing in the progenies of the corrected lines. All F1 to F3 plants originating from each corrected plant looked phenotypically alike. Reciprocal crosses between corrected mutants showed that the correction could be transmitted through the male as well as the female, which excludes maternal cytoplasmic inheritance.

1

It appears that the foreign DNA was translocated long distances across cell membranes to the nucleus where it became covalently linked to the recipient DNA as double-stranded material.[1] Other evidence suggested that the thiamine-synthesizing information was added to the genome without replacing the mutant gene. It is too early to decide whether the foreign DNA remained outside the chromosomes as a very stable form of exosome or was integrated within a chromosome, but there is nothing yet to suggest that this would be relevant to the propagation of homozygous plants.

Ledoux et al introduced foreign heterologous DNA from bacteria into a flowering plant. There is also evidence for the incorporation of homologous DNA into the genome of a flowering plant. Hess (5) prepared DNA from plants of *Petunia hybrida* with colored flowers and transferred it to plants with white flowers. By treating seeds, seedlings, or embryos of a white-flowered line with this DNA he obtained plants with colored flowers, some of them homozygous for the "corrected" color and breeding true on selfing. Other examples of DNA-mediated corrections have been reported in eukaryotes other than flowering plants; Hess, and Ledoux et al have reviewed the literature. For prokaryotes the story is an old one; it has long been known that purified DNA can be taken up by bacterial cells and incorporated into their chromosomes.

To return to the man treating oat cells with DNA from corn, why was he using DNA from a gene for susceptibility instead of one for resistance? The answer is, it would not matter which allele he used. The one is as good as the other. The DNA difference between susceptibility and resistance at a locus is possibly no more than can produce a different side chain of a single amino acid in the specified protein. It is a difference easily bridged by a change to virulence in *P. sorghi*. The man would not rely on puny defense by a missense mutation, but would aim at blocking the gene-for-gene systems of *P. graminis* and *P. coronata*. The corn DNA, he believes, is similar enough to home in on appropriate oat loci for susceptibility to these two fungi and cause host-parasite incompatibility, but dissimilar enough not to be bridged by changes to virulence in them. He could try DNA from several corn genes, singly or in combination, because corn has several genes for susceptibility to *P. sorghi*. Or, if he wants greater dissimilarity, he could try corn DNA for susceptibility to *Angiopsora zeae,* or sunflower DNA for susceptibility to *P. helianthi*, or flax DNA for susceptibility to *Melampsora lini,* or pine or currant DNA for susceptibility to *Cronartium ribicola*. . . . To dissimilarity there would be no limit.

He had two pieces of knowledge not available in 1976. He knew how to extract and purify gene-long DNA intact from corn, and he knew what a gene for susceptibility was.

---

[1] It is to reduce these distances that the man is treating wall-less oat cell protoplasts instead of seeds with the corn DNA. The techniques of tissue culture are advancing fast; and the development of flowering tobacco plants from single protoplasts (13) is now an experimental commonplace. When tobacco men lead, can oat men be far behind?

WHAT IS A GENE FOR SUSCEPTIBILITY?

It is basically incorrect to talk of a gene for susceptibility. The term implies self-inflicted harm as a primary function; it implies that the gene is a gene of welcome to the parasite. This welcome, it is reasonable to suggest, is a secondary function forced on the host by the parasite. The counter view which I put to you is that the gene is a gene essential to the host and existing for reasons other than to make the host susceptible to infection. The gene has a primary function which it carries on irrespective of whether there is a parasite in the offing or not. We must think of the gene as a plant gene and not particularly as a host plant gene.

Consider *Phytophthora infestans.* It devastated the potato fields of western Europe in 1845. Those fields were of *Solanum tuberosum.* In the 1920s, work began in Germany with the resistant W-potatoes (W for wild), which had the resistance gene later called $R1$ and probably derived from *S. demissum.* These W-potatoes crossed easily with *S. tuberosum,* and the progeny segregated roughly 1:1 for resistance:susceptibility. Thus we came to recognize $R1$ and its allele for susceptibility $r1$.

All the evidence agrees that $r1$ was universal in *S. tuberosum* in Europe before the 1920s when $R1$ was brought in, because potatoes were universally susceptible to blight, which they would not have been had $R1$ been present. At that time virulence on $R1$ was so rare in populations of *P. infestans* in Europe that, despite extensive searches by plant pathologists and breeders, it was not discovered until the 1930s.

This universal occurrence of $r1$ means that it existed primarily for reasons other than to make potatoes susceptible to *P. infestans.* Genetics recognizes that harmful recessives may occur in a population as a genetic load, but not at a frequency of 100%. The frequency of a harmful allele is maintained by a balance between selection and mutation, from which it follows that an allele universal in the population cannot be just harmful. The hypothesis, that $r1$ is a useful and normal constituent of *S. tuberosum,* is a straight deduction from a general proposition in genetics.

What about $R1$? Primarily it substitutes for $r1$. In the absence of *P. infestans,* as in a dry climate or in fields fully protected with fungicide, potato plants with $R1$ and those with $r1$ are alike. Even an expert could not tell them apart unless he happened to know their parentage. Whatever it is that $r1$ does, that too $R1$ does. Mutation from $r1$ to $R1$ is neutral or nearly neutral in its effects on primary functions.[2]

Neutral or nearly neutral effects of mutation on the phenotype are recognized in modern genetics as the rule rather than the exception. Mutation seldom has a

---

[2] The argument might be objected to, because a potato variety is a tetraploid clone, and for it to be resistant, only one of the four alleles at a locus needs to be a resistance allele. But the objection can be ignored because a parallel essay could have been written about (say) a rust disease of wheat.

great effect. Great effects, as in Mendel's tall pea plants and short pea plants, and correspondingly clear concepts about dominance are just the tip of the iceberg. Below and inconspicuous are a mass of neutral or nearly neutral effects. Langridge (7), studying the enzyme function of β-galactosidase from *Escherichia coli,* estimated that only 11 out of 733 mutations involving amino acid substitutions strongly reduced enzyme activity. In quoting this result as relevant to our topic, I am assuming for various reasons that $r1$ and $R1$ (and corresponding genes in other diseases) specify a metabolic enzyme, and are not genes whose products form part of a structural complex or are involved in a nucleic acid—protein complex. To turn to other evidence about neutral mutation, those who follow the literature of population genetics will know that Kimura (6) caused a stir by proposing that different alleles at a given locus produce the same phenotype, so that there would be relatively little genetic variation which could be affected by evolutionary processes other than genetic drift within a finite population. Heritable variation in most characters has been the foundation of Darwinian evolution, and Kimura's hypothesis has inevitably been challenged on various grounds. What matters to us here as a guide to modern thinking is that the challenges have been about population size, migration, inbreeding patterns, and other parameters concerned with the sexual breeding structure; the foundation of Kimura's proposal, that different alleles at a given locus mostly produce the same phenotype, has not been seriously questioned.

It is in a secondary function that $R1$ contrasts sharply with $r1$, and the terms *gene for resistance* and *gene for susceptibility* apply. In the presence of a population of *P. infestans* avirulent on $R1$, potatoes with $r1$ are susceptible and those with $R1$ are resistant.

There is no essential improbability in postulating two separate functions carried out by $r1$ and $R1$ differently in the healthy host and in the host-parasite combination. One might, for example, suppose that the enzymes specified by $r1$ and $R1$ differ by an amino acid not in contact with the substrate for binding or catalysis, but involved during parasitism in the host enzyme—parasite enzyme association, possibly dimerization, which is the essence of our hypothesis for diseases with gene-for-gene relations and a topic of the next essay.

It may have been noticed that when referring to the genes' primary functions I mentioned neutral or nearly neutral effects. A near, but not complete, neutrality is possibly involved in vertical resistance to disease; but this is another story outside the scope of this essay. I shall not digress further.

A gene for susceptibility is a normal useful gene in the healthy host. But what sort of gene is it? We can perhaps narrow the field. At least in diseases in which there is a gene-for-gene relation it seems likely that the gene in the host and the gene in the parasite have similar primary functions. Doubly et al's (3) demonstration of antigenic affinities suggests this. On this hypothesis of similar functions one can exclude host genes involved in photosynthesis or in any other metabolic or structural processes not shared by the parasite. Even with this exclusion, trying to identify the gene among the thousands involved in basic common metabolism is like looking for a needle in a haystack. Perhaps we should turn to trying to identify genes in the parasite in the hope that it would tell us what sort of gene to look for in the host. Since the genome of fungi is smaller than that of flowering plants, the haystack would be smaller.

CIRCLES OF INTIMACY

Every now and then plant pathologists are taken by an orismological urge, and reassess parasites and pathogens, parasitism and pathogenism, obligate parasites and facultative parasites, biotrophs and necrotrophs, mutualism and commensalism, and other terms. I notice that the British Mycological Society's (1) definition of a parasite is widely accepted (4, 10): a parasite is an organism or virus existing in intimate association with a living organism from which it derives an essential part of the material for its existence while conferring no benefit in return. This definition is all very well, but it glosses over the question: How intimate is the association?

Genes characterize an organism, and the only universal difference between different species of plants or animals is in their genes. One might therefore theorize that the most intimate association is between the DNA of the host and the DNA of the parasite. For plant viruses, substitute RNA. The further away from association at nucleic acid level, the less is the intimacy.

Following this line of thought, I suggest three widening circles of intimacy: an inner circle with direct association between host nucleic acid and parasite nucleic acid; a middle circle with association between the proteins (enzymes) specified by the DNA of the host and those specified by the DNA of the parasite; and an outer circle with association between products catalyzed directly or indirectly by the specified proteins.

In the inner circle, in the innermost part, are associations as in citrus exocortis disease. The virus (or viroid or metavirus) is a naked strand of infectious RNA too small to code for replication on its own. Presumably it relies on host mechanisms for even this most elementary requirement for existence. The host cell's nucleus remains intact. Confirmation of my suggestion comes from the finding by Semancik & Geelen (9) that citrus exocortis RNA interacts with host DNA to form an RNA-DNA hybrid within the nucleus.

In the middle circle are associations commonly regarded as gene-for-gene associations. The associations are in fact not at gene level; light and electron microscope studies make this clear. The principal associations, I suggest, are between the primary coded proteins. This was the theme of the previous essay, the proteins being identified as metabolic enzymes and not as structural proteins or proteins involved in repressor protein-operator nucleic acid interactions. Enzymes easily permeate membranes, and their tertiary and quaternary structure gives them great versatility.

On this theory of the middle circle, it would benefit the parasite to keep the host nucleus alive. This is the goose that lays the golden eggs. The general facts fit this deduction. In the rusts, downy mildews, powdery mildews, smuts, and gall diseases like potato wart, for which gene-for-gene relations have been suggested, the parasite tends to keep the host nucleus alive at least until sporulation begins and often until later. I am of course referring to susceptible (compatible) host plants. With the rust fungi, for example, the haustorium or occasional intracellular mycelium approaches the host nucleus or draws it towards its own, so that the nuclei of host and pathogen often lie close together across the membrane. Good management (some call it traumatotaxis) makes the goose nest near the fence across which her eggs can easily be reached. With wart disease of potatoes and club root disease of brassicas the parasite is purely intracellular; it stays near the goose.

The enlargement of the host nucleus and the apparent increase of host ribosome concentration and protein synthesis, common features of infection belonging to the middle circle of intimacy, are in line with the other evidence.

It is in the middle circle that disease is most often controlled by vertical resistance. However, vertical resistance and a gene-for-gene effect need not always be manifest, even though they are present. Consider *Phytophthora infestans*. *Solanum tuberosum* has no vertical resistance; had this been the only known host, no gene-for-gene relation would have been suggested. It needed the gene $R1$ from *S. demissum* to show up $r1$ in *S. tuberosum*, and start thoughts about genes-for-genes.

There are not only genes-for-genes but also unmatched genes. *Phytophthora infestans* uses (among other foods) carbohydrates in the host. However, it seems unlikely that the genes in the parasite governing carbohydrate digestion are matched, gene for gene, with host genes governing the photosynthesis that makes the carbohydrate. It is these unmatched genes that determine whether parasitism is obligate or not; in our scheme of things obligate parasitism, although of importance, is only of secondary importance.

In the outer circle are diseases in which the pathogen consumes pectin, cellulose, and other direct and indirect products of the primary enzymes of the host, as well as these enzymes and the host nucleus itself. Unmatched genes preponderate. At the extreme are wound parasites of food-rich storage organs and fruits, like *Erwinia carotovora* and *Botrytis cinerea*, which efficiently macerate cell walls. It has been known for more than 60 years (2) that macerating enzymes swiftly kill the protoplast; they probably kill it even while much of the cell wall remains undigested. It seems that the living host nucleus and the direct products of its genes have no special part in the parasitism except as food for the parasite. The parasite can afford to cook its goose.

To return to a matter discussed in the previous essay, most mutations are neutral or nearly neutral in their effect. The mutants produce the same, or nearly the same, phenotype. Electrophoresis detects many enzyme mutants (isozymes) that catalyze normally. Qualitatively, primary coded enzymes may vary much, but their products little. On the theory I have proposed, there should be many variants at the middle circle of intimacy, the circle of coded enzyme interaction, but fewer at the outer circle of interaction between products of these enzymes. Differential interactions between host and parasite, which are the interactions involved in physiologic races and vertical resistance, require that there be matching variation in both host and parasite. In the middle circle, host enzyme-parasite enzyme interaction would allow full play to most mutation, except samesense mutation. In the outer circle, in a reaction between, for example, host pectins and parasite pectic enzymes, the scope for differential interaction would be reduced, because, on the evidence about neutral mutations, the host pectins would reflect little or none of the mutations of their parent enzymes. The facts fit. At the middle circle, wheat and *Puccinia graminis* have an abundance of differential interactions, expressed as an abundance of races of *P. graminis* interacting with standard differential wheat lines, and reflecting neutral or nearly neutral mutation in both wheat and *P. graminis*. With this goes ample scope for trying to find vertical resistance in wheat against stem rust. At the outer circle, differentially interacting races of *Erwinia carotovora, Botrytis cinerea,* or their like are scarce or absent, and nobody has yet found vertical resistance against them, reflecting that mutation involving coded enzymes is relatively seldom carried through to their products.

The theory that the abundance of differential races of rust (or other diseases at the middle circle) derives from an abundance of neutral or nearly neutral mutations, that is, from an abundance of enzyme variations not carried further, demands that the host enzyme and parasite enzyme associate directly without their products being intermediaries. The simplest form of direct assocation would be as dimers. In enzyme (or, more generally, polypeptide) dimerization, or higher polymerization, I believe host-parasite specificity to lie at the middle circle. (Dimerization, or higher polymerization, probably proceeds largely on the host side of the interface; this could explain why the host cell dies before the parasite in incompatible combinations.) Although I have cited dimers as being the simplest form of direct host enzyme-parasite enzyme association, it would be more logical to consider tetramers when the mycelium is dikaryotic; and varying degrees of polymerization could give the simplest explanation of the dose effect implicit in heteroecism.

As a general proposition in plant pathology, widening circles of intimacy reduce differential interactions between host and parasite. Differential interactions (if there are any) involving samesense mutations must stop at the inner circle; and those involving missense mutations must fade in number beyond the middle circle. (Samesense, also called degenerate, mutations affect the DNA and RNA, but not the coded amino acid. Missense mutations affect the amino acid as well.)

As another general proposition, there is special importance in the survival of the nucleus in the diseased parenchymatous cell of a susceptible (compatible) host plant, at least until the fungus begins to sporulate. It determines the intimacy of host and parasite. It determines whether systemic infection is possible. It determines whether a gene-for-gene relation is likely to occur. It determines the probability that vertical resistance against the parasite will be found. This proposition must be seen against the background that in explosive epidemics sporulation begins early; early sporulation is a prerequisite for an explosive epidemic. For the fungus the need for early sporulation can override the need to prolong the survival of the nucleus, as it does in potato late blight epidemics.

On purpose I have oversimplified. There is every gradation from an indefinitely preserved nucleus to one killed ahead of the parasite, and circles of intimacy are often smudged.

AUTOBIOGRAPHY AND RE-VIEWING

In Romsey Abbey, Hampshire, in the floor of the central aisle of the nave near the choir is a slab inscribed: "In memory that here lyeth the body of John Vanderplank Senr. who departed this life ye 2nd day of October in ye year of our Lord 1717 and in ye 69th year of his age." His son, also John, is buried there too, the slabs lying side by side. The names were already anglicized. The next we know of is another John Vanderplank, died 1767 and buried in the church of St. Bartholomew the Great in the City of London. The family must have stayed in London, because my grandfather[3] was born there. My mother was from rural Aberdeenshire. Myself was born in Eshowe, Zululand, in the days of no motor cars and few fences.

My education was in the pure sciences, botany and chemistry, and I have never listened to a classroom lecture in plant pathology in my life. Teachers and students who think, "That explains it," possibly have different thoughts in mind. For many years down-to-earth potato breeding has been my main job. I think I have been successful, and predict that within a few years 75% of the potatoes in South Africa will be of my varieties, despite there being no restriction on the use or importation of other varieties (apart of course from normal and not very restrictive phytosanitary requirements). Any local reputation I may have is as potato breeder and seed grower, and I have no doubt that it would come as a surprise to most of my fellows in the Department of Agriculture to know I had any other professional interest. It is also on record that my son, when asked by his girl friend in Australia what his father did, replied, "Pa? He grows spuds." To end this synopsis, I have never attended a classroom lecture on genetics or plant breeding.

My best-known teacher was V. H. Blackman, the botanist. But the most news worthy was J. L. B. Smith, who taught me when I was working for an MSc degree in organic chemistry. This may have had a traumatic effect on him, because soon afterwards he switched from chemistry to fish. Within a few years he became world famous by identifying and naming the first living coelacanth fish to be caught. It was described in the news media as a living fossil. It was bone for bone, fiber for fiber almost identical with those that abounded in the seas of 300 to 70

[3] The book *Sisters of the South* about South Africa, Australia, and New Zealand, written by C. Lighton (Howard Timmins Publishers, Cape Town, 1958), tells something of grandfather John's doings. He brought seed of the wattle tree *Acacia mearnsii* from Australia to South Africa. Tens of thousands of acres have since been planted for the tannin in the bark.

million years ago, then vanished except in the fossil record. The special importance to biologists of the coelacanth is that it is the nearest surviving relative of the extinct rhipidistian fishes, which for some reason left the water to conquer the land as the ancestors of the amphibia, reptiles, birds, and mammals. Smith had discovered living in the Indian Ocean the nearest thing to the missing link in the descent of land vertebrates. He was a busy taxonomist, and on a tally made 25 years ago he had identified and named more than one percent of the total named sea fish species of the world.

My brainiest teacher was J. W. Bews, whose biography has been written by Gale (University of Natal Press, Pietermaritzburg, 1954). His books, *Plant Forms and their Evolution in South Africa* (1925) and *Studies in the Ecological Evolution of the Angiosperms,* published as *New Phytologist Reprint No 16* (1927) appeared while I was a student of his. Half a century later I retain some of that ecological background (12, Chapter 5). I am one of the lucky ones who looking back can think happily of all their teachers.

Consider books and students. There is a limit to the bulk of undigested detail a student can be expected to absorb. Fortunately, and especially in the older and more exact sciences, experience shows that what must be memorized need not increase in proportion to accumulated experimental fact. As new facts to be remembered are discovered, many old facts are taken up in generalizations easily learned; and many old concepts and controversies simply fade away. But the process is not automatic. Synthesis and generalization must be as brisk as the churning out of new experimental data. My purpose in writing books has been just this, to give new ideas and in the process to fit together loose pieces and to allow old concepts to lapse. It is my boast that in none of my books have any experimental data of my own collecting appeared. My reliance on the experimental findings of others has been deliberately complete. But nine tenths or more of the discussions of these findings have been my own.

My task has been to look at published data, assess them, and see how they fit together. Often one comes to conclusions different from the original author's. Perhaps my most profitable scoop was to unravel the data on the artificial physiologic races of *Puccinia graminis tritici* in Canada, transform them into data on virulence frequencies in the fungus population, and compile a table (12, p. 154) that must be one of the most remarkable in all population genetics. It showed beyond question the reality of stabilizing selection acting against virulence, a matter disputed on the same, but untransformed, data by their original author. It set (I hope) a new pattern of analysis based on virulence frequencies instead of races, which should bring this branch of plant pathology into line with population genetics generally. But my favorite scoop was from a German potato trade journal *Kartoffelbau.* From an informal article—it did not even give the author's initials—of less than two pages entitled "Ergebnisse von Krautfäule-Spritzversuchen" which did not mention disease resistance, I derived a full-page table (11, p. 186) showing that horizontal resistance to potato blight had been badly eroded by potato breeders in the course of breeding for vertical resistance. What I specially like about this scoop is that I was in an agricultural library in Wageningen playing truant from a conference when I saw *Kartoffelbau*—which I had never seen before and have not seen since—which proves that reading is better than attending lectures.

I have been a re-viewer of evidence. Dictionaries are helpful. Among various other meanings the verb *review* has two: view again, and survey. The *Oxford English Dictionary* says that when the verb has the first meaning it should be pronounced with the accent on the first syllable and spelled with a hyphen. *Webster's Third New International Dictionary* agrees about the accent. The invaluable journal, the *Review of Plant Pathology,* formerly *RAM,* reviews—surveys—the literature, but can give little space to re-viewing evidence, although fortunately it does give some. The *Annual Review of Phytopathology* is where plant pathologists look for re-viewings. We know how very good much of the reviewing has been in the past. But the *Annual Review's* greatest tasks are yet to come. The twin features of science, the publication explosion and ever-increasing specialization, are going to make us need the *Annual Review* more and more. It will not fail us; of that I am confident.

Literature Cited

1. Anonymous. 1950. Definition of some terms used in plant pathology. *Trans. Br. Mycol. Soc.* 33:154−60
2. Brown, W. 1915. Studies in the physiology of parasitism. 1. The action of *Botrytis cinerea. Ann. Bot. London* 29: 313−48
3. Doubly, J. A., Flor, H. H., Clagett, C. O. 1960. Relation of antigens of *Melampsora lini* and *Linum usitatissimum* to resistance and susceptibility. *Science* 131:229
4. Hall, R. 1974. Pathogenism and parasitism as concepts of symbiotic relationships. *Phytopathology* 64:576−77
5. Hess, D. 1972. Transformationen an höheren Organismen. *Naturwissenschaften* 59:348−55
6. Kimura, M. 1968. Evolutionary rate at the molecular level. *Nature* 217:624−26. *Also* Kimura, M., Ohta, T. 1971. *Theoretical Aspects of Population Genetics.* Princeton, NJ: Princeton Univ. Press. 219 pp.
7. Langridge, J. 1974. Mutation spectra and the neutrality of mutations. *Aust. J. Biol. Sci.* 27:309−19
8. Ledoux, L., Huart, R., Jacobs, M. 1974. DNA-mediated genetic correction of thiamineless *Arabidopsis thaliana. Nature* 249:17−21
9. Semancik, J. S., Geelen, J. L. M. C. 1975. Detection of DNA complementary to pathogenic viroid RNA in exocortis disease. *Nature* 256:753−56
10. Thrower, L. B. 1966. Terminology for plant parasites. *Phytopathol. Z.* 56: 258−59
11. Vanderplank, J. E. 1963. *Plant Diseases: Epidemics and Control.* New York: Academic. 349 pp.
12. Vanderplank, J. E. 1975. *Principles of Plant Infection.* New York: Academic. 216 pp.
13. Watts, J. W., Motoyoshi, F., King, J. M. 1974. Problems associated with the production of stable protoplasts of cells of tobacco mesophyll. *Ann. Bot.* 38:667−71

Copyright © 1976 by Annual Reviews Inc.
All rights reserved

CONTRIBUTIONS TO THE HISTORY OF PLANT PATHOLOGY IN SOUTH AMERICA, CENTRAL AMERICA, AND MEXICO

♦3630

J. A. B. Nolla

University of Puerto Rico, Mayaguez, Puerto Rico

Manuel V. Fernandez Valiela

Delta Agricultural Experiment Station, Cas. Correos 14 Campana, Argentina

INTRODUCTION

Reviewing the growth of the science of plant pathology in Mexico, Central America, and South America is difficult because of the lack of sources of information from some countries. As more students become interested in the subject, information may be unearthed from periodicals or other written vehicles unknown nationally. In a few decades we may have a more precise and well-documented history than is now possible. This review emphasizes the disease rather than the pathogen or author although they are closely united.

EARLY HISTORY

All of us relish being able to present a subject from its clear beginnings. We agree with the late Professor Whetzel (105) of Cornell University that "phytopathology, like all natural sciences, had its beginnings in the dawn of man's civilization." Until man began to record his observations, our science passed by word-of-mouth among peoples of various cultures. When man picked the fruit or dug the root of wild plants, disease, if present, may have passed unnoticed. Awareness of disease in plants must have come when early peoples began to cultivate plants in an endeavor to produce food in abundance for their subsistence. Whether individually or as a tribe they first cleared just enough forest for seasonal plantings. Come the following season they would clear more forest, and in this way established a system of agriculture. Continued cultivation created the conditions under which pathogens and pests were more apt to appear and increase. One day, a man or woman observed that young plants were blighting or wilting, or fruit

11

rotted or the grains were malformed or displaced by abnormal growths. These were the symptoms of disease for which they must have had names. Migratory tribes and birds carried seeds to other regions and thus transported disease agents. In the absence of migration or communication with other people, diseases were limited to local areas because man himself was not an agent of transportation. Thus, the concept of disease in plants was born in different areas of the world. The etiology was to remain shrouded in mystery. So it was with the old civilizations of Europe, Asia, and Africa.

Popenoe (82) mentioned a decline in maize production in Maya country in the period 400–800 AD which Wellman (104) suggests may be attributable to disease. This may well be a good starting point in a chronological discussion of the events that make history in plant pathology in this area.

Cordero (28) in 1882 suggested that the people inhabitating Mexico before the discovery of America had some knowledge of plant diseases as shown by their name *Chahuiztli* for a rust, supposedly corn rust. To this day the word *Chahuixtle* designates all rusts, but may also apply to other diseases.

IMPORTANT DISEASES OR PATHOGENS

In his *Natural and Moral History of the Indies* (1590), the eminent Father Joseph De Acosta (30), who traveled widely in South America, wrote, "and, if the year is plentiful of these [potatoes], they [Indians of Perú] are happy, because during many years they [potatoes] blight or freeze in the very ground." He was referring, according to most writers, to late blight disease, caused by *Phytophthora infestans* (Mont.) de Bary. This then was the first recorded epidemic of late blight of potato. The second recorded epidemic of blight in Perú was referred to by García Merino (44) in 1878. In Chipaque, Colombia, it appeared as a severe epidemic in 1861 (81) and extended through several municipalities; in 1867 it appeared in the Bogotá savanna. [See (106).] In Brazil, Puttemans (85) reported late blight in 1892. Mexico in 1898 (70) suffered from an epiphytotic of late blight known there as *niebla rocio* (mildew or fog, dew). Its occurrence in Argentina was reported by Spegazzini in 1898 (96) in epidemic form in the southeast of the province of Buenos Aires.

Linnaeus (62) as cited by Wellman (104) classified some dried specimens received from South America as species of three genera of phanerogamic parasites of other higher plants, i.e. *Loranthus* spp. in 1730, and *Cassytha filiformis* and *Cuscuta americana* in 1753. By 1892 the Mexican farmers became aware of the menace of *Cuscuta* to crop plants (69). An outbreak of "polvillo" [*Puccinia graminis tritici* Eriks. & Henn., *P. glumarum* (Schm.) Eriks. & Henn.] was reported to occur in the vicinity of Ecuador in 1808, as reported by Caldas (20) according to Orjuela (80). A very severe epiphytotic of *P. graminis* was recorded in Mexico in 1884–1885 (66).

The first mention of diseases of crop plants in Argentina was by de Moussy (31) in 1860. He described rust, smut, and blight of cereals, leaf-curl ("torque") of peach, and sooty mold of oranges.

Moko disease of bananas was described by Rorer (87) from Trinidad in 1911. According to Buddenhagen (19) a bacterial disease of banana was recorded by Schomburg in 1840 in Guyana, which Stover (101) believes may be the same disease. If this is confirmed Schomburg's record would become the first report of a bacterial disease, Draenert's sugarcane gumming, second, and Burrill's pear blight, third. A severe epidemic about 1890 is reported to have wiped out the Moko fig variety of plantain which had been used to provide shade for young cacao trees. One striking difference from the Panama wilt noted by Rorer was the resistance of the Gros Michel, a banana variety very susceptible to the Panama wilt organism. The causal agent has been established as *Pseudomonas sólanacearum* E. F. Sm. This organism is a special form of the bacterium that is a very common pathogen of many solanaceous hosts and has exhibited great variation. The disease is now common in Central America.

In 1869, a serious epidemic of bacterial disease of sugarcane was described in Bahia, Brazil, by Draenert (32, 33). The causal agent is now known as *Xanthomonas vasculorum* (Cobb) Dows. This disease, which is the second bacterial disease reported in the world, is known as gommosis or gumming. It has devastating effects on susceptible clones but disappears when a resistant cane is planted as a substitute, to reappear if susceptible canes are returned to the area of the epidemic.

The first coffee disease in America was reported by Saenz (89) from Colombia in 1876, the American leaf disease or "viruela" of coffee. The cause was attributed to *Stilbum flavidum* by Cooke in 1880, reconfirmed by Ellis from leaves sent by Saenz [Michelsen (71)], and now known in its perfect stage as *Mycena citricolor* (Berk. & Curt.) Sacc. It is widespread in the Antilles and in the coffee growing countries of continental America. Puttemans (85) reported it in Brazil in 1904 and Hecq (55) in Perú in 1906.

Le Feuvre reported two important diseases on grapes in Chile, the powdery mildew (*Oidium tuckeri* Berk.) (60) in 1877 and anthracnose (*Elsinoe ampelina* Shear) (61) in 1878.

French zoologist Jobert (57) in 1878 made the discovery in Brazil of the first nematode disease reported in this hemisphere. He observed considerable injury to the roots of coffee and proved that the disease was caused by an *anguillula* or nematode (85). Goeldi (49) in 1887 commissioned by the National Museum of Brazil to make a detailed study of the new disease agreed with Jobert's conclusions and described the genus and species *Meloidogyne exigua*. This is the first nematode reported in Latin America.

Brown-eye spot of coffee was reported from Jamaica in 1881 and its causal agent described as *Cercospora coffeicola* by Berkeley & Curtis (12). Later, in that decade, Goeldi (49) reported brown-eye spot from Brazil, but thought it to be caused by a *Ramularia* which Saccardo named *Ramularia goeldiana*. Puttemans (85) regarded this as a synonym of *C. coffeicola*. Goeldi's microscopic illustrations were considered by Puttemans as the first published in Brazil.

A very serious epiphytotic of wheat rust occurred in Mexico (68) in 1885, which destroyed 25−50% of the wheat crop; in the same year and country García

Díaz (43) called attention to the seriousness of an epidemic of *Plasmopara viticola* (Berk. & Curt.) Berl. & de Toni on grape vines. In 1887 Ernst (36) studied corn smut [*Ustilago maydis* (D.C.) Corda] in Venezuela but did not go into detail about its distribution there.

Cacao witches'-broom was first recorded in the Saramacca district of Surinam in or about 1895. Its cause was determined by Stahel (97) as *Marasmius perniciosus* Stahel. From 1908 to 1912 a severe epidemic of cacao witches'-broom reduced production in this area by approximately 60%. The same fate was met by the crop in Ecuador (88) and Perú (8).

Nematodes in coffee was the subject of study by D'Utra (35) about or before 1900 in Brazil. Various papers on plant pathology in Argentina were written by Huergo (56) between 1897 and 1908. And from 1899–1902, Lavergne (58, 59) wrote on diseases of potato, grape, watermelons, apples, and pears in Chile.

Hevea leaf spot or South American leaf disease of *Hevea* was reported in 1901 by Ernst Heinrich Ule, a German botanist who collected specimens in Brazil. When the perfect stage was described it was classified as *Dothidella ulei;* but in 1917 Stahel (98), who studied the disease and fungus in Surinam, named the sexual stage *Melanopsammopsis ulei* Stahel. The disease has been considered responsible for the failure of rubber production on a large scale in South and Central America.

The first banana disease to attract the attention of growers and scientists was the Panama disease or wilt which McKenny (64) believed was first observed in epidemic form in 1904 in Panama and Costa Rica and even as early as 1890 in the former. Stover (101) gave credit to Bancroft (7) for reporting from Queensland, Australia in 1876 a disease on the sugar banana which may be the same as Panama disease. It spread rapidly and was soon found in other areas of America. Smith (93) studied samples sent from Cuba to Washington DC in 1910 and named but did not describe the causal fungus which is named *Fusarium oxysporum* Schlecht f. sp. *cubense* (E.F. Smith) Snyder and Hansen. Brandes (17) established the pathogenicity of the incitant; much of the work was done at USDA Experiment Station in Puerto Rico. The disease has brought ruinous losses to the banana growing and exporting interests in Panama, Costa Rica, Honduras, Nicaragua, British Honduras, Colombia, Venezuela, Surinam, and Guyana. The great losses suffered from this disease have come not only from the crop itself, but also from the need of abandoning thousands of acres in the affected areas and relocating in new fungus-free areas.

Red ring of coconuts was first seen by Hart (Nowell, 79) in Trinidad in 1905, and Nowell (78) proved the pathogenicity of the nematode, now designated as *Rhadinaphelenchus cocophilus* Cobb. It occurs in Panama, Venezuela, Colombia, British Honduras, Surinam, etc. This disease is difficult to control and is possibly one of the greatest menaces to coconut production in America.

The ubiquitous fungus *Pellicularia rolfsii* (Curzi) West (*Sclerotium rolfsii* Sacc.) was recognized in tropical America in 1903 (3). It is the cause of root and foot rot of young plants in the nursery or field, and even a seedling blight.

In 1906 Puttemans (83) reported the bean disease caused by *Isariopsis griseola*

Sacc. in Brazil. This is important because beans constitute an important item in the diet of many South American and Central American countries and Mexico.

By 1907 a disease, later to be identified with mosaic or yellow leaf stripe of sugarcane, was laying waste the fields of sugarcane in the province of Tucumán, Argentina (Matz, 67). This was the first epidemic of sugarcane mosaic in America. The transmission of this disease by the corn aphid was shown by Brandes (18) in 1920. In 1923, Chardón & Veve (26) demonstrated transmission by the same vector under field conditions in Puerto Rico.

Mycosphaerella linorum (Wr.) García Rada (*Phlyctaena linicola*) is the cause of Pasmo, a serious disease of flax in Argentina, described by Spegazzini (96) in 1909. It has gained in importance because of its worldwide distribution.

In 1911 Puttemans (84) reported a disease of the inflorescence of cauliflower which he attributed to the fungus *Alternaria brassicae* Sacc., new to science. This disease occurs also in the United States of America.

Tristeza, one of the most serious virus diseases of citrus, was first observed in 1930 in Argentina (Bella Vista, Corrientes Province). Seven years later it appeared in Pariba Valley, Brazil, where it was given its present name, which means sadness or prostration. It had been recognized in 1896 in South Africa as an "incompatibility." A similar observation was made in Java in 1928. The principal symptoms consist of a decay of the rootlets and a resulting top deterioration which leads to the characteristic "tristeza." These symptoms are brought about by the girdling effect of sieve tube collapse in the bud-union region. The causal virus is transmitted by the aphid vector *Toxoptera citricida* and several other aphids. The introduction of the virus and the insect vector in Argentina is thought to have occurred in orange grafted on the tolerant rough lemon rootstock. Losses in Argentina alone have been as high as 20 million bearing trees, worth approximately 500 million dollars.

Citrus fruit culture is of much importance in the agricultural economy of the southern nations of South America. One of the factors that play a definite role in its progress is the control of diseases such as tristeza.

The Sigatoka disease of banana is a leaf spot disease so fast spreading that Stover (99) in 1962 was led to regard it as transported by air currents from the East over land and sea to Africa and thence to the Caribbean region, South America, Central America, and Mexico in a relatively short time. The epidemic of 1934–1935 inflicted such heavy losses upon banana plantations that several Central American countries saw their exports reduced considerably. Variation of the causal fungus in pathogenicity and morphology was shown by Stover (101a) who suggested the following nomenclature: *Mycosphaerella fijiensis* Morelet var. *musicola* (Sigatoka), *M. fijiensis* var. *difformis* (Black Sigatoka) from Honduras, and *M. fijiensis* (black leaf streak). The name *Sigatoka* is now less prominent than *black leaf streak*.

The sugarcane industry in Tucumán, Argentina, was menaced again in 1941 by a second foreign intruder reminiscent of the days when mosaic was advancing rapidly during the first decade of this century. We are now concerned with the smut, *Ustilago scitaminea* Syd. (39), a fungus originally observed in India on wild cane, *Saccharum spontaneum* L., which on its westward march reached Natal by 1877 [McMartin (65)]. A severe epidemic destroyed tens of thousands of hectares of cane in Argentina (39). The disease has also been reported from Paraguay, Bolivia, and Brazil.

The golden nematode of potato (*Heterodera rostochiensis* Wollem.) was reported in 1953 from the Andean region of Bolivia and Perú (9) by Bazán de Segura.

Radopholus similis (Cobb) Thorne, first considered of importance in 1957, now has displaced *Fusarium oxysporum* f. sp. *cubense* as the major banana root pathogen (101). The Cavendish varieties are generally infected except in some Mexican districts. The very susceptible Gros Michel appears somewhat free of the nematode. Two special forms of the nematode are recognized, one on bananas and the other on citrus. The banana form has shown some variation in races. The citrus form has not been clearly demonstrated to be the only cause of spreading decline in citrus. Thorne (102) agrees that another organism may be working in association with the nematode to produce the characteristic symptoms.

Coffee rust, an Oriental disease discovered and described in Ceylon in 1869, which is caused by *Hemileia vastatrix* Berk. & Br., was transported to Brazil by way of Africa, presumably by natural forces across the ocean. First found and reported in 1970 by Gómez Medeiros (50) near the city of Itabuna, Bahia, it has spread to neighboring states of Brazil and to Paraguay and Argentina. Presently it constitutes the most feared plant disease in South and Central America, where coffee has reigned supreme in their economy (15, 21–23, 86, 90).

CONTROL

The European colonizers did not bring much to improve the conditions they found in America. The land was plentiful. As soon as plants became diseased or the soil was depleted of its nutrients the colonizers moved up or down the land to clear new sites for their crops. As commercial agriculture took its first steps, growers began to realize that losses from disease were increasing. They became more careful about selecting, cleaning, and storing their seed; they thought of cultural practices to prevent losses; they burned the dead dried plants and as plows were made available some fields were plowed under. Plowing alone would not control disease when the pathogen was probably a soil inhabitant requiring a more sophisticated treatment. Drainage of the land helped in cases of low fields, and rotation of crops was beneficial especially in controlling nematodes. Growers had to wait until the early part of this century to begin to change the pattern of their agriculture.

With the advent of Bordeaux mixture (72) in Europe, the discovery of lime sulfur in both Europe and North America (29), and the help of the few scientists available, some progress was made. Scientific agriculture has taken us from the relatively small farm planting to large-scale cropping. The new system has made possible the application of new techniques on a large scale. It has required costly equipment and greater skill on the part of personnel.

Panama Disease

The story of the control of Panama disease of banana by the British in Jamaica and Trinidad and later in Central and South America by the United Fruit Company and its subsidiaries is an example difficult to match in any other disease. Of the great number of banana cultivars in the world it seems that only one, the Gros Michel, suited the taste of the consumers. But this was the most susceptible of the dessert bananas and had to be protected at all costs. Here then began a never-ending battle. All cultural practices known were tried with varied results. At first drainage was tried. Then in an effort to isolate the pathogen, a trench was dug around groups consisting of one diseased plant and eight apparently healthy plants. This slowed down the advance of the fungus but did not stop it, because for one thing not all the surrounding plants were really healthy. Then roguing and destroying by fire diseased plants as well as the healthy plants immediately surrounding them also met with failure. Chemical treatment also was not effective.

After the great epidemic of the disease in the early part of this century in Panama and Costa Rica the disease spread so rapidly that vast tracts of land had to be abandoned and areas of virgin lands were opened to cultivation. This seemed to be the only alternative, but an expensive one. Laboratory tests indicated flood fallowing was logical and worthy of a trial. Drost (34) had attempted it in Surinam in 1912. [V. C. Dunlap made the first large field experiment of flooding in Honduras in 1939 in an abandoned area of extremely high infestation. Maintaining the water at a constant depth for varying periods of time resulted in good control. The treated fields remained productive for 11 years (R. H. Stover, personal communication). By 1956 some 25,000 acres in Panama and Honduras were flood fallowed.] The use of chemicals in the water had no appreciable effect (100). We have come almost to the end of the road. Remarkable as these accomplishments have been, their application is limited by economic realities. The cost of labor, materials, and equipment has risen. Then, Stover (100) showed with detailed studies why flooding became ineffective in control. Few other avenues remain to be explored.

In 1912 Drost (34) found resistance to the pathogen in Congo (Robusta) banana in Surinam. The United Fruit Company started a collection of bananas throughout the world (100) and by 1930 had accumulated 157 cultivars. In 1947 Jamaica began changing from Gros Michel to Lacatan, a resistant Cavendish cultivar from the East. This banana was not accepted in the North American markets. But survival in the future depended on resistant varieties with the good qualities of the Gros Michel. A breeding program requires a long time and while we wait for the results the industry must carry on.

In order to ensure as healthy crops as possible, corms were selected from symptomless clumps and given a preplanting treatment which included paring off all leaf sheaths and lesions on corms, and dipping in a nematicide 1,2-dibromo-3-chloropropane (DBCP) solution. A hot-water treatment has also been used. A soak for 10 minutes in a concentration of 1000 ppm of Mocap® has been recommended (101). The treated corms are planted in a nursery in nematode- and

Panama disease—free lands; they are completely isolated and all necessary phytosanitary precautions are taken to avoid introduction of the pathogens. This healthy planting system will, if permanently continued, reduce infection to a minimum.

Moko

Sequeira (91, 92) developed the first effective control measures against Moko disease. Economical control depends on inherent resistance of certain cultivars as in Pelipita, the first bacterial and fusarial resistant plantain reported. Because the disease is transmitted by knives and other tools that may injure plant parts, disinfection of these is an essential preventive measure.

Sigatoka

The fight against this disease started with Bordeaux mixture. [Dunlap revived the collapsing banana industry in Central America in the years 1937—1940 with his development of a permanent pipeline spray system for the application of Bordeaux mixture, installed in more than 100,000 acres (R. H. Stover, personal communication).] This system remained in use until the discovery of low volume (1 gallon/acre) petroleum oil sprays in Guadeloupe (52, 53) in the midfifties. The next step was an oil-in-water emulsion with a dithiocarbamate fungicide. Aircraft took the place of most of the laborious hand application. After 1970 oil-in-water emulsions alone or with a systemic fungicide (benomyl) and maneb came into use.

Science has solved the problems with which the banana industry has contended; research seems to have planned the future with intelligence and foresight. The industry seems to be secure, so let us examine what the future has in store.

Breeding for resistance to disease in bananas was first tried in Trinidad and Jamaica beginning in the first quarter of this century. As far as we have been able to determine none of the artificially bred bananas has shown superiority to the natural cultivar Lacatán, with respect to resistance to the Panama disease pathogen.

The United Fruit Company or its subsidiaries began a breeding project in 1958, sending out in the succeeding years expeditions to Asia for the collection of breeding stocks. All of these have been a substantial addition to the gene pool from which the plant breeder will draw materials for the development of new cultivars tailored to the demands of the market.

TEACHING AND RESEARCH

Argentina

In 1879, the Italian biologist, Carolus Spegazzini, who was destined to become a prominent figure in the sciences of mycology and plant pathology in Latin America, settled in Argentina. He was preeminently a mycologist but made important contributions to phytopathology. Before 1900, he published on diseases of sugarcane in Argentina and on coffee in Costa Rica. In the next three decades his phytopathological interest spread to alfalfa, fig, cacao, chestnut, "mate," *Ligustrum* sp. in Argentina and to citrus in Paraguay.

Huergo (56) is recognized as the first Argentinian phytopathologist, with a number of papers on plant disease problems written between 1897 and 1908.

RESEARCH   The Office of Chemistry and Agriculture in the province of Buenos Aires was established in 1897, and Spegazzini became its first botanist. In 1899 he was appointed director of ·the Vegetable Biology Section of the Ministry of Agriculture. This section included phytopathology. He can rightly be considered the first federal phytopathologist in that country. During his professional life the number of species of fungi he studied was close to 5000. When he started, only 39 species had been described (96). (See also 94, 95.)

Spegazzini was the pioneer who laid the groundwork of the development of mycology and plant pathology in South America. The Museo Spegazzini in Buenos Aires stands out as a silent tribute of praise and gratitude from his adopted country.

In 1902, Girola (48) studied the status of agriculture under the guidance of Spegazzini and in 1904 published the first paper of importance on diseases of cultivated plants in Argentina and their control.

In answer to a request from growers an experiment station was established in Tucumán in 1907. It dealt mainly with diseases of sugarcane. Here G. L. Fawcett worked, beginning in 1917. He is regarded as one of the pioneers in the development of phytopathology in Argentina. Zeman (107) and Backhouse (4) were government pathologists until 1922.

After Spegazzini, Juan B. Marchionatte has been the central figure in the development of phytopathology. He joined the Ministry of Agriculture and Animal Husbandry in 1923 and was promoted to various important positions until he reached the post of Director General of Plant Protection. He established the Laboratory of Plant Pathology and Zoology, Insecticides, and Fungicides and regional laboratories of Phytopathology and Entomology; a plant quarantine station and port of entry inspection of plants and plant parts; he organized the diagnostic and disease control service, laid the foundation for a sound plant protection legislation, and took part in the signing of international agreements and treaties in plant protection. He surpassed his Argentinian contemporaries as an organizer and writer.

Fernández Valiela (41) mentions 29 crops suffering from 77 diseases of economic importance in 1964. These are all subtropical and temperate zone diseases.

TEACHING   The first course in plant pathology in Argentina was given in 1904 by Hauman (54), a Belgian, at the Faculty of Agronomy and Veterinary Science of the University of Buenos Aires. He taught for 21 years. His successor was Marchionatto, who in addition to his other duties in the Ministry of Agriculture, found time to teach phytopathology at the Universities of Buenos Aires and La Plata. Many of his students have held important positions in teaching or research. Teaching is also done in the Agronomy Faculty of the National Universities at Corrientes, Córdoba, Tucumán, La Pampa del Sur; at the Catholic University de Mar de Plata; and at the Faculty of Agricultural Science, University of Cuy.

In 1964 an international postgraduate course in phytopathology was developed under the auspices of the Organization of American States and South Zone [OAS (OEA)], the Faculty of Agronomy of La Plata, and the National Institute of Agricultural Technology (INTA). As a contribution to the knowledge of phytopathology Fernández Valiela (40) and Marchionatto (66) have each published a book.

Bolivia

TEACHING   Martin Cárdenas, as president of the University of San Simon de Cochabamba, organized the Superior School of Agronomy in 1939 and became the first professor of phytopathology. He taught until 1951. Teaching has continued from 1951 until the present time under Herbas.

RESEARCH   Cardenas made extensive collections of fungi in cooperation with John A. Stevenson, Curator, National Fungus Collections, USDA and the Smithsonian Institution, Washington, DC. Stevenson made the taxonomic determinations.

The only agricultural experiment station in Bolivia, La Tamborada, which was established in 1945, was transferred to the Interamerican Agricultural Service, and the Phytopathological Service was placed under Bell, in the USA. Alandia became Bell's assistant.

In Bolivia the diseases that caused problems in 1964 were late blight (P. infestans), mosaic and leaf roll of potato; rusts of wheat (Puccina graminis and P. glumarum); Septoria tritici Rob. and S. nodorum Berk. and the smuts [Ustilago tritici (Pers.) Rostr., Tilletia sp.; U. hordei (Pers.) Lagerh. and Rhynchosporium secalis (Oud.) J. J. Davis] of barley.

Brazil

RESEARCH   The first position of state phytopathologist was created in 1888 by Dafert (85) at the Agronomical Institute of Campinas, Sao Paulo. The position was filled by Benecke, a German, who held it for only a short term. Actually the first phytopathologist to hold this position was Noack who described 23 diseases of orchard and garden (75) besides contributing papers on diseases of grape vines, wheat, and citrus trees (85).

The federal government organized a laboratory of plant pathology in the National Museum in Rio de Janeiro in 1910 and Arsène Puttemans became the first director. The Instituto Biológico de Defensa Agrícola (Biological Institute of Crop Protection) was organized in the Ministry of Agriculture at Rio de Janeiro in 1920, and eight years later the Biological Institute of Sao Paulo was founded.

Phytopathology has made extraordinary progress in Brazil, especially since the beginning of the fourth decade of this century. Many workers have shared in this development with diligence and competence, among them, Bitancourt, Chaves

Batista, Santos Costa, Moreira, Victoria Rossetti, Viegas, Silverschmidt, and da Costa Neto. If one is to give special mention to a living pathologist it should be A. A. Bitancourt who recently retired as head of the section of phytopathology in the Museo Biológico de Sao Paulo.

In 1964 there were 96 diseases reported as requiring some study in 29 crops (41). The crops requiring more attention appeared to be potatoes, beans, citrus, sugarcane, tomatoes, and grapes.

TEACHING   Interest in the teaching of plant pathology in Latin America was first shown in Brazil. The first teacher was Garcia Redondo (85) who taught plant pathology in the general course in agriculture at the Polytechnical School, State of Sao Paulo, from 1893 to 1899. Arsene Puttemans taught this course from 1903 to 1910.

An agricultural college was founded at Piracicaba in 1901 with a course in botany which included plant pathology. According to Grillo (51) the department of plant pathology here was organized by Honey and that in Viçosa by Müller.

In 1964 phytopathology was being taught at the undergraduate level in 12 universities and schools of agriculture. Postgraduate courses are offered in the schools of agronomy at Piracicaba and Viçosa, in the Biological Institute of Sao Paulo, and in the Virology Section, Institute of Agronomy of Campinas.

Chile

RESEARCH   Early papers were published in 1877 and 1878 by Le Feuvre (60, 61). In 1896, Chile established a phytopathological station at the request of the National Society of Viticulturists. Lavergne, a French phytopathologist, was chosen to organize it and became the first director in 1897. Camacho, who succeeded Lavergne (58, 59) in 1906 as director of the experiment station, was instrumental in the organization of laboratories for seed testing, and foreign and domestic plant quarantines. Graf Marín took over in 1932 and introduced new techniques emphasizing modern chemical control. Even at this stage there is emphasis on mycology (73).

TEACHING   The teaching of plant pathology in Chile began in 1904 in the Faculty of Agronomy of the University of Chile with Camacho, who taught until 1928. Others were Larenas from 1928 to 1931, Graf Marín from 1931 to 1954, and Tartakowsky since 1954. A special course in advanced plant pathology was inaugurated in 1956 under Caglevic. The Catholic University of Santiago started teaching plant pathology in 1950 under Arentsen, and Vergara since 1957. The Universidad Austral included phytopathology in its curriculum, taught by Oherens Hertossi since 1958. Likewise, the University of Concepción began teaching phytopathology in 1959 with Arentsen as professor.

Colombia

Interest in phytopathology may have been enhanced by the visits of a small group of Puerto Rican scientists to Colombia under the leadership of Chardón. In 1926 after his first visit Chardón wrote on coffee diseases (24) and published the first record of the occurrence of sugarcane gumming in Colombia (25). The papers by Chardón and Toro are contributions to our knowledge of the diseases of Colombia. Toro's papers (103) written while at Medellin are important additions (3).

Nolla (76) was the plant pathologist of the Chardón mission to the Cauca Valley in 1929 (27). Forty-four diseases on 22 crop plants were diagnosed and control measures recommended; at least 10 had been reported in 1929.

TEACHING   Plant pathology became a subject of study in the Department of Antioquía by Law 38 and Official Order 11 of 1914. It was first taught at the School of Agriculture (now National University, Faculty of Agronomy). Now it is part of the curriculum in the Faculty of Agronomy (National University) at Bogotá and at the Universities of Manizales, Ibagué, Tunja, and Pasto.

RESEARCH   The Agricultural Experiment Station was established at Palmira, near Cali, in 1929. Durán Castro was appointed director. The research complex in 1957 consisted of the laboratories at Tibaitatá (Bogotá), Palmira, and Medellín. Presently postgraduate courses are offered in these centers.

Orjuela (80), the principal phytopathologist at Tibaitatá, regards Mejía Franco as the potent factor in the development of plant pathology. He is the first Colombian phytopathologist, also the first Colombian to teach plant pathology. At the present time Gálvez, the phytopathologist at CIAT (Palmira), is engaged in the study of virus diseases of legumes. Garcés is director of the graduate school at ICA. There is a long list of young, well-trained women and men doing good work in research.

Costa Rica

TEACHING   The National School of Agriculture established in 1927 is now the Faculty of Agronomy of the University of Costa Rica, in San Jose. Phytopathology is part of the curriculum.

RESEARCH   The Instituto Interamericano de Ciencias Agrícolas of the Organization of American States (OAS), which was established in Turrialba in 1944, has been very helpful in training young Costa Ricans who have contributed interesting and valuable research papers. In 1964 32 diseases on 12 crops were studied. Leaders in teaching and research in Costa Rica are L. C. González, Gámez, and Bianchini.

Two Costa Ricans are on the teaching staffs of northern universities: Eddie Echandi is at North Carolina State University, and Luis Sequeira, a Fellow of the American Phytopathological Society, is at University of Wisconsin.

El Salvador

TEACHING   An elementary course is offered in the Faculty of Agronomy of the University.

RESEARCH   In 1948, phytopathology was organized as a section in the National Center of Agronomy and has now been transferred to the Dirección General de

Investigaciones Agronómicas. The first chief of the section was Abrego, who was succeeded by Alas López. In 1956 the Instituto Salvadoreño de Investigaciones del Cafe (Coffee Research Institute) was established with a section of phytopathology dealing exclusively with diseases of coffee. Other problems are in tobacco, cotton, maize, beans, rice, and vegetables.

Guatemala

TEACHING   The Universidad de San Carlos de Guatemala and National Agricultural School have courses in phytopathology in their curriculum. A book on coffee culture was published by Alvarado (1) in 1935, with 194 pages devoted to plant diseases. Müller, director of the National Agricultural School between 1943 and 1948, was the first teacher in mycology and phytopathology. The position of head pathologist was held by Darley for a year in 1949, Le Beau from 1950 to 1956, Flores from 1956 to 1959, and Schieber for several years. All of them were engaged in research, especially in coffee, maize, potato, beans, avocado, citrus, and cucumber.

Honduras

RESEARCH   The important research in Honduras is done on bananas by the pathologists of the United Fruit Company and Standard Fruit Company.

Mexico

TEACHING   The teaching of phytopathology in Mexico started in the National School of Agriculture at Chapingo in 1935. Growth has been at a rapid pace. In 1959, a graduate school was organized in Chapingo, offering a master's degree in phytopathology. Undergraduate courses are offered in three other universities and schools of agriculture.

RESEARCH   With the establishment of the Office of Special Studies, a cooperative effort between the Ministry of Agriculture and the Rockefeller Foundation, research in agriculture started on the road to many successes. Goals have been established, and attained. The battle against hunger has been almost won. New very high yielding cultivars of wheat, potatoes, and maize resistant to disease have replaced or are rapidly replacing the susceptible and low-yielding native cultivars of three decades ago.

Galindo A., head of the Department of Plant Pathology in the Graduate School at Chapingo, has done outstanding research on late blight of the potato. He is a Fellow of the American Phytopathological Society.

Prominent phytopathologists who have contributed to the development of the cooperative Mexico-Rockefeller program are E. C. Stakman, George Harrar, E. J. Wellhausen, Norman Borlaug, and John Niederhauser.

Gene pools in potato, wheat, and corn have been established which have become international centers for basic studies in which many other countries participate.

Nicaragua

TEACHING   Presently only a semester course in phytopathology is taught at the School of Agriculture near Managua.

RESEARCH   The first Agricultural Experiment Station was organized as a cooperative project between Nicaragua and the USA in 1952. Phytopathological studies were included in the Department of Agronomy headed by phytopathologist S. C. Litzenberger (63).

Panama

TEACHING   A course is offered at the National Institute of Agriculture and in the Faculty of Agronomy of the National University. Ocaña is professor of phytopathology.

RESEARCH   The diseases of 12 crop plants are the subject of study. Emphasis has been placed on banana, cacao, and yuca (*Manihot*).

Paraguay

TEACHING AND RESEARCH   The National School of Agriculture (now Botanical Garden) was established in 1895. Swiss botanist Bertoni was director until 1905 (13, 14) and also taught plant pathology. Winkelricd taught phytopathology at the National School of Agriculture at Trinidad from 1931 to 1938. This school became the Mariscal Estigarribia National School of Agronomy in 1939 and in 1957 was incorporated into the new Faculty of Agronomy and Veterinary Medicine.

In 1952 the Interamerican Technical Cooperative Agricultural Service came into being, and Ana Michalowski, a Pole, was the phytopathologist for eight years; during the last two she also taught at the Faculty of Agronomy of the National University in Asunción. Alvarez García, a Puerto Rican phytopathologist, served in the Interamerican Technical Service since 1952 and in 1957 became consultant to the Faculty of Agronomy. He published *Course in Phytopathology and Entomology* with Michalowski and Chunchuy in 1955. Doffield, from the USA became consultant for the Faculty of Agronomy in 1962.

Peru

Dongo (personal communication, Bazán de Segura, Perú) associates the beginning of interest in phytopathology with the publication by García Merino (44) in 1878 of the first book in this field in Latin America, *Plant Disease Epidemics on Coastal Lands of Peru*. Diseases were then considered simply as botanical abnormalities. A more modern approach is considered to begin with the establishment of the experiment station at La Molina in 1927. Here Abbott was in charge of phytopathology for three years and laid the foundation of formal research. He diagnosed diseases of crop plants and published many papers. His associate García Rada continued the study of diseases and published a book in 1947 (45). Two other books, one in 1965 (10) and another in 1975 (11), are by Bazán de

Segura. Dongo became head of the department of phytopathology of La Molina in 1968. Another experiment station was opened in Tingo María with a Department of Plant Pathology.

TEACHING   Phytopathology was part of the course in botany offered in 1902 at the National School of Agriculture and Veterinary Medicine. Early teachers were Leopoldo Hecq, Carlos Deueumostier (to 1912), and Julio Gaudron (1912—1950). At the end of this period phytopathology became a separate course taught by García Rada until 1957. The Agricultural University of La Molina opened in 1957 and phytopathology was included in the curriculum as a separate course.

Uruguay

The faculty of agriculture was founded in 1908. The first teacher in plant pathology was Gustavo Gassner, who was succeeded by Augusto Rimbach in 1910. Gassner paid special attention to the cereal rusts of Uruguay and Argentina. Arturo Montero Guarce succeeded Rimbach in 1917.

Papers of general interest have been published: Girardi on bacterial canker of peach (47), Giacconi on diseases of the grape vine, Báez (5) on wheat rust, Bianchi on potato diseases (16), Fantini (37) on control of anthracnose of the grape, and Fischer (42) on chemical treatment of seed.

Venezuela

In Fernández Valiela's (41) account of 1964, Malaguti established the starting point of phytopathology in this country as the organization of the Faculty of Agronomy in 1937. Actually, Nolla (77) in 1936, in his report to the Minister of Agriculture, recommended mycology, elementary plant pathology, and advanced plant pathology as required courses in the Department of Botany; a Department of Phytopathology in the Experiment Station, an Extension phytopathologist; and a phytopathologist at the Practical School of Agriculture. Phytopathologist Müller was one of the first appointees. He became the first phytopathologist to engage in teaching and research. His list of the diseases of cultivated plants in Venezuela (74) in 1937—1938 is probably the first contribution to phytopathology in this century.

The College of Agriculture was transferred to the Central University, Ministry of Education, in 1948, and phytopathology has been taught there ever since. It is also a subject of study in the Universities of Zulia and Oriente.

Research is carried out on many crop plants, mainly in sesame (*Sesamum indicus* L.), cotton, coffee, cacao, beans, corn, coconut, and sugarcane.

SPECIAL EVENTS

One may mention the organization of meetings and societies as a measure of the increasing interest of phytopathology as a profession. In 1936 the pathologists of Brazil held their first meeting at Rio de Janeiro; in 1949 the Central American pathologists met in Guatemala; the Latin American pathologists met in Mexico City; the Caribbean Division of the American Phytopathological Society was founded in 1960–1961 and meets every year in different countries of the area, in rotation. The Asociación Latinoamericana de Fitotecnia was organized with a section of Phytopathology (1964–1966), and the Sociedad Mexicana de Fitopatologia was host to the American Phytopathological Society in 1972 at Mexico City.

In 1943, the Rockefeller Foundation, at the request of the Mexican government, entered into an agreement with that government's Ministry of Agriculture which made possible a joint program in research and teaching, aimed at crop improvement as a solution to the food shortage. A similar program was subsequently organized in Colombia and Chile.

ACKNOWLEDGMENTS

We have drawn freely from Wellman's *Tropical American Disease*. We are indebted to Guillermo Gálvez, CIAT, Colombia; to María de Lourdes de la I de Bauer, and Jorge Galindo, Graduate School, Chapingo, Mexico; to E. R. French, International Potato Center, and Consuelo Bazán de Segura, both of Lima, Perú; to Gino Malaguti, Centro Investigaciones, Maracay, Venezuela; to R. H. Stover, United Fruit Company, La Lima, Honduras; and B. H. Waite, American Embassy, El Salvador, for valuable information. We are indebted also for library facilities to Chancellor R. Pietri, Puerto Rico University at Mayaguez; Director Frank W. Martin, Institute of Tropical Agriculture, Mayaguez, Puerto Rico; and Director M. Perez Escolar and Miss Joan Hayes, Librarian of the Agricultural Experiment Station, Rio Piedras.

Literature Cited

1. Alvarado, J. A. 1935. Enfermedades del cafeto. *In Tratado de caficultura práctica,* 176–369. Guatemala: Impreso Tipogr. Nac. 524 pp.
2. Deleted in proof
3. Aycock, R. 1966. Stem rot and other diseases caused by *Sclerotium rolfsii;* or, the status of Rolf's fungus after 70 years. *N. C. Agric. Exp. Stn. Tech. Bull.* 174:1–202
4. Backhouse, W. O. Mejoramiento de trigos. *Argent. Dir. Ensén. Agric. Inv. Agric. Publ.* 73
5. Báez, H. 1922. La rulla o polvillo de trigo. *Uruguay Def. Agric.* 3:163–67
6. Deleted in proof
7. Bancroft. J. 1876. *Report, 1st, of the Board Appointed to Inquire into the Causes of Disease Affecting Livestock and Animals.* Brisbane
8. Bazán de Segura, C. 1952. Escoba de brujas del cacao en Bagua. *Cent. Nac.*

Inv. Exp. Agric. La Molina, Perú
9. Bazán de Segura, C. 1953. More about the golden nematode in Perú. *Plant Dis. Reptr.* 37(5):326
10. Bazán de Segura, C. 1965. *Enfermedades de cultivos tropicales y subtropicales.* Lima: Editorial Jurídica Lima. 439 pp.
11. Bazán de Segura, C. 1975. *Enfermedades de cultivos frutícolas y hortícolas,* 3–276. Lima: Editorial Juridica
12. Berkeley, M. J., Curtis, A. M. 1881. *Grevillea* 9:99
13. Bertoni, M. S. 1911. Contribución al estudio de la gomosis del naranjo. *Paraguay Puerto Bertoni Agron.* 5: 77–89
14. Bertoni, M. S. 1919. La gomosis de los citrus y un nuevo medio preventivo y curativo. *Ann. Cient. Paraguayos.* Ser. 2:408–21
15. Bettencourt, A. J., Carvalho, A. 1968.

Melhoramento visando a resistencia do cafeeiro a ferrugem. *Bragantia* 27:35−68

16. Bianchi, A. T. 1920. Enfermedad de la papa. *Argent. Sucor* 1(6):8−9
17. Brandes, E. W. 1919. Banana wilt. *Phytopathology* 9:339−89
18. Brandes, E. W. 1920. Artificial and insect transmission of sugar cane mosaic. *J. Agric. Res.* 19:131−38
19. Buddenhagen, I. W. 1961. Bacterial wilt of bananas: history and known distribution. *Trinidad Trop. Agric.* 38:107−21
20. Caldas, F. J. 1803. Memoria sobre la nivelación de las plantas que se cultivan en la vecindad del Ecuador. In *Obras de Caldas por Eduardo Posada*. Bogotá: Imprenta Nacional
21. Carvalho, A., Monaco, L. C. 1972. Melhoramento do cafeeiro visando a resistencia a ferrugem alaranjada. *Cienc. Cult. Sao Paulo* 23(2):141−46
22. Carvalho, A., Monaco. L. C. 1972. Adaptaçãõ e productividade de cafeeiros portadores de factores para resistencia a *Hemileia vastatrix. Cienc. Cult. Sao Paulo* 24:924−31
23. Castillo, J., López, S., Torres, E. 1972. Comportamiento de introducciones de café con resistencia a *Hemileia vastatrix* en Colombia. *Cent. Nac. Inv. Café, Chinchina, Colombia.* 31 pp.
24. Chardón, C. E. 1926. Observaciones sobre las enfermedades del café. *Colomb. Medellín Inf. Esc. Sup. Agric. Med. Vet.*, pp. 43−52
25. Chardón, C. E. 1926. La gomosis, una epidemia grave de la caña en Antioquía. *Gob. Dep. Antioquía, Col. Agric. Vet. Colombia Circ.* 1:1−22
26. Chardón, C. E., Veve, R. A. 1923. The transmission of sugar cane mosaic by *Aphis maidis* under field conditions in Puerto Rico. *Phytopathology* 3:24−29
27. Chardón, C. E. et al 1929. Reconocimiento agropecuario del Valle del Cauca. *Informe de la Misión Agrícola Puertorriqueña.* Ed. in San Juan, Puerto Rico, 1930
28. Cordero, M. D. 1882. Apuntes sobre el Chahuistle. *Mex. Soc. Agric. Bol.* 4:283
29. Cordley, A. B. 1908. Lime-sulfur spray as a preventive of apple scab. *Rural New Yorker.* March 1, 1908:26
30. De Acosta, P. J. 1590. *Historia Natural y Moral de las Indias.* Edición primera de O'Gorman, Edmundo, 1962. *Fondo de Cultura Económica,* Mexico. 444 pp.

31. de Moussy, V. M. 1860. *Description Géogr. Statist. Conféd. Argent.* 1:531−33
32. Draenert, F. M. 1869. Bericht über die Krankheit des Zuckerrohres. *Z. Parasitenkd.* 1:13−17
33. Draenert, F. M. 1869. Weitere notizen über die Krankheit des Zuckerrohre. *Z. Parasitenkd.* 1:212
34. Drost, A. W. 1912. De surinaamsche panama-siekte in the gros michel bacoven. *Dep. Landb. Surinam Bull.* 26. Trans. 1913, S. F. Ashby, in *Jam. Dep. Agric. Bull.* 2
35. D'Utra, G. 1900. Molestias vermiculares do cafeeiro. *Bol. Agric. Sao Paulo* L:1−16
36. Ernst, A. 1891. El tizón de maíz. *Venez. Caracas Min. Obras Publ.* 4 pp.
37. Fantini, N. 1920. La antracnosis y medios de curación. *Uruguay Def. Agric.* 1:179−80
38. Fawcett, G. L. 1924. Enfermedades de la caña de azúcar en Tucumán. *Argent. Est. Exp. Agric. Bol.* 1. 47 pp.
39. Fawcett, G. L. 1941. El carbón o tizón de la caña de Tucumán. *Ind. Azucar.* 46:80−81
40. Fernández Valiela, M. V. 1955. *Introducción a la Fitopatología.* Talleres Graficos Gadola, Buenos Aires
41. Fernández Valiela, M. V. 1964. Desenvolvimiento de la fitopatología en los diferentes países latinoamericanos. Mesa Redonda de Fitopatología. *VI Reunión Latinoam. Fitotec. Lima, Perú*, pp. 1−75 (mimeo)
42. Fisher, F. E. 1968. Timing of spray applications for control of *Cercospora* diseases of citrus. *Phytopathology* 58: 553
43. García Díaz, C. 1885. Influencia del invierno sobre los organismos. *Soc. Agric. Mex. Bol.* 9:84
44. García Merino, M. 1878. *Las Epidemias de las Plantas en la Costa de Perú.* 192 pp.
45. García Rada, G. 1947. *Fitopatología Agrícola del Perú.* 423 pp.
46. García Redondo. 1936. See Ref. 85
47. Girardi, J. 1920. Tumor bacteriano del duraznero. *Uruguay Def. Agríc.* 1:300−2
48. Girola, C. D. 1904. Investigación agrícola de la República Argentina. *Minist. Agric. Nac. An.* 1:217−38
49. Goeldi, E. A. 1892. Relatorio sobre a molestia do cafeeiro na Provincia do Rio de Janeiro 1887. *Arch. Mus. Nac. Rio de Janeiro* 8:8−123
50. Gómez Medeiros, A. 1970. Observing

on 17 January, occurrence of *Hemileia vastatrix* on coffee in Fazenda Asunciao, Municipio Aurelino Leal, Bahia. Processed for president of CEPLAC, Jan. 26, 1970. Rio de Janeiro, Brazil. 1 p.
51. Grillo, H. V. S. 1935. A evolucao da Phytopathología. *Rodriguésia* 1(3): 2−11
52. Guyot, H. 1953. La lutte contre *Cercospora musae* dans les bananerais de Guadeloupe. Essais de nebulisation (fogging). *Fruits Outre Mer* 8:525−32
53. Guyot, H., Cuillé, J. 1954. Les formules fongicides huileuses pour le traitement des bananerais. *Fruits Outre Mer* 9:289−92
54. Hauman, L., Parodi, L. 1921. Los parásitos vegetales de las plantas cultivadas en la República Argentina. *Rev. Fac. Agron. Vet. Univ. Rio Grande do Sul* 3:234
55. Hecq, L. 1906. Una enfermedad del café en Perú. *Minist. Fom.* 4(9):30−39
56. Huergo, J. M. Enfermedades criptogámicas en la zona norte y oeste de la provincia de Buenos Aires y el broussin. *Ofic. Nac. Agric. Argent.* 21:575
57. Jobert, C. 1878. Sur une maladie du caféier observée au Brésil. *Paris Acad. Sci. C. R.* 87:941−43
58. Lavergne, G. 1899. Una enfermedad criptogámica de la papa. *Chile Bol. Soc. Nac. Agric.* 30:667−74
59. Lavergne, G. 1901. La antracnosa de la vid *(Gloeosporium ampelophagum)*. *Chile Estac. Pat. Veg. Cartilla* 8:1−8
60. Le Feuvre, R. 1877. *El Oidium tuckeri, enfermedad de las viñas y medio de curarlas.* Imprenta Nac. Santiago Cartilla. 22 pp.
61. Le Feuvre, R. 1878. La antracnosis de la vid. *Chile Soc. Agric. Bol.* 9:490−95
62. Linnaeus, C. 1753. *Species Plantarum.* Holmiae, L. Salvatii, Paris
63. Litzenberger, S. C., Stevenson, John A. 1957. *Plant Dis. Reptr. Suppl.* 243:3−19
64. McKenney, R. E. B. 1910. The Central American banana blight. *Science N.S.* 31:750−51
65. McMartin, A. 1945. Sugar cane smut: reappearance in Natal. *S. Afr. Sugar J.* 29:55−57
66. Marchionatto, J. B. 1948. *Tratado de fitopatología, Ed. Sudam., Buenos Aires, Argentina*
67. Matz, J. 1922. Gumming disease of sugar cane. *P. R. Dep. Agric. J.* 6(3):5−21

68. *Mex. Bol. Soc. Agric.* 1884−1885. ¿Que haremos? 8:658
69. *Bol. Soc. Agric. Mex.* 1892. Destrucción de la Cúscuta. 16:366
70. *Mex. Bol. Soc. Agric.* 1898. Enfermedades de las papas. 22:814−815
71. Michelsen, C. 1893. Enfermedad del café. *Agric. Soc. Agric. Colomb.* 9: 415−18
72. Millardet, P. M. A. 1885. Traitement du mildiou par le mélange de sulphate de cuivre et de chaux. *J. Agric. Prat.* 2:707−10
73. Mujica, R., Fernando, Vergara, C. 1945. *Flora Fungosa Chilena. Indice Preliminar de los Huéspedes de los Hongos Chilenos y sus Referencias Bibliográficas, Minist. Agric.* 199 pp.
74. Müller, A. S. 1950. A preliminary survey of plant diseases in Guatemala. *Plant Dis. Reptr.* 34:161−64
75. Noack, F. 1898. Cogumelos parasitas das plantas de pomar, horta e jardim. *Braz. Inst. Agron. Estado Sao Paulo Campinas Bol.* 9:75−88
76. Nolla, J. A. B. 1929. Enfermedades de las plantas. In *Reconocimiento Agropecuario del Valle del Cauca, Informe de la Misión Agrícola Puertorriqueña,* ed. C. E. Chardón et al. San Juan: Chardón
77. Nolla, J. A. B. 1936. Informe al Doctor Alfonso Mejía, Ministro de Agricultura de la República de Venezuela Relacionado con el Proyecto de Organización de un Colegio de Agricultura, una Estación Experimental y el Fomento Agrícola, Caracas
78. Nowell, W. 1919. The red ring or root disease of coconut palms. *West Indian Bull.* 17(4):180−202
79. Nowell, W. 1923. *Diseases of Crop Plants in the Lesser Antilles.* London: West India Comm. 383 pp.
80. Orjuela N. J. 1964. Desarrollo de la fitopatologia en Colombia. *Agric. Trop.* 20:541−49
81. Pardo, P. 1883. Contestaciones a la circular sobre enfermedades de la papa. *Agric. Soc. Agric. Colomb.* 5:251−54
82. Popenoe, H. 1963. The pre-industrial cultivator in the tropics. *Int. Union Conserv. Natl. Res. Proc.,* pp. 66−73
83. Puttemans, A. 1906. Sobre una molestia de los feijoeros *(Isariopsis griseola)* e sus synonymos. *Braz. Rev. Agric.* 130:1−7
84. Puttemans, A. 1911. Nouvelles maladies des plantes cultivées. *Soc. Bot. Belg. Bull.* 48:235−47

85. Puttemans, A. 1936. *Alguns Dados para Servir a Historia da Phytopathología no Brasil e as Primeiras Notificacoes de Doencas de Vegetaes Neste País*. Presented at 1st Meet. Phytopathol. Braz., Jan. 1937. *Rodriguesia*. 1936 (1937). 2 (Spec. no.):17—36. *P. R. Univ. J. Agric.* 1940. 24(3):77—108

86. Rodrigues, C. J. Jr., Bettencourt, A. J., Rijo, L. 1975. Races of the pathogen and resistance to coffee rust. *Ann. Rev. Phytopathol.* 13:49—70

87. Rorer, J. B. 1911. A bacterial disease of banana and plantains. *Phytopathology* 1(2):45—49

88. Rorer, J. B. 1918. Enfermedades y plagas del cacao en el Ecuador, y métodos modernos apropiados al cultivo del cacao. *Asoc. Agric. Ecuador, Ambato, Ecuador.* 80 pp.

89. Saenz, N. 1895. *Memoria Sobre el Cultivo del Cafeto.* Bogota: Casa Ed. Pérez. 185 pp. 3rd ed.

90. Schieber, E. 1975. Present status of coffee rust in South America. *Ann. Rev. Phytopathol.* 13:375—82

91. Sequeira, L. 1958. Bacterial wilt of banana; dissemination of the pathogen and control of the disease. *Phytopathology* 48:64—69

92. Sequeira, L. 1962. Control of bacterial wilt of bananas by crop rotation and fallowing. *Trinidad Trop. Agric.* 38:211—17

93. Smith, E. F. 1910. A Cuban banana disease. *Science* N.S. 31:754—55 (Abstr.)

94. Spegazzini, C. 1909. Mycetes Argentinensis. *An. Mus. Nac. Buenos Aires Ser.* 3:82—287

95. Spegazzini, C. 1909. *An. Sociedad Cient. Argent.* 83:231; 1929. 108:7

96. Spegazzini, C. 1911. Mycetes Argentinenses. *An. Mus. Nac. Buenos Aires* 3:389

97. Stahel, G. 1915. *Marasmius perniciosus* nov. sp. de verovzaker der Krullotenziekte van de cacao en Surinam. *Dep. Landb. Surinam Bull.* 33:1—27

98. Stahel, G. 1917. De zuid-amerikaansche hevea—bladziekte veroorzaakt door *Melanopsammopsis ulei* nov. gen. (= *Dothidella ulei* P. Hennings). *Dep. Landb. Surinam Bull.* 34:1—111

99. Stover, R. H. 1962. Intercontinental spread of banana leaf spot. *Trop. Agric. Trinidad* 39:327—38

100. Stover, R. H. 1962. Fusarial wilt (Panama Disease) of bananas and other *Musa* species. *Commonw. Mycol. Inst. Phytopathol. Pap.* 4:1—117

101. Stover, R. H. 1972. *Banana, Plantain and Abacá Diseases. Commonw. Mycol. Inst.*, Kew, Surrey, England. 316 pp.

101a. Stover, R. H. 1974. *Pathogenic and Morphologic Variation in Mycosphaerella fijiensis (M. musicola).* Presented at Ann. Meet. Caribb. Div., Am. Phytopathol. Soc., Trinidad

102. Thorne, G. 1961. *Principles of Nematology.* New York: McGraw-Hill. 333 pp.

103. Toro, R. A. 1930. Colombia: crop disease and pests. *Int. Plant Prot. Bull.* 4(1):3—4

104. Wellman, F. L. 1972. *Tropical American Plant Disease.* Metuchen, NJ: Scarecrow. 989 pp.

105. Whetzel, H. H. 1918. *An Outline of the History of Phytopathology.* Philadelphia: Saunders. 130 pp.

106. Zapata, F., Cardona, C. 1965. *Agric. Trop. Columbia* 21(11):750—76

107. Zeman, V. 1921. Bacteriosis del bananero (una nueva enfermedad). *Univ. Nac. Rev. Fac. Agron. Argentina* 14:17—30

Copyright © 1976 by Annual Reviews Inc.
All rights reserved

DISEASES AS A FACTOR IN PLANT EVOLUTION ♦3631

Jack R. Harlan

Agronomy Department, University of Illinois, Urbana, Illinois 61801

INTRODUCTION

Disease epidemics are among the most spectacular of biological phenomena. Not only can mortality be enormous but it is selective, sparing the occasional resistant or escaped individual and destroying the susceptible ones. Crops, domesticated animals, and man himself have suffered repeated epidemic disasters throughout recorded history. Of the many plagues that raged across Europe and Asia, the one in the winter and spring of 1347−1348 was especially virulent and the death rate exceptional. In many communities, less than 10% of the population survived and there were not enough healthy people to bury the dead (16). In 1633, a smallpox epidemic swept through the New England Indian population with such virulence that some villages were carried off entirely (14). Even measles, a comparatively mild disease in Europe, took a devastating toll of Polynesians in Hawaii (61). From the havoc wrought on the forces of Sennacherib (2 Kings 19:35) and the plagues visited on Egypt to the 20th century, history has been punctuated by the dread and terror of epidemic disease.

Most of us have seen towns and cities stripped of their best shade trees by Dutch elm disease, and some of us can recall the great, stark, grey hulks of dead chestnut trees that once dotted our eastern forests. We can, somehow, get by without shade and timber trees, but when epidemics strike the food plants we live by, famine and starvation may result. We all know of the great potato famines of 1846−1851 which had such devastating effects on the population of Ireland. Perhaps we are not so well informed on the great famine in Bengal of 1943 (65) caused primarily by *Helminthosporium oryzae (Cochliobolus miyabeanus)*. Even a nation producing substantial food surpluses like the United States can suffer heavy financial loss in epidemic years. In countries that normally import some food and that have limited exchange resources, epidemics of crop disease are utter disaster. Plant diseases can cause humans to die (of starvation) as well as the black death or cholera. One must suppose that any agent that *selectively* destroys genotypes, populations, or whole species as efficiently as epidemic disease surely has played a role in the evolution of modern faunas and floras.

31

Haldane (33, 34), Motulsky (61), Fenner (20, 21), Black (7), and others have indicated that "infectious diseases have exerted some of the strongest of the pressures that shaped the development of modern man" (20). I attempt in this essay to evaluate plant disease as a factor in evolution. This, I must do as an evolutionist, not as a plant pathologist, plant pathology being an area in which I have little knowledge or expertise. I attempt to avoid tedious descriptions of situations familiar to all plant pathologists, but at the same time am compelled to use examples common in phytopathological literature. This will not be a literature review but rather a personally biased assessment of the subject.

At the outset I make two presuppositions that may or may not be correct, but which appear to limit the operation of plant diseases in evolution. First, the extreme susceptibility and virulence of epidemic situations are not the norm over long-term evolution and second, diseases have had much less influence in the drier and colder regions of the world than in the warm wet ones. The first presupposition calls for some sort of understanding of the kinds of disturbances or imbalance that would generate epidemics, and both presuppositions call for an analysis of possible roles of disease in the absence of epidemics.

ENDEMIC BALANCE

In Vanderplank's terminology (87) a disease is endemic when it is always present. The normal situation is characterized by a high degree of horizontal resistance in the host, relatively low levels of virulence of the pathogen, or both. The disease is always present but does little damage. It is not good evolutionary strategy for a pathogen to kill off its host; indeed the best strategy for the pathogen is one in which the host flourishes. The area occupied by an endemic disease is described by Vanderplank as "protean," bulging here and receding there like an amoeba (87). Over a period of time the disease may flare up and become more destructive, then subside again in cyclic fashion. The recent history of oak wilt, caused by *Ceratocystis fagacearum,* is cited as an example. The disease is endemic in Appalachia with a center in West Virginia. In the 1950s and 1960s it became epidemic, spreading into new areas in Pennsylvania and Ohio, then it became anti-epidemic and receded again. Oak wilt is spread by the oak bark beetle which in turn spreads or recedes according to its own parasites, predators, and other factors (87).

Genetic Cost of Resistance

In an endemic situation, host and pathogen accommodate to each other in a dynamic balance not unlike a climax flora. The strains of pathogen that appear to have a low level of virulence may turn out to be highly virulent on hosts well removed from the endemic center which have not previously been subject to attack by the pathogen. Similarly, if the host is moved to an area where the disease does not occur, it may lose its resistance rather promptly.

An example often cited is that of southern maize rust, *Puccinia polysora,* in Africa. The endemic situation is described eloquently by Borlaug (8): "Although one or the other of these rusts *(P. sorghi* and *P. polysora)* is commonly found infecting nearly every plant of maize throughout its natural range in Mexico, Central America, and northern South America, the infection seldom occurs in sufficient intensity to cause appreciable damage, except rarely and locally, where some farmer (which is uncommon) or some scientist (which is more common) has upset the balanced biotic system which exists between host and pathogen. *Indeed there is no well-documented report indicating extensive epidemics of rust on maize in its native Latin American habitat either in colonial or recent times.* Undoubtedly, the near perfect biotic balance between host and parasite has probably existed from the time maize was first domesticated, and almost certainly even prior to that time on its now-extinct wild ancestors." (Borlaug's italics)

When maize was introduced into Africa, *P. sorghi* was brought along and continued to be docile; *P. polysora* was left behind and was not noticed in Africa until 1949. By this time, African maize had lost its resistance, and the epidemic was extremely severe, causing losses of 50% or more, and plants frequently died before maturity. In three years the rust spread across Africa from west to east, and on to Mauritius, Réunion, and Madagascar. Asian maize had also become susceptible and another focus was established in Malaysia in 1950, spreading eastward across the Philippines, Indonesia, and Oceania, northward into Indo-China and, eventually, westward to India. Meanwhile, in West Africa the epidemic abated as the local maize populations became more resistant (32). In a similar way, 70 years of breeding and selection for high-yielding clones of rubber in the Orient resulted in loss of resistance to leaf blight, *Dothidella ulei* (87).

The implication of these and other examples is that horizontal resistance is developed at a cost of fitness. In the absence of the disease, the host quickly loses resistance. The implication is supported by the "Vertifolia effect," i.e. the decrease in horizontal resistance during the process of breeding for vertical resistance. As Vanderplank (86) pointed out, the Vertifolia effect may not always be a loss in horizontal resistance in the hands of the plant breeder, but sometimes could be due to selection for horizontal susceptibility under the protection of vertical resistance. But, why would a plant breeder select susceptible lines unless they had some superior fitness qualities? At any rate, it seems clear that high levels of horizontal resistance evolve at some cost in fitness.

The nature of the cost in fitness is not known in detail, is undoubtedly complex, and no doubt varies with host, pathogen, and the nature of resistance. If, for example, the defense strategy calls for the production of an array of fungitoxic or fungistatic compounds or of enzymes that detoxify pathotoxins, it would seem likely that such productions would compete with other metabolic processes and would require continuous selection for maintenance. Resistance by hypersensitivity would also appear to be a highly evolved state that could be lost readily in the absence of disease. The loss of resistance alleles in interbreeding landrace populations is readily achieved through simple genetic recombination where susceptibility has an advantage.

Since evolutionary theory has a broad biological base, it might be appropriate to point out that in humans, genes that protect against infectious disease may lower general fitness in dramatic fashion. Thalassemia, the sickling trait, and glucose-6-phosphate dehydrogenase deficiency owe their present frequencies and distribution to selection by falciparum malaria (61). Deleterious genes for cystic fibrosis, phenylketonuria, spastic diplegia, schizophrenia, and hyperuricemia, among others, exist in populations at such frequencies they could not be sustained by mutation, and were probably dispersed by past epidemics (61). Defense against disease may require a heavy genetic load if selection pressures are strong.

Fitness of Pathogens

Pathogens have their fitness problems, too. Most of our information comes from studies involving vertical resistance because it is seldom possible to identify a gene unless it has a specific effect. For example, Frandsen (24) surveyed the races of late blight, *Phytophthora infestans,* in Europe and found all races virulent in R_4 hosts. Graham (28) found that about half of the isolates tested in Canada also attacked R_4 plants. This particular virulence gene apparently confers fitness to the pathogen even in the absence of R_4 potato genotypes. The R_4 gene, of course, is useless for breeding purposes and would not be used at all unless it conferred some fitness advantage to the potato. On the other hand, the allele for avirulence at R_1 must also confer fitness to the pathogen since R_1 potato genotypes introduced into cultivation in 1935 were almost completely protected for a number of years (87).

Similarly, stem-rust resistance provided by Sr6 plus Sr9d gave almost complete protection to Canadian wheats over a period of 20 years, even though the same combination failed in Texas. Since rusts work their way from Texas to Canada yearly, it seems that the subraces of race 15B that can attack cultivars with the Sr6 gene are poorly fitted for Canadian conditions (87). Some highly specialized races of pathogen can attack only a few genotypes of the host, and if these are not grown the race could disappear. Races of rust have been monitored for many years in Australia (90), United States, and Canada (30), and the evidence is clear that different races not only have very specific host ranges but differ in adaptation to climatic and ecological conditions as well (70). Mutations from avirulence to virulence may be common, but the virulent genotype is not always well adapted (90).

One must visualize the endemic balance developing over a period of time with continuous fluctuations, adjustment, and accommodation by both host and pathogen. Both increase in resistance in the host and decrease in virulence of the pathogen are likely, but the latter is often difficult to detect. Both phenomena have been demonstrated experimentally in the artificially induced epidemic of infectious myxomatosis in Australian rabbits. Rabbits were infected in 1950, and the initial impact was devastating as hoped for; rabbits died by the millions (20). Within a few years, somewhat attenuated strains of the virus became dominant since they were transmitted more readily by mosquitoes than the original, highly virulent strain (57). The attenuation was rather slight at first, killing 90% of the rabbits challenged instead of 99%, but other much less virulent strains appeared, including one that does not even make a rabbit sick. At the same time, resistance in the rabbit population was demonstrably increasing. The disease is still serious for rabbits, but the trend is clearly moving toward an endemic balance (22).

One may also visualize the evolution of endemic balance as a form of coevolution, perhaps not as spectacular as that between plants and animals (26), and much more difficult to document, though real enough. Good examples of biochemical warfare have, so far, been demonstrated in only a few cases, but the evolutionary strategy is probably common. Romanko (71) showed that the toxin victorin produced by *Helminthosporium (Cochliobolus) victoriae* is inactivated by resistant plants. Toxin formation by *Verticillium albo-atrum* is reduced in wilt-resistant tomatoes (82). Several fungitoxic substances have been isolated and characterized chemically, including α-solanine and α-chaconine of potato, tomatine of tomato, pisatin of pea, trifolirhizin from red clover roots, orchinol from orchid tubers, ipomeamarone from sweet potato, phaseolin from common bean, and viciatin from horse bean. Allen & Kuc (2) showed the first two to be highly toxic on *Helminthosporium carbonum,* a pathogen that does not attack potato. Pathogens that do attack the potato are more resistant. Thus, two can play the game. Either host or pathogen can produce a toxin; the other can detoxify it. The production of phytoalexins and their breakdown, and of pathotoxins and their breakdown, could develop into an elaborate interplay. The same would be true for other kinds of attack and defense mechanisms. Gallegly (25) indicated, for example, that after selection for increased horizontal resistance, the pathogen may respond with increased "aggressiveness."

Cross protection may be one mechanism in the evolution of endemic balance. It has been shown that infection by one disease may protect the host from another (94, 95). As in animals, a mild form of the disease may also protect against a virulent strain (74). Even the exceptionally virulent *Endothia parasitica,* that has threatened extinction of the American chestnut, may be transformed into a much less injurious form by transmission of a cytoplasmic determinant through hyphal anastomosis (85). By whatever means, the trend over time appears to be an adjustment of virulence and resistance toward a dynamic balance in which disease does not threaten the host, and the host supports the pathogen indefinitely.

Linkage

Genes for fitness tend to be linked (29). Indeed, genetic elements that control developmental processes must be protected against recombination either by homozygosity throughout the entire population, redundancy, tight linkage, or combinations of all three. Genes for disease resistance are among the most important in conferring fitness, so important in fact that they are accumulated at the expense of fitness in other respects. The selection pressures are powerful and the deployment of genetic defenses is likely to involve genetic linkage with an array of genes involved in general fitness. The arrangement of genes on the chromosome and their allelic states is far from random, and is subject to precise selection pressures. In Dobzhansky's terminology (18), they became *co-adapted,* or develop *relational balance* according to Mather (59). The evolutionary result is an integrated, balanced genome in which genetic defenses against disease are enmeshed in a matrix of genes controlling other fitness characters.

It is said that vertical resistance has little effect on general fitness because genes can be inserted into cultivars without affecting performance. But, if genes for vertical resistance were linked to genes for good or acceptable fitness, we might not be able to detect such linkages. We can demonstrate that genes for vertical resistance are linked to genes conferring *adverse* fitness for cultigens when they come from distantly related sources. For example, Sears (75) used X rays to induce translocations between an *Aegilops umbellulata* chromosome carrying leaf-rust resistance and *Triticum aestivum* chromosomes. At least 17 different translocations were identified. One of them, analyzed in detail, carried deleterious genes distal to the gene for leaf-rust resistance, which was located near the centromere. Another translocation, however, showed no deleterious effects and differed from normal wheat only in being slightly later and carrying the gene for leaf-rust resistance. The translocation could not be detected cytologically and presumably involved only a small segment of chromosome.

The difficulty of extracting an individual gene from the background matrix is illustrated in breeding tomatoes resistant to root-knot nematode, *Meloidogyne incognita* (69). High resistance was detected in *Lycopersicon peruvianum* by Bailey in 1941 in Tennessee. The difficult cross was made by Smith in 1944 in California. The equally difficult backcross was made by Watts in 1946 in Arkansas. Many more backcrosses followed without much difficulty, but tight linkages prevented development of good fruit types. Finally, these linkages were broken by Gilbert and McGuire in 1956 in Hawaii. This isolated the gene *Mi* which can now be inserted into cultivars by routine backcross procedures (69).

A. L. Hooker (personal communication) had a similar experience in transferring resistance to northern corn leaf blight, *Helminthosporium turcicum,* from *Tripsacum floridanum.* For a number of generations all the resistant plants in the population gave an agronomically unacceptable performance, while the susceptible plants were normal for corn belt maize. Eventually, the linkages were broken and resistant plants with adequate agronomic fitness were recovered. The implication is that disease resistance genes are, in fact, linked to other genes conferring fitness for specific ecological niches. Linkage, in a general sense, is not rigid and can be broken up by crossing-over; however, it does give a considerable inertia to the breakup of adaptive genetic complexes (59).

Studies by Qualset (66) and associates have shown a highly nonrandom association among morphological and developmental characters in Ethiopian barley and incidence of resistance to barley yellow dwarf virus. Resistance is associated with white and purple grain color, short-haired rachilla, long glume awn, lax spike, covered caryopsis, and late flowering time. Except for late flowering, which might be expected in high elevation barleys, there is no known selective advantage for this array of characters. It should be emphasized that close association of characters can be easily maintained in self-fertilizing species without physical linkage of genes on the chromosomes.

Disruptive Selection

Fitness under the threat of disease is different from fitness in the absence of disease. The difference is sufficient that one would expect disruptive selection to operate between the two types of populations. In one population, resistance genes accumulate at the cost of general fitness; in the other population general fitness is enhanced at the expense of resistance. It is usual in disruptive selection for the intermediates to have low survival value, and this would probably be the case for individuals along the fluctuating margins of an area of endemic disease. Loss of general fitness imposed by high levels of resistance may restrict the host species geographically and in ecological amplitude.

We would expect that disruptive selection (between populations under the threat of disease and those not threatened) would generate a considerable genetic cohesion within populations and a considerable differentiation between populations (58). At the same time, the various races and ecotypes of the host will deploy different genetic defenses because they are adapted to different environments and are subjected to a genetically different array of pathogens. The fragmentation of a species into distinct populations with different peaks of fitness has long been considered a classic strategy for efficient evolutionary change (89, 92). Where several important diseases are operating independently, they could have a profound effect on the population structure of the host and, over evolutionary time, the effect could be beneficial. The cost in fitness caused by disease resistance is a kind of genetic load which may have long-term advantages for the host (89).

Genetic Load

In fact, according to Wallace (89), "the ability to evolve requires a genetic load." Populations of outbreeding species normally carry a remarkable number of genes that in homozygous condition are subvital, semilethal, or lethal in their effects. This is the genetic load referred to by geneticists and evolutionists. As pointed out earlier, some of these genes are known to confer some measure of protection against disease in humans. The heterozygotes must be favored in order to keep the deleterious genes in the population. Heterozygotes are favored in inbreeding species as well, to the point that homozygosity is reached much more slowly than predicted by calculation. Balanced polymorphisms are characteristic of natural populations because heterozygosity confers a fitness advantage (18, 93). Evolutionary advance of the species may require the genetic sacrifice of many individuals.

Under epidemic conditions selection pressures are extremely strong. Thoday & Boam (83) in 1959 were the first to show in a carefully designed experiment that disruptive selection of sufficient intensity can separate populations even against a gene flow of 50% per generation. Since then, there have been a number of studies of both natural and artificial populations to confirm the general principle (10, 45, 77). Under extreme conditions, adjacent populations of outbreeding species can evolve strikingly different genetic constitutions despite repeated crossing between them. Some of these models apply rather well to epidemic conditions.

Under conditions of endemic balance, selection continues, but tends to be stabilizing within the populations subject to disease. The races of pathogen may be attenuated, but any individual host plant that lowers its genetic defenses too far will be pruned out. Vanderplank (87) suggested that the host could evolve over-resistance, but this would lead to lowered general fitness and would not be to the advantage of the host. Thus, stabilizing selection maintains the balance on both sides. The point is that selection pressures due to disease are present even if the disease is not causing much damage. Stabilizing selection does not require a raging epidemic to operate. Indeed, the apparently comfortable accommodation of host and parasite in endemic balance requires constant selection for maintenance. The principle is of general evolutionary application, and investigations of predator-prey relationships show many similarities to host-pathogen interactions (27).

Mather (58, 59) is inclined to believe that disruptive selection, once started, is a self-stimulating drive, generating polymorphisms, barriers to gene flow, and eventual speciation. Genetic barriers prevent wasteful production of nonadopted hybrids and, consequently, have a selective advantage. This may well be true if disruptive selection is long sustained, but short-term differentiation can set the stage for the operation of differentiation-hybridization cycles, and introgressive hybridization between distinct populations (93). Hybrids between populations are not always at a disadvantage, especially in a changing environment or under the artificial conditions of agriculture. It may be going too far to suggest that diseases may lead species down proscribed evolutionary pathways, but the genetic effects on the host could be profound and should not be ignored.

At any rate, we may conclude that the evolutionary endpoint is an endemic balance in which the disease is always present but does either little or only occasional damage. Long-term selection pressures are in the direction of such a balance. Gene complexes that permit the host to avoid epidemics have an advantage over those that allow a fraction of the host population to survive them. The evolutionary strategy is to pay the cost of resistance rather than suffer the consequences of vulnerability. Plant breeders and pathologists may find the natural strategy instructive.

EPIDEMIC IMBALANCE

Perhaps the most extreme cases of imbalance are epidemics in which the host is so susceptible and the pathogen so virulent that, after the epidemic subsides, all host individuals are either dead or immune and the pathogen fails to survive, or if there is no immunity, both fail to survive. Epidemics of this kind have been well documented in small, isolated human populations (7). Some diseases require sizable host populations for maintenance, and it is thought that most of the human diseases caused by virus (e.g. measles, influenza, smallpox, poliomyelitis, and some intestinal disorders like typhoid or cholera) could not have been supported by the small isolated populations of paleolithic times, and that they become important only after the evolution of agricultural societies with larger and more concentrated populations (7, 20, 61). Livingstone (56) pointed out that falciparum malaria could hardly have been a significant disease of man in Africa until agricultural peoples opened up the forest and became the most widespread large mammal. The importance of selection pressures caused by disease can hardly be illustrated more dramatically than by the insertion into the human population of genes modifying hemoglobin that confer some resistance to malaria at the cost of near lethality when the genes are homozygous.

Stevens (81) complained that plant pathologists have not studied epidemiology much, and the literature is scanty compared to that for human and animal epidemiology. He is correct, but the biological models of the black death (plague), tuberculosis, smallpox, syphilis, yellow fever, and so on do not appear to be very different in basic character from chestnut blight, Dutch elm disease, late blight of potato, stem rust of wheat, witches' broom of cacao, Panama disease of banana, mosaic of sugarcane, coffee rust, rice blast, or leaf blight of rubber. Most evolutionary principles apply to both plants and animals.

Cases of extreme susceptibility and extreme virulence most often occur when host and pathogen are introduced for the first time or for the first time in a long time (81, 87). The host has little in the way of genetic defenses to deploy and all races of the pathogen are likely to be virulent. The imbalance is usually brought about by (a) introduction of the disease agent into areas where it had not occurred (e.g. white pine blister rust, Dutch elm disease, chestnut blight, coffee rust to the New World, late blight of potato, powdery mildew of grape, maize rust to the Old World) (60) or (b) introduction of the host into areas where it had no opportunity to develop resistance to local diseases. Fire blight of pears and apples in America (88) is one example. Asian rice grown in West Africa may suffer severely from local races of blast, while African rice is so sensitive to virus diseases that it can be grown in the Philippines only under a canopy of insecticides to prevent infection (T. T. Chang, personal communication).

Man is the direct or indirect cause of most (possibly all) of the epidemic imbalances we know about. The disturbances caused by human activity are relatively recent in evolutionary time, but so profound and far-reaching that they have probably caused the most intensive evolutionary crisis in the history of the earth. It is the short time span that is critical here. Changes in climate that brought about the deposition and advance of glaciers over large segments of continental land masses and their subsequent retreat also profoundly influenced evolutionary history all over the globe, but these events took place over a much longer time span. There was more time for evolutionary balances to be maintained. Even so, the aftermath can be detected biologically in definable suture zones of biome interaction (67). On the other hand, man has not only altered the terrestrial biosphere fundamentally; he has done it in only a few thousand years. It is pertinent, then, to analyze briefly some of his activities that are implicated in recent evolution.

MAN-MADE IMBALANCE

Evolution of Agriculture

The effect of hunting-gathering man on the world ecological balance must have been relatively minor. It is true, he set fires with the intention of altering vegetation, and he did develop hunting skills that may have had an impact on the large mammalian fauna. But, the total global human population before agriculture is estimated to have been only about 10 million (52), and the cumulative disturbance of fauna and flora was probably not much more than that of any other large mammal (41). It was with the evolution of effective agriculture that man began to churn up the landscape, destroying the natural flora and inserting an artificial agricultural flora. According to current archaeological evidence, the process began slowly and probably in a number of widely separated parts of the world more or less simultaneously. We shall always be ignorant of the first attempts at plant cultivation, but we can detect signs of effective agriculture or plant manipulation emerging between eight and nine thousand years ago in the Near East, Southeast Asia, Meso-America, and South America (40).

Agriculture did not burst full-blown upon the world; it evolved over a period of several thousand years and so did the host-pathogen relationships of the crops. Basic subsistence agriculture with a typical small-village farming pattern has survived here and there in various parts of the world into modern times, and has been studied by many people (40). Whether in Asia, Africa, Oceania, or the Americas, these agricultural economies have several features in common: (a) a large number of plants are harvested in the wild so that the crops provide only a portion of the diet; (b) crops are grown in mixtures on the same piece of ground; (c) there are usually several landraces or cultivars of each crop; (d) the landraces themselves are genetically heterogeneous; (e) fields are small and often scattered; (f) villages are small and often widely separated; (g) the populations are likely to be mobile and the site of the village is shifted from time to time; (h) in the wet tropics especially, bush-fallow rotations evolve in which new land is brought into cultivation every 1−3 years and abandoned for 10−20 years.

Each of these features provides some defense against epidemics and, taken together, tend to maintain a stable endemic balance. Plant populations are often small enough that inoculum buildup is not sufficient to generate an epidemic (51, 87). The same is true of the human population, and many diseases cannot be sustained under these conditions (61).

With an increase in population and the evolution of more complex social and political structures, towns inevitably appear. These may be economically based market towns or religiously based ceremonial centers, or both. Defenses against disease begin to erode. The villagers cultivate larger fields so they will have surpluses to sell or barter. The villages are more permanent, being tied to the town economically, socially, religiously, and politically. The number of villages increase and more of the land is put under cultivation. Sources of wild plant foods decline, placing still more demands on the crops. The landscape begins to change and the flora to be impoverished.

The emergence of city-states intensifies the process. A cash economy develops. More and more land must be planted to support the urban population. The fields become larger, the number of crops fewer. The agricultural flora is extended at the expense of the natural flora. There may be considerable trade between cities, with an expanded opportunity for exchange of hosts and pathogens. The consolidation of large territories into nations and empires simply accelerates the processes long under way. The only new feature added is long-distance trade on a sufficient scale that the introduction of entirely new and virulent pathogens is likely, and the chances of serious epidemics are enhanced. At this stage, in several parts of the world, literature developed and we find the first historical documentation of plagues, diseases, and pestilence of plants (50).

At the same time that agricultural land use is intensified, successful agricultural systems also expand into new areas. The Near Eastern agricultural complex based on barley, wheat, sheep, and goats can be traced in some detail by archaeological evidence. The time is significant in indicating the approximate number of generations available to move toward a balance between host and pathogen. The system was effectively established in the nuclear area (Palestine, southeastern Turkey, northern Iraq, western Iran) by 7000 BC (40). It spread westward across Anatolia to Greece by 6000 BC, along the shores of the Mediterranean to Spain by about 5000 BC, up the Danube and down the Rhine to the North Sea by about 4000 BC. Herding peoples at least occupied the Sahara in the fifth millennium BC when the rainfall was considerably greater than now. Slowly, the system moved eastward across Iran to the Indus by 3000 BC and southward across Arabia to the Ethiopian plateau (time unknown). Wheat, barley, and goats apparently reached China during the Shang period in the second half of second millennium BC (43).

The settlement of Europe by farmers is well documented archaeologically (62, 68, 84). Expansion of Asian, African, and American agricultural systems can also be traced, but with much less evidence available. Plant domestication and the development of farming systems may have been as early in southeast Asia and the Americas as in the Near East, but major expansion possibly occurred somewhat later (40). There are few significant dates for Africa as yet.

There are two obvious consequences of expanded agricultural systems: (a) vast continuous areas are established yearly to a few select species, and (b) these species must adapt to new environments. Wheat and barley originated in an area of winter rainfall and summer drought. They were moved into regions of summer rainfall. Adaptation on the Eurasian continent was feasible where summer temperatures were not too high, but in India they could not tolerate summer monsoons and had to be grown in the cool, winter, dry season. In the Nilgiris of south India, temperatures were low enough, but conditions were especially favorable for rusts. Traditionally, emmers have been grown there because of greater disease resistance, and in modern times the "Khapli" emmers have been used as sources of rust resistance for bread wheats.

The Ethiopian plateau also has a cool climate and conditions favoring leaf and stem diseases. Both wheat and barley were able to respond by deployment of genetic defenses. In recent years, plant breeders have found Ethiopian cultivars to be enormously useful sources of resistance to an array of diseases. For example, in one test 113 of 117 cultivars resistant to barley yellow dwarf virus were from Ethiopia. High resistance to powdery mildew, loose smut, leaf rust, net blotch, septoria, scald, yellow dwarf virus, and barley stripe mosaic virus has been found in Ethiopian barleys. Resistance is common; many lines have resistance to several diseases and some to all races of a disease tested (39). The Ethiopian accession CI 5791, for instance, has shown resistance to all races of net blotch so far tested. We do not know when wheats and barleys were brought to Ethiopia, but apparently the time has been sufficient for hosts and pathogens to adjust reasonably well.

Qualset (66) has pointed out that the distribution of resistance to barley yellow dwarf virus is nonrandom in Ethiopia, being concentrated in one or two detectable localities, especially at high elevations. A thorough analysis of Ethiopian barleys based on a more complete sampling, together with surveys of incidence of the disease could be rewarding for an understanding of host-pathogen interactions in an endemic center of a disease.

Some stabilizing features of subsistence agriculture have persisted until very recently, and still provide some protection where agriculture has not been too strongly influenced by modern technologies. One concrete example is PI 178383, a wheat collection I made in eastern Turkey in 1948. It is reported to have near immunity to four races of stripe rust, *Puccinia striiformis* (1); good resistance to 35 races of common bunt, *Tilletia caries* and *T. foetida;* 10 races of dwarf bunt, *T. controversa;* and usable tolerance to snow molds, *Fusarium* spp. and *Typhula* spp. (12, 15). Kendrick & Hoffman (48) reported that it carried three genes, each conferring resistance to all races of common bunt. On the other hand, the accession is highly susceptible to leaf rust, which is not an important disease in the remote mountainous province of Hakkari. Several other collections from the area showed somewhat similar genetic protection. The assembly of this unusual array of genetic defenses evidently resulted from long culture of local landraces under specific selection pressures by disease.

The techniques of subsistence agriculture select continuously for a quality the French call *rusticité* and for which there is no appropriate English word. Cultivars or landraces with rusticité yield something despite flood, drought, disease, insects, nematodes, birds, monkeys, or witchweed. In basic subsistence agriculture there is little pressure for high yields, but a crop failure means famine and starvation. Dependability is essential, a matter of life or death to the cultivator. Crops developed under these conditions are hardly ever high-yielding, but they have excellent general fitness for local conditions and good wide-spectrum resistance to the diseases of the region (36, 37). Specialized, derived cultivars developed by modern breeding methods may be high-yielding, but this is usually achieved at the cost of restricted and specialized fitness.

In a general way, there seems to be some positive correspondence between centers of origin or centers of diversity and frequencies of genes for disease resistance (23, 55). The longer a crop is grown in a particular region, the more likely that it will develop genetic protection against the prevalent diseases. Of course, sometimes the crop originates in one area and the disease in another [e.g. potato from the Andes and *Phytophthora infestans* from Mexico (55, 64)], but ancient centers of culture are good places to look for resistant germ plasm. When crops expand into new regions and new environments the balance is frequently upset. Recent trends in agriculture have brought about cases of the most extreme imbalance.

Modern Times

The Columbian discovery was something of a watershed in many ways. Columbus was obviously not the first man to reach the New World, but it was only after 1500 AD that transoceanic exchanges of crops and diseases became common. These have provided some of the most spectacular epidemics in the history of crops and man. The global population at the time is thought to have been about 300 million (52). In four hundred and fifty years the figure increased tenfold to three billion. All of the most suitable land and a good deal of marginal land filled up with farmers by the early decades of the 20th century.

An unprecedented ecological situation has developed. Wheat is grown in a continuous carpet from northern Mexico far into Canada. The vast plains of Argentina and southern Brazil, large areas of Australia, and enormous stretches across the Soviet Union are devoted to a near monoculture of wheat. At the same time an enormous expansion of rice took place in Asia and Oceania, largely in the great river deltas and mostly in the last hundred years (40). It is said that over half the people on earth survive because of these two crops, wheat and rice. The trend has in no way been reversed and is, in fact, accelerating as ever more millions depend on ever fewer species for survival.

Even the crops we have left are becoming steadily more uniform as old landraces give way to modern-style cultivars (13, 38). This is a tale all too familiar to plant pathologists, and need not be elaborated here. The point should be made, however, that selection for disease resistance is no longer made in the production fields; it is done in the plant breeders' nurseries. We are risking our food supply on the experience, skill, and judgment of a comparative handful of people. It would appear that the ecological imbalance is about as extreme as it can get. There are not many important crops left to be abandoned. The challenge is to look to our defenses and see if we are doing the right things.

Would it be possible, in time, to generate some sort of endemic balance for the few major crops that have become absolutely essential for human survival? Probably not; the situation is too artificial to stabilize completely, but we could do more to move in that direction.

Several strategies have been proposed that might improve our defenses against epidemic outbreak (5, 49, 63, 76). Multiline synthetic cultivars have received considerable comment but little action (11, 53, 54). Other breeding systems are faster and more convenient. The general subject is treated in detail by Baker & Cook

(5). Duvick's 1975 paper (19) is of interest in describing the strategies actually used by his seed company in producing maize hybrids. Inbreds are monitored for some 24 diseases. Some can be used without problems even when known to be susceptible simply by growing hybrid derivatives in parts of the country where a particular disease is not serious, or by combining with a resistant inbred. In fact, a good many of the inbreds contributing to the US hybrid seed industry are susceptible to a number of diseases, but they have not caused serious problems. The procedures for selecting inbreds for combining ability are designed to maximize heterozygosity and in successful single-cross hybrids one inbred is likely to cover the susceptibilities of its mate. Three-way and four-way crosses would probably be even safer.

Vertical resistance is used very little in maize breeding (19). A few genes are known (e.g. *Ht* on chromosome 2, *Rp* on chromosome 10), but acceptable levels of horizontal resistance to most diseases of the corn belt can be developed. Generally, maize is pretty healthy in the region, and poor yields are more likely to be due to adverse weather than to disease. The notable exception in 1970 had the character of a failure of vertical resistance since susceptibility to southern corn leaf blight can be attributed to a single source (T cytoplasm). Diseases appear every year in the corn belt and do some damage, but severe epidemics are uncommon. One need only grow some nonadapted exotic material or genetic stocks near a pathologist's nursery to appreciate the degree of tolerance of the better corn-belt materials.

Vertical resistance has been much used in wheat breeding (44). In recent decades the crop has been generally protected most of the time, but occasional severe epidemics have occurred, and losses in yield can be extremely high (72). Resistance genes work fine when they work, but would not make so striking a difference if horizontal resistance had not been neglected. Vanderplank (86) suggests that in some cases "plant breeders and pathologists are busy patching their mistakes and bragging about how big the patches are." The question is whether it is possible to move in the direction of endemic balance when genes for vertical resistance are used as defense. Possibly, but not without dedicated effort to that end and, perhaps, not without some cost in agronomic fitness. One must wonder what the situation would be now if the same effort had been invested in raising levels of horizontal resistance as has been invested in incorporation of genes conferring vertical resistance.

In potato, both approaches have been used with respect to late blight. Some cultivars have good levels of field resistance with no R genes introduced (86). Other cultivars have been developed with an array of R genes. Despite many years of intensive plant breeding, late blight is still with us and losses can still be heavy (46). On the other hand, we no longer suffer epidemics like those of 1846, and the world produces over 300 million metric tons of potatoes yearly.

Since I am not a plant pathologist I am not qualified to enter the debate, but as an evolutionist it seems to me that stabilizing strategies that tend in the direction of balanced host-pathogen relationships are much to be desired. The pressure on the world food supply is such that modest yearly losses would be far better than occasional disastrous epidemics.

North American Wheat Belt

The North American wheat belt may be singled out as one of the most artificial and unnatural disturbances in all of agriculture. The annual epidemics of rusts moving over 4000 km along the Puccinia Path from a region where they cannot oversummer to a region where they cannot overwinter and back again, must be included among the marvels of the biological world. Obviously, it would be impossible if wheat were a minor crop, grown here and there in a few isolated patches. It is the enormous dependence of the human species on wheat that has caused this distortion of nature.

It might be instructive to investigate a natural solution to the same biological problem. Blue grama, *Bouteloua gracilis,* is distributed, in part, over the drier sections of the high plains wheat belt from northern Mexico well into Canada. Stands are often continuous over many thousands of square kilometers. For the more humid parts of the wheat belt, and especially along the water courses, little bluestem *(Andropogon scoparius)* and western wheatgrass *(Agropyron smithii)* have somewhat similar north-south ranges although stands are more interrupted. All three species are infected by rusts (4): blue grama by *Puccinia bartholomaei,* little bluestem by *P. andropogonis,* while western wheatgrass, being related to wheat, can be infected by *P. graminis* and *P. rubigo-vera.* The usual situation is for most plants in a population to carry a few pustules sometime during the year and senescent leaves may show quite a few. In some years, under special conditions, some genotypes can be rather severely damaged for part of the growing season. Selection, however, is minimal, and I have never seen a plant killed by rust.

The genetic defenses of these three native grasses have not been studied in detail, partly because rusts are not a serious problem, but we have investigated the genetic variability of the hosts enough to know that these species are enormously complex. In one study of blue grama over a small part of its range (35, 78), we found chromosome races of $2n = 20, 40, 60, 42,$ and 84. Many thousands of single-plant isolates have been grown in grass-breeding nurseries at several stations over many years. Viewing the array of genetic diversity displayed, we must conclude that it is on an entirely different order of magnitude than that deployed by the wheat crop over the same geographic range. Genetic diversity and coevolution over time have generated an endemic balance.

Man-made imbalances are by no means confined to his cultivated crops. Enormous losses have been sustained among forest trees as well. Bingham et al (6) report that most forest tree pathological work is "devoted to salvaging what man has brought about." They refer not only to introduced diseases such as white pine blister rust and chestnut blight but also point out that native diseases can do great damage when man disturbs the natural balance, e.g. Scots pine blister rust in Europe and fusiform rust in the southern United States.

NATURAL IMBALANCE

The world is so dominated by man-made disturbances that we have little experience with the natural order. Do epidemics occur in the absence of man? Did imbalance between host and pathogen develop before man began to dominate the scene? These questions are almost impossible to study today because the human influence is so pervasive that it can hardly be excluded. Most of what follows must, consequently, be speculation.

We know that long-distance dispersal of viable spores is possible (42). Viable spores have been found in upper air currents even over the Arctic and Antarctic and in midocean (31, 32). They could, conceivably, be carried from one continent to another. Whether this actually happens is subject to differences of opinion. It may be significant that coffee rust did not cross the south Atlantic until it reached Angola and the southern trade winds (9, 73). On the other hand, this may have simply made it more convenient to transport by airplane. We do not know. In any case, if it happens at all, it should be a rare event and, therefore, the more devastating. The spores must not only survive long-distance transport but must alight on a congenial host. For specialized pathogens, the new host would have to be related to the species normally attacked in the endemic source. If all these contingencies come about, a natural imbalance could result and epidemic conditions be established.

Population Fragmentation

If the host is extremely susceptible, an epidemic may occur in such severity that only remnants of the host population remain after the disease has run its course. Small, fragmented populations have particular evolutionary characteristics (93). They are subject to differentiation by genetic drift and gene fixation by inbreeding. If widely separated, they are subject to differentiation for fitness to local conditions (77, 79, 80). Some populations may behave like founder colonies and spread back over the original range of the species; others may not be able to do so. Again, the stage is set for operation of classical differentiation-hybridization cycles, introgressive hybridization, or other evolutionary strategies based upon the occasional encounter of divergent populations within the total gene pool. Evolutionary efficiency may be enhanced by the process.

As an illustration, Diller (17) reported in 1965 that a survey had located over 500 chestnut trees eight inches or more in diameter breast high. These were challenged by *Endothia* and over 90% succumbed. He was unwilling to identify the survivors as resistant, but suppose a few resistant trees could be found. Consider the genetic effect of funneling the entire species through a handful of survivors. The genetic base would be narrow to begin with, and could be lowered still more by drift and inbreeding effects (91). On the other hand, some computer simulation studies on bottle-neck populations or founder-colony populations suggest that loss of genetic variability is not as great as one might expect, provided there is a rapid increase in the population after the event (92, 93). If several widely separated colonies of trees survived, maintained isolation for some generations, and eventually made genetic contact, genetic diversity could even be enhanced, as suggested above.

If the model has validity and if epidemics of this sort should happen repeatedly with different diseases attacking different components of the natural flora over evolutionary time, one might predict a response of the flora as a whole. Vast, uniform stands of one or a few dominant species would be especially vulnerable. Where such plant formations occur in nature, the epidemic model just described probably does not apply. The taiga or most other coniferous forests, desert shrub, grassland steppes, and tropical savanna show no evidence that such natural imbalances have ever occurred.

Tropical Forests

The forests of the wet tropics are of a different order. As a rule, they are immensely rich in species deployed, it seems for maximum protection against disease epidemics. Individual specimens of a given species are likely to be widely separated, rarely occurring in groves or colonies. The whole flora is such a heterogeneous mixture that it would appear difficult for any pathogen to build up sufficient inoculum to increase the population of lesions very much. Endemic diseases are numerous, but natural epidemics unlikely.

Janzen (47) put forward an interesting suggestion that the unexpectedly uniform dispersal and low density of tropical forest species may largely be due to insects and other animals that eat seeds or seedlings. Most of the seeds fall near a parent tree, but most offspring to reach maturity are at a considerable distance. No doubt herbivores are a factor, but diseases could well be as important or more so.

The dispersal strategy is a good one, and it should be noted that cultivators in subsistence agriculture of the wet tropics have adopted the same defense. I have charted gardens with over 25 species of food plants and mixed orchards with over 30 species of fruits and nuts (see also Anderson, 3). Conversely, it is easy to see why tropical plantation crops have repeatedly suffered disastrous epidemics in the past and, no doubt, will in the future. As suggested in the introduction, plant diseases are likely to have had the most impact on host evolution in the warm wet parts of the world.

CONCLUSIONS

Disease has had a powerful effect on plant evolution even under conditions of endemic balance where diseases do little damage. Epidemics are spectacular results of disturbing the balance, and man is the usual cause of disturbance. Epidemics threaten his food crops, his forests, his livestock, and himself. They stem primarily from the enormous populations, unnatural population densities and distributions, the disturbed habitats, and artificially extended geographic ranges of the threatened species.

The human population continues to grow, and somehow continues to be fed. The fact that most people on earth are poorly nourished and some are starving suggests a·chronic imbalance with no relief in sight. If the human population could be stabilized, there would, perhaps, be a chance of working the plants man depends on into a reasonably stable relationship with their pathogens. In time, generalized field resistance might be developed to the point that losses due to disease would be an acceptable price to pay. This is the direction that evolution would take over time if selection were continuous and directional.

Under conditions of modern agriculture, however, host genotypes are deployed and removed so rapidly that selection occurs primarily in breeding nurseries and not in field populations. Miscalculations by plant breeders and plant pathologists could mean disaster. Failure of politicians and administrators to allocate adequate resources for critical work could have the same effect. The latter would appear the more likely, considering the fact that governmental decision-makers are generally ignorant of agriculture and have little interest in it (38).

Literature Cited

1. Allan, R. E., Purdy, L. H. 1970. Reaction of F_2 seedlings of several crosses of susceptible and resistant wheat selections to *Puccinia striiformis*. *Phytopathology* 60:1368−72
2. Allen, E. H., Kuć, J. 1968. α-Solanine and α-chaconine as fungitoxic compounds in extracts of Irish potato tubers. *Phytopathology* 58:776−81
3. Anderson, E. 1952. *Plants, Man and Life*. Boston: Little, Brown. 245 pp.
4. Anonymous. 1960. *Index of Plant Diseases in the United States, US Dep. Agric. Handb.* 165:1−531
5. Baker, K. F., Cook, R. J. 1974. *Biological Control of Plant Pathogens*. San Francisco: Freeman. 433 pp.
6. Bingham, R. T., Hoff, R. J., McDonald, G. I. 1971. Disease resistance in forest trees. *Ann. Rev. Phytopathol.* 9:433−52 ·
7. Black, F. L. 1975. Infectious diseases in primitive societies. *Science* 187: 515−18
8. Borlaug, N. E. 1972. A cereal breeder and ex-forester's evaluation of the progress and problems involved in breeding rust resistant forest trees: moderator's summary. *US Dep. Agric. Misc. Publ.* 1221:615−42. Washington DC
9. Bowden, J., Gregory, P. H., Johnson, C. G. 1971. Possible wind transport of coffee leaf rust across the Atlantic Ocean. *Nature* 229:500−1
10. Bradshaw, A. D. 1971. Plant evolution in extreme environments. In *Ecological Genetics and Evolution,* ed. R. Creed, 20−50. Oxford: Blackwell
11. Browning, J. A., Frey, K. J. 1969. Multiline cultivars as a means of disease control. *Ann. Rev. Phytopathol.* 7:355−82
12. Burgess, S., ed. 1971. *The National Program for Conservation of Crop Germ Plasm*. Athens, Ga.: Univ. Georgia Press. 73 pp.
13. Committee on Genetic Vulnerability of Major Crops. 1972. *Genetic Vulnerability of Major Crops*. Washington DC: Natl. Acad. Sci. 307 pp.
14. Cook, S. F. 1973. The significance of diseases in the extinction of the New England indians. *Hum. Biol.* 45:485−508
15. Creech, J. L., Reitz, L. P. 1971. Plant germ plasm now and for tomorrow. *Adv. Agron.* 23:1−49
16. Deaux, G. 1969. *The Black Death, 1347.* London: Hamish Hamilton. 229 pp.
17. Diller, J. D. 1965. Chestnut blight. *US Dep. Agric. For. Pest Leafl.* 94:1−7
18. Dobzhansky, Th. G. 1970. *Genetics of the Evolutionary Process*. New York: Columbia Univ. Press. 505 pp.
19. Duvick, D. N. 1975. Using resistance to manage pathogen populations: A corn breeder's commentary. *Iowa State J. Res.* 49:505−12
20. Fenner, F. 1959. Myxomatosis in Australian wild rabbits—evolutionary changes in an infectious disease. *Harvey Lect.* 1957−1958:25−55
21. Fenner, F. 1970. *The Impact of Civilization on the Biology of Man,* ed. S. V. Boyden, 48−68. Toronto: Univ. Toronto Press. 233 pp.

22. Fenner, F., Marshall, I. D. 1957. A comparison of the virulence for European rabbits *(Oryctolagus cuniculus)* of strains of myxoma virus recovered in the field in Australia, Europe and America. *J. Hyg.* 55:149—91

23. Flor, H. H. 1971. Current status of the gene-for-gene concept. *Ann. Rev. Phytopathol.* 9:275—96

24. Frandsen, N. O. 1956. Rasse 4 von *Phytophthora infestans* als Feld rasse in Deutschland. *Phytopathol. Z.* 26:124—30

25. Gallegly, M. E. 1968. Genetics of pathogenicity of *Phytophthora infestans. Ann. Rev. Phytopathol.* 6:375—96

26. Gilbert, L. E., Raven, P. H. 1975. *Coevolution of Animals and Plants.* Austin, Tex.: Univ. Texas Press. 246 pp.

27. Gilpin, M. E. 1975. *Group Selection in Predator-Prey Communities.* Princeton: Princeton Univ. Press. 110 pp.

28. Graham, K. M. 1955. Distribution of physiological races of *Phytophthora infestans* (Mont.) de Bary in Canada. *Am. Potato J.* 32:277—82

29. Grant, V. 1975. *Genetics of Flowering Plants.* New York: Columbia Univ. Press. 514 pp.

30. Green, G. J. 1971. Physiologic races of wheat stem rust in Canada from 1919 to 1969. *Can. J. Bot.* 49:1575—88

31. Gregory, P. H., Monteith, J. L., eds. 1967. *Airborne Microbes.* London: Cambridge Univ. Press. 385 pp.

32. Gregory, P. H. 1973. *The Microbiology of the Atmosphere.* Aylesbury, England: Leonard Hill. 377 pp. 2nd ed.

33. Haldane, J. B. S. 1949. Disease and evolution. *Suppl. La Ricerca Sci.* 19:68—76

34. Haldane, J. B. S. 1957. Natural selection in man. *Acta Genet. Stat. Med.* 6:321—32

35. Harlan, J. R. 1958. Blue grama types from west Texas and eastern New Mexico. *J. Range Manage.* 11:84—87

36. Harlan, J. R. 1969. Evolutionary dynamics of plant domestication *Jpn. J. Genet.* 44:*(Suppl.)*337—43

37. Harlan, J. R. 1970. The evolution of cultivated plants. In *Genetic Resources in Plants: their Exploration and Conservation,* ed. O. H. Frankel, E. Bennett, 19—32. Oxford: Blackwell. 554 pp.

38. Harlan, J. R. 1972. Genetics of disas-

ter. *J. Environ. Qual.* 1:212—15

39. Harlan, J. R. 1973. Barley genetics and breeding (Summary). *East Afr. Agric. For. J.* 39(Spec. issue 6):21

40. Harlan, J. R. 1975. *Crops and Man.* Madison, Wis.: Crop Sci. Soc. 295 pp.

41. Harlan, J. R., de Wet, J. M. J. 1973. On the quality of evidence for origin and dispersal of cultivated plants. *Curr. Anthropol.* 14:51—62

42. Hirst, J. M., Stedman, O. J., Hogg, W. H. 1967. Long-distance spore transport: methods of measurement, vertical spore profiles and the detection of immigrant spores. *J. Gen. Microbiol.* 48:329—55

43. Ho, P-T. 1969. The loess and the origin of Chinese agriculture. *Am. Hist. Rev.* 75:1—36

44. Hooker, A. L. 1967. The genetics and expression of resistance in plants to rusts of the genus *Puccinia. Ann. Rev. Phytopathol.* 5:163—82

45. Jain, S. K., Bradshaw, A. D. 1966. Evolutionary divergence in adjacent plant populations. I. The evidence and its theoretical analysis. *Heredity* 21:407—41

46. James, W. C., Shih, C. S., Hodgson, W. A., Callbeck, L. C. 1972. The quantitative relationship between late blight of potato and loss in tuber yield. *Phytopathology* 62:92—96

47. Janzen, D. H. 1970. Herbivores and the number of tree species in tropical forests. *Am. Nat.* 104:501—28

48. Kendrick, E. L., Hoffman, J. A. 1963. Reactions of wheat varieties and selections to pathogenic races of *Tilletia controversa. Plant Dis. Reptr.* 47:736—38

49. Knott, D. R. 1972. Using race-specific resistance to manage the evolution of plant pathogens. *J. Environ. Qual.* 1:227—31

50. Kramer, S. N. 1959. *History Begins at Sumer.* Garden City, NY: Doubleday. 247 pp.

51. Kranz, J., ed. 1974. *Epidemic of Plant Diseases: Mathematical Analysis and Modeling.* New York: Springer. 170 pp.

52. Lee, R. B., DeVore, I., eds. 1968. *Man the Hunter.* Chicago: Aldine. 415 pp.

53. Leonard, K. J. 1969. Factors affecting rates of stem rust increase in mixed plantings of susceptible and resistant oat varieties. *Phytopathology* 59:1845—50

54. Leonard, K. J. 1969. Selection in

heterogeneous populations of *Puccinia graminis* f. sp. *avenae*. *Phytopathology* 59:1851−57

55. Leppik, E. E. 1970. Gene centers of plants as sources of disease resistance. *Ann. Rev. Phytopathol.* 8:323−44

56. Livingstone, F. B. 1958. Anthropological implications of sickle cell gene distribution in West Africa. *Am. Anthropol.* 60:533−62

57. Marshall, I. D., Fenner, F. 1958. Studies on the epidemiology of infectious myxomatosis in rabbits. V. Changes in the innate resistance of Australian wild rabbits exposed to myxomatosis. *J. Hyg.* 56:288−302

58. Mather, K. 1955. Polymorphism as an outcome of disruptive selection. *Evolution* 9:52−61

59. Mather, K. 1973. *Genetical Structure of Populations.* London: Chapman & Hall. 197 pp.

60. McNew, G. L. 1960. The nature, origin, and evolution of parasitism. In *Plant Pathology; an Advanced Treatise,* ed. J. G. Horsfall, A. E. Dimond, 2:19−69. New York: Academic. 715 pp. 3 vols.

61. Motulsky, A. G. 1960. Metabolic polymorphisms and the role of infectious diseases in human evolution. *Hum. Biol.* 32:28−62

62. Murray, J. 1970. *The First European Agriculture.* Edinburgh: Edinburgh Univ. Press. 380 pp.

63. Nelson, R. R. 1972. Stabilizing racial populations of plant pathogens by use of resistance genes. *J. Environ. Qual.* 1:221−27

64. Niederhauser, J. S., Cervantes, J., Servian, L. 1954. Late blight in Mexico and its implications. *Phytopathology* 44:406−86

65. Padmanabhan, S. Y. 1973. The great Bengal famine. *Ann. Rev. Phytopathol.* 11:11−26

66. Qualset, C. O. 1975. Sampling germplasm in a center of diversity: an example of disease resistance in Ethiopian barley. *Crop Genetic Resources for Today and Tomorrow,* ed. O. H. Frankel, J. G. Hawkes, 2:81−96. London: Cambridge Univ. Press. 492 pp.

67. Remington, C. L. 1968. Suture-zones of hybrid interaction between recently joined biotas. *Evol. Biol.* 2:321−428

68. Renfrew, J. 1973. *Paleoethnobotany: The Prehistoric Food Plants of the Near East and Europe.* New York: Columbia Univ. Press. 248 pp.

69. Rick, C. M. 1967. Exploiting species hybrids for vegetable improvement. *Proc. XVII Int. Hortic. Congr.* 3:217−29

70. Roelfs, A. P. 1974. Evidence for two populations of wheat stem and leaf rust in the U.S.A. *Plant Dis. Reptr.* 58:806−9

71. Romanko, R. R. 1959. A physiological basis for resistance of oats to victoria blight. *Phytopathology* 49:32–36

72. Romig, R. W., Calpouzos, L. 1970. The relationship between stem rust and loss in yield of spring wheat. *Phytopathology* 60:1801−5

73. Schieber, E. 1975. Present status of coffee rust in South America. *Ann. Rev. Phytopathol.* 13:375−82

74. Schnathorst, W. C. 1966. Cross protection in cotton with two strains of *Verticillium albo-atrum*. *Phytopathology* 56:151 (Abstr.)

75. Sears, E. R. 1956. The transfer of leaf-rust resistance from *Aegilops umbellulata* to wheat. *Brookhaven Symp. Biol.* 9:1−22

76. Simons, M. D. 1972. Polygenic resistance to plant disease and its use in breeding resistant cultivars. *J. Environ. Qual.* 1:232−40

77. Snaydon, R. W. 1970. Rapid population differentiation in a mosaic environment. I. The response of *Anthoxanthum odoratum* populations to soil. *Evolution* 24:257−69

78. Snyder, L. A., Harlan, J. R. 1953. A cytological study of *Bouteloua gracilis* from western Texas and eastern New Mexico. *Am. J. Bot.* 40:702−7

79. Stebbins, G. L. 1966. *Processes of Organic Evolution.* Englewood Cliffs, NJ: Prentice-Hall. 191 pp.

80. Stebbins, G. L. 1969. *The Basis of Progressive Evolution.* Chapel Hill, NC: Univ. North Carolina Press. 150 pp.

81. Stevens, R. B. 1974. *Plant Disease.* New York: Ronald. 459 pp.

82. Tarr, S. A. J. 1972. *The Principles of Plant Pathology.* London: Macmillan. 632 pp.

83. Thoday, J. M., Boam, T. B. 1959. Effects of disruptive selection. II. Polymorphism and divergence without isolation. *Heredity* 13:205−18

84. Tringham, R. 1971. *Hunters, Fishers and Farmers of Eastern Europe 6000−3000 B.C.* London: Hutchinson Univ. Library. 240 pp.

85. Van Alfen, N. K., Jaynes, R. A.,

Anagnostakis, S. L., Day, P. R. 1975. Chestnut blight: Biological control by transmissible hypovirulence in *Endothia parasitica. Science* 189:890−91

86. Vanderplank, J. E. 1968. *Disease Resistance in Plants.* New York: Academic. 206 pp.

87. Vanderplank, J. E. 1975. *Principles of Plant Infection.* New York: Academic. 216 pp.

88. Van der Zwet, T. 1968. Recent spread and present distribution of fire blight in the world. *Plant Dis. Reptr.* 52:698−702

89. Wallace, B. 1970. *Genetic Load: Its Biological and Conceptual Aspects.* Englewood Cliffs, NJ: Prentice-Hall. 116 pp.

90. Watson, I. A. 1970. Changes in virulence and population shifts in plant pathogens. *Ann. Rev. Phytopathol.* 8:209−30

91. Wright, S. 1931. Evolution in mendelian populations. *Genetics* 16:97−159

92. Wright, S. 1932. The roles of mutation, inbreeding, crossbreeding and selection in evolution. *Proc. Sixth Int. Congr. Genet.,* 1:356−66. Brooklyn, NY: Brooklyn Bot. Garden. 396 pp.

93. Wright, S. 1968−1969. *Evolution and the Genetics of Populations; A Treatise.* Chicago: Univ. Chicago Press. 2 vols. 469 pp., 511 pp.

94. Yarwood, C. E. 1956. Cross protection with two rust fungi. *Phytopathology* 46:540−44

95. Yarwood, C. E. 1967. Response to parasites. *Ann. Rev. Plant Physiol.* 18:419−38

Copyright © 1976 by Annual Reviews Inc.
All rights reserved

ETHYLENE IN SOIL BIOLOGY   ✦ 3632

A. M. Smith

Biological and Chemical Research Institute, New South Wales Department of Agriculture, Rydalmere, NSW 2116, Australia

INTRODUCTION

Ethylene, a simple unsaturated hydrocarbon gas, is well known in biology for its profound and often spectacular effects on plant growth. It is an endogenous plant growth regulator functional at concentrations as low as 0.04 ppm, with most growth responses saturated at about 1 ppm. Ethylene is produced by most plant parts, and directly or indirectly affects the physiology of most organs and virtually all developmental stages of plants. It is also involved in some host-pathogen interactions. These aspects of ethylene in plant biology have been thoroughly reviewed (1, 17, 18, 46, 87).

More recently, ethylene has been identified as a common constituent of the soil atmosphere of both aerobic and anaerobic soils. This ethylene is produced by microbial activity and concentrations are high enough to be biologically active. Important, far-reaching claims have been made about its role in soil biology. This review concentrates on ethylene in this context. Most emphasis is placed on ethylene as a critical regulator of microbial activity in soil with its implications in the important soil processes of rates of turnover of organic matter, availability of essential plant nutrients, and the incidence and control of soil-borne plant diseases. Cursory attention is paid to the significance of this exogenous source of ethylene on plant root growth and seed germination.

IDENTIFICATION OF ETHYLENE IN SOIL

Definitive research on ethylene production in soil was hampered until the advent of flame ionization gas chromatography, which enabled the detection and measurement of the trace concentrations that occur (110). The first authentic reports of ethylene production in soil came from geochemical studies (106, 110), but it was difficult to determine whether the source was the soil or contaminating plant material. Research, started in 1968 by K. A. Smith and colleagues at Letcombe Laboratory in the United Kingdom, showed unequivocally that ethylene is common in waterlogged anaerobic soil (111, 113). They found that the ethylene was probably of microbial origin (111) and showed that it may adversely affect plant growth (25, 110, 112, 113).

In 1973, my research showed that ethylene was common in Australian soils, that it was produced in appreciable quantities in open, aerobic systems, most probably by microbial activity and, more importantly, that it is a cause of soil fungistasis (102). Ethylene has been measured in field soils at concentrations ranging from a trace to 30 ppm or more, depending on the origin and history of the soil and the prevailing climatic conditions (34, 103–105, 109, 110). The methods used to record and quantify ethylene production in soil under laboratory and field conditions have been published (34, 104, 108, 111).

A special warning: great care should be exercised when testing soil or microorganisms for ability to produce ethylene because the gas can be liberated from virtually any organic substance subjected to heat or oxidation (1). In our laboratory, contamination of this type has come from rubber stoppers and tubing, plastic tubing, heated soil and plant material, trace amounts in commercial cylinders of oxygen, nitrogen, carbon dioxide, etc, slow leaks of liquid petroleum and coal gas supplies, and unburnt residues of these gases remaining in glassware after flaming to sterilize.

ETHYLENE AND SOIL FUNGISTASIS

Soil fungistasis is the term applied to the widespread inability of fungal propagules to germinate in soil under apparently favorable conditions unless supplied with exogenous nutrients (31). This phenomenon has considerable ecological significance in soil microbiology, and especially in plant pathology, and has been reviewed extensively (32, 58, 59, 65, 121). In the 20 years since its discovery, none of the alternative hypotheses proposed to explain soil fungistasis has been entirely satisfactory. Watson & Ford (121) attempted to reconcile the different hypotheses and proposed an inhibitor-stimulator model as the best explanation of the available data. They argued, however, that many different biotic and abiotic factors were likely to be inhibitors and stimulators of germination. I subscribe more to Lockwood's view (66) that a few, rather than many factors are probably responsible for soil fungistasis.

Ethylene has been shown to be an inhibitor causing soil fungistasis (102–104). Preliminary evidence that a volatile factor caused fungistasis was obtained with water-saturated airstreams passed over soil. This treatment annulled fungistasis. More than 50 soils with widely different origins and histories were studied, and ethylene, identified by gas chromatography, was the only material found in all soils at concentrations likely to be biologically active. Results with sterilized soil indicated that this ethylene was formed by microbial activity.

Volatiles have unique properties (40, 54, 64) that favor them as causal agents of fungistasis. They were implicated previously in soil fungistasis (11, 12, 23, 49, 50, 92–94), and, in fact, the results of Balis & Kouyeas (12) suggested that ethylene or a closely related material may be responsible. Many volatiles with biological activity, however, are produced by microorganisms (40, 53, 54, 73) and by plant material (64).

Ethylene was established as a cause of fungistasis by comparing the germination and subsequent growth of propagules of test fungi (including sclerotia of *Sclerotium rolfsii* and conidia of *Helminthosporium sativum*) placed on unsterile soil, wetted to about −0.5 bars, and exposed to water-saturated airstreams containing 0, 0.5, 0.7, and 1.0 ppm ethylene (102, 103; A. M. Smith, unpublished). Propagules of test fungi germinated and grew vigorously only when soils were exposed to ethylene-free airstreams (Figure 1). Propagules of the natural soil mycoflora including species of *Pythium, Fusarium, Trichoderma, Penicillium, Aspergillus, Cephalosporium,* and many unidentified fungi, behaved similarly. The inhibitory effects of ethylene were reversed when ethylene-free airstreams were swept over all treatments at the end of the experiments (102). Similar results were obtained with a range of soils including rainforest, grassland, plantation, and intensively cultivated soil.

Figure 1 Ethylene inhibiting the germination and growth of sclerotia of *Sclerotium rolfsii* and propagules of natural soil mycoflora in an avocado soil. *Top,* treated with water-saturated airstream + 1 ppm ethylene. *Bottom*, treated with water-saturated airstream only.

Hyphal growth of the fungi tested was also sensitive to ethylene. If propagules were germinated in an ethylene-free environment and then exposed to ethylene at the concentrations described, the hyphae lysed prematurely.

It should be emphasized that, if air-dried soils are used in these experiments, the soils must be wetted for at least two days before propagules of test fungi are placed on the soil and the airstream treatments started. If this is not done, most propagules germinate in all treatments. This is interpreted as the ethylene effect being overridden by excess organic nutrient liberated on wetting a dried soil (117). Such treatments are well known to annul fungistasis. Similarly, amendment of soil with simple carbohydrates or organic residues including ground wheat straw, alfalfa meal, and oat grain overrides the inhibitory effects of ethylene on most fungal activity (102).

These experiments clearly show that trace concentrations of ethylene effectively inhibit most fungal activity even in soils naturally rich in organic nutrients. Of paramount importance, however, is the interaction between the concentration of ethylene required in soil to maintain fungistasis and the organic nutrient levels available for microbial growth. Soils rich in organic matter produce most ethylene (102, 104, 105, 110, 111), but such soils also require higher concentrations of ethylene to inhibit germination of fungal propagules. Thus, organic nutrient is the main stimulator of propagule germination, and ethylene the inhibitor. The balance between the two determines whether a soil is fungistatic at any given time (102, 103). Appreciation of the importance of this balance helps reconcile my claims for ethylene as an explanation for fungistasis with Lockwood's (65, 66) nutrient-depletion hypothesis. The nutrient-depletion hypothesis may have relevance in intensively cultivated soils but is difficult to apply in rich forest soils where fungistasis was first described (31).

It is impossible to specify a particular concentration of ethylene that will cause fungistasis in all soils because this should vary from soil to soil and within the one soil, depending on organic nutrient status. Practical implications are that it is useless testing sensitivities of microorganisms to ethylene in pure culture. Results obtained in rich media and in the absence of competitive pressures cannot be extrapolated to unsterilized soil. Failure to appreciate these aspects leads to invalid conclusions about the sensitivity of particular species of soil fungi to ethylene.

This does not necessarily mean that all soil fungi will be inhibited by ethylene, or are equally sensitive to ethylene. Preliminary evidence indicates that some basidiomycetes and phycomycetes are more tolerant of ethylene than most fungi (104). In fact, mycelial growth of *Agaricus bisporus*, the cultivated mushroom, has been stimulated in our experiments by low concentrations of ethylene. The greater tolerance of these fungi may reflect their long evolutionary association with elevated concentrations of ethylene: many basidiomycetes are litter decomposers inhabiting forest and grassland soils, while phycomycetes are common in wet soil. Ethylene concentrations are usually high in these habitats (103–105). A similar association occurs with higher plants where root growth of lowland rice is unaffected by ethylene concentrations in excess of 40 ppm, whereas tomato, tobacco, and barley are sensitive to trace amounts (110). Differences in sensitivity of fungi to ethylene, however, will have important ecological implications in soil microbiology.

Research on ethylene has not eliminated other volatiles found in soil as possible causal agents of fungistasis. To warrant serious consideration, however, a volatile should occur in most soils most of the time at concentrations high enough to be biologically active. Ethylene fulfills these criteria but, as far as I am aware, they have not been met yet by other candidate compounds. Experiments made to test fungistatic effects of other volatiles in soil require the rigorous exclusion of ethylene now that we know that ethylene is common in soil and active at trace concentrations. Residual fungistasis (30) is a distinct form of fungistasis and does not appear to be caused by ethylene (62, 66).

Ethylene and Inhibition of Other Soil Microorganisms

Observations made in experiments with water-saturated airstreams passed over soil indicated that actinomycetes, bacteria, and some nematodes were also inhibited by ethylene (102–105).

Actinomycetes are almost certainly sensitive to ethylene in soil. In our fungistasis studies, actinomycete colonies grew prolifically after about seven days on soil subjected to ethylene-free airstreams but were absent in all ethylene treatments. Fungi partially overgrow actinomycete colonies in ethylene-free treatments making it difficult to obtain quantitative data. Steam-air treatment of soil at 60°C for 30 min to kill most fungi but not some actinomycetes (9) has not resolved this problem. Actinomycetes grow well on steamed soil even in 5 ppm ethylene presumably because the treatment liberates organic nutrients and removes some of the competition. In other experiments with untreated soils in closed vials, we obtained additional evidence that actinomycetes are sensitive to ethylene, but are more tolerant than most fungi. (104, 105).

Attempts to show quantitatively that ethylene affects bacterial activity in soil have been singularly unsuccessful. Ethylene concentrations up to 20 ppm have been imposed on soils without any measurable effects on bacterial activity as measured by oxygen consumption, carbon dioxide production, and total counts of bacteria in soil. Soils leached free of soluble carbohydrates in boiling 0.01 M calcium chloride (115) were also tested without effect (A. M. Smith, D. H. Pike, and W. L. Morrison, unpublished). On the data available, we must conclude that bacteria are not directly inhibited by ethylene at the concentrations likely to occur in soil. The ecological implications are discussed later.

Quantitative work on the effect of ethylene on soil nematodes is lacking. Marked increases in populations of saprophytic soil nematodes occurred at sites of low ethylene, but not where ethylene was high in experimental systems in soil (102). Fungi multiplied in the same sites, and bacteria accumulated on and around the fungal hyphae. Perhaps the nematodes simply responded to these microorganisms as a food source, rather than to the ethylene gradient. Careful experimentation is required to separate these effects.

MICROBIAL PRODUCTION OF ETHYLENE IN SOIL

All evidence points to microorganisms as the major source of ethylene in soil (102, 105, 111, 113) although the particular group or species responsible is in dispute. Many microorganisms, including soil inhabitants, produce ethylene in culture under certain conditions (1, 14). Ethylene is a common metabolite of fungi (14, 55, 56, 67, 70), is produced by some yeasts (14, 67), but is recorded from only one species of bacteria (38).

Initially, K. A. Smith & Restall (111) suggested that facultative anaerobic microorganisms were the most likely producers of ethylene in soil. Later research in the same laboratory led Lynch & Harper (70) to propose that *Mucor hiemalis* is a major producer. They used a medium enriched with methionine to selectively isolate aerobic ethylene-producing microorganisms. They reconciled their conclusion with earlier findings that ethylene production occurs under anaerobic or microaerobic conditions (111) by arguing that substrate for ethylene production by this aerobic fungus was mobilized anaerobically (70). We concluded (104, 105) from our studies that the main production of ethylene in soil is by spore-forming bacteria in anaerobic microsites. This conclusion was based on experiments that showed that ethylene was not produced in soil unless, or until, most of the oxygen was depleted, that production was most rapid in soils incubated under oxygen-free atmospheres, that it was produced in soil at water potentials of -5 bars or wetter but only slightly or not at all at -15 bars or drier, irrespective of whether incubated under air or nitrogen. Also, in our studies using aerated-steam treatments (9), production of ethylene was reduced significantly only when soils were heated above 80°C for 30 min, and was prevented only when soils were autoclaved. Heating soil to 60°C and 80°C enhanced initial production of ethylene, indicating that heat activation of spores may be involved (41). We argued that these results ruled out *M. hiemalis* as a major producer of ethylene in our soils because most fungi are killed by aerated steam at 60°C or above (9). However, Lynch (69) has challenged our conclusions.

Park's (77, 78) observation that the behavior of fungi in soil and in staled fungal cultures have much in common may be relevant in this controversy. He postulated that staled cultures formed a chemical inhibitor that caused autolysis of aged hyphae and that increased in concentration as cultures aged. His model system reproduced many of the features of behavior of fungi in soil, including fungistasis and lysis of mycelium (65). A number of volatiles, although not ethylene, have been involved in the phenomenon of staling (92–94). Ethylene, which I showed causes fungistasis and lysis of fungal hyphae in soil, is produced in pure culture by fungi, mainly during their death phase and immediately before autolysis (55, 68, 70, 71). Perhaps there are two sources of ethylene regulating fungal growth in soil—(*a*) an exogenous source from spore-forming bacteria in anaerobic microsites from which it diffuses and inhibits germination of fungal propagules and promotes premature lysis under limiting nutrient conditions, and (*b*) an endogenous source produced by fungi as hyphae age and that promotes their autolysis. There are precedents from plant research for both endogenous and exogenous sources of ethylene influencing physiology, particularly in the ripening of fruit (1, 17, 18, 46, 87).

Similarly, Lynch's finding (68) that Fe^{2+} (200 ppm) stimulated ethylene production from some chemical intermediate released into culture medium by *M. hiemalis,* may be relevant. Quite independently, we found (A. M. Smith, P. J. Milham, D. H. Pike, and W. L. Morrison, unpublished) that Fe^{2+}, at concentrations of about 100–1000 ppm in the soil solution, markedly stimulated ethylene production in soil. The onset of ethylene production after the Fe^{2+} treatment is so rapid that it is conceivable that a strictly chemical reaction is involved. Significant concentrations of Fe^{2+} are ephemeral in most soils because Fe^{2+} exists in soil only under reduced conditions resulting from intense microbial activity

(83, 85). It is possible that a chemical precursor of ethylene forms in soil as a result of aerobic microbial growth, but is not converted to ethylene until the microbial activity is intense enough in microsites to reduce Fe^{3+} to Fe^{2+}. The Fe^{2+} then chemically catalyzes ethylene formation. Although this remains a feasible interpretation of the data, all our results (103–105) indicate that anaerobic bacteria are primarily responsible for ethylene production in soil, but depend on activity of aerobes or aerobic conditions for production of an essential substrate or precursor.

OXYGEN-ETHYLENE CYCLE IN SOIL

In collaboration with R. J. Cook, an oxygen-ethylene cycle that has far-reaching implications for the biological balance of soil was described (103–105). This cycle has great practical potential for agriculture because its successful manipulation may lead to regulation of organic matter turnover in soil, better availability of essential plant nutrients, and development of effective biological control of soilborne plant pathogens.

There is general agreement that ethylene is produced in soil only under anaerobic conditions (70, 103–105, 110, 111). Its relevance as a critical regulator of microbial activity in field soils, therefore, may be challenged because most soils are considered to be adequately supplied with oxygen, at least in the upper layers (42). Elevated concentrations of ethylene have been recorded in field soils in both England and Australia (34, 103–105, 109, 110). Most ethylene occurred in wetter soils especially in warmer parts of the year, but concentrations high enough to be biologically active were recorded in apparently well-aerated soils.

Recognition of the presence of numerous anaerobic microsites in soil, especially in the rhizosphere is the key to resolving the apparent contradiction of ethylene production in field soils (104). Surprisingly, this concept causes much scepticism, misunderstanding, and argument even among soil microbiologists who have readily accepted that soil is composed of an innumerable variety of microhabitats (118). The evidence for the presence of anaerobic microsites in soil seems irrefutable. For example, strictly anaerobic bacteria are common in the surface layers of most soils (100, 101) ranging from the arctic to the tropics (60, 72), and have been recorded even in extremely sandy soils (79). Also, on purely physical considerations, anaerobic microsites are predicted to form continually in soil as a result of aerobic organisms utilizing oxygen, especially near organic matter (26, 42–44, 48).

Anaerobic microsites in soil are most likely to occur in the rhizosphere of plants (104) because this is the site of intense microbial activity (95–97), this is a source of electron donors (123), and the roots are additional sinks for oxygen (42). The rhizosphere's environment is dominated by the root exudates which, during the life of the plant, may account for from 2 to 20% or more of the total carbon fixed by the plant (13, 21). Marked increases in populations of strict anaerobes (72) and denitrifying bacteria (61, 122, 123) in the rhizosphere, as compared to the soil mass, confirm this conclusion. Exudate patterns alter markedly from site to site on a root system, vary with age of roots, and depend on environmental conditions (95–97). It follows, therefore, that individual anaerobic microsites are not stable during the life of the plant but are transitory and dynamic in location and size.

It is suggested (103–105) that anaerobes proliferate in these microsites and produce ethylene, which diffuses through the soil. Sensitive species of aerobic microorganisms will be inactivated first, but eventually, if ethylene concentrations become high enough in the soil atmosphere, most if not all aerobic microbial growth will be arrested, at least in the vicinity of organic nutrient sources. As a result, a self-regulating microbial cycle is established in soil. As aerobic growth slows, oxygen demand falls and oxygen will diffuse back into the microsites and prevent anaerobic ethylene production. This will permit resumption of aerobic growth. This cycle is summarized (105).

Oxygen Supply

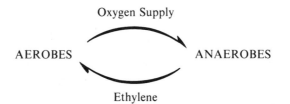

AEROBES   ANAEROBES

Ethylene

Verification of the existence of such a cycle in soil comes from numerous observations made in laboratory tests on ethylene production in soil (103–105). Prolific hyphal growth was observed under a dissecting microscope on the moist soil crumbs when ethylene was low but not when ethylene was high. In tests continued for over 80 days, ethylene concentrations reached a peak, dropped, then eventually rose again in cyclic fashion. Fungal hyphae appeared, lysed, and reappeared in response to the fluctuations. Actinomycetes behaved similarly except that colonies became evident only at intermediate concentrations of ethylene and after most fungal hyphae had lysed. As concentrations of ethylene increased further, these colonies also disappeared. These observations led us to conclude that actinomycetes, as a group, are more tolerant of ethylene than fungi.

It could be inferred that bacteria will proliferate in soil when ethylene concentrations are elevated because apparently they are unaffected by this gas. Similarly, it seems unlikely that soil bacteriostasis (16, 27, 28) is caused directly by ethylene, although as a phenomenon it closely resembles fungistasis. Ethylene, however, may indirectly limit aerobic bacterial activity in soil by restricting the growth of fungi and actinomycetes. There is good evidence that the fungal biomass dominates the metabolism of most agricultural and forest soils (6, 7, 22, 98). Fungi are better adapted to degrade the resistant carbonaceous materials that constitute the bulk of plant residues in soil, whereas bacteria proliferate most rapidly on simple breakdown products of these residues (3, 5). Also, growth of some bacteria is restricted in soil until fungal hyphae have developed (99). Therefore, ethylene may act indirectly as a pacesetter for bacterial activity in soil.

FACTORS AFFECTING THE OXYGEN-ETHYLENE CYCLE IN SOIL

Inhibition by Nitrate Nitrogen

Concentrations of nitrate nitrogen (NO_3^--N) in soil have a marked influence on ethylene production. K. A. Smith & Restall (111) first showed that NO_3^--N retarded ethylene production, but they amended soils with 2,000 ppm and 10,000 ppm NO_3^--N, which are unrealistically high concentrations to expect in soil.

Later, it was found (103–105) that soils under different systems of management produce vastly different concentrations of ethylene. Bare cultivated soil always produced less ethylene than one under crop, which in turn produced less than one left undisturbed. There was a negative correlation between ability of these soils to produce ethylene and the concentrations of NO_3^--N present. Detailed chemical analysis of these soils confirmed other published data that bare, cultivated soil often has from 20 ppm to over 200 ppm NO_3^--N whereas cropped soils commonly have much lower amounts and undisturbed soils under grass or forest contain only trace amounts (4, 19, 22, 51, 63, 90, 91). These variations in concentrations of NO_3^--N are reflected by differences in populations of autotrophic nitrifying bacteria (19, 114).

We studied ethylene production in six soils amended with different forms of nitrogen ranging from 20 to 200 ppm N. A chemical inhibitor of nitrification [2-chloro-6 (trichloromethyl) pyridine] was added to all treatments after preliminary experiments showed that the inhibitor did not directly affect ethylene production. Amendment of soil with NO_3^--N always delayed production of ethylene in proporation to the amount added whereas NH_4^+-N stimulated production, or had no effect, depending on the soil used (103–105). Ethylene production was best in soils leached free of NO_3^--N, but when these soils where reamended with nitrate at the concentrations described, production started only after denitrification, or not at all. Denitrification was assumed to be completed when nitrous oxide evolution ceased. A minimum concentration of NO_3^--N was required in soil to inhibit ethylene production, and this varied from soil to soil depending on the microbial nutrient status (104). However, concentrations of NO_3^--N in the range expected to occur in most agricultural soils were sufficient to retard or inhibit ethylene production in the soils tested.

Nitrate nitrogen acts as an efficient terminal electron acceptor for a range of facultative anaerobic bacteria in the absence of oxygen (15, 33, 123). In microbiologically active soils adequately supplied with NO_3^--N, hydrogen donors, and suitable species of bacteria, the redox potential will be poised sufficiently high to prevent activity of strict anaerobes or to stop other electrochemical reactions that occur at lower redox potentials (80, 81, 83, 119, 120). This seems to explain how NO_3^--N prevents ethylene production in soil. This hypothesis is supported by our results (unpublished) on ethylene production in soil following addition of excess Mn^{2+} and Fe^{2+} to activate the manganous—manganic and ferrous—ferric redox couples. Mn^{2+} always inhibited ethylene production whereas Fe^{2+} markedly enhanced production. Mn^{2+} will poise the redox potential of soil slightly lower than NO_3^--N, but Fe^{2+} poises it at a considerably lower potential (83, 84, 119, 120).

In soil, NO_3^--N will interfere more with ethylene production than manganese because NO_3^--N is in a high state of oxidation, is soluble in water, is used as an electron acceptor by many microorganisms, and occurs in appreciable concentrations in agricultural soils (57). Manganese, on the other hand, is insoluble in the oxidized form and is a minor constituent of most soils. This unbalancing effect of NO_3^--N on ethylene production has important significance for the biological balance of soil because it leads to more or less unrestricted activity of aerobic microorganisms until it is leached, denitrified, or used by plant roots.

Persistence of Ethylene in Soil

Our concept of an oxygen-ethylene cycle in soil and its critical role in regulating microbial activity depends on ethylene persisting at biologically active concentrations in aerobic or, at least, microaerobic sites in soil. Microbial degradation of ethylene in aerobic soils is well established (1, 2, 24, 107). Cornforth (24) claims that ethylene is microbially decomposed in aerobic soils much faster than it is produced in anaerobic soils. He concludes that biologically active concentrations of ethylene will occur in soil for only short periods associated with waterlogging. The concept that soil is composed of a mosaic of anaerobic microsites centered on the sites of availability of organic nutrients overcomes most of Cornforth's implied criticisms of an oxygen-ethylene cycle. Microbial activity in soil is restricted to the isolated islands of organic nutrient, and these are the only sites where a regulator of activity, such as ethylene, is required. Thus, an intimate relationship exists between the sites where ethylene is produced and where it is required to act. Away from the nutrient sources, ethylene may be degraded rapidly and completely but this in no way invalidates the oxygen-ethylene cycle. Furthermore, restriction of oxygen partial pressures and elevation of carbon dioxide partial pressures increase persistence of ethylene in soil (2, 24). These are the conditions most likely to exist in the sites where ethylene concentrations need to be highest to limit microbial activity.

The identity of the microorganisms mainly responsible for ethylene degradation in soil has not been established although methane-oxidizing bacteria have this capability (29, 37). We have found that soils that are poor ethylene producers and low in organic matter, degrade artificially imposed ethylene best.

Cornforth (24) questioned the significance of ethylene as a regulator of microbial activity in soil because he recorded healthy fungal colonies on the surface of soil in an ethylene atmosphere and could not detect any inhibitory effect of ethylene on soil respiration. Such vigorous fungal growth does not occur on soil, however, unless it is pretreated or amended to increase nutrient availability or decrease microbial competition, or both (65, 121). It has already been discussed how treatments of this type annul any inhibitory effects of ethylene on fungal activity in soil. Also, soil respiration may remain relatively high in the presence of ethylene because soil bacteria do not appear to be directly inhibited by ethylene.

Effects of Temperature

Temperature has a marked effect on ethylene evolution by soil in both laboratory and field studies (109, 111). The rate of production is greatest at higher temperatures, and below 10°C evolution is considerably reduced (111). In the field, there was a twentyfold increase in concentration of ethylene over the range 4–11°C (109).

The effect of temperature on ethylene production is a reflection of the microbial activity in soil. It is the intensity of the microbial activity that ultimately determines the number, size, and persistence of anaerobic microsites in soil. K. A. Smith & Dowdell (109) pointed out that the volume of soil in which ethylene can be produced increases as the cube of the radius of the anaerobic zones. Cold temperatures in winter reduce ethylene production in soil, but by also directly restricting microbial activity the requirement for additional regulators such as ethylene is diminished.

Influence of Organic Nutrients

The oxygen-ethylene cycle in soil is temporarily unbalanced by large inputs of organic nutrients because these inputs override the inhibitory effects of ethylene (103–105). A similar mechanism has been described for fungistasis (31, 65). The balance is rapidly reestablished because the resultant increase in microbial activity associated with the additional nutrient expands the volume of anaerobic microsites leading to enhanced ethylene production. This temporary unbalancing is important because it ensures that large seasonal inputs of organic matter are degraded rapidly enough to prevent problems of accumulation in soil. A similar disturbance of the oxygen-ethylene cycle may partly explain the priming effect of additions of fresh organic matter on rates of decomposition of native soil organic matter (20).

There is evidence that organic nutrients in soil derived from plant residues directly affect ethylene production by acting as substrates (109). Also, these substrates appear to be limiting for ethylene production under certain conditions (109). There is evidence that anaerobes depend in part on aerobes or aerobic conditions for production of substrate for ethylene formation (103–105). Prolonged and maximum production of ethylene occurs when a mosaic of microsites with anaerobic-aerobic interfaces forms in soil (104).

IMPLICATIONS OF THE OXYGEN-ETHYLENE CYCLE

Modern conventional agriculture is an exploitive system leading inevitably to a decline in soil organic matter, soil structure, availability of essential plant nutrients, and to an increase in the incidence of plant disease. It contrasts sharply with natural undisturbed ecosystems such as forests and grasslands where organic matter and plant nutrients are slowly recycled, nitrate nitrogen does not accumulate, and root diseases are rare. Soil fertility, as measured by organic carbon content and availability of plant nutrients, increases continually in undisturbed ecosystems until a climax equilibrium, dictated by the prevailing climatic conditions, is established. The highest concentrations of ethylene are consistently recorded from soil under such ecosystems, indicating that the oxygen-ethylene cycle is in balance. Conversely, in most agricultural soils ethylene is barely detectable because the oxygen-ethylene cycle is impaired or destroyed (103–105).

We think (103–105) that this is a major factor determining the inefficiency of agriculture and, that if the oxygen-ethylene cycle was restored, many of the exploitive aspects of modern agriculture could be reversed. Conversion of a natural ecosystem to agricultural usage usually requires destruction of existing vegetation, and tillage of the soil. Tillage breaks soil aggregates, aerates the soil, and raises the redox potential of microsites, causing a temporary destruction of the oxygen-ethylene cycle. This increases the rate of organic matter breakdown and mineralization of nitrogen, which is rapidly nitrified because plants are not present to utilize the NH_4^+-N released. Even when intense microbial activity recreates oxygen-free microsites in the soil, ethylene production will not resume because the NO_3^--N poises the redox potential at an unfavorably high potential unless it is removed by plant roots of a subsequent crop or, less desirably, by leaching or denitrification. The unbalancing of the oxygen-ethylene cycle becomes more intense, prolonged, and difficult to reverse the longer the soil is denuded of vegetation and the more it is cultivated.

Soils stored in the laboratory at warm temperatures and moderate water potentials rapidly lose their ability to produce ethylene. Ability to produce ethylene is regained after growing plants in the soil for some weeks (111). We have made similar observations that emphasize that plants are essential to maintain ethylene production in soil. They act not only as sinks for the nitrogen released from organic matter decomposition, and so prevent excessive nitrification, but also supply a continual source of high energy substrate essential for ethylene formation. Thus, excessive removal of plants by any agricultural operation, such as forestry or overgrazing of pastures, unbalances the oxygen-ethylene cycle even in the absence of any form of tillage and will lead to some reduction in soil fertility. Studies over several years have shown that this cycle can be unbalanced temporarily even in undisturbed ecosystems if extremes in climatic conditions, such as excessive rainfall or drought, occur (104). The cycle is rapidly reestablished when the climatic conditions return to normal. In contrast, such rapid reversals are rare when the cycle is disturbed by agricultural practices.

It should be noted also that excessive use of nitrogenous fertilizers on arable or pasture systems will unbalance the oxygen-ethylene cycle by enhancing the rate of nitrification. This effect will be most pronounced on arable land (52).

Nitrification is so universal and rapid in agricultural systems that NO_3^--N is often considered synonymous with available nitrogen (51). This stimulation in the rate of nitrification upsets the biological balance of soil by destroying the oxygen-ethylene cycle. Excessive nitrification has other disadvantages because, in addition to converting nitrogen into a leachable and denitrifiable form, it also leads to acidification and a loss of cations from soil (4, 63, 75). The nitrification reaction is proton producing (4), and the hydrogen ions liberated displace cations from the base-exchange complex into the soil solution (63, 75). The classical studies on the Hubbard Brook watershed-ecosystem (63) showed clearly that the cations and nitrogen may then be leached into waterways and cause problems with eutrophication of streams, rivers, and lakes.

AVAILABILITY OF PLANT NUTRIENTS IN REDUCED MICROSITES IN SOIL

Microbial activity in soil is dominated by plants because they supply organic nutrients as crop residues and root exudates. A major limitation to plant growth in most agricultural soils is an inadequate supply of cations and anions. There are large reserves of these essential plant nutrients in soil, but they occur in highly insoluble forms as crystals, salts, or attached to the base-exchange complex. Their high degree of insolubility prevents loss by leaching but has the major disadvantage that plants cannot obtain a satisfactory supply from the equilibrium solution in soil.

To obtain an adequate supply of these inorganic nutrients, plants must change the environment around their roots to increase the solubility of these cations and anions (104). Research on the chemistry of submerged soils has shown that when the redox potential of soil falls the concentration of cations and anions in the soil solution increases dramatically (80, 81, 83). Research on ethylene production in soil reveals that reduced microsites are common in drained, adequately aerated soil. Furthermore, these reduced microsites are most likely to occur in the rhizosphere of plants because of root exudation. Availability of cations and anions should alter as markedly in these microsites as in flooded soils. This concept ascribes a positive, beneficial, and essential role to root exudation and the rhizosphere effect which, previously, has often been viewed as neutral or even harmful to the plant.

Extrapolating from the chemistry of submerged soil (80, 81, 83, 119, 120), it is possible to predict the sequential series of reactions in the anaerobic microsite that will lead ultimately to an increase in availability of inorganic nutrients. Presuming that the organic nutrient supply is adequate in the microsite, aerobic microbial growth will exhaust the oxygen supply, increase the carbon dioxide concentration, and cause the redox potential to fall. Any nitrate in the microsite will be denitrified and, at about the same time, manganese will be reduced, to be immediately followed by reduction of iron. Iron and, to a lesser extent manganese, dominate the chemical reactions in the microsite and are the key to availability of other cations and anions, including minor elements. Iron and manganese usually occur in soil as minute crystals of the respective oxides (74). Because of their small particle size, these oxides have a high surface area and contribute significantly to the surface activity of soil. Other elements are adsorbed on the surface or, perhaps are built into the crystal lattice. As a result the oxides play an important role in retaining and supplying trace elements (74). In most soils a significant portion of the phosphate is similarly associated with iron oxides (K. Norrish, personal communication).

Reduction of iron and manganese in the anaerobic microsites, therefore, releases bound phosphate and minor elements into the soil solution. Also, the reduced iron and manganese undergo base exchange with other cations such as calcium, magnesium, potassium, and ammonium, which will increase their concentration in the soil solution in the microsite (83).

Elevated partial pressures of carbon dioxide in the microsite are important in all these reactions because carbon dioxide dominates the pH of the microsite (57, 82). The oxidation-reduction reactions of iron and manganese are extremely sensitive to changes in pH, e.g. for $Fe^{3+} \rightleftarrows Fe^{2+}$ couple, $dEh/dpH = -180$ mV (83).

Provided the redox potential drops sufficiently to reduce iron and manganese, plant roots or root hairs in or at the edges of the reduced microsites have a ready supply of essential cations and anions. Furthermore, any of these cations or anions that are not utilized by plants will not be leached from soil because, as soon as they migrate to the edge of the reduced microsite, reoxidation will occur and their solubility is again restricted. Thus, a mosaic of reduced microsites in the rhizosphere of plants will ensure a ready supply of essential nutrients for plants, especially since the microsites are dynamic in location and size. Plant roots function imperfectly in the absence of oxygen (42), but it must be remembered that, in the system described, the bulk of the root system is aerobic and it is probable that internally the roots remain adequately aerated even directly adjacent to the reduced microsites.

It is clear from the discussion on ethylene production that, in most agricultural soils, the concentrations of NO_3^--N are elevated sufficiently to prevent or restrict the occurrence of reduced microsites. This, in turn, prevents the reduction of iron and manganese and will severely restrict the availability of plant nutrients.

The presence of reduced microsites in the rhizosphere of plants may also have important implications for nonsymbiotic fixation of nitrogen. To operate efficiently, free-living nitrogen fixation requires anaerobic conditions and substantial quantities of carbohydrate, as well as suitable species of microorganisms and high concentrations of nitrogen gas (5, 60, 72, 86, 88, 89). These are exactly the conditions that should exist in the reduced microsites in the rhizosphere. Nonsymbiotic nitrogen fixation is usually considered of limited importance in agricultural soils (5, 60, 72, 88), but this is probably explained by the elevated concentrations of NO_3^--N preventing formation of reduced microsites.

Ability of a soil to fix nitrogen nonsymbiotically is often measured with the acetylene reduction technique (47). This technique relies on the reduction of acetylene to ethylene by the nitrogenase enzyme, and the ethylene produced is quantified by gas chromatography. Exogenous soil ethylene will obviously interfere with this assay and must be considered whenever this technique is used. Nitrogenase is not involved in exogenous ethylene production in soil (A. M. Smith, D. F. Herridge, and D. H. Pike, unpublished).

In summary, stimulation of the rate of nitrification by conventional agriculture not only upsets the biological balance of soil by destroying the oxygen-ethylene cycle, but it may also directly limit availability and uptake of plant nutrients.

SOIL ETHYLENE AND PLANT GROWTH

The significance and value of ethylene as a critical regulator of microbial activity in soil is in jeopardy if it directly and adversely affects growth and development of plants. K. A. Smith & Jackson (110) recently reviewed this subject. Clearly, the high concentrations of exogenous ethylene produced in waterlogged soils are detrimental to plant growth although the degree of damage varies with plant species, type of organ, and stage of development, as well as the growing conditions (110).

This does not mean, however, that the concentrations of ethylene in soil under more normal moisture regimes will be detrimental to plants. In such soils, much of the root system occupies the intercrumb macropores where aeration is high (26) and, hence, ethylene is low (24). High concentrations of ethylene occur only in localized, scattered microsites within crumbs, at some crumb-root interfaces, or associated with fresh organic residues. The elevated carbon dioxide concentrations and low oxygen concentrations in these microsites will tend to protect plant roots from any adverse effects of ethylene (1, 110). Roots are also important sites for production of cytokinins and gibberellins (10) and their presence can reduce sensitivity of plant parts to ethylene (110). Furthermore, exposure of one portion of a root system to elevated concentrations of ethylene does not cause abnormal growth of unexposed roots (25).

Not all effects of exogenous ethylene are harmful to plant growth, nor did K. A. Smith & Jackson (110) claim this was the case. A relationship almost certainly exists between the much branched, thickened, and shallow system of feeder roots found under litter layers and in surface soils in forests and grasslands, and the elevated concentrations of ethylene recorded in such sites. This form of root development is ideally suited to the thorough exploration and absorption of nutrients by roots from the zone highest in available plant nutrients (110). It seems no coincidence that similar root patterns occur in surface layers under minimum tillage (8, 36) where higher concentrations of ethylene also could be expected. Finally, there are some reports that applications of ethylene to soil under growing crops even stimulate yields (39).

Seed germination in many plant species is either stimulated or inhibited by ethylene (76). Thus, the concentration of exogenous ethylene in soil will have important effects on establishment of crop plants and weeds. Many seeds leak considerable quantities of organic nutrients (116) which increase the likelihood of elevated concentrations of ethylene in their vicinity. These aspects must be considered when attempts are made to increase ethylene concentrations in agricultural soils. In North Carolina, ethylene is injected directly into soil to control witchweed [*Striga asiatica* (L.) O. Kuntze]. Seed of this phanerogamic parasite remains dormant in soil for up to 20 years, and will not germinate until it comes into intimate contact with the roots of a host plant. Research by Eplee (35) showed that ethylene at about 1 ppm in the soil atmosphere triggered suicidal germination of up to 90% of these seeds in the absence of hosts. This eradicant treatment is now used on about 6000 ha in North Carolina. Eplee has applied ethylene to soil under maize crops at rates twenty times more concentrated than his recommended treatment without causing injury to the crop.

Ethylene has been reported to inhibit nodulation of leguminous plants (45) and this is another area where elevated concentrations of exogenous soil ethylene may cause problems. Nitrification, however, is enhanced in the rhizosphere of legumes (4), and this will restrict ethylene production and may permit unimpaired nodulation under most field conditions.

CONCLUSION

Ethylene is produced microbially in anaerobic microsites in soil and directly regulates activity of fungi and actinomycetes and probably indirectly affects bacteria and other soil microorganisms. It has a central role in a self-regulating microbial cycle in soil, with important implications for rates of turnover of organic matter, availability of plant nutrients, and incidence and control of soil-borne plant pathogens. This exogenous source of ethylene also may directly affect the morphology and function of plant roots and the germination of seeds. Excessive amounts produced in waterlogged soils may damage plant growth.

The key to ethylene production in field soils is the formation of numerous anaerobic microsites, especially in the rhizosphere of plants. Availability of essential plant nutrients also is altered markedly in these microsites. Individual microsites are ephemeral and dynamic both in location in soil and in size. Formation of microsites depends primarily on intensity of aerobic microbial activity and, therefore, factors such as supply of readily available organic nutrients, temperature, and water potential in soil are important influences. Nitrate nitrogen stops formation of reduced microsites in soil by poising the redox potential: this prevents ethylene production and inhibits the reductive processes that lead to availability of plant nutrients. Nitrification is much enhanced in agricultural soils and this may explain why conventional agriculture is such an exploitive system. Consequently, the role of nitrification in agriculture requires urgent reappraisal.

Successful manipulation of the oxygen-ethylene cycle in soil will demand alterations to certain established concepts in agriculture. Techniques aimed at increasing aeration and oxidation states of soil give short-term increases in plant growth but rapidly create long-term problems of nutrient depletion and plant-disease incidence. To improve the efficiency of agriculture, these practices should be replaced by techniques of minimum tillage, better use of organic amendments, and chemical inhibitors to suppress nitrification to ensure the establishment of a balanced oxygen-ethylene cycle in soil. Precise manipulation of the oxygen-ethylene cycle has greatest potential benefit for tropical agriculture where conventional attempts at mechanized agriculture lead to a rapid depletion of nutrient reserves in the soil.

Manipulation of the oxygen-ethylene cycle in plant pathology may lead to the development of effective methods of biological control of soil-borne plant pathogens. Identification of the species responsible for ethylene production will make it possible to add suitable microorganisms to soil or, better still, modify soil conditions to favor increases in their population in soil. Research on ranking pathogens and saprophytes for sensitivity to ethylene in soil will be profitable, and may give important insight into the ecology of particular species. Chemicals such as 2-chloroethylphosphonic acid (1) which degrade slowly in soil to release ethylene are valuable experimental tools in research on effects of ethylene on microorganisms. They may even have a role in controlling certain plant pathogens in the field.

Ethylene seems destined,‘ through the oxygen-ethylene cycle, to play an increasingly important role in soil biology. Being such a new field, there is a limited amount of definitive research. As a result this review has attempted a broader than usual appraisal of the potential significance of ethylene and reduced microsites in soil to agriculture. Undoubtedly, some of the views expressed will require modification in the light of future research but, hopefully, the approach outlined here will attract increased attention to this exciting field of research.

Literature Cited

1. Abeles, F. B. 1973. *Ethylene in Plant Biology.* New York: Academic. 302 pp.
2. Abeles, F. B., Craker, L. E., Forrence, L. E., Leather, G. R. 1971. Fate of air pollutants: removal of ethylene, sulfur dioxide, and nitrogen dioxide by soil. *Science* 173:914–16
3. Alexander, M. 1961. *Introduction to Soil Microbiology.* New York: Wiley. 472 pp.
4. Alexander, M. 1965. Nitrification. In *Soil Nitrogen,* ed. W. V. Bartholomew, F. E. Clark, 307–43. Madison, Wis.: Am. Soc. Agron. 615 pp.
5. Allison, F. E. 1973. *Soil Organic Matter and its Role in Crop Production.* Amsterdam: Elsevier. 637 pp.
6. Anderson, J. P. E., Domsch, K. H. 1973. Quantification of bacterial and fungal contributions to soil respiration. *Arch. Microbiol.* 93:113–27
7. Anderson, J. P. E., Domsch, K. H. 1975. Measurement of bacterial and fungal contributions to respiration of selected agricultural and forest soils. *Can J. Microbiol.* 21:314–22
8. Baeumer, K., Bakermans, W. A. 1973. Zero-tillage. *Adv. Agron.* 25:77–123
9. Baker, K. F. 1970. Selective killing of soil microorganisms by aerated steam. In *Root Diseases and Soil-Borne Pathogens,* ed. T. A. Toussoun, R. V. Bega, P. E. Nelson, 234–39. Berkeley: Univ. Calif. Press. 252 pp.
10. Baker, K. F., Cook, R. J. 1974. *Biological Control of Plant Pathogens.* San Francisco: Freeman. 433 pp.
11. Balis, C. 1975. The nature of volatile inhibitors involved in soil fungistasis. In *Biology and Control of Soil-Borne Plant Pathogens,* ed. G. W. Bruehl, p. 197. St. Paul, Minn.: Am. Phytopathol. Soc. 216 pp.
12. Balis, C., Kouyeas, V. 1968. Volatile inhibitors involved in soil mycostasis. *Ann. Inst. Phytopathol. Benaki,* n.s. 8: 145–49
13. Barber, D. A., Martin, J. K. 1976. The release of organic substances by cereal roots into soil. *New Phytol.* In press
14. Bird, C. W., Lynch, J. M. 1974. Formation of hydrocarbons by microorganisms. *Chem. Soc. Rev.* 3:309–28
15. Broadbent, F. E., Clark, F. E. 1965. Denitrification. See Ref. 4, pp. 344–59
16. Brown, M. E. 1973. Soil bacteriostasis limitation in growth of soil and rhizosphere bacteria. *Can. J. Microbiol.* 19:195–99
17. Burg, S. P. 1962. The physiology of ethylene formation. *Ann. Rev. Plant Physiol.* 13:265–302
18. Burg, S. P. 1968. Ethylene, plant senescence and abscission. *Plant Physiol.* 43:1503–11
19. Chase, F. E., Corke, C. T., Robinson, J. B. 1968. Nitrifying bacteria in soil. In *The Ecology of Soil Bacteria,* ed. T. R. G. Gray, D. Parkinson, 593–611. Liverpool: Liverpool Univ. Press. 681 pp.
20. Clark, F. E. 1968. The growth of bacteria in soil. See Ref. 19, pp. 441–57
21. Clark, F. E. 1972. Soil microor-

ganisms. In *McGraw-Hill Yearbook Science and Technology,* 1972: 380–83. New York: McGraw-Hill. 450 pp.

22. Clark, F. E., Paul, E. A. 1970. The microflora of grassland. *Adv. Agron.* 22:375–435

23. Coley-Smith, J. R., Cooke, R. C. 1971. Survival and germination of fungal sclerotia. *Ann. Rev. Phytopathol.* 9:65–92.

24. Cornforth, I. S. 1975. The persistence of ethylene in aerobic soils. *Plant Soil* 42:85–96

25. Crossett, R. N., Campbell, D. J. 1975. The effects of ethylene in the root environment upon development of barley. *Plant Soil* 42:453–64

26. Currie, J. A. 1961. Gaseous diffusion in the aeration of aggregated soils. *Soil Sci.* 92:40–45

27. Davis, R. D. 1975. Bacteriostasis in soils sterilized by gamma irradiation and in reinoculated sterilized soils. *Can. J. Microbiol.* 21:481–84

28. Davis, R. D. 1975. Soil bacteriostasis: inhibition of spore germination and microcolony development in agar discs incubated on nonsterile soils. *Can. J. Microbiol.* 21:1270–72

29. De Bont, J. A., Mulder, E. G. 1974. Nitrogen fixation and co-oxidation of ethylene by a methane-utilizing bacterium. *J. Gen. Microbiol.* 83:113–21

30. Dobbs, C. G., Gash, M. J. 1965. Microbial and residual mycostasis in soils. *Nature* 207:1354–56

31. Dobbs, C. G., Hinson, W. H. 1953. A widespread fungistasis in soils. *Nature* 172:197–99

32. Dobbs, C. G., Hinson, W. H., Bywater, J. ,1960. Inhibition of fungal growth in soils. In *The Ecology of Soil Fungi,* ed. D. Parkinson, J. S. Waid, 130–47. Liverpool: Liverpool Univ. Press. 324 pp.

33. Doelle, H. W. 1969. *Bacterial Metabolism.* New York: Academic. 486 pp.

34. Dowdell, R. J., Smith, K. A., Crees, R., Restall, S. W. F. 1972. Field studies of ethylene in the soil atmosphere—equipment and preliminary results. *Soil Biol. Biochem.* 4:325–31

35. Eplee, R. E. 1975. Ethylene: a witchweed seed germination stimulant. *Weed Sci.* 23:433–36

36. Finney, J. R. 1973. Root growth in relation to tillage. *Chem. Ind.* London 1973:676–78

37. Flett, R. J., Rudd, J. W. M., Hamilton, R. D. 1975. Acetylene reduction assays for nitrogen fixation in fresh-

waters: a note of caution. *Appl. Microbiol.* 29:580–83

38. Freebairn, H. T., Buddenhagen, I. W. 1964. Ethylene production by *Pseudomonas solanacearum. Nature* 202:313–14

39. Freytag, A. H., Wendt, C. W., Lira, E. P. 1972. Effects of soil-injected ethylene on yields of cotton and sorghum. *Agron. J.* 64:524–26

40. Fries, N. 1973. Effects of volatile organic compounds on the growth and development of fungi. *Trans. Br. Mycol. Soc.* 60:1–21

41. Funke, B. R., Harris, J. O. 1968. Early respiratory responses of soil treated by heat or drying. *Plant Soil* 28:38–48

42. Greenwood, D. J. 1969. Effect of oxygen distribution in the soil on plant growth. In *Root Growth,* ed. W. J. Whittington, 202–23. London: Butterworths. 450 pp.

43. Greenwood, D. J., Goodman, D. 1967. Direct measurements of the distribution of oxygen in soil aggregates and in columns of fine soil crumbs. *J. Soil Sci.* 18:182–96

44. Griffin, D. M. 1972. *Ecology of Soil Fungi.* London: Chapman & Hall. 193 pp.

45. Grobbelaar, N., Clarke, B., Hough, M. C. 1971. The nodulation and nitrogen fixation of isolated roots of *Phaseolus vulgaris* L. III. The effect of carbon dioxide and ethylene. *Plant Soil* (special volume):215–23

46. Hansen, E. 1966. Postharvest physiology of fruits. *Ann Rev. Plant Physiol.* 17:459–80

47. Hardy, R. W. F., Burns, R. C., Holsten, R. D. 1973. Applications of acetylene-ethylene assay for measurement of nitrogen fixation. *Soil Biol. Biochem.* 5:47–81

48. Hattori, T. 1973. *Microbial Life in the Soil; An Introduction.* New York: Dekker. 427 pp.

49. Hora, T. S., Baker, R. 1970. Volatile factor in soil fungistasis. *Nature* 225:1071–72

50. Hora, T. S., Baker, R. 1972. Soil fungistasis: microflora producing a volatile inhibitor. *Trans. Br. Mycol. Soc.* 59:491–500

51. Huber, D. M., Watson, R. D. 1974. Nitrogen form and plant disease. *Ann. Rev. Phytopathol.* 12:139–65

52. Huntjens, J. L. M. 1972. Does nitrogen fertilization of grassland lead to eutrophication of surface water? *Stikstof* 15:52–55

53. Hutchinson, S. A. 1971. Biological ac-

tivity of volatile fungal metabolites. *Trans. Br. Mycol. Soc.* 57:185–200

54. Hutchinson, S.A. 1973. Biological activities of volatile fungal metabolites. *Ann. Rev. Phytopathol.* 11:223–46

55. Ilag, L. L. 1972. Periodicity of ethylene production in fungi. *Philipp. Agric.* 56:9–15

56. Ilag, L. L., Curtis, R. W. 1968. Production of ethylene by fungi. *Science* 159:1357–58

57. *Int. Rice Res. Inst. Ann. Rept.,* 1964, pp. 200–40

58. Jackson, R. M. 1960. Soil fungistasis and the rhizosphere. See Ref. 32, pp. 168–76

59. Jackson, R. M. 1965. Antibiosis and fungistasis of soil microorganisms. In *Ecology of Soil-Borne Plant Pathogens,* ed. K. F. Baker, W. C. Snyder, 363–73. Berkeley: Univ. Calif. Press. 571 pp.

60. Jensen, H. L. 1965. Nonsymbiotic nitrogen fixation. See Ref. 4, pp. 436–80

61. Katznelson, H., Rouatt, J. W., Payne, T. M. B. 1956. Recent studies on the microflora of the rhizosphere 1. *Trans. 6th Int. Congr. Soil Sci., Paris,* C: 151–56

62. Ko, W. H., Hora, F. K. 1974. Factors affecting the activity of a volatile fungistatic substance in certain alkaline soils. *Phytopathology* 64:1042–43

63. Likens, G. E., Bormann, F. H., Johnson, N. M., Fisher, D. W., Pierce, R. S. 1970. Effects of forest cutting and herbicide treatment on nutrient budgets on the Hubbard Brook watershed-ecosystem. *Ecol. Monogr.* 40: 23–47

64. Linderman, R. G., Gilbert, R. G. 1975. Influence of volatiles of plant origin on soil-borne plant pathogens. See Ref. 11, pp. 90–99

65. Lockwood, J. L. 1964. Soil fungistasis. *Ann. Rev. Phytopathol.* 2:341–62

66. Lockwood, J. L. 1975. Colloquium on soil fungistasis and lysis, summary and synthesis. See Ref. 11, pp. 194–97

67. Lynch, J. M. 1972. Identification of substrates and isolation of microorganisms responsible for ethylene production in the soil. *Nature* 240:45–46

68. Lynch, J. M. 1974. Mode of ethylene formation by *Mucor hiemalis. J. Gen. Microbiol.* 83:407–11

69. Lynch, J. M. 1975. Ethylene in soil. *Nature* 256:576–77

70. Lynch, J. M., Harper, S. H. T. 1974. Formation of ethylene by a soil fungus. *J. Gen. Microbiol.* 80:187–95

71. Lynch, J. M., Harper, S. H. T. 1974. Fungal growth rate and formation of ethylene in soil. *J. Gen. Microbiol.* 85:91–96

72. Mishustin, E. N., Shil'nikova, V. K. 1971. *Biological Fixation of Atmospheric Nitrogen,* transl. A. Crozy. London: Macmillan. 420 pp.

73. Moore-Landecker, E., Stotsky, G. 1972. Inhibition of fungal growth and sporulation by volatile metabolites from bacteria. *Can. J. Microbiol.* 18:957–62

74. Norrish, K. 1975. Geochemistry and mineralogy of trace elements. In *Trace Elements in Soil—Plant—Animal,* ed. D. J. D. Nicholas, A. R. Egan, 55–81. New York: Academic. 417 pp.

75. Nye, P. H., Greenland, D. J. 1960. The soil under shifting cultivation. *Commonw. Bur. Soils, Tech. Commun.* 51:1–156

76. Olatoye, S. T., Hall, M. A. 1973. Interaction of ethylene and light on dormant weed seeds. In *Seed Ecology,* ed. W. Heydecker, 233–49. London: Butterworths. 578 pp.

77. Park, D. 1960. Antagonism—the background to soil fungi. See Ref. 32, 148–59

78. Park, D. 1961. Morphogensis, fungistasis and cultural staling in *Fusarium oxysporum* Snyder & Hansen. *Trans. Br. Mycol. Soc.* 44:377–90

79. Parker, C. A. 1954. Non-symbiotic nitrogen-fixing bacteria in soil. I. Studies on *Clostridium butyricum. Aust. J. Agric. Res.* 5:90–97

80. Patrick, W. H. Jr., Mahapatra, I. C. 1968. Transformation and availability to rice of nitrogen and phosphorus in waterlogged soils. *Adv. Agron.* 20: 323–59

81. Patrick, W. H., Mikkelsen, D. S. 1971. Plant nutrient behaviour in flooded soil. *Fertilizer Technology and Use,* ed. R. A. Olsen, 187–215. Madison, Wis.: Am. Soc. Soil Sci. 2nd ed. 611 pp.

82. Ponnamperuma, F. N. 1967. A theoretical study of aqueous carbonate equilibria. *Soil Sci.* 103:90–100

83. Ponnamperuma, F. N. 1972. The chemistry of submerged soils. *Adv. Agron.* 24:29–96

84. Ponnamperuma, F. N., Castro, R. U. 1964. Redox systems in submerged soils. *Trans. 8th Int. Congr. Soil Sci., Bucharest* 3:379–86

85. Ponnamperuma, F. N., Tianco, E. M., Loy, T. 1967. Redox equilibria in flooded soils: 1. The iron hydroxide systems. *Soil Sci.* 103:374–82

86. Postgate, J. R. 1974. New advances and future potential in biological nitrogen fixation. *J. Appl. Bacteriol.* 37:185–202

87. Pratt, H. K., Goeschl, J. D. 1969. Physiological roles of ethylene in plants. *Ann. Rev. Plant Physiol.* 20:541–84

88. Rice, W. A., Paul, E. A. 1972. The organisms and biological processes involved in asymbiotic nitrogen fixation in waterlogged soil amended with straw. *Can. J. Microbiol.* 18:715–23

89. Rice, W. A., Paul, E. A., Wetter, L. R. 1967. The role of anaerobiosis in asymbiotic nitrogen fixation. *Can. J. Microbiol.* 13:829–36

90. Richardson, H. L. 1938. The nitrogen cycle in grassland soils: with especial reference to the Rothamsted Park grass experiment. *J. Agric. Sci.* 28:73–121

91. Robinson, J. B. 1963. Nitrification in a New Zealand grassland soil. *Plant Soil* 19:173–83

92. Robinson, P. M., Garrett, M. K. 1969. Identification of volatile sporostatic factors from cultures of *Fusarium oxysporum. Trans. Br. Mycol. Soc.* 52:293–99

93. Robinson, P. M., Park, D. 1966. Volatile inhibitors of spore germination produced by fungi. *Trans. Br. Mycol. Soc.* 49:639–49

94. Robinson, P. M., Park, D., Garrett, M. K. 1968. Sporostatic products of fungi. *Trans. Br. Mycol. Soc.* 51:113–24

95. Rovira, A. D. 1965. Plant root exudates and their influence upon soil microorganisms. See Ref. 59, pp. 170–86

96. Rovira, A. D. 1965. Interactions between plant roots and soil microorganisms. *Ann. Rev. Microbiol.* 19:241–66

97. Rovira, A. D., McDougall, B. M. 1967. Microbiological and biochemical aspects of the rhizosphere. *Soil Biochemistry,* ed. A. D. McLaren, G. F. Peterson, 1:417–63. New York: Dekker. 509 pp.

98. Shields, J. A., Paul, E. A., Lowe, W. E., Parkinson, D. 1973. Turnover of microbial tissue in soil under field conditions. *Soil Biol. Biochem.* 5:753–64

99. Siala, A., Hill, I. R., Gray, T. R. G. 1974. Populations of spore-forming bacteria in an acid forest soil, with special reference to *Bacillus subtilis. J. Gen. Microbiol.* 81:183–90

100. Skinner, F. A. 1968. The anaerobic bacteria of soil. See Ref. 19, pp. 573–92

101. Skinner, F. A. 1971. The isolation of soil Clostridia. *Isolation of Anaerobes,* ed. D. A. Shapton, R. G. Board, 57–80. London: Academic. 270 pp.

102. Smith, A. M. 1973. Ethylene as a cause of soil fungistasis. *Nature* 246:311–13

103. Smith, A. M. 1975. Ethylene as a critical regulator of microbial activity in soil. *Proc. 1st Intersect. Congr. Int. Assoc. Microbiol. Soc., Tokyo* 2:463–73

104. Smith, A. M. 1976. Ethylene production by bacteria in reduced microsites in soil and some implications to agriculture. *Soil Biol. Biochem.* In press

105. Smith, A. M., Cook, R. J. 1974. Implications of ethylene production by bacteria for biological balance of soil. *Nature* 252:703–5

106. Smith, G. H., Ellis, M. M. 1963. Chromatographic analysis of gases from soils and vegetation, related to geochemical prospecting for petroleum. *Bull. Am. Assoc. Pet. Geol.* 47:1897–1903

107. Smith, K. A., Bremner, J. M., Tabatabai, M. A. 1973. Sorption of gaseous atmospheric pollutants by soils. *Soil Sci.* 116:313–19

108. Smith, K. A., Dowdell, R. J. 1973. Gas chromatographic analysis of the soil atmosphere: automatic analysis of gas samples for O_2, N_2, Ar, CO_2, N_2O and C_1-C_4 hydrocarbons. *J. Chromatogr. Sci.* 11:655–58

109. Smith, K. A., Dowdell, R. J. 1974. Field studies of the soil atmosphere. 1. Relationships between ethylene, oxygen, soil moisture content, and temperature. *J. Soil Sci.* 25:217–30

110. Smith, K. A., Jackson, M. B. 1974. Ethylene, waterlogging and plant growth. *Agric. Res. Counc. Letcombe Lab. Ann. Rept.,* 1973:60–75

111. Smith, K. A., Restall, S. W. F. 1971. The occurrence of ethylene in anaerobic soil. *J. Soil Sci.* 22:430–43

112. Smith, K. A., Robertson, P. D. 1971. Effects of ethylene on root extension of cereals. *Nature* 234:148–49

113. Smith, K. A., Russell, R. S. 1969. Occurrence of ethylene, and its significance, in anaerobic soil. *Nature* 222:769–71

114. Smith, W. H., Bormann, F. H., Likens, G. E. 1968. Response of

chemoautotrophic nitrifiers to forest cutting. *Soil Sci.* 106:471–73

115. Stanford, G., Vander Pol, R. A., Dzienia, S. 1975. Denitrification rates in relation to total and extractable soil carbon. *Proc. Soil Sci. Soc. Am.* 39:284–89

116. Stanghellini, M. E., Hancock, J. G. 1971. Radial extent of the bean spermosphere and its relation to the behavior of *Pythium ultimum*. *Phytopathology* 61:165–68

117. Stevenson, I. L. 1956. Some observations on the microbial activity in remoistened air-dried soils. *Plant Soil* 8:170–82

118. Stotsky, G. 1972. Activity, ecology, and population dynamics of microorganisms in soil. *CRC Crit. Rev. Microbiol.* 2:59–137

119. Takai, Y., Kamura, T. 1966. The mechanism of reduction in waterlogged paddy soil. *Folia Microbiol. Prague* 11:304–13

120. Turner, F. T., Patrick, W. H. 1968. Chemical changes in waterlogged soils as a result of oxygen depletion. *Trans. 9th Int. Congr. Soil Sci., Adelaide* 4:53–65

121. Watson, A. G., Ford, E. J. 1972. Soil fungistasis—a reappraisal. *Ann. Rev. Phytopathol.* 10:327–48

122. Woldendorp, J. W. 1962. The quantitative influence of the rhizosphere on denitrification. *Plant Soil* 17:267–70

123. Woldendorp, J. W. 1968. Losses of soil nitrogen. *Stikstof* 12:32–46

Copyright © 1976 by Annual Reviews Inc.
All rights reserved

EPIDEMIOLOGY AND CONTROL OF TOMATO MOSAIC VIRUS

♦3633

Leonard Broadbent

School of Biological Sciences, University of Bath, Claverton Down, Bath, England

INTRODUCTION

The cultivated tomato *(Lycopersicon esculentum)* is subject to a range of diseases resulting from infection with certain strains of tobacco mosaic virus (TMV). The mosaic disease in tomato was described in the Netherlands in 1910 (158), and in the USA in 1916 (9). Since then considerable efforts have been made to understand its epidemiology and to devise control measures. The fact that the disease remains widespread suggests that the efforts to date have not been entirely successful.

TMV is the most infectious plant virus known; young tomato plants are infected when rubbed with infected tomato sap diluted one part in 5 million parts of water (27). It is also the most persistent virus in terms of its ability to survive outside plant cells and in dead tissues (38). Its numerous strains infect a very wide range of plant species, although there is considerable specificity between strain and host. It is not surprising, therefore, that the disease is ubiquitous wherever tomatoes are grown under glass and in many areas of outdoor cultivation. Tomatoes are not a cheap crop to produce, particularly under glass, and until recently about 20% of world production probably was lost because of TMV. The literature on TMV is voluminous; so also is the literature on the disease in tomato. Many of the papers mentioned in my 1960 review (25) are omitted from this paper.

THE DISEASE IN TOMATO

Causal Viruses

Early investigators reported that ordinary or mild tomato mosaic virus was similar to common TMV. In tobacco of the White Burley type, TMV was known to develop systemically but did not cause necrotic local lesions (2, 9). In tomato,

75

TMV caused a light and dark green mottle, slight plant stunting, and sometimes leaf distortion. Since then many serologically related strains of TMV have been isolated from tomatoes, and the problem is to decide what was meant by tomato mosaic. By 1927 stripe or streak disease of tomato was often associated with "mosaic," plants occasionally developing necrotic areas on leaves, brown necrotic stripes on the stems and petioles, and sometimes brown pits on the fruits (15). This disease was called also single virus streak to distinguish it from that caused by a mixture of TMV and potato virus X.

Inoculation of tomato with sap from stripe plants often caused mosaic only (1) and in 1940 it was demonstrated that the strain of TMV common in German tomato crops was not common tobacco virus, but caused only necrotic local lesions when inoculated into certain *Nicotiana* species (97). Occasionally the common tobacco strain occurred as well, causing a systemic infection. Experimentally, it is difficult to reproduce and distinguish single virus streak from double virus streak—the symptoms may result from a complex of factors, including environmental shocks, root damage, and possibly a mixture of established or unstable tomato strains of TMV (124). New strains of TMV with increased infectivity, or extended host range, or both have resulted from passage through resistant hosts. This suggests that protracted periods of evolution are not required for changes in strains (50, 91, 92).

A review of the TMV strains and isolates would require a paper by itself. The literature is confused because different workers have used symptoms to distinguish strains, or indicator plant reactions, host range, genetic or biochemical techniques. A paper on this subject by M. Hollings is to be published in 1976.

Tomato mosaic is now known to be caused by a range of strains, allied to those causing streak, most of which cause local lesions only when inoculated into appropriate test plants, such as the necrotic line of White Burley tobacco selected by workers in the Netherlands (143). Unfortunately the reactions of such test plants are not entirely consistent, and several tomato mosaic serotypes invade them systemically. Nevertheless, surveys of commercial tomato crops in several countries, using different indicator plants such as the necrotic strain of White Burley tobacco, *Nicotiana sylvestris* and Petunia, have confirmed that tomato and tobacco crops tend to be infected by different viruses. In the UK during 1960−61 only 4 out of a total of 187 crops that were infected with the tomato strains of TMV carried a tobacco strain as well; these 4 consisted of grafted plants and the workers had been smoking while grafting (26). Other surveys have shown that the tomato strains prevail in Canada (56), Korea (17), Japan (83), Belgium (151), the Netherlands (124), and elsewhere, although Dutch, Korean, and Japanese workers found some tomatoes infected with tobacco strains of TMV. Even when tomatoes were grown contiguous to tobacco in Canada, the TMV strains infecting both crops were distinct (90). In the Netherlands, work based largely on symptom expression distinguished eight strains in tomato, of which the normal mosaic was the most common (124). The common tobacco strains were seldom found, and the other six were more closely related to the tomato than to the tobacco type. The common tomato strain not only more easily infected tomato than the tobacco strain but also multiplied faster after invasion (83). The two strains do not cross-protect and they can coexist in tomato (26, 93, 124).

Symptoms

The reaction of the tomato plant to infection with TMV depends upon cultivar, virus strain, time of infection, soil nutrient, water status, day length, light intensity, and temperature (55). In addition to the leaf mottle or mosaic, leaves tend to be elongated or otherwise distorted in appearance and smaller and more pale than those on healthy plants. Leaves of infected plants sometimes are scorched in sunlight, particularly soon after infection and after wetting with water, and lower leaves die earlier than those on healthy plants (28).

Infected seedlings are stunted, with smaller, less fibrous root systems as well as less foliage; late infection does not affect root growth significantly (32, 152). The smaller root systems in young plants are caused by a failure to grow, rather than by necrosis.

During the incubation period before symptoms show, growth slows considerably (4). There is a sharp drop in transpiration rate coincident with the appearance of symptoms but then the rate increases gradually, and usually TMV-infected leaves transpire faster than healthy ones. The water content of infected plants is lower than that of healthy ones during the early stages of disease, but later it is higher (68, 132). Although TMV can seriously affect quantities of root, shoot, and fruit produced, depending mainly on the age of the plant at the time of infection, it does not affect the general pattern of growth and development (32).

Leaf symptoms often become less severe as infected plants age. TMV concentration in young leaves, as well as symptom severity, increases with increasing temperature from 16 to 28°C, but once the maximum concentration is reached, symptom severity starts to decrease (11). It has been suggested that virus multiplication in the shoot may be controlled by two distinct root factors, the supply of cytokinins and the amount and nature of the nitrogenous compounds present (135). A tolerant cultivar Virocross contained more gibberellins, more cytokinins, and a lower concentration of amino compounds than a susceptible one (135). Because the sugars available for growth are less than in healthy plants, it has been recommended that infected ones be grown at somewhat lower temperatures, with free ventilation, adequate root moisture, and frequent water sprays for the foliage (67). There is no doubt that experienced growers can do much to alleviate the effects of infection, but it does not seem possible to greatly minimize them by altering nutrition. Early experiments on the effects of nutrients on symptoms showed that lack of potash but not of any other major nutrient increased the incidence of necrotic streak (1) and that the addition of lime or sulfate of potash retarded symptom development (131). On the other hand, excess potash in relation to growth increased susceptibility to infection. Attempts to mitigate the effects of virus infection by weekly foliar sprays with an aqueous solution of urea, with or without solutions of monoammonium phosphate and potassium sulfate, failed to affect fruit yield or quality sufficiently to warrant commercial use (36).

Fruit Quality and Yield

Tomato fruits can react to stresses in the plants in a limited number of ways, so it is not surprising that the quality defects that occur have been attributed to several causes. The yellow patches associated with the aucuba strain of TMV, and the severe blistering and necrosis caused by double virus streak were recognized, but for a long time there was considerable confusion over the very common unevenness of ripening referred to as blotchy ripening or severe mottle, graywall, cloud, internal browning or bronzing of the fruits.

The suggestion that internal browning or bronzing might be caused by TMV infection was made as early as 1941 (69) but it was 1956 before it was established that it can be induced as a shock reaction some three to four weeks after invasion by the common tomato strains of TMV, affecting only fruits that were already set when the virus became systemic (19, 22, 23, 28, 76, 140). TMV isolate, high soil moisture, low nitrogen and boron in the soil, and cultivar markedly affected the incidence and severity of this disease and of blotchy ripening (13, 21, 141). Internal browning is particularly evident in unripe fruit and usually is restricted to the fleshy parenchyma whose cells often collapse. It is most prominent near the shoulders of the fruit at the stem end, and most severe between the vascular bundles, which are not affected. White tissue as well as necrosis occurs frequently in the pericarp (85).

For a period there was considerable controversy over the causes of internal browning, some workers asserting that nutritional imbalance affected its incidence (127, 139); others thought its cause was environmental (62). The superficially similar graywall symptom was usually induced in the absence of TMV especially when plants were grown at low light intensity, low air temperature, and high soil moisture (100, 101, 136), but these noninfected fruits frequently have prominent necrotic streaks in the vascular region in the lower half extending to the stylar scar. However, the subject is again confused because some workers now attribute both graywall and internal browning to TMV infection (20, 43); it has been suggested that TMV and bacteria enhance the development of graywall, bacteria multiplying more in TMV-infected than in TMV-free fruits (138). Experiments under glass in the UK showed clearly that necrotic pitting, especially around the calyx, bronzing or internal browning, and severe mottle affect some but not all fruits that are already fairly mature when TMV invades them (28, 76). Many fruits with necrotic pitting fall off prematurely, and calices of affected fruits are dry and crinkled. The severe shock reaction lasts up to almost eight weeks after systemic invasion by the virus. The proportion of seriously affected fruits gradually decreases with successive harvests (28). The reason that most growers had not previously attributed these symptoms to TMV infection was that virus invading via the roots makes its way into the lower fruit trusses before penetrating into the young shoots; thus, fruit defects may be visible several days before leaves show symptoms. When leaf infection occurs, trusses show symptoms in a sequence that depends both on leaf position and the number of trusses already formed (40). While internal browning or bronzing, necrotic pitting, and severe mottle usually are caused by TMV infection, especially when the plants are lush in growth, one cannot always be sure of this; superficially similar

fruit defects also can be caused by adverse environments and nutritional imbalance in the absence of virus (79).

In addition to poor quality, fruit yield can be adversely affected by TMV in two
ways—fewer fruits or smaller ones (28). Many estimates of loss have been made
on both outdoor and indoor crops but most have been underestimates because
the control plants became infected during the experiments. The experiments were
then comparing early with late-infected plants, rather than infected and noninfected ones. Loss differed with cultivar, virus strain, growing conditions, and
time of infection (5, 94). Some workers found that the earlier the plants were infected the greater was the loss (1, 18, 44), whereas others found that early inoculation decreased yields most in the early and midharvest pickings but in later
pickings yields were less in late- than in early-infected plants (157). Experiments
on the seasonal pattern of fruit production in tomato showed that infected plants
always produced less than healthy ones but the magnitude of the loss varied with
time of year and age of plant, being greatest in spring and in young plants (32).
Losses have varied considerably in the different trials reported, but most fell
within the range of 5 to 50%, and most losses have been due to fewer fruits rather
than smaller ones (18, 130).

Experiments under glass in the UK, in which the control plants remained noninfected, showed similar losses with widely different dates of infection, from
before any fruit had set to after it had all done so; with early infection there were
fewer fruits, with late infection smaller ones (28, 29). Trusses coming into flower
or setting at the time the virus became systemic, suffered the most in terms of
amount of fruit set (28). Desiccation affected both stigmas (which were less
sticky) and pollen (which germinated poorly) for about six weeks after inoculation. Plants inoculated in the seedling stage produced better pollen than plants inoculated later (120).

The factors affecting crop losses are so numerous and varied that it is impossible to more than guess at an average figure: somewhere between 15 and 25% loss
in yield is probably about right. On a worldwide scale this forms a tremendous
loss, whether in fruit tonnage, wasted manpower, or monetary value. In addition
to losses in yield, many growers have suffered financially because badly blem-
ished fruits were unsalable early in the season when prices are high. Before quality defects were recognized as a result of TMV infection, growers maintained that
the greatest financial losses are incurred when TMV affects a crop before the
fourth truss is flowering and that later infection was of little consequence.
However, if quality defects make much of the early fruit unsalable, the total
financial loss is often greater when infection occurs later (28). These considerations led to the development of deliberate early inoculation (see below).

SOURCES AND TRANSMISSION OF VIRUS

The Tomato Crop

SEEDS   There was controversy for many years between growers, who maintained
that TMV was seed-borne in tomato, and plant pathologists who said it was not,

because they were unable to obtain any infected plants by sowing seeds from infected fruits and growing the seedlings in isolation (25, 80, 130). Early workers who claimed that virus was in the embryo used relatively few seeds with inadequate controls; however, when 2000 tomato seeds were dissected in Australia, virus was found in the testa but not in the endosperm or embryo (45). Other workers also found some testae infected but showed, in addition, that the endosperm sometimes carried TMV (30, 142). Infected seeds were sometimes necrotic, especially those from early-infected plants; such seeds tended to contain virus in the endosperm (30). Severely necrotic seeds often failed to germinate. About half the seeds from infected fruits carried TMV, the proportion differing with tomato cultivar, time of infection, truss position, and method of cleaning (30). Seeds usually carried the virus externally, depending on the method of cleaning, but about a quarter also carried it within the testa or the endosperm. High virus concentrations sometimes occurred internally when the endosperm was infected but no evidence of embryo infection has ever been obtained. Endosperm infection occurred mainly in fruits that set after the plants were systemically infected, and decreased with increasing age at the time of infection. Thus there would seem to be a critical period during which the endosperm can become infected.

TMV persists in or on the testae for several months or even years after harvest. At first this was thought to depend on the way in which the seeds were stored (103); but this was not substantiated when seeds were stored in open dishes or in closed glass tubes (30). In both, virus concentration remained high in stored seeds for nearly a year and then fell and remained at a constant low level for five years; presumably the virus on or in the testa (but not that in the endosperm) was gradually inactivated. Some seeds extracted from fruits by fermentation remained infected internally when stored in paper packets for nine years (30). Some TMV strains survived only a few days on germinating tomato seeds, and it was suggested that these contained substances that inhibited TMV, differing with tomato cultivar (125). However, other workers grew tomato embryos in media containing virus but found no evidence of inactivation (46, 153). Rapid inactivation of virus was observed in seeds after sowing, perhaps caused by soil microflora, but testae remained infective for at least ten days. This was long enough for contamination during transplanting (30, 150, 153).

It is clear that TMV is seed-borne in tomato, but because it does not occur in the embryo, is this a source of virus for the crop? It has been shown that seedlings are not infected when left undisturbed, but growers often transplant the young seedlings a few days after germination. Loosely held TMV on the surface of seeds may be the most likely source, contaminating the seedling and entering through mechanical abrasions made during transplanting; in such cases, most infections were presumed to occur after the testae were carried above ground on the cotyledons (142). However, other work indicated that infection usually occurred below the soil surface when testae or the remaining endosperm were pressed against the seedlings during transplanting (30). Normally very few seedlings were infected in this way because the testae often were left behind in the compost when the seedlings were transferred. Also much of the virus carried externally was inactivated after a week in the soil, and very young seedlings resisted infection. Seedlings were not infected by absorbing TMV from the endosperm during germination because none was infected unless transplanted.

Only a few seedlings need to be infected in a tomato crop for the virus to spread rapidly. A weekly survey of 374,000 seedlings from thirteen different sources of seed on ten commercial holdings found 0.05% of them infected and although these were removed, virus had been spread to other young plants that formed sources of TMV for 7 of 22 growing houses to which the seedlings were transplanted (34). Direct seeding has been adopted by some growers to eliminate pricking out; this method is easier now that it is possible to obtain pelleted seeds.

SHOOTS AND ROOTS   Other plant parts that can act as virus sources are the leaves and shoots, and the roots, or the remains of both. Handling the plants while tying them, removing side shoots, or picking fruits is undoubtedly the most important method of spreading TMV within tomato crops (34). Many growers attempt to delay virus spread by handling last all obviously infected plants; but this is seldom effective because virus is in the leaves some days before they show symptoms. The longevity of the virus in dried plant debris poses a special problem for tomato growers. Plant debris is usually cleared up from the soil surface between crops, but small scraps of infected leaves may adhere to parts of the glasshouse structure. Root debris left in the soil is a different matter and one that was long overlooked by growers. This is surprising because in 1928 it was shown that TMV remained active for at least 70 days in glasshouse soils that had carried infected crops, and tomato plants were infected when planted into such soils. Soil conditions affect virus survival, TMV remaining active longer in heavy than in light soils (78). Powdered leaf debris lost infectivity within a month in moist soil but remained infective for two years in dry; also, root debris remained infective for six months even in moist soil. Tomatoes were infected when planted into glasshouse soils within five weeks of clearing infected plants but not when a different crop was grown for six months in between the tomato crops (78, 80). Root debris from previous crops is undoubtedly the most important source of virus for tomatoes under glass (29, 53). Earlier workers were correct in stating that the proportion of plants that became infected in this way is usually small (but see 137) but later spread within the crop is so rapid that only a few foci are needed to start an epidemic (53). It makes no difference if the plants are set out into soil containing root debris or into a layer of sterilized soil covering it; once the roots penetrate into contaminated soil they may be infected (29). The virus was shown to persist in tomato root debris at a depth of at least 120 cm for at least 22 months in fallow soil, and for over two years in soil under black polyethylene that carried a layer of clean compost above it (35). Transmission presumably occurs when root hairs are damaged in the presence of infected debris or free virus leached from it into the soil (29, 128). TMV has even been detected in surface water used for watering the tomato crop in the Netherlands (148). There is no evidence to suggest transmission to neighboring plants via root contact or anastamosis (29), although it may occasionally occur from virus exuded into the rhizosphere from infected plants (98).

INSECT AND BIRD TRANSMISSION   Some early work on insect transmission of TMV is suspect because of inadequate controls and the ease with which the virus is transmitted by contact (87, 105). However, recent work has shown that glandular leaf hairs are susceptible to injury by clawing. Aphids can pick up TMV from tomato hairs and subsequently inoculate healthy plants by clawing them (24, 64). By feeding aphids (*Myzus persicae*) through a membrane on a solution containing TMV to minimize the chances of their claws being contaminated with virus, it was shown that they could acquire TMV only when it had been mixed with poly-L-ornithine or poly-L-lysine, although the reasons for this are not clear (116, 117).

TMV-like particles were found in the guts of some sucking insects fed on infected plants (107) and in aphid honeydew (119), but TMV could not be detected in droplets emerging from the cut stylets of *M. persicae* which had been feeding on infected plants (149). If insect transmission of TMV occurs in tomato crops it must be of minor importance relative to other methods.

It seems obvious that a virus so readily spread by contact might be carried within crops, and from one crop to another, by birds on their feet, plumage, or beaks, or by mammals on their fur. Wild animals occur in outdoor crops and occasionally birds enter glasshouses, fly among the plants, and even peck the fruits. These can undoubtedly transmit TMV within tomato crops (31).

Alternative Host Plants

WEEDS   Viruses of the TMV group occur in a wide range of cultivated plants and weeds, but it has seldom been determined if these are sources of virus for tomato crops. TMV is commonly found in *Plantago* (70) but it differs from the common tomato and tobacco strains (65). TMV overwintered in rootstocks of *Physalis subglabrata* in Indiana (60), and field investigations suggested that this weed was a source of virus for outdoor crops. Apart from this there is little evidence that weeds play much part in the epidemiology of tomato mosaic.

SMOKING TOBACCO   The relative importance of smoking tobacco as a source of TMV in tomato crops used to be as controversial a subject as seed transmission. It is only during the last fifteen years that it has been realized that the virus strains common in tomato are different from those in tobacco. It is possible that the strain prevalent in tomato in the 1920s was the tobacco form which later mutated after frequent passage through tomato; but the latest research does not support this view (124).

Many workers found most cigarettes, cigars, pipe and chewing tobaccos infected (3, 26). The habit of chewing tobacco was incriminated as a common cause of infection for many tobacco and tomato crops, because of the infective saliva on the hands. Pruning knives, also, were often kept in pockets containing tobacco dust (71). There is no doubt that tomato can be infected with the tobacco form (3, 26). The speculation was removed from the subject when forms of White Burley tobacco were selected, which reacted to inoculation with most tomato strains of TMV with necrotic local lesions only, whereas the common tobacco strains caused systemic mosaic and no local lesions (97, 143). These and other test plants made it possible to survey tomato crops for various strains of TMV. As a result, it is now known that the tobacco strains of TMV can and occasionally do infect tomato, but in general tobacco is not an important source of virus for the tomato crop (17, 26, 55, 82, 83, 124, 151).

Clothing

Although some workers had shown that TMV remained active on clothing and could be transmitted also by machinery passing through crops, the importance of contaminated clothing was long overlooked by growers and plant pathologists. A survey of clothing worn by workers in a major English tomato growing area in 1962 indicated that 42% of outer garments had not been cleaned during the year; 24% of them had been worn the previous year and two thirds of these had not been cleaned between seasons (33). That infected clothing is a potent source of virus was shown by a worker wearing a cotton overall for two days while working in an infected crop and then walking between rows of healthy plants for 10 min so that the overall brushed the leaves: 19 of the 20 plants were infected but none of the controls.

Washing the infected clothes with detergents in hot water, or dry cleaning them, decreased the TMV content. Virus on clothing stored in sunlight was inactivated within a few months. TMV-impregnated clothing stored in a shady place remained infective throughout a winter, and when stored in the dark remained so for over three years. There is no doubt that nursery workers, extension advisers, and firms' representatives can easily carry virus from one crop to another as well as spread it within the crop. However, it is doubtful if clothing forms a major reservoir of overwintering virus.

Structures

It used to be assumed that TMV could survive between crops on glasshouse structures but tests showed that infective sap was rapidly inactivated in daylight, occasionally persisting at low concentrations for about two months (33). A survey confirmed that TMV seldom overwintered on commercial glasshouse structures although it might persist in small fragments of dry tomato leaves if the houses were not washed. The wires used to support crops were occasionally still contaminated after four months in storage, but could be cleaned by heat or chemical inactivators (103).

VIRUS MOVEMENT IN TOMATO

The development of disease in the tomato plant depends not only on the virus strain, cultivar, and cultural conditions but also on movement of the virus once it has effected entry. This in turn is influenced by the site of infection, and a knowledge of this movement is important in understanding the epidemiology of the disease. The development of symptoms in the young leaves around the growing points following leaf inoculation depends upon the age of the plant and time of year, young plants often showing symptoms in five days in summer but taking three weeks in winter, and taking several days longer in older plants than in seedlings (2, 9). When infection occurs through a leaf, the virus multiplies therein for a few days and then moves into the phloem; from there it is carried rapidly first to the roots and then to the shoot apices, or to both simultaneously, bypassing many parts of the plant en route (37, 129). Large plants may never become completely invaded. Because the tomato plant grows by a succession of axillary meristems, which terminate as fruit trusses, lower trusses often are infected before virus passes to the growing point (133) and so may show severe fruit blemishes before leaves show symptoms.

When plants were infected via the roots, the virus often failed to invade the tops especially when growth was slow (59, 128), although young seedlings inoculated in the roots usually became systemically infected (122). The virus sometimes is localized in the root around the initial entry point. Systemic invasion can vary with individual plants and with the time of year; it is common in summer but rare in winter (128). Experiments to test the frequency of root infection and the time taken for virus to become systemic showed that TMV could be detected in young shoots before symptoms appeared. The periods between inoculation of roots and symptom development in the shoots vary from three weeks in young plants to six months in older plants (the average periods being between ten and sixteen weeks). Such periods are shorter in summer than in spring or winter (29). In some older plants the virus did not leave the roots during the experimental period of about six months. Seedlings showed leaf symptoms an average of five weeks after inoculation in the roots, although some did not do so in ten weeks. Symptoms appearing soon after planting out could well result from infection of the roots during transplanting (122). The lower fruit trusses often showed severe blemishes before the foliage showed mosaic, being invaded by the virus before the upper shoots. This undoubtedly contributed to the failure of growers to attribute such fruit symptoms to TMV.

The frequency of infection by root contact with debris from the previous crop was low, but a few primary infections can lead quickly to a serious epidemic of the disease. The sudden development of symptoms in previously healthy commercial crops in the UK during May and June probably results from spread by handling during tying and sideshooting, following a few initial infections through the roots.

CONTROL

Prevention of Infection

Knowledge of the epidemiology of the disease now makes it much easier than before to grow a healthy crop, but it usually is impossible to prevent infection under commercial growing conditions.

CLEAN SEEDS OR DIRECT SEEDING   Seeds taken from healthy plants can be sown, and the seedlings planted into clean soil but it is essential to test the mother plants and the seed batches by inoculation into *Nicotiana glutinosa* or some other test plant to ensure that fruits are not harvested from late-infected plants. In the Netherlands, some growers use an electronic scanning device to sort out necrotic seeds, because these germinate poorly and are often infected in the endosperm (30, 124).

If infected seeds have to be used, infection of the crop can be prevented by direct sowing into the final positions, so that the seedlings are not transplanted (30, 52, 61). Recent cultural changes, such as seed pelleting and the development of plastic containers and compost blocks, have made direct seeding a commercially viable practice.

ISOLATION   The crop should be kept isolated from sources of infection. This means planting outdoor crops into soil that has not carried tomatoes, tobacco, or other plants susceptible to TMV for one or preferably two seasons. If this cannot be achieved under glass, the contaminated soil can be isolated with sheets of polyethylene and the plants grown in a suitable compost above this (159). This method was used by several UK growers for a time but has been superseded by the bag method—isolated plastic bags, or troughs being filled with peat compost (156). This is now the predominant method of culture in Guernsey and in a few other areas of the UK.

The latest method to isolate the roots from contaminated material is the waterbed method (41), a British hydroponic technique in which the tomato roots are immersed in a flowing stream of water containing dissolved nutrients in lay-flat plastic tubing. If clean seeds are used, the crop ought to be kept free from TMV, but in view of the Dutch observation of TMV in irrigation water (148), if a plant did become infected by contact with a worker's hands or clothing there would be the danger of rapid spread throughout the house via the water. Another technique that proved successful is to grow the plants on straw bales isolated from the soil by polyethylene and watered with nutrient solutions, but this did not become popular. Isolation by distance from infected crops is also desirable to minimize the transmission of virus by birds, mammals, or insects.

HYGIENE   The following measures are desirable but it is unrealistic to expect them to be practiced by commercial growers and their employees. All tomato crop debris should be disposed of at a sufficient distance from growing crops to prevent it becoming a source of virus. Workers should not be asked to tend infected as well as healthy crops, and should not wear uncleaned outer clothing that had been worn in infected crops. If these measures are not possible, it is little use working symptomless areas first, which is a common practice. Visits by managers, foremen, advisory officers, or commercial representatives should be minimized. All persons entering a healthy crop should be obliged to put on clean overalls and to wash their hands; visits should be restricted to one pathway in a crop. Workers should be made aware that TMV can be carried on their footwear and hair as well as on clothing and hands.

Although smoking tobacco is only a minor source of infection, smoking should be forbidden in tomato crops and smokers should wash their hands before tending the crop. They should not keep tools in pockets that might contain tobacco debris.

Virus Inactivation

IN PLANTS AND SEEDS   Attempts to inactivate TMV in infected plants or decrease the effects of disease have not proved practicable (14, 88, 160), although β-3-indolylpropionic acid or α-naphthalene sprayed on field-grown tomatoes in India were reported to alleviate symptoms (16). Much more work has been done with infection inhibitors or virus inactivators on the surface of plants, particularly cow's milk. Following an initial report that spraying tomato foliage every ten days with milk, diluted with water, eliminated TMV spread (102), numerous tests have shown that skimmed milk sprayed on to the plants, and used to wet the worker's hands shortly before tending them, have prevented or delayed TMV spread (63). However, there are several reports of poor results from such treatments (47, 66); they delayed but did not prevent the spread of TMV in the crop. Such treatments would not be worthwhile where late infection causes internal browning or other fruit defects; but they have been shown to increase significantly yields of field crops (51, 75). The casein in milk is the strong inhibitor, lactalbumin and α-lactglobulin being weak ones (74).

If possible, tomato seeds should be harvested only from plants shown by testing to be virus-free. Such plants are seldom available, and if they are not, seeds should be harvested from the lower trusses of late-infected plants so that there is a good chance that they will be free from internal virus. In the past, most commercial seed stocks have been infected. Several attempts have been made to inactivate virus on or in seeds by chemical means. After extraction from the fruit, most tomato seeds are cleaned by fermentation or by a chemical treatment such as adding sodium carbonate solution or concentrated hydrochloric acid to the pulp. They are then washed in water and dried. Many investigators claim that these treatments, especially the HCl, free contaminated seeds from TMV (6, 39, 58, 104) but others, who probably worked with seeds that contained virus internally, were unable to free all batches of seeds completely (30, 47, 72). Similarly, treatments of seeds with trisodium orthophosphate or sodium hydroxide, or with ultraviolet light have sometimes been successful, sometimes not (6, 30, 103).

Heat treatment (three days at 70°C or one day at 80°C) of dried seeds has also been used successfully; it is the only method that is successful for seeds infected internally, although three weeks at 70°C did not free those infected in the endosperm (30, 73, 84, 155). Heat treatments may delay germination but no deleterious effects occur if seeds that are newly extracted are adequately dried first; such seeds will withstand at least 70 days at 70°C (126). Heat-treated seeds are now readily available commercially.

TMV INACTIVATION IN DEBRIS   TMV in soil and in plant debris can be destroyed by heat. Virus in undiluted plant sap is inactivated in 10 min at 93°C or in 80 min at 90°C (118); but in dried debris it may survive higher temperatures (12, 35).

The length of exposure necessary at different temperatures depends on the concentrations of virus and its location—whether in sap or in leaf or root debris (35).

Steaming of glasshouse soils, as done commercially, may not eradicate TMV even from the upper soil layer because an adequate temperature is not maintained long enough to penetrate thick roots or to reach every part of the soil, especially that near brickwork, water pipes, or other structures (53, 154). It is extremely difficult to inactivate TMV in thick roots by steaming the soil; as many large roots as possible should be removed before steaming (53). Steam penetration is best in dry soil but dry roots are more difficult to free from TMV than moist ones. Sheet steaming gives satisfactory results, but the effectiveness of all methods depends upon the structure, texture, and moisture of the soil. Even if all the TMV in the upper 30 cm or so of soil is inactivated, however, debris at lower depths can remain infective for several months, particularly in fallow soil (35).

Although formaldehyde inactivated purified TMV in laboratory tests (81) and considerably decreased virus concentration in debris (35), treatment of debris-containing soil with formalin did not prevent infection of tomato seedlings (86).

When concentrations of five times the recommended doses of chloropicrin, metham sodium, methyl isothiocyanate, D-D, and nabam were tested, these soil fumigants did not inactivate TMV in plant debris or in soil or when they were mixed with infective leaf sap. Indeed, the main effect of treating buried roots with chloropicrin and metham sodium was to increase the proportion of root fragments that remained infective over a period up to 13 months after treatment, from 33% in the untreated controls to 61% and 73% respectively (35, 77, 89). The consequence of such chemical sterilization was an earlier attack of TMV in the crop through root infections. There was no clear indication of a causal relation between the persistence of TMV and the microorganisms colonizing roots although the population of fungi and bacteria, many of them involved in the breakdown of vegetation, were significantly affected by the treatments (35). Only methyl bromide destroyed all the TMV in root fragments in closed containers but it might not be effective under glasshouse conditions because, although it inactivated TMV in thin roots, it did not do so completely in thick ones (77, 108, 147).

TMV INACTIVATION ON HANDS, TOOLS, AND STRUCTURES   It is not easy to eliminate TMV from the hands after working in an infected crop, and just washing them with soap and water is inadequate. One of the best methods is to wash in a 3% solution of trisodium orthophosphate and then scrub well with soap and water, although even this does not always eliminate virus in the areas under finger nails (27, 99).

Tools can be heat sterilized, or dipped in 3% trisodium orthophosphate (27, 99, 103). Detergents and washing-soap solutions only inactivate virus on tools, wires, etc if these are soaked for long periods in high concentrations of the solutions. Although light soon inactivates TMV adhering to glasshouse structures, small fragments of leaf debris may carry the virus, and so thorough washing of the houses with a disinfectant, when they are empty between crops, is to be recommended.

Minimizing Fruit Losses

ALTERING THE ENVIRONMENT   Although growing conditions affect the plant's response to virus infection, there is no concensus as to the optimal environment to minimize loss. Many reports are concerned with leaf symptom severity, whereas all that matters is salable fruit yield. Because the sugars available for growth in infected plants are less abundant than in healthy ones, it has been suggested that infected plants under glass should be grown at less than 27°C by day and a few degrees lower than healthy plants by night (67). Free ventilation, frequent water sprays, and adequate root moisture also have been recommended to aid the growth of infected plants in sunny weather.

Some workers have suggested that such symptoms as internal browning and blotchy ripening may be enhanced by heavy watering, or inadequate nitrogen and boron (134, 139); others did not find the incidence of internal browning affected by nitrogen, although blotchy ripening was enhanced by high nitrogen (42). These severe fruit reactions to virus infection are much worse when vegetative growth is abundant and when watering is excessive. When growth is better balanced, leading to a high dry matter ratio, they occur less frequently or not at all (10).

DELIBERATE INOCULATION WITH TMV   The discovery that late infection with TMV often causes severe fruit blemishes which are not present when the plants become infected before any fruit has set, led to experiments which confirmed that deliberate early infection would be worthwhile for a grower who always suffered quality loss (28). Plants infected before the first truss set often suffered less weight loss than plants infected later. Thus, on the average, the financial returns were greater from inoculated crops than from noninfected ones that suffered late infection and defects in fruit quality.

Growers in the Isle of Wight (UK) were the first to adopt deliberate inoculation in 1964, although a mild strain of TMV was not available and the normal, fairly severe strain of tomato mosaic was used for inoculation in the seedling stage. The proportion of unsalable and poor quality fruit was reduced from about 30% to under 3% in 1965. The benefit to be gained from seedling infection would be greater if a mild strain of tomato TMV could be used that would protect the plants against severe strains, and attenuated strains that did this were obtained by prolonged heat treatment (106, 109, 110). Then in 1968 an almost symptomless mutant (MII-16) which gave good cross-protection was isolated in the Netherlands by the mutagenic action of nitrous acid (121, 124).

Tests on commercial crops proved very satisfactory in decreasing crop loss, and a spray gun method of inoculation was found to be better than manual inoculation because of the risk of contamination with other strains (121). Such inoculation with the mild strain also prevented the development of severe strains acquired through seedling root contact with inoculum (123). However, it may not give complete protection against all naturally occurring strains of TMV, either of the tomato or tobacco forms (57, 96).

The mild strain, when inoculated into seedlings, caused a temporary inhibition of growth and delayed flowering and fruit set, so it was necessary to advance the sowing date by about a week to compensate for this (124). Seedling inoculation with this strain resulted in negligible internal infection of seeds in subsequent fruits, in contrast to plants inoculated with the parent strain when endosperm infection occurred, so deliberate inoculation with the mild strain could be used by seed growers together with subsequent heat treatment, when seed plants cannot be kept healthy.

The mild strain, MII-16, is now widely used by commercial growers in the UK (57) and other countries, as well as in the Netherlands, and apart from the breeding of resistant cultivars, its development has done more than any other factor to decrease crop losses from TMV in tomato. Largely as a result of such inoculation, average yields of early heated crops in the Netherlands increased by 15% between 1971 and 1973, and trials in the UK in 1971 gave increased yields of 7% (146). The very success of the method may well make growers reluctant to accept resistant cultivars unless these can be shown to have considerable advantages over existing ones (124).

Cross-protection between virus strains is dependent on temperature; a mild strain that prevented the multiplication of a severe one at 15 or 20°C did not do so at 25 or 30°C (145), probably because high temperature has a detrimental effect on the multiplication of mild strains (112). The mild strain MII-16 was inactivated when stored in dried leaves for a long period, and it was found necessary to freeze concentrated suspensions for commercial use (124).

Most of the work on inoculation of seedlings with attenuated strains of TMV has been commercially successful, but phenotypic mixing of strain characteristics can occur with TMV strains, producing new, recombinant strain types. M. Hollings has warned *(in litt.)* that if MII-16 is widely used over a lengthy period, it may saturate the soil in tomato houses and may combine with more virulent strains. Because MII-16 is derived from a strain 1 type, any new recombinants will stand a good chance of breaking resistance in Tm-1 plants. Also the attenuated strain can increase considerably in virulence during passage through some test plants. Further, MII-16 multiplies much more slowly than typical TMV strains in tomato, thus offering a poor chance of this strain surviving in competition with more virulent strains against which it does not protect.

Breeding Resistant or Tolerant Cultivars

An adequate review of such a large subject as tomato breeding for disease resistance would be out of place in one on epidemiology, and the subject has been well reviewed elsewhere (111, 113, 124). However, because the use of resistant cultivars is now the most widely practiced means of preventing losses from TMV infection in several countries, a brief account of recent work is essential.

The success achieved so far suggests that the development of commercially acceptable resistant cultivars is not only possible but also may result in long-lived resistant or immune cultivars, despite the apparent ease with which TMV can mutate. The several strains of TMV that occur in different parts of the world are not identical and this complicates the problem (7, 95). Field resistance was sought without much success, so the breeders have relied on three major gene factors for resistance, Tm-1, Tm-2, and Tm-2^2. Five strains of TMV have been identified in tomato in Europe: strain 0, the commonest, induces no symptoms on plants with any of the resistance factors; strain 1 causes symptoms on Tm-1 plants, strain 2 on Tm-2 plants; strain 1:2 causes symptoms on both and also on Tm-1 Tm-2 plants. Strain 2^2 induces symptoms on Tm-2^2 plants, but it is rarely found. Only strains 0 and 1 are common (113, 114, 125). This work is based on the resistance that exists in wild *Lycopersicon* species, e.g. *L. peruvianum*. The mechanisms of resistance are not fully understood, but virus multiplies very slowly in resistant plants and concentration remains low in contrast to susceptible plants (7, 49, 115).

There are differences of opinion on the plant breeding policy to be followed, some breeders thinking it worthwhile to release cultivars with single gene resistance even if they last only for limited periods (8). In the UK, tomato crops sampled prior to the introduction of cultivars with monogenic resistance to TMV yielded only strain 0, but within two years of the introduction of cultivars heterozygous for Tm-1, strain 1 was found with increasing frequency (114). Strain 1 continued to increase following the introduction of cross-protection with the avirulent TMV strain MII-16 (originally derived from a strain 1 isolate) and two years later 94% of isolates sampled from nurseries using the technique were of strain 1 in contrast to 39% from other nurseries (54). It was suggested that this increase in the incidence of strain 1 increases the risk of breakdown of resistant cultivars and that it is desirable to achieve multigene resistance, preferably with all the three known factors, in a cultivar before it is released into commercial production (114, 144).

Unfortunately, when some of the resistant cultivars are grafted on to TMV-susceptible rootstocks, the virus invades the shoots and causes severe fruit necrosis and fruit drop. Recently a rootstock has been introduced with single-gene resistance (Tm-2^2) but again there is the danger of selection of new TMV strains if a three-gene scion is grafted on to a one-gene stock; attempts are now being made to produce a rootstock carrying Tm-1, Tm-2, and Tm-2^2 (48).

CONCLUSIONS

Almost all aspects of the epidemiology of tomato mosaic have been reexamined during the last twenty years: The work is being done in several countries, which illustrates the importance attached to the disease. Many myths resulting from inadequate early research have been dispelled, and the sources of virus and its transmission are now well understood.

Infected seeds and root debris in the soil are the two most important sources of TMV. The virus is easily carried from one crop to another on clothing as well as on hands and tools. Hygiene measures necessary to prevent TMV introduction and spread are too uncertain and involve complications which make them commercially unacceptable, so the development of a mild-strain inoculation technique that minimizes fruit losses and protects the plants against invasion by more damaging strains has been of great importance. However, virus epidemiologists have never been very happy about deliberately disseminating viruses, because of the risks of selection of virulent strains, and would much prefer to try to eliminate the virus. Breeding for resistance to TMV is going ahead in several countries and so far several resistant cultivars have been accepted in commerce. It remains to be seen if the plant breeders can keep ahead of the selection pressures on the viruses. If they can, the great reservoir of tomato strains of TMV should gradually be reduced, and healthy crops become the norm.

ACKNOWLEDGMENTS

I am most grateful to Dr. J. T. Fletcher and Dr. M. Hollings for criticizing the draft of this review.

Literature Cited

1. Ainsworth, G. C. 1932. Mosaic disease of the tomato. *Exp. Res. Stn. Cheshunt Ann. Rept., 1931,* pp. 42–43
2. Ainsworth, G. C. 1933. An investigation of tomato virus diseases of the mosaic "stripe", streak group. *Ann. Appl. Biol.* 20:421–38
3. Ainsworth, G. C. 1935. Cigarettes as a possible source of tomato mosaic. *Exp. Res. Stn. Cheshunt Ann. Rept., 1934,* pp. 65–66
4. Ainsworth, G. C., Selman, I. W. 1936. Some effects of tobacco mosaic virus on the growth of seedling tomato plants. *Ann. Appl. Biol.* 23:89–98
5. Alexander, L. J. 1952. Effect of the tobacco mosaic disease on the yield of unstaked tomatoes. *Phytopathology* 42:463
6. Alexander, L. J. 1960. Inactivation of TMV from tomato seed. *Phytopathology* 50:627
7. Alexander, L. J. 1962. Strains of TMV on tomato in The Netherlands and in Ohio, USA. *Meded. Landbouwhogesch. Opzoekingsstn. Staat Gent* 27:1020–30
8. Alexander, L. J. 1971. Host-pathogen dynamics of tobacco mosaic virus on tomato. *Phytopathology* 61:611–17
9. Allard, H. A. 1916. The mosaic disease of tomatoes and petunias. *Phytopathology* 6:328–35
10. Anonymous. 1965. The effect of tobacco mosaic virus infection on yield and quality of tomatoes. *Rept. Lee Valley Exp. Hortic. Stn., 1964,* pp. 26–31
11. Bancroft, J. B., Pound, G. S. 1956. Cumulative concentrations of tobacco mosaic virus in tobacco and tomato at different temperatures. *Virology* 2:29–43
12. Bartels, W. 1955. Untersuchungen über die Inaktivierung des Tabakmosaikvirus durch Extrakte und Sekrete von höheren Pflanzen und einigen Mikroorganismen. Ein Beitrag zur Frage der Kompostierung tabakmosaikvirushaltiger Pflanzenmaterials. *Phytopathol. Z.* 25:72–98, 113–52
13. Bergman, E. L., Boyle, J. S. 1965. Symptoms, yield and leaf mineral composition of greenhouse tomatoes as affected by soil moisture and tobacco mosaic virus. *J. Pa. Agric. Exp. Stn.* 3016:96–99
14. Bergmann, L. 1958. Uber den Einfluss von Thiouracil und Cytovirin auf das Wachstum und die Virusproduktion isolierter Tomatenwurzeln. *Phytopathol. Z.* 34:209–20
15. Bewley, W. F., Corbett, W. 1929. 'Mosaic' disease of the tomato. *Exp. Res. Stn. Cheshunt Ann. Rept., 1927,* pp. 51–59
16. Bhatt, J. G., Verna, S. S. 1958. Effect of growth regulating substances on virus affected tomato plants. *Sci. Cult.* 23:610–11
17. Bong Cho Chung. 1969. An investigation of strains and viruses isolated from mosaic and streak types of tomato. *Bull. Aphid Lab. Korea* 1:7–12
18. Bovey, R., Canevascini, V., Mottier,

P. 1957. Influence du virus de la mosaique du tabac sur le rendement de tomates infecteés à différentes dates. *Rev. Romande Agric. Vitic. Arboric.* 13:36–39

19. Boyle, J. S. 1956. The nature of the internal browning disease of tomato. *Phytopathology* 46:7

20. Boyle, J. S. 1970. Evidence that Pennsylvania tobacco mosaic virus induced internal browning and Florida graywall are the same. *Phytopathology* 60:571

21. Boyle, J. S., Bergman, E. L. 1967. Factors affecting incidence and severity of internal browning of tomato induced by tobacco mosaic virus. *Phytopathology* 57:354–62

22. Boyle, J. S., Wharton, D. C. 1957. The role of tobacco mosaic virus in abnormal ripening disorders of tomato. *Phytopathology* 47:4

23. Boyle, J. S., Wharton, D. C. 1957. The experimental reproduction of tomato internal browning by inoculation with strains of tobacco mosaic virus. *Phytopathology* 47:199–207

24. Bradley, R. H. E., Harris, K. F. 1972. Aphids can inoculate plants with tobacco mosaic virus by clawing. *Virology* 50:615–18

25. Broadbent, L. 1961. The epidemiology of tomato mosaic. 1. A review of the literature. *Glasshouse Crops Res. Inst. Ann. Rept., 1960*, pp. 96–116

26. Broadbent, L. 1962. The epidemiology of tomato mosaic. 2. Smoking tobacco as a source of virus. *Ann. Appl. Biol.* 50:461–66

27. Broadbent, L. 1963. The epidemiology of tomato mosaic. 3. Cleaning virus from hands and tools. *Ann. Appl. Biol.* 52:225–32

28. Broadbent, L. 1964. The epidemiology of tomato mosaic. 7. The effect of TMV on tomato fruit yield and quality under glass. *Ann. Appl. Biol.* 54:209–24

29. Broadbent, L. 1965. The epidemiology of tomato mosaic. 8. Virus infection through tomato roots. *Ann. Appl. Biol.* 55:57–66

30. Broadbent, L. 1965. The epidemiology of tomato mosaic. 9. Transmission of TMV by birds. *Ann. Appl. Biol.* 55:67–69

31. Broadbent, L. 1965. The epidemiology of tomato mosaic. 11. Seed-transmission of TMV. *Ann. Appl. Biol.* 56:177–205

32. Broadbent, L., Cooper, A. J. 1964. The epidemiology of tomato mosaic. 6. The influence of tomato mosaic virus on root growth and the annual pattern of fruit production. *Ann. Appl. Biol.* 54:31–43

33. Broadbent, L., Fletcher, J. T. 1963. The epidemiology of tomato mosaic. 4. Persistence of virus on clothing and glasshouse structures. *Ann. Appl. Biol.* 52:233–41

34. Broadbent, L., Fletcher, J. T. 1966. The epidemiology of tomato mosaic. 12. Sources of TMV in commercial tomato crops under glass. *Ann. Appl. Biol.* 57:113–20

35. Broadbent, L., Read, W. H., Last, F. T. 1965. The epidemiology of tomato mosaic. 10. Persistence of TMV-infected debris in soil, and the effects of soil partial sterilization. *Ann. Appl. Biol.* 55:471–83

36. Broadbent, L., Winsor, G. W. 1964. The epidemiology of tomato mosaic. 5. The effect on TMV-infected plants of nutrient foliar sprays and of steaming the soil. *Ann. Appl. Biol.* 54:23–30

37. Caldwell, J. 1931. The physiology of virus diseases in plants. 2. Further studies on the movement of mosaic in the tomato plant. *Ann. Appl. Biol.* 18:279–98

38. Caldwell, J. 1959. Persistence of tomato aucuba mosaic virus in dried leaf tissue. *Nature* 183:1142

39. Canova, A. 1957. Richerche intorno ad una virosi del Pomodoro II. *Phytopath. Z.* 28:415–22

40. Cocking, E. C., Pojnar, E. 1968. A study of the infection of tomato fruit by tobacco mosaic virus. *Phytopathol. Z.* 63:364–72

41. Cooper, A. J. 1973. Rapid crop turnround is possible with experimental nutrient film technique. *Grower* 79:1048–52

42. Cotter, D. J. 1961. The influence of nitrogen, potassium, boron and tobacco mosaic virus on the incidence of internal browning and other fruit quality factors of tomatoes. *Proc. Am. Soc. Hortic. Sci.* 78:474–79

43. Cox, R. S. 1970. Observations on the cause of graywall of tomato in Florida. *Plant Dis. Reptr.* 54:678

44. Crill, P., Burgis, D. S., Jones, J. P., Strobel, J. W. 1973. Effect of tomato mosaic virus on yield of fresh-market, machine-harvest type tomatoes. *Plant Dis. Reptr.* 57:78–81

45. Crowley, N. C. 1957. The effect of developing embryos on plant viruses. *Aust. J. Biol. Sci.* 10:443–48

46. Crowley, N. C. 1957. Studies on the

seed transmission of plant virus diseases. *Aust. J. Biol. Sci.* 10:449−64

47. Crowley, N. C. 1958. The use of skim milk in preventing the infection of glass-house tomatoes by tobacco mosaic virus. *J. Aust. Inst. Agric. Sci.* 24:261−63

48. Darby, L. A. 1975. Letter in *Grower.* 84:312

49. Dawson, J. R. O. 1965. Contrasting effects of resistant and susceptible tomato plants on tomato mosaic virus multiplication. *Ann. Appl. Biol.* 56: 485−91

50. Dawson, J. R. O. 1967. The adaptation of tomato mosaic virus to resistant tomato plants. *Ann. Appl. Biol.* 60:209−14

51. Denby, L. G., Wilks, J. M. 1963. The effect of tobacco mosaic on the yield of field tomatoes as influenced by sprays of milk and DOSS. *Can. J. Plant Sci.* 43:457−61

52. Doolittle, S. P. 1948. Tomato diseases. *Farm. Bull. US Dep. Agric., 1934,* pp. 36−50

53. Fletcher, J. T. 1969. Studies on the overwintering of tomato mosaic in root debris. *Plant Pathol.* 18:97−108

54. Fletcher, J. T., Butler, D. 1975. Strain changes in populations of tobacco mosaic virus from tomato crops. *Ann. Appl. Biol.* 81:409−12

55. Fletcher, J. T., MacNeill, B. H. 1971. Influence of environment, cultivar and virus strain on the expression of tobacco mosaic virus symptoms in tomato. *Can. J. Plant Sci.* 51:101−7

56. Fletcher, J. T., MacNeill, B. H. 1971. The identification of strains of tobacco mosaic virus from tomato crops in Southern Ontario. *Can. J. Microbiol.* 17:123−28

57. Fletcher, J. T., Rowe, J. M. 1975. Observations and experiments on the use of an avirulent mutant strain of tobacco mosaic virus as a means of controlling tomato mosaic. *Ann. Appl. Biol.* 81:171−79

58. Fry, P. R., McCallum, D. W. 1960. Tobacco mosaic virus in tomato glasshouses. Influence of extraction method on seed transmission and seed production. *N.Z. Comm. Grow.* 15: 9−10

59. Fulton, R. W. 1941. The behavior of certain viruses in plant roots. *Phytopathology* 31:575−98

60. Gardner, M. W., Kendrick, J. B. 1923. Field control of tomato mosaic. *Phytopathology* 13:372−75

61. Gol'din, M. I., Jurcenko, M. A. 1959. Direct sowing as an anti-virus method in the control of mosaic and streak of tomatoes. *Trans. Inst. Microbiol. Virusol. Kaz. SSR* 3:166−68 (Orig. in Russian)

62. Hall, C. B., Dennison, R. A. 1955. Environmental factors influencing vascular browning of tomato fruits. *Proc. Am. Soc. Hortic. Sci.* 65:353−56

63. Hare, W. W., Lucas, G. B. 1959. Control of contact transmission of tobacco mosaic virus with milk. *Plant Dis. Reptr.* 43:152−56

64. Harris, K. F., Bradley, R. H. E. 1973. Importance of leaf hairs in the transmission of tobacco mosaic virus by aphids. *Virology* 52:295−300

65. Harrison, B. D. 1956. A strain of tobacco mosaic virus infecting *Plantago* spp. in Scotland. *Plant Pathol.* 5:147−48

66. Harrison, B. D. 1957. Tobacco mosaic in the tomato crop. *Ann. Rept. Scott. Hortic. Res. Inst., 1956−57,* pp. 31−32

67. Harrison, D. J. 1960. Tomato mosaic. *Guernsey Growers Assoc. Rev.* 24: 27−49

68. Heuberger, J. W., Norton, J. B. S. 1933. Water loss in tomato mosaic. *Phytopathology* 23:15

69. Holmes, F. O. 1941. A distinctive strain of tobacco-mosaic virus from Plantago. *Phytopathology* 31:1089−98

70. Holmes, F. O. 1950. Internal-browning disease of tomato caused by strains of tobacco-mosaic virus from Plantago. *Phytopathology* 40:487−92

71. Howles, R. 1950. Tobacco as a source of virus infection in the tomato crop. *Ann. Rept. Exp. Res. Stn. Cheshunt, 1949,* pp. 33−34

72. Howles, R. 1957. Chemical inactivation of virus in tomato seed. *Glasshouse Crops Res. Inst. Ann. Rept., 1954/1955,* p. 45

73. Howles, R. 1961. Inactivation of tomato mosaic virus in tomato seeds. *Plant Pathol.* 10:160−61

74. Hsieh, S. T., Hsu, Y. L., Lai, T. C. 1967. Study on milk and its effective constituents as TMV virus inhibitors. *Ann. Rept. Tobacco Res. Inst. Taiwan, 1967,* pp. 171−76

75. Jaeger, S. 1966. Milch als Infektionshemmstoff mechanisch ubertragbarer Viren in Tomaten-und Gurkenkulturen. *Phytopathol. Z.* 56:340−52

76. Jenkins, J. E. E., Wiggell, D., Fletcher, J. T. 1965. Tomato fruit bronzing. *Ann. Appl. Biol.* 55:71−81

77. Johnson, E. M., Chapman, R. A.

1958. Resistance of tobacco and tomato virus to heat and chemicals. *Bull. Kyoto Agric. Exp. Stn. 23*

78. Johnson, J., Ogden, W. B. 1929. The overwintering of the tobacco mosaic virus. *Bull. Wis. Agric. Exp. Stn. 95:* 1—25

79. Jones, J. P. 1958. The relation of certain environmental factors, tobacco mosaic virus strains, and sugar concentration to the blotchy ripening disease of tomato and the inheritance of the tendency to the disease. *Diss. Abstr.* 19:650—51

80. Jones, L. K., Burnett, G. 1935. Virous diseases of greenhouse-grown tomatoes. *Bull. Wash. State Agric. Exp. Stn.* 308:1—36

81. Kassanis, B., Kleczkowski, A. 1944. The effect of formaldehyde and mercuric chloride on tobacco mosaic virus. *Biochem. J.* 38:20—24

82. Komuro, Y., Iwaki, M. 1968. Strains of tobacco mosaic virus contained in cigarettes, with special reference to the infection source for mosaic disease of tomato. *Ann. Phytopathol. Soc. Jpn.* 34:98—102

83. Komuro, Y., Iwaki, M., Nakahara, M. 1966. Viruses isolated from tomato plants showing mosaic and/or streak in Japan, with special reference to tomato strain of tobacco mosaic virus. *Ann. Phytopathol. Soc. Jpn.* 32: 130—38

84. Laterrot, H., Pecaut, P. 1965. Tomato seed production. i. Rapid cleansing using pectolytic enzymes. ii. Decreasing the content of tobacco mosaic virus by dry-heat treatment. *Ann. Epiphyt.* 16: 163—70

85. Lewis, G. D., Taylor, G. A. 1967. Association of tomato fruit abnormalities with tobacco mosaic virus infection. *Hortscience* 2:163—64

86. Limasset, M. P. 1957. Observations relatives au role des terreaux et des tabacs manufactures dans la transmission du virus de la mosaique du tabac. *C. R. Seances Acad. Agric. Fr.* 43: 749—52

87. Lojek, J. S., Orlob, G. B. 1972. Transmission of tobacco mosaic by *Myzus persicae. J. Gen. Virol.* 17:125—27

88. Lucas, G. B., Winstead, N. N. 1958. Control of tobacco mosaic with an antiviral agent. *Phytopathology* 48:344

89. McKeen, C. D. 1962. The destruction of viruses by soil fumigants. *Phytopathology* 52:742—43

90. MacNeill, B. H. 1962. A specialized tomato form of the tobacco mosaic virus in Canada. *Can. J. Bot.* 40: 49—51

91. MacNeill, B. H., Boxall, M. 1974. The evolution of a pathogenic strain of tobacco mosaic virus in tomato: a host-passage phenomenon. *Can. J. Bot.* 52:1305—7

92. MacNeill, B. H., Fletcher, J. T. 1971. Changes in strain characteristics of tobacco mosaic virus after passage through grafted tomato hosts. *Phytopathology* 61:130—31

93. MacNeill, B. H., Ismen, H. 1960. Studies on the virus-streak syndrome in tomatoes. *Can. J. Bot.* 38:9—20

94. McRitchie, J. J., Alexander, L. J. 1957. Effect of strains of tobacco mosaic virus on yields of certain tomato varieties. *Phytopathology* 47:24

95. McRitchie, J. J., Alexander, L. J. 1963. Host-specific Lycopersicon strains of tobacco mosaic virus. *Phytopathology* 53:394—98

96. Marrou, J., Migliori, A. 1971. Essai de protection des cultures de tomate contre le virus de la mosaique du tabac: mise en evidence d'une specificite etroite de la premunition entre souches de ce virus. *Ann. Phytopathol.* 3:447—59

97. Melchers, G., Schramm, G., Trurnit, H., Friedrich-Freksa, H. 1940. Die biologische, chemische und elektronenmikroskopische Untersuchung eines Mosaikvirus aus Tomaten. *Biol. Zentralbl.* 60:524—56

98. Migliori, A., Marrou, J. 1970. Le virus de la mosaique du tabac dans le sol: influence sur la dissemination de ce virus dans les cultures de tomates. *Ann. Phytopathol.* 2:669—80

99. Mulholland, R. I. 1962. Control of the spread of mechanically transmitted plant viruses. *Commonw. Phytopathol. News* 8:60—61

100. Murakishi, H. H. 1960. Comparative incidence of graywall and internal browning of tomato and sources of resistance. *Phytopathology* 50:408—12

101. Murakishi, H. H. 1960. Diagnostic aids in distinguishing internal browning and graywall of tomatoes. *Phytopathology* 50:648

102. Newell, J. 1954. Milk spray cured tomato mosaic. *Grower* 41:1409

103. Nitzany, F. E. 1960. Tests for tobacco mosaic virus inactivation on tomato trellis wires. *Ktavim Rec. Agric. Res. Stn.* 10:59—61

104. Nitzany, F. E. 1960. Transmission of tobacco mosaic virus through tomato seed and virus inactivation by

methods of seed extraction and seed treatments. *Ktavim Rec. Agric. Res. Stn.* 10:63−67

105. Orlob, G. B. 1963. Reappraisal of transmission of tobacco mosaic virus by insects. *Phytopathology* 53:822−30

106. Oshima, N., Komoti, S., Goto, T. 1965. Study on control of plant virus diseases by vaccination of attenuated virus. (i) Control of tomato mosaic disease. *Bull. Hokkaido Agric. Exp. Stn.* 85:23−33

107. Ossiannilsson, F. 1958. Is tobacco mosaic virus not imbibed by aphids and leafhoppers? *K. Lantbrukshoegsk. Ann.* 24:369−74

108. Pallett, I. H. 1969. *The survival of tomato mosaic virus in root debris.* PhD thesis. University of London

109. Paludan, N. 1968. Tomato mosaic virus. Investigation concerning TMV in different plant genera, the virulence of TMV strains, virus attenuation by heat treatment, cross protection and yield. *Tidsskr. Planteavl.* 72:69−80 (In Danish, Engl. summ.)

110. Paludan, N. 1973. Tobak-mosaik-virus. Infektionsforsog, krydsbeskyttelse, smittetidspunkt og udbytte med tomatlinier af TMV hos tomat. *Tidsskr. Planteavl* 77:495−515

111. Pelham, J. 1966. Resistance in tomato to tobacco mosaic virus. *Euphytica* 15:258−67

112. Pelham, J. 1968. TMV resistance. *Ann. Rep. Glasshouse Crops Res. Inst., 1967,* pp. 45−48

113. Pelham, J. 1972. Strain-genotype interaction of tobacco mosaic virus in tomato. *Ann. Appl. Biol.* 71:219−28

114. Pelham, J., Fletcher, J. T., Hawkins, J. H. 1970. The establishment of a new strain of tobacco mosaic virus resulting from the use of resistant varieties of tomato. *Ann. Appl. Biol.* 65:293−97

115. Phillip, M. J., Honma, S., Murakishi, H. H. 1965. Studies on the inheritance of resistance to tobacco virus in the tomato. *Euphytica* 14:231−36

116. Pirone, T. P., Kassanis, B. 1975. Aphid transmission of TMV. *Rothamsted Exp. Stn. Rept., 1974,* p. 213

117. Pirone, T. P., Shaw, J. G. 1973. Aphid stylet transmission of poly-L-ornithine treated tobacco mosaic virus. *Virology* 53:274−76

118. Price, W. C. 1933. The thermal death rate of tobacco-mosaic virus. *Phytopathology* 23:749−69

119. Proeseler, G., Karl, E. 1973. Pflanzenpathogene Viren im Honigtau der Blattläuse. *Biol. Zentralbl.* 92:357−60

120. Rast, A. T. B. 1967. Yield of glasshouse tomatoes as affected by strains of tobacco mosaic virus. *Neth. J. Plant Pathol.* 73:147−56

121. Rast, A. T. B. 1972. M11−16, an artificial symptomless mutant of tobacco mosaic virus for seedling inoculation of tomato crops. *Neth. J. Plant Pathol.* 78:110−12

122. Rast, A. T. B. 1973. Systemic infection of tomato plants with tobacco mosaic virus following inoculation of seedling roots. *Neth. J. Plant Pathol.* 79:5−8

123. Rast, A. T. B. 1974. TMV in tomaat. *Groenten Fruit* 29:1581

124. Rast, A. T. B. 1975. Variability of tobacco mosaic virus in relation to control of tomato mosaic in glasshouse tomato crops by resistance breeding and cross protection. *Agric. Res. Rept., 834, Wageningen.* 76 pp.

125. Raychaudhuri, S. P. 1952. Studies on internal browning of tomato. *Phytopathology* 42:591−95

126. Rees, A. R. 1970. Effect of heat treatment for virus attenuation on tomato seed viability. *J. Hortic. Sci.* 45:33−40

127. Rich, S. 1958. Fertilizers influence the incidence of tomato browning in the field. *Phytopathology* 48:48−50

128. Roberts, F. M. 1950. The infection of plants by viruses through roots. *Ann. Appl. Biol.* 37:385−96

129. Samuel, G. 1934. The movement of tobacco mosaic virus within the plant. *Ann. Appl. Biol.* 21:90−111

130. Selman, I. W. 1941. The effects of certain mosaic-inducing viruses on the tomato crop under glass. *J. Pomol. Hortic. Sci.* 19:107−36

131. Selman, I. W. 1943. The influence of lime and potash on mosaic infection in the tomato (var. Potentate) under glass. *J. Pomol. Hortic. Sci.* 20:89−106

132. Selman, I. W. 1945. Virus infection and water loss in tomato foliage. *J. Pomol. Hortic. Sci.* 21:146−54

133. Selman, I. W. 1946. The localisation of tobacco mosaic virus in tomato fruits. *J. Pomol. Hortic. Sci.* 22:226−30

134. Selman, I. W. 1947. Resistance to mosaic infection in the tomato in relation to soil conditions. *J. Pomol. Hortic. Sci.* 23:71−79

135. Selman, I. W., Yahampath, A. C. I. 1973. Some physiological characteristics of two tomato cultivars, one tolerant and one susceptible to tobacco mosaic virus. *Ann. Bot.* 37:853−65

136. Smith, P. R., Stubbs, L. L., Sutherland, J. L. 1965. Internal browning of tomato fruit induced by a late infection of tobacco mosaic virus. *Aust. J. Exp. Agric. Anim. Husb.* 5:75−79

137. Smyglya, V. A. 1963. On root virus infection of tomatoes and potatoes. *Dokl. Mosk. Skh. Akad.* 89:393−99

138. Stall, R. E., Alexander, L. J., Hall, C. B. 1970. Effect of tobacco mosaic virus and bacterial infections on occurrence of graywall of tomato. *Proc. Fla. State Hortic. Soc. 1969*, pp. 157−61

139. Taylor, G. A. 1957. The influence of some environmental and nutritional factors on the incidence of internal browning of tomato. *Diss. Abstr.* 17:211−12

140. Taylor, G. A., Lewis, G. D., Rubatzky, V. E. 1969. The influence of time of tobacco mosaic virus inoculation and stage of fruit maturity upon the incidence of tomato internal browning. *Phytopathology* 59:732−36

141. Taylor, G. A., Smith, C. B., Fletcher, R. F. 1958. Influence of some environmental and nutritional factors on incidence of internal browning of tomato. *Bull. Pa. State Univ. Agric. Exp. Stn.* 629:1−21

142. Taylor, R. H., Grogan, R. G., Kimble, K. A. 1961. Transmission of tobacco mosaic virus in tomato seed. *Phytopathology* 51:837−42

143. Termohlen, G., van Dorst, H. J. M. 1959. Verschillen tussen *Lycopersicum* virus 1 en *Nicotiana* virus 1. *Jversl. Proefstn. Groenten en Fruitteelt Glas, Naaldwijk, 1958.* pp. 125−26

144. Thomas, B. J. 1973. Tomato mosaic virus. *Ann. Rept. Glasshouse Crops Res. Inst. 1972*, pp. 30−31

145. Thomas, B. J. 1974. Tobacco (tomato) mosaic virus. *Ann. Rept. Glasshouse Crops Res. Inst., 1973*, pp. 42−43

146. Upstone, M. E. 1974. Effects of inoculation with the Dutch mutant strain of tobacco mosaic virus on the cropping of commercial glasshouse tomatoes. *Rept. Agric. Dev. Adv. Serv., 1972*, pp. 162−65.

147. Van den Broeck, H., Vanachter, A.,

van Winckel, A. 1967. De inactiviering van Tabaksmozaiekvirus met chemische grondontsmettingsmiddelen. *Agricultura Louvain* 15:120−26

148. Van Dorst, H. J. M. 1970. *Jversl. Proefstn. Groenten Fruitteelt Glas, Naaldwijk, 1969*, p.75

149. van Soest, W., de Meester-Manger Cats, V. 1956. Does the aphid *Myzus persicae* (Sulz.) imbibe tobacco mosaic virus? *Virology* 2:411−14

150. van Winckel, A. 1965. Tabakmosaiekvirus op Tomatenzaad. *Agricultura Louvain* 13:721−29

151. van Winckel, A. 1967. Bijdrage tot de studie van het tabakmozaiekvirus (TMV) bij tomaten. *Thesis Kath. Univ. Leuven.* 120 pp.

152. van Winckel, A. 1967. Invloed van het TMV op de vegetatieve ontwikkeling van tomatenplanten. *Agricultura Louvain* 15:67−75

153. van Winckel, A. 1968. Natuurlijke inactivering van het tabakmozaiekvirus (TMV) op tomatenzaad. *Parasitica* 24:1−9

154. van Winckel, A., Geypens, M. 1965. Inactivering van Tabakmosaiekvirus in de grond. *Parasitica* 21:124−28

155. Vovk, A. M. 1961. Inaktivatsiya virusa mosaiki Tabaka v sememakh Tomatov pri raznykh Srokakh khraneniya. *Tr. Inst. Genet. Akad. Nauk SSSR* 28:269−76

156. Wall, E. T. 1973. Isolated growing systems. *The UK Tomato Manual*, 94−99. London: Grower Books

157. Weber, P. V. V. 1960. The effect of tobacco mosaic virus on tomato yield. *Phytopathology* 50:235−37

158. Westerdijk, J. 1910. Die Mosaikkrankheit der Tomaten. *Meded. Phytopathol. Lab. Willie Commelin Scholten,* 1

159. Wheeler, G. F. C. 1961. Glasshouse crops in soilless composts. *Comm. Grower* 3436:920−23

160. Wittman, H. G. 1958. Untersuchungen über die wirkung des Cytovirins auf Virusvermehrung und Wirtswachstum. *Phytopathol. Z.* 34:221−27

Copyright © 1976 by Annual Reviews Inc.
All rights reserved

SPORE YIELD OF PATHOGENS IN INVESTIGATIONS OF THE RACE-SPECIFICITY OF HOST RESISTANCE

♦ 3634

R. Johnson and A. J. Taylor
Plant Breeding Institute, Trumpington, Cambridge, England

Many methods have been developed for measuring the susceptibility of host plants to disease. Especially where rapid assessments must be made, as in plant breeding, susceptibility frequently is estimated visually, often with the aid of diagrams indicating the appearance of various amounts or types of infection. Zadoks has suggested (98) that "the counting of propagules is an alternative for, or a complement to, disease assessment" and that "in mathematical analyses, cumulative spore counts are analogous to disease severity." Unfortunately the measurement of spore production is laborious by all methods devised so far. Consequently, it has not yet been used as a selection method in plant breeding and has been used only to a very limited extent in the genetic analysis of resistance. Nevertheless, it is one of the most accurate and least subjective ways of assessing the growth of pathogens and the susceptibility of hosts. In this review we draw attention to numerous examples which show that differences in host resistance and pathogen growth have been detected by measuring spore production where other methods have not been sufficiently sensitive. Many measurements of spore production by pathogens on living hosts have been for pathogens that show race-specificity on host cultivars. These studies have special importance, therefore, in the discussion of race-specificity and the distinction between resistance known to be race-specific and that which is thought not to be.

Outlined below are some of the many techniques that have been used to measure spore production. In later sections we refer to the techniques only when they are relevant to interpretations of the results.

METHODS OF MEASURING SPORE PRODUCTION

Spore production can be measured by collecting spores directly from infected leaves or disease lesions and counting them or otherwise assessing the quantities. It may also be assessed indirectly by sampling spores released into the air over in-

97

fected crops or by testing the progeny of samples of spores collected in a previous generation.

Direct Methods

Two operations, collection and assessment, are needed to measure spore production and many methods have been used for each. The references are given only as examples.

COLLECTION OF SPORES   In many studies the spores or sporangia have been collected in liquid by washing infected leaves or leaflets with water or other fluids (e.g. 29, 59, 71, 90, 97). With such pathogens as *Phytophthora infestans,* or *Bremia lactucae* with easily wettable sporangia or spores, pure water has been used to produce a suspension (e.g. 4, 22, 90) but for hydrophobic spores, such as those of *Puccinia* or *Erysiphe* species, wetting agents have often been used (29, 91, 97). In many cases spores have been suspended in liquid after collection by various methods, such as shaking or enclosing whole leaves in glass tubes (16, 35, 49), by vacuum collection on Millipore filters® (66), or by means of cyclone spore collectors (44). The efficiency of cyclone spore collectors has been discussed by Mehta & Zadoks (57). Shaner (72) used a specially designed spore sampler to collect spores from single pustules of *Erysiphe graminis* and to impact them on slides coated with silicone grease. Schwarzbach (71) claimed that it was simpler to collect spores of *E. graminis* in liquid. Yarwood (97) allowed spores of *Uromyces phaseoli* to fall directly onto glass slides to study the degree of clumping.

Samples of spores have been collected from parts of leaves, sometimes for the convenience of using detached leaves (14), sometimes to investigate the production of spores from lesions only, or from parts of lesions (24, 28, 30, 45).

ASSESSMENT OF QUANTITIES OF SPORES   The quantities of spores present in liquid suspensions have, in most cases, been determined by counting small samples with a hemocytometer or other counting chamber. In some cases spores have been counted automatically by Coulter counters (66, 91) or relative proportions of spores have been estimated by determining the turbidity of spore suspensions by nephelometer (35, 83). In several experiments with pathogens having hydrophobic spores, the quantities collected have been assessed by weighing (23, 38, 56, 41, 76).

Results of such investigations have been expressed as quantities of spores per pustule or per lesion, per unit area of infected tissue, per unit area of tissue exposed to infection, or per unit of dry weight (56) or fresh weight (71) of host tissue.

Methods such as measurement of turbidity of spore suspensions, or weighing, are less laborious than those in which spores are counted (35). It would be an advantage, however, when such methods are used, to include data relating the unit of measurement to the number of spores.

Indirect Methods

Two types of technique may be considered to sample a proportion of the spores produced by pathogens growing on host plants.

PROGENY TESTS   This technique is illustrated by numerous investigations into the competitive ability of different pathogen strains (4, 19, 62, 79, 92). The usual procedure in such experiments has been to grow, together on a susceptible host cultivar, two races of a pathogen that can be distinguished by their interaction with selected differential cultivars. The mixture of races is maintained for a number of generations on the susceptible host cultivar and samples from each generation are transferred to the differential cultivars. The frequencies of the differing types of infection are used to assess the proportions of the different races occurring in the previous generation on the susceptible host. In some experiments color mutations have been used to identify different strains of a pathogen, and the need for a progeny test on differential varieties has been avoided (8, 19). By these methods the relative ability of races to reproduce in competition on a susceptible cultivar is assessed and, conversely, the relative susceptibility of that cultivar to the two races is measured.

SPORE SAMPLING IN THE FIELD   A proportion of spores produced on infected plants can be collected by trapping spores over infected plots. Such techniques can be used to assess the relative resistance of both pure line and multiline cultivars (9, 10, 85). A refinement proposed by Wolfe & Schwarzbach (94) is to expose seedlings of groups of cultivars with known resistance in plots for brief periods and then to count the number of infections on the different cultivars. An estimate of the frequencies of certain virulence characteristics in the pathogen population is thus obtained.

SPORE PRODUCTION AND RACE-SPECIFIC RESISTANCE

It is sometimes assumed that disease resistance that will allow some sporulation (often described by such terms as incomplete, partial, or slow rusting) will not show differential interactions with strains of a pathogen and thus may be described as non-race-specific or horizontal (87). Consequently it is sometimes suggested that non-race-specific resistance can alter the infection rate in a disease epidemic whereas race-specific resistance cannot. This view is strongly expressed by Vanderplank (88) in such comments as "The infection rate and therefore horizontal resistance are the main theme of this chapter" (p. 164) or "Horizontal resistance is a slowing down of an epidemic" (p. 169). Similar ideas have been expressed by other authors (e.g. 25). Cases in which resistance can be expressed in terms of reduced spore production per lesion or increased time between infection and sporulation have been interpreted as evidence for non-race-specific resistance (e.g. 64, 74). In work on late blight of potatoes a distinction has been made between race-specific genes for hypersensitive resistance (R genes), which prevent sporulation of *Phytophthora infestans,* and all other forms of inherited resistance to this pathogen (5). It is commonly assumed that forms of resistance which allow sporulation are non-race-specific.

In contrast to these views, there is much evidence that race-specific resistance occurs in many host-pathogen interactions in which sporulation is not inhibited completely. For example, all the infection types higher than 0 used in the classification of races of *Puccinia graminis* permit some sporulation (77), and many, if not all, are known to be race-specific. Toxopeus (82) noted that even the *R* genes for resistance to *Phytophthora infestans* could occasionally allow the development of lesions in which some sporulation occurred.

Using the technique of Johnson & Bowyer (35) we have assessed the relationship between infection type and spore production per unit area of leaf for *Puccinia striiformis* (Figure 1). In two experiments, the infection types were classified from 0 to 10 where 0 is any nonsporulating resistance reaction and 10 is the most susceptible reaction, with heavy sporulation and no chlorosis. Intermediate values from 1 to 9 indicate increasing sporulation and decreasing chlorosis. The graphs are nonlinear and can be fitted to a quadratic equation, the probabilities being < 0.001 for graph a and < 0.01 for graph b assuming that, by definition, there is a point at the origin in which infection type 0 produces no spores. The shape of the graphs indicates that, at the higher end of the infection-type scale, small differences in classification are associated with large differences in sporulation.

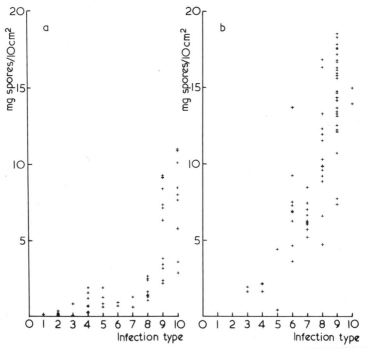

Figure 1 Spore yield (mg/10cm^2) from single wheat seedlings infected with *Puccinia striiformis* plotted against infection type. Equivalence of our 0–10 scale to standard 0–4 scale:

0–10:	0	1	2	3	4	5	6	7	8	9	10
0– 4:	00–0	1−	1	1+	2−	2	2+	3−	3	3+	4

Sporulation in Differential Interactions

Because they provide a sensitive method for determining resistance, measurements of spore production are useful in studying differential, or race-specific, interactions between hosts and pathogens. With these techniques, differential interactions have been shown to occur in cases where they would not have been detected with visual techniques of assessment. Two types of experimental method have been used; in the first, sporulation of single pathogen isolates on individual host cultivars is assessed and in the second, pairs or mixtures of pathogen isolates are grown on individual host cultivars and their relative competitive ability is noted.

SINGLE PATHOGEN ISOLATES ON INDIVIDUAL HOST CULTIVARS   Johnson & Taylor (38) showed that slight differences in infection types caused by isolates of *Puccinia striiformis* on seedlings of wheat cultivars Joss Cambier and Hybrid 46 were related to large differences in spore production. The differential interaction between two pathogen isolates and two wheat cultivars could be demonstrated more clearly by measuring spore production than by visual assessment of infection types. This result indicated that the wheat cultivars each possessed different factors for race-specific resistance and that the pathogen isolates separately possessed corresponding factors for virulence. The differential interactions were more marked in field trials (33) showing that the race-specific genetic factors in the wheat cultivars were more strongly expressed at later growth stages than in seedlings.

Using potato cultivars that lack *R* genes, Jeffrey, Jinks & Grindle (32) noted that isolates of *Phytophthora infestans* sporulated more rapidly on leaves or tubers of cultivars from which they were collected than on the tissues of other cultivars; similar results were reported by Caten (12). Denward (20) observed that more sporangia were produced by races of this pathogen on hosts that contained *R* genes exactly corresponding with the virulence genes of the pathogen than on hosts possessing fewer or no *R* genes. Clifford & Clothier (17) measured sporulation of isolates of *Puccinia hordei* on various barley cultivars, including those from which the isolates were collected, none of which contained genes for hypersensitive resistance. There was a highly significant interaction between isolates and varieties in the statistical analysis of sporulation per unit area of infected leaf, because isolates tended to sporulate best on the cultivar from which they originated.

It is evident from these experiments that measurements of spore production provide a sensitive test for differential interactions, and that such interactions often occur due to genes for race-specific resistance that are unrecognized because the interactions are not detectable visually.

In interpreting these experimental results, we have assumed that statistical interactions in an analysis of variance are indicative of differential interactions as suggested by Vanderplank (88, p.19) and Robinson (67). Whether this would always be true could depend on the correctness of assumptions that can be made about the effects of aggressiveness (88) in the pathogen (P. R. Scott, personal communication; 35). Vanderplank (88) defined aggressiveness as a component of pathogenicity that does not interact with individual host cultivars; increases in aggressiveness would decrease the resistance of cultivars with non-race-specific resistance. Assuming that a doubling of aggressiveness in the pathogen would double the amount of disease on a host, we have constructed an artificial example in which cultivar A is twice as resistant as cultivar B; this difference is expressed equally to two isolates of the pathogen and is therefore non-race-specific. One of the isolates (X) is twice as aggressive as the other (Y), and X would therefore infect both cultivars twice as heavily as Y. Imaginary data have been created and analyzed for such an experiment with six replicates (Table 1).

Although we have assumed that there is no differential interaction for specific virulence of the pathogen with the host cultivars there is a highly significant interaction in the statistical analysis. This statistical interaction could be eliminated by transforming the data to logarithms. If, however, an increase of aggressiveness in the pathogen resulted in an arithmetically equal increase in susceptibility for each cultivar, it would be more appropriate to use the untransformed data. Thus it is important to investigate experimentally the effects of increased aggressiveness on cultivars with different levels of resistance.

The assessment of differential interactions is simpler if a significant change in ranking of the susceptibility of cultivars results from interactions with pathogen isolates (67).

Table 1 Milligram spores per unit leaf area produced by two pathogen isolates, X and Y, growing on two host cultivars, A and B, and analysis of variance. Imaginary data.

				Replicate				
Isolate	Host	1	2	3	4	5	6	Mean
X	A	156	157	158	162	163	164	160
X	B	78	76	82	77	84	83	80
Y	A	82	83	77	76	84	78	80
Y	B	36	38	42	44	43	37	40

Analysis of variance

Source	df	ms	P
Replicates	5	15.1	NS[a]
Isolates	1	21600	***[b]
Hosts	1	21600	***
Isolates × hosts	1	2400	***
Error	15	10.4	
Total	23		

[a]NS = Not significant.
[b]*** = significant at P < 0.001.

Competition Between Pathogen Isolates on Individual Host Cultivars

Several studies on competition between races of pathogens provide further evidence of differential interactions between the races and cultivars that do not possess recognized genes for resistance. In such experiments, pairs of races have been grown for several generations on cultivars that were believed to be equally susceptible to both races (40, 52, 92). Frequently one race of the pair becomes dominant in the mixture, as shown by the relative numbers of spores produced of each race, assessed by the infection types they produce on a differential variety. Where more than one susceptible cultivar has been used in such experiments the same pair of races may respond differently on the different cultivars. Thus Loegering (52) showed that, in a mixture of races 17 and 56 of *Puccinia graminis*, race 17 predominated strongly on the wheat cultivars Fulcaster and Little Club but only weakly on Ceres. Kak et al (40) applied a mixture of races 21 and 40 to Agra local wheat and four selected wheat lines, all of which were described as producing a type 4 (susceptible) infection to both races. On Agra local wheat the races remained in equal proportions over several generations, while on three lines (NP4, C591, and NP718) race 40 and on one line (NP 710) race 21 predominated. Watson (92) showed that at 31.6°C (89°F) race 34 of *P. graminis* became dominant in a mixture of five races on Little Club wheat but that race 19 predominated on the cultivar Soft Federation.

In all these cases a differential interaction is shown between the races and the cultivars, affecting the ability of the races to reproduce in competition with each other. Thus they provide further examples of differential interactions where there are susceptible infection types and where known genes for resistance are absent. It is probable that such competition studies provide one of the most sensitive methods of detecting differential interactions because the effect of differential reproduction is allowed to accumulate over several generations rather than being assessed in one generation as with most other tests. Watson (92) noted that although races could survive when separate from each other, they often failed to compete in mixtures. In a mixture, spores can only be used in the ratio in which they occur; a less competitive race will be rapidly diluted in such a mixture, eventually becoming so rare that there are too few spores in a given sample to ensure further infection.

Vanderplank (89) criticized the findings of Jeffrey, Jinks & Grindle (32) who stated that isolates of *Phytophthora infestans* grew best on potato cultivars from which they were isolated, because he considered that even if such specially adapted isolates existed, many of the colonies growing on a cultivar would not be of the most adapted type; it was therefore unlikely that all the colonies they collected would be specially adapted to the cultivar from which they were collected. However, evidence from competition studies indicates that, in mixtures of pathogen isolates, one type can become predominant in very few generations. For example, Watson (92) showed that race 19 of *Puccinia graminis* declined from 50% to 2% of a mixture with race 56 in four generations at 22.7°C (73°F) on Soft Federation wheat. Thus, if specially adapted pathogen strains do occur they could be expected to predominate rapidly, making it likely that they would be frequent in random samples collected after several generations of multiplication. The results of Paxman (65) showed that no change occurred in adaptation of isolates of *Phytophthora infestans* after repeated subculturing on potato cultivars; this suggests that the adaptations shown by Jeffrey, Jinks & Grindle (32) were the results of selection occurring among the variants existing in the pathogen population prior to collection of their isolates.

Our conclusions from this survey are that there are differential interactions of varying magnitude, indicating large and small steps of adjustment of pathogens to individual host cultivars. The smaller interactions can be more readily detected by measurements of spore production than by visual assessments. Race-specific resistance often does not provide complete resistance but can cause a quantitative reduction in production of spores; it could therefore slow down the rate of development of epidemics (6) in a way similar to that attributed to horizontal resistance (88).

HOST-SPECIFICITY IN THE PATHOGEN

The counterpart of race-specific resistance in the host has been called *virulence* in the pathogen (88) or *vertical pathogenicity* (67) while the counterpart of non-race-specific resistance in the host has been called *aggressiveness* in the pathogen (88) or *horizontal pathogenicity (67)*. There is no general agreement about the use of these terms (78, 84); our choice here is to use *virulence* and *aggressiveness* as defined by Vanderplank (88) and discussed by Caten (11). By this definition, an increase in aggressiveness in the pathogen would result in a corresponding decrease in resistance of cultivars that had non-race-specific resistance, whereas an increase in virulence would be detected as a specific differential interaction with an individual or a group of cultivars.

Variation in either of these pathogen characters would affect the resistance expressed by host cultivars and the degree to which it is race-specific. Because the characters in the pathogen are, by definition, counterparts of those in the host, the quantitative aspects of race-specific resistance in the host, discussed above, reflect the quantitative nature of host-specific virulence in the pathogen. Other variables in the pathogen population which have been investigated, and which could influence the expression of host resistance, are the distribution of virulence and the accumulation of virulence in single races of the population. If the geographic distribution of virulence in the pathogen population is controlled by environmental factors it might be possible for a race-specific gene for resistance in the host to be exploited successfully to control disease in an area where a corresponding gene for virulence could not survive. If the accumulation of genes for virulence in single races of the pathogen population resulted in decreased fitness or aggressiveness of those races, the production of host cultivars with many genes for race-specific resistance might attenuate the development of disease epidemics.

It is not within the scope of this review to discuss these theories in the detail that has been done elsewhere (7, 9, 27, 42, 50, 88). Nevertheless they have been investigated by assessments of spore production of different races, and we examined some of these experiments to determine whether the results have helped to justify these theories.

Prevalence of Pathogen Races

Some experiments have been carried out in attempts to reveal causes for the predominance of certain races in surveys of pathogen populations, especially when the predominance was not explained by selection due to widespread use of cultivars with race-specific resistance (41, 52, 55, 68, 92). As already noted, such studies have sometimes revealed unexpected differential interactions which could themselves be causes of changes in the frequencies of races. In other studies, environmental factors such as temperature have appeared to influence competitive ability (92).

Katsuya & Green (41) suggested that race 56 of *Puccinia graminis* predominated in competition with race 15B-1(Can) at 20°C and 25°C whereas race 15B-1(Can) predominated at 15°C. They suggested that the rise in frequency of race 15B-1(Can) during the stem rust epidemics of 1953 and 1954 in Canada, was partly related to lower than average temperatures in those years, and the subsequent increase of race 56 in later years to higher temperatures. Critical examination of their extensive data suggests an alternative interpretation.

In experiments on separate plants, pustules of race 56 opened sooner and expanded more rapidly than those of race 15B-1(Can), but pustules of 15B-1(Can) later became larger than those of 56. Two methods used to determine areas of pustules gave different times for pustules of the two races to reach equal size, but it may have been between 12 and 15 days at 20°C and between 17 and 26 days at 15°C. In the competition experiments spores were harvested after 14 days at 20°C and after 18 days at 15°C. Thus the apparent differential effect of temperature on competitive ability may have depended on the critical timing of harvests with respect to pustule size. Other data in the paper suggest that, in the early stages of infection, seedlings of all three cultivars Little Club, Red Bobs, and Marquis are considerably more susceptible to 56 than to 15B-1(Can) at 15°C. For example, based on their data, total spore production per 100 seedlings at a single collection 13 days after infection on Little Club was, for race 56, 13.7 mg and for 15B-1(Can), 2.0 mg. Corresponding figures for Red Bobs are 4.8 mg for race 56 and 3.6 mg for race 15B-1(Can), and for Marquis 42.9 mg for race 56 and 19.7 mg for race 15B-1(Can). Similar results were obtained at 20°C. Cooler seasons may not, therefore, have favored the development of race 15B-1(Can) more than that of race 56, since race 56 seems to possess the advantage of earlier development under all the conditions tested. Changes of frequency recorded for the two races may have been influenced more than was suggested, by the manner of sampling, the cultivars from which collections were made in the surveys, and possible unnoticed differential resistance to the two races, of cultivars grown elsewhere in North America.

Several laboratory experiments have given results not corresponding with field observations. Loegering (52) noted that the predominance of race 56 of *Puccinia graminis* in field plots of the wheat cultivar Ceres did not correspond with results of laboratory experiments in which race 17 predominated in a mixture with race 56. He suggested that virulence alone may not be sufficient to explain prevalence of races, being only one of several factors influencing survival. Martens (55) found that race C1 of *Puccinia graminis* var *avenae* tended to predominate in mixtures with several other races in glasshouse tests on susceptible seedlings. In the field, plots of Victory oats were inoculated with mixtures of three races to which they were susceptible, including C1, and after allowing a few generations of multiplication, samples of rust were collected. Single-pustule isolates were tested on differential varieties and race C1 was found to be rare, corresponding with its low frequency in nature. Again it may be assumed that environmental factors influenced the difference between glasshouse and field results, especially as the same pathogen isolates were used in both environments. It is also possible that differential resistance could develop in cultivars at later growth stages, similar to that observed in some competition studies with pathogens on seedlings; this might not be easily detectable other than by studies of competition between isolates, but could influence survival of races in the field.

Virulence and Aggressiveness

Several authors have suggested that pathogen strains carrying unnecessary genes for virulence tend to have reduced aggressiveness and thus reduced competitive ability, compared with strains with few or no genes for virulence, on host cultivars lacking genes for race-specific resistance (87). Black (4) reported that a strain of *Phytophthora infestans* carrying three recognized genes for virulence was eliminated in four generations in competition with a strain carrying only two genes for virulence. Similar results were found with several other pairs of races, by assessing relative proportions of each race in samples tested on differential cultivars. Thurston (79) found that, in most cases, isolates of race 0, with no genes for specific virulence, predominated in competition with race 1, possessing one gene for virulence. With *Puccinia graminis* f. sp. *avenae* Leonard (50) showed that races with more genes for virulence were eliminated in competition with races with less genes for virulence. Day (19) reported that race 2 of *Cladosporium fulvum,* virulent on tomato cultivars possessing resistance gene *Cf2,* could not compete with race 0 on cultivars without known genes for resistance.

By contrast, in a number of experiments on competition between races in the laboratory, races with more genes for virulence have predominated over those fewer genes (1, 60−62). Brown & Sharp (8) showed that a yellow-spored (wild-type) race of *Puccinia striiformis* possessing several genes for virulence, predominated in competition with a white-spored mutant of a race possessing fewer genes for virulence. They stated that the white-spored mutant was an equal competitor with the yellow race from which it was derived and that it possessed the same range of virulence; they thus concluded that additional genes for virulence did not reduce aggressiveness. The data presented show, however, that the white-spored mutant fell from 47% to 8% of the mixture in seven generations in a warm environment on the wheat cultivar Lemhi in competition with the wild-type culture from which the mutant originated, and from 47% to 24% in five generations in a cool environment. Similar results were obtained on the cultivar Itana, but none of the differences was declared to be significant. When the white-spored mutant was in competition with the yellow-spored wild type with wider virulence, decreases in frequency of the white-spored mutant from 39% to 9% on Lemhi and from 38% to 8% on Itana were found to be significant. It would appear that the white-spored mutant was not competitive with either of the yellow-spored wild types; this makes it difficult to assess the effect of additional virulence on competitive ability.

Day (19), on the other hand, found that a pigment mutation in *Cladosporium fulvum* decreased its competitive fitness but that the fitness of a wild-type race with an extra gene for virulence was still less. The rarity of pigment mutants under natural conditions, however, suggests that most of them would have so little competitive ability that they could not compete with any wild-type races.

The frequent divergence of laboratory and field observations on competition and prevalance of races could be attributed to several causes. Among them are the difficulties of selecting races or isolates of pathogens that are representative of the field population, and testing them on suitable cultivars, at relevant growth stages and under suitable environments. It is likely that pathogen isolates usually will be adapted to the environment in which they are collected, in addition to being adapted to their host cultivar. Thus if isolates are collected at, say, different temperatures and from different cultivars it is possible that they would belong to different races and possess different temperature optima. This need not imply any fixed relationship between the temperature optima and the array of genes for virulence.

It is also difficult to show by laboratory experiments that there is any fixed relationship between virulence and aggressiveness. There is considerable evidence that the aggressiveness of individual isolates of pathogens varies within as well as between races. Thurston (79) showed that one isolate of race 0 of *Phytophthora infestans* was less aggressive than other isolates of the same race. The results of numerous other investigations lead to similar conclusions (11, 24, 26, 32, 93).

Our conclusions from this survey are that it is difficult to obtain, from laboratory and glasshouse experiments of this type, data that correspond with, and therefore help to explain, results obtained in the field. Much remains to be learned about the relationships between genes for virulence and fitness in the pathogen populations. It may be that the development of improved sampling techniques in field experiments, such as those proposed by Wolfe & Schwarzbach (94) will prove more relevant than laboratory tests for investigating these problems.

COMPONENTS OF RESISTANCE

It is evident from our discussion so far that there is great variation within pathogen populations for many characters including the interaction of individual pathogen isolates with host cultivars. Therefore, it is difficult to test any cultivar with a representative range of pathogen isolates. Add to this the wide variation in magnitude of differential interactions and the problems of statistical interpretation and it becomes difficult to confirm experimentally that any host cultivar possesses resistance that can be described as non-race-specific or horizontal in the strict sense. It is equally difficult experimentally to separate aggressiveness from virulence in the pathogen.

As pointed out by Caten (12), however, it is not simply the presence or absence of differential interactions that matters but their magnitude relative to the differences in resistance between cultivars. Our observation of the common occurrence of small as well as large differential interactions does not conflict with the observation that some cultivars can be widely grown for long periods, without succumbing to new highly virulent strains of pathogens, while others cannot. The most easily distinguished feature of such resistance is that it is durable (36, 37) or long-lasting (75). Resistance can be considered to be durable without knowledge of its mechanism, its genetic control, or whether it shows any race-specificity; it needs only to have stood the test of time while being widely exposed to the pathogen under conditions favoring the development of disease.

In cultivars that have been thought to display stability or durability of resistance, experiments have been conducted to identify components of the resistance mechanism. In some cases the cultivars have been selected on various criteria, such as slow disease development, intermediate level of resistance (e.g. incomplete or partial resistance), mechanism of resistance (e.g. nonhypersensitive), or genetic control of resistance (e.g. polygenic). We suggest that none of these criteria constitutes, by itself, a guarantee of durability or stability of resistance, and that adequate tests of durability are essential in addition.

The components of resistance usually examined have been (a) resistance to penetration by the pathogen, (b) resistance to growth of mycelium in the host tissue, and (c) resistance to production of spores or sporangia. In our view all three components, and especially b and c, could be displayed by the incomplete types of race-specific resistance so that their recognition is not a guarantee of non-race-specificity.

These categories were first used to investigate "field" or "nonhypersensitive" resistance of potato cultivars to *Phytophthora infestans* (21, 58, 70, 80, 90). Objectives of such experiments have included attempts to obtain satisfactory correlations between laboratory and field tests (43, 46, 47, 90). These have been used to identify the component of greatest effect which could be used as a selection criterion in potato breeding (86). Also the stages of the infection process at which the resistance mechanisms operate have been studied (28, 31). Experiments with similar objectives have been carried out more recently on cereals (16, 29, 44, 63, 64, 72–74).

Relationships Between the Components of Resistance

All these proposed components of resistance have been found to contribute to resistance, and frequently they occur together. It may be pertinent to ask how independent they are from each other. It is known that increasing infection density, by application of higher doses of inoculum, shortens the latent period from infection to sporulation (41, 48, 49, 56, 63, 97). Heagle & Moore (29) reported that *Puccinia coronata* produced fewer pustules, had retarded hyphal growth and a longer latent period, and produced smaller pustules with fewer spores on the oat cultivar Portage than on the susceptible cultivar Coachman. Parlevliet & van Ommeren (64) and Clifford (16) found that *Puccinia hordei* produced fewer pustules and had a longer latent period on the barley cultivar Vada than on more susceptible varieties. Similar results were described for potatoes and *Phytophthora infestans* (48). In each case at least part of the increase in latent period could have been due to the reduced infection density and reduced competition between pathogen colonies. Some variation in latent period may, however, occur independently of infection density (64).

It seems likely that resistance to growth of mycelium through the host tissues and resistance to production of spores or sporangia could be closely related—both resulting from slow growth of the pathogen in the host tissue from whatever cause. Results of Deshmukh & Howard (21) and Lapwood (47) indicate correlations between these two components, and in other studies where the components of resistance include small lesions with reduced sporulation, it can be inferred that a similar correlation would be found (29, 74). In some experiments similar growth rates of the pathogen through tissues have been reported for cultivars on which rates of sporulation differed. Rate of advance of *Phytophthora infestans* through leaf tissues was reported to be equal in potato cultivars King Edward, Majestic, Arran Viking, and Up-to-Date, but the number of spores per lesion was greatest in King Edward and decreasing in Majestic, Up-to-Date, and Arran Viking (45). The amount of sporulation corresponded, more closely than the growth rate of mycelium, with resistance of these cultivars in the field. Knutson (43) reported that the potato cultivar Sebago had larger lesions but produced fewer spores per lesion of *P. infestans* than the cultivar Pontiac. In the work of Lapwood (45) determination of the rate of sporulation perhaps was a more sensitive test than measurement of lesion size. This would not explain the contrast between Sebago and Pontiac potatoes, however. It is possible that growth patterns of a pathogen on a leaf could vary and that measurements of linear growth might not detect variation involving depths of penetration or degree of colonization of infected tissue. Estimation of fungal mass in host tissues by chemical tests (53) could be useful in comparing growth patterns, and might help to explain the occurrence of different rates of sporulation where growth rates of mycelium through the host tissues are apparently similar.

From this evidence it may be concluded that the degree of independence between the components of resistance may be difficult to determine but that some independence does exist. Where components are not independent, a useful consequence would be that selection for one component in breeding programs would be adequate to conserve the others and to retain all the available resistance. Where they are independent it may be possible to recombine them, perhaps thereby increasing the level of resistance, but this might necessitate separate tests for each component, which could be a formidable task in some breeding programs. In such a case, spore production, which represents the total effect of all the components of resistance, may be the most useful criterion upon which to base selection. Further development of techniques for relating quantitative assessment of sporulation to other simpler methods of assessment would improve the possibilities of exploiting this component in selection programs.

Sporulation on Single Lesions and on Whole Leaves

It was noted in our brief survey of methods of determining spore production that many variations are possible in the proportion of spores collected. Total quantities of spores produced throughout the period of sporulation may be collected from infections on a whole organ, such as a leaf. Alternatively spore production may be determined only at certain selected intervals. In some cases, measurements have been made of sporulation from single lesions, or even from parts of lesions (45). Results have been given on the basis of sporulation occurring on whole infected plant organs, or per unit of host tissue that has been initially exposed to infection, or per disease lesion. Interpretation of results may be affected by which of these methods are used.

It was noted by Yarwood (97) and by Mehta & Zadoks (56) that within wide limits the total quantity of spores per infected bean or wheat leaf was independent of the number of rust pustules occurring on a leaf. This was evidently because, with few pustules per unit area of leaf, the pustules were larger and sporulation continued longer than with many pustules per unit area. If results were given according to the production of spores per pustule, under these circumstances a plant might be thought to be more susceptible only because it had received a small amount of inoculum or because fewer pustules had formed on it. Torres & Browning (81) measured the size of 50 crown-rust pustules on each of the oat cultivars Clinton and Cherokee. Those on Cherokee were 0.0332 mm^2 larger than those on Clinton, but the total yield of spores per cm of leaf length was greater on Clinton than on Cherokee. Presumably the number of pustules per unit length of leaf was lower on Cherokee, so that they were able to develop to a slightly larger size, but not sufficiently to compensate for the production of the more densely distributed pustules on Clinton. In several cases, not only were there fewer pustules but also spore production per pustule was less in resistant than in susceptible cultivars (16, 29, 44, 74). The data of Shaner (74) show that a high density of pustules may minimize differences in the area of individual pustules on resistant and susceptible cultivars. Nevertheless, even at high inoculum densities it seems likely from his data that fewer spores of *Erysiphe graminis* would be produced per unit area of leaf on the wheat cultivar Knox than on the more susceptible Vermillion, due to the smaller number of pustules occurring on Knox. In all these cases it may be concluded that differences between resistant and susceptible host cultivars for sporulation per unit area of host leaf would be greater than differences shown by either of the two components separately. Determination of the total spore production of a pathogen per unit area of host tissue exposed to infection may be the most sensitive method of detecting differences in resistance, but that other ways of presenting data may be useful in subdividing the resistance into components.

Spore Production on Cultivars Classified as Disease-Tolerant

The term *tolerance* is sometimes used to describe the ability of cultivars to yield well despite apparently heavy disease infections. Tolerance is evidently considered to be a non-race-specific character. A full discussion of uses of the word in plant pathology was given by Schafer (69); Robinson (67) defined as tolerant a cultivar that develops as severe disease symptoms as another but suffers less damage in terms of agricultural yield. Torres & Browning (81) measured sporulation of race 216 of *Puccinia coronata* on the tolerant oat cultivar Cherokee (25) and the susceptible cultivar Clinton, and showed that total spore production per linear centimeter of leaf was 444 μg on Cherokee compared with 715 μg on Clinton. Chakravarti (13) reported that *P. graminis* produced fewer spores on wheat cultivars that were described as tolerant. In these cases, the cultivars described as tolerant, are in fact resistant and the resistance is detectable by measuring spore production although, presumably, not in visual assessments.

Seedling and Adult Plant Resistance

In studies on the components of resistance and the genetic control of durable resistance it is advantageous if resistance of any type that operates in the field can also be detected in seedlings. Isolates of pathogens can be handled more easily in the laboratory or glasshouse, environments can be more easily controlled, and more plants can be examined.

Resistance of the barley cultivar Vada to *Puccinia hordei* was observed in seedlings when the number of spores per pustule was assessed (16). Resistance of the oat cultivar Portage was detected by measurement of the number of spores per pustule on seedlings, but it was noted that the effect was less pronounced than in adult plants (29). Differential resistance of wheat cultivars to *Puccinia striiformis* was detected by measuring spore production in seedlings and was correlated with resistance in the field (33).

Similarly, we have observed that *P. striiformis* produced 105 mg/100 cm^2 of seedling leaf on the wheat cultivar Nord-Desprez and infected 60% of its leaf area at heading time in the field. On the cultivars Vilmorin 27 and Cappelle-Desprez *P. striiformis* produced 59 and 64 mg spores/100 cm^2 of seedling leaf respectively and infected 2% and 11% respectively of the leaf area in the field. In many of these cases the resistance of the seedlings could not be assessed visually but could be detected by the measurement of spore production.

Johnson & Law (37) were able to show that a chromosome of the wheat cultivar Hybride de Bersée controlled the extent of adult plant resistance to *P. striiformis* for those races that could sporulate on seedlings of the cultivar. By measuring spore production they were also able to detect this resistance in seedlings and to show that its expression was strongly influenced by environmental factors. The use of measurements of spore production in this genetic experiment was made possible by the availability of genetically homozygous host lines differing in chromosome complement. Spore production has not yet been used to measure susceptibility of plants in genetically segregating populations, probably because of the great amount of labor required.

In some cases, no resistance is expressed at the seedling stage despite its expression at later growth stages. The wheat cultivar Little Joss possesses resistance to *P. striiformis* at post-seedling stages (54), but is susceptible as a seedling. In an experiment with *P. striiformis* we have recorded 114 mg spores/100 cm^2 of leaf on seedlings of both Little Joss and Nord Desprez, but in field trials with the same isolate of *P. striiformis* the percentage of leaf tissue infected at heading time was 1.2 on Little Joss and 46.0 on Nord-Desprez. In such cases, it is necessary to work with older plants in the laboratory or glasshouse to obtain results corresponding with those in the field, as in the experiments described by Shaner (72–74).

MEASUREMENTS OF SPORE PRODUCTION IN FIELD EXPERIMENTS

As already described, an indirect method of measuring spore production, and thereby estimating disease intensity, is by trapping spores released from infected crops. While many such experiments have been carried out in epidemiological investigations few have been related to cultivars with different levels of resistance. Advantages of such field investigations include the ability to follow events over periods of time and under natural environments. Epidemiological effects of partial resistance can be assessed and the effects of complex interactions between pathogen populations, host cultivars, and environments, which can occur in multiline cultivars, can be monitored. Results of such investigations may help to provide a basis for strategies of deploying host resistance to obtain minimum loss due to disease.

Spore Production in Pure-Line Crops

It was shown by Turner & Hart (85) using rotorod spore samplers (3) that populations of spores of *Helminthosporium turcicum* were smaller in the air above plots of maize possessing resistance gene *Ht-1* than over plots of nearly isogenic lines lacking this resistance gene. Such a result could depend on the relative frequencies of pathogen races able to infect plants carrying the gene *Ht-1* as well as on the extent of resistance of the cultivars.

Wolfe & Schwarzbach (94) described procedures for obtaining estimates of the frequencies of different virulence characteristics in populations of *Erysiphe graminis* by counting pustules of powdery mildew developing on seedlings of selected barley cultivars which had been exposed in crops of barley. Based on the relative frequency of individual factors for virulence, their expected frequencies in combination could be estimated. These could be compared with observed frequencies of mildew pustules, either on cultivars in which appropriate genes for resistance were present together, or by sequentially retesting pathogen cultures, taken from the primary test cultivar, on cultivars with different resistance genes. It was reported that the frequency of colonies developing on the cultivar Maris Mink, with resistance genes *Ml-g* and *Ml-as,* was always less than expected on the basis of observed frequencies of colonies on host cultivars carrying these genes separately. It was suggested that Maris Mink may possess resistance in addition to that provided by *Ml-g* and *Ml-as,* although it could also be suggested that the virulence genes in the pathogen occurred together less often than expected (95).

In our discussion of components of resistance we noted that equal amounts of inoculum often lead to the formation of different numbers of disease lesions. This factor must therefore be taken into account when trying to assess whether the frequency of lesions occurring on cultivars with different genes for resistance is due to host or to pathogen population characteristics. The use of a number of different cultivars believed to have similar genes for resistance could provide useful supplementary data in this case. Competition of the type discussed above, between different races of the pathogen population, might also become important and lead to changed frequencies if sequential tests were used to establish the frequency of combinations of genes for virulence in the pathogen population. These factors emphasize the importance of the additional technique proposed by Wolfe & Schwarzbach (94) for estimating total spore production from leaves of selected cultivars infected in the field. This would provide an estimate of the effect of the whole pathogen population on the selected host cultivar, including effects of unrecognized factors for resistance in the host cultivars, although providing less information about the frequencies of specific genes for virulence in the pathogen.

Spore Production in Multiline Crops

It has been suggested that disease will develop more slowly in a mixture of host lines or cultivars possessing different race-specific genes for resistance to disease. This would indicate a means of deploying race-specific resistance so as to enhance its effect and possibly, depending on the ability of the pathogen to accumulate genes for virulence, to simulate effects of non-race-specific resistance. This topic has been reviewed by Browning & Frey (9). In some experiments on the effectiveness of multiline cultivars, spore production has been measured to assess disease intensity.

Cournoyer et al (18) grew plots of a multiline oat cultivar consisting of seven component lines and infected these either with race 264 of *Puccinia coronata* var *avenae* which was virulent on all the component lines, or with a mixture of six races, to one or more of which each component line was susceptible. These plots were compared with those of a pure line, susceptible both to race 264 and also to all six races of the mixture. Spores were trapped with a rotorod sampler downwind of each plot. Maximum spore production occurred on the pure-line cultivar infected with race 264, and slightly less production was recorded for the mixture of six races on the pure line. Spore production for the mixture of six races on the multiline was about half as abundant; this result was attributed to the operation of mechanical interference, or "spore-trapping," resulting from a proportion of spores landing on resistant lines in the host mixture (9, 25). The data given by Browning & Frey (9) indicate, however, that even fewer numbers of spores resulted from infection of the multiline with race 264, which was virulent on all components of the multiline. It follows that this result cannot be explained by spore-trapping due to spores landing on resistant lines. Nor does it appear to be due to reduced aggressiveness of race 264 if the development of this race on the pure line is compared with that of the mixture of races. It could be suggested that, although race 264 was virulent on all the component lines of the multiline, these lines were in fact more resistant to race 264 than was the susceptible pure line, and that the resistance was more easily detectable by measuring spore production than by visual assessment. This explanation is speculative and perhaps a different one would emerge from appropriate experiments including investigations of sporulation of race 264, and of each of the individual races of the mixture, on pure stands of each of the components of the multiline and on the susceptible pure line.

In other investigations of multiline cultivars the increase of rust diseases has been assessed by counting pustules (9, 49). Results supported the idea that mechanical effects of spore-trapping contributed to delaying the development of disease.

In one experiment Leonard (49) reported that, in a mixture of oat cultivars Clintland D and Garry, Clintland D was susceptible to races 6F and 7A of *P. graminis* var *avenae* while Garry was resistant to both. Fewer rust pustules developed on Clintland D in such a mixture when races 6F and 7A were applied together than when either race was applied separately. Watson (92) reported results of Y. S. Tsiang, in which it was shown that less rust developed when a mixture of races of *P. graminis* var *tritici* was used to infect three wheat cultivars than when pure races were used. It is difficult to imagine the mechanism of such interference between races, unless an interaction occurs between the host cultivar and each pathogen race which is inhibitory to the development of another race, but not to itself. When a race grew alone it would not, therefore, be self-inhibiting.

These field experiments contain some features that are difficult to explain. Perhaps this is not surprising where complex interactions between host, pathogen, and environment are possible and the techniques of investigation are new. The results suggest that the value of race-specific resistance may be enhanced by appropriate use, but much remains to be learned about the behavior of pathogen populations when such methods of disease control are applied. The various techniques of spore collection and assessment will play an important role in elucidating these aspects of disease epidemiology and the relationships between race-specific and non-race-specific components of disease resistance.

SPORE PRODUCTION AND INDUCED RESISTANCE

Several authors have reported that inoculation of plants with a nonvirulent strain of a pathogen caused them to become resistant to a virulent strain of the same pathogen to which they would normally be susceptible (15, 33, 34, 51, 96). In none of these cases was an accurate assessment made of the numbers of effective infection points by the nonvirulent strain, but estimates were given of the degree of control achieved compared with test plants inoculated only with the virulent strain. In flax a maximum of 68% fewer pustules developed when a virulent race of *Melampsora lini* was applied 24 hr after a nonvirulent race than when a virulent race alone was applied (51). Cheung & Barber (15) recorded an 80% reduction in the number of pustules of a virulent race of *Puccinia graminis* applied three days after inoculation of a wheat cultivar with a nonvirulent race. Johnson & Allen (34) noted a reduction in the total quantity of spores of *P. striiformis* produced on a wheat cultivar when a nonvirulent race was applied simultaneously with a virulent strain.

Similar observations have been reported when hosts have been inoculated with alien rusts, for example, maize rust on beans (2), oat crown rust on wheat (39), wheat rusts on flax (51), bean rust on sunflowers, and sunflower rust on beans (96). Evidently the resistant response triggered by the incompatible strain of the pathogen is effective against any normally virulent strain. It may therefore be regarded as a non-race-specific mechanism elicited by a race-specific interaction. Yarwood (96) calculated that an average dose of *Uromyces phaseoli* at 1.3 mg spores/dcm^2 of sunflower leaf gave 50% control of *Puccinia helianthi* and 4 mg of *P. helianthi* spores/dcm^2 of bean leaf gave 50% control of *U. phaseoli*. He compared these results with the effectiveness of fungicides and concluded that application of spores would be uneconomical for disease control. Nevertheless, it has been suggested that this type of induced resistance might contribute to disease control in the field in multiline cultivars or mixed crops (2, 7, 34). Little experimental evidence from the field has yet been presented to support this hypothesis. However, the data of Leonard (49) and that reported by Watson (92), in which mixtures of races of *Puccinia graminis* resulted in less disease than pure races, may indicate the operation of mechanisms related to induced resistance.

CONCLUSIONS

The measurement of spore production of pathogens growing on infected host plants provides an accurate method of measuring the pathogenicity of the pathogen and the resistance of the host. It measures the sum of the effect of all components of resistance mechanisms in the host. It is less subjective than visual estimation of disease intensity and has revealed differences that were not detected by visual methods. The data indicate the widespread occurrence of differential interactions and their wide range in magnitude in host-parasite relationships; they show that incomplete resistance can be race-specific. This makes it difficult to identify non-race-specific resistance but does not conflict with the observation that disease resistance can be durable.

It is evident from these results that pathogen populations are highly variable. When pathogen isolates are selected from the population they may be adapted to many unrecognized components in the particular environment from which they are collected. These adaptations can have an important and often unrecognized influence on the results of experiments.

Our purpose in this review has been to identify the positive contribution to our knowledge of plant disease made by the use of accurate techniques for measuring spore production. There are numerous other ways of estimating disease intensity, and although they may in some cases be less accurate, they are nevertheless essential. Especially in plant breeding, large populations of plants must be examined and very rapid methods of disease assessment are required. Measurements of spore production are not rapid enough for such purposes, but they can be useful in assessing the accuracy of more rapid methods.

Literature Cited

1. Allen, D. J. 1975. Variation in pathogenicity of *Uromyces appendiculatus*. *Ann. Rept. Bean Improv. Co-op.* 18: 13−14
2. Allen, D. J. 1975. Induced resistance to bean rust and its possible epidemiological significance. *Ann. Rept. Bean Improv. Co-op.* 18:15−16
3. Asai, G. N. 1960. Intra- and interregional movement of uredospores of black stem rust in the Upper Mississippi River Valley. *Phytopathology* 50:535−41
4. Black, W. 1952. A genetical basis for the classification of strains of *Phytophthora infestans*. *Proc. R. Soc. Edinburgh Sect. B* 65:36−51
5. Black, W., Gallegly, M. E. 1957. Screening of *Solanum* species for resistance to physiologic races of *Phytophthora infestans*. *Am. Potato J.* 34:273−81
6. Browder, L. E. 1973. Specificity of the *Puccinia recondita* f. sp. *tritici: Triticum aestivum* 'Bulgaria 88' relationship. *Phytopathology* 63:524−28
7. Brown, J. F. 1975. Factors affecting the relative ability of strains of fungal pathogens to survive in populations. *J. Aust. Inst. Agric. Sci.* 41:3−11
8. Brown, J. F., Sharp, E. L. 1970. The relative survival ability of pathogenic types of *Puccinia striiformis* in mixtures. *Phytopathology* 60:529−33
9. Browning, J. A., Frey, K. J. 1969. Multiline cultivars as a means of disease control. *Ann. Rev. Phytopathol.* 7:355−82
10. Browning, J. A., Cournoyer, B. M., Jowett, D., Mellon, J. 1970. Urediospore and grain yields from interacting crown rust races and commercial multiline oat cultivars. *Phytopathology* 60:1286 (Abstr.)
11. Caten, C. E. 1970. Spontaneous variability of single isolates of *Phytophthora infestans*. II. Pathogenic variation. *Can. J. Bot.* 48:897−905
12. Caten, C. E. 1974. Intra-racial variation in *Phytophthora infestans* and adaptation to field resistance for potato blight. *Ann. Appl. Biol* 77:259−70
13. Chakravarti, B. P. 1972. Infection processes in wheat stem rust. *Proc.* *Eur. Mediterr. Cereal Rusts Conf. Prague, 1972* 1:63−67
14. Chaudhary, R. C., Schwarzbach, E., Fischbeck, G. 1976. Differences in sporulation as a quantitative measure of the mildew resistance in barley. *Proc. Third Int. Barley Genet. Symp. Munich, 1975.* In press
15. Cheung, D. S. M., Barber, H. N. 1972. Activation of resistance of wheat to stem rust. *Trans. Br. Mycol. Soc.* 58: 333−36
16. Clifford, B. C. 1972. The histology of race non-specific resistance to *Puccinia hordei* Otth. in barley. *Proc. Eur. Mediterr. Cereal Rusts Conf. Prague, 1972* 1:75−79
17. Clifford, B. C., Clothier, R. B. 1974. Physiologic specialisation of *Puccinia hordei* on barley hosts with non-hypersensitive resistance. *Trans. Br. Mycol. Soc.* 63:421−30
18. Cournoyer, B. M., Browning, J. A., Jowett, D. 1967. Crown rust intensification within and dissemination from pure line and multiline varieties of oats. *Phytopathology* 58:1047 (Abstr.)
19. Day, P. R. 1968. Plant disease resistance. *Sci. Prog. London* 56:357−70
20. Denward, T. 1967. Differentiation in *Phytophthora infestans*. I. A comparative study of eight different biotypes. *Hereditas* 58:191−220
21. Deshmukh, M. J., Howard, H. W. 1956. Field resistance to potato blight *(Phytophthora infestans). Nature* 177: 794−95
22. Dickinson, C. H., Crute, I. R. 1974. The influence of seedling age and development on the infection of lettuce by *Bremia lactucae. Ann. Appl. Biol.* 76:49−61
23. Eyal, Z., Peterson, J. L. 1967. Uredospore production of five races of *Puccinia recondita* Rob ex. Desm. as affected by light and temperature. *Can. J. Bot.* 45:537−40
24. Fowler, A. M., Owen, H. 1971. Studies on leaf blotch of barley *(Rhynchosporium secalis). Trans. Br. Mycol. Soc.* 56:137−52
25. Frey, K. J., Browning, J. A., Simons, M. D. 1973. Management of host re-

sistance genes to control diseases. *Z. Pflanzenkr. Pflanzenschutz* 80:160−80

26. Fuchs, E. 1966. Physiologic races of *Puccinia striiformis* on wheat identified in the glasshouse. *Rept. Cereal Rusts Conf. Cambridge* 1964:39−46

27. Green, G. J. 1971. Hybridization between *Puccinia graminis tritici* and *Puccinia graminis secalis* and its evolutionary implications. *Can. J. Bot.* 49:2089−95

28. Guzmann, J. 1964. Nature of partial resistances of certain clones of three *Solanum* species to *Phytophthora infestans. Phytopathology* 54:1398-1404

29. Heagle, A. S., Moore, M. B. 1970. Some effects of moderate adult resistance to crown rust of oats. *Phytopathology* 60:461−66

30. Hilu, H. M., Hooker, A. L. 1963. Monogenic chlorotic lesion resistance to *Helminthosporium turcicum* in corn seedlings. *Phytopathology* 53:909-12

31. Hodgson, W. A. 1962. Studies on the nature of partial resistance in the potato to *Phytophthora infestans. Am. Potato J.* 39:8−13

32. Jeffrey, S. I. B., Jinks, J. L., Grindle, M. 1962. Intraracial variation in *Phytophthora infestans* and field resistance to potato blight. *Genetica* 32:323−38

33. Johnson, R. 1976. Genetics of host-parasite interactions. In *Specificity in Plant Disease, ed. A. Graniti, R. K. S. Wood.* Sardinia: NATO Adv. Stud. Inst. In press

34. Johnson, R., Allen, D. J. 1975. Induced resistance to rust diseases and its possible role in the resistance of multiline varieties. *Ann. Appl. Biol.* 80:359−63

35. Johnson, R., Bowyer, D. E. 1974. A rapid method for measuring production of yellow rust spores on single seedlings to assess differential interactions of wheat cultivars with *Puccinia striiformis. Ann. Appl. Biol.* 77:251−58

36. Johnson, R., Law, C. N. 1973. Cytogenetic studies on the resistance of the wheat variety Bersée to *Puccinia striiformis. Cereal Rusts Bull.* 1:38−43

37. Johnson, R., Law, C. N. 1975. Genetic control of durable resistance to yellow rust *(Puccinia striiformis)* in the wheat cultivar Hybride de Bersée. *Ann. Appl. Biol.* 81:385−91

38. Johnson, R., Taylor, A. J. 1972. Isolates of *Puccinia striiformis* collected in England from the wheat varieties Maris Beacon and Joss Cambier. *Nature* 238:105−6

39. Johnston, C. O., Huffman, M. D.

1958. Evidence of local antagonism between two cereal rust fungi. *Phytopathology* 48:69−70

40. Kak, D., Joshi, L. M., Prasada, R., Vasudeva, R. S. 1963. Survival of races of *Puccinia graminis tritici* (Pers.) Eriks & J. Henn. *Indian Phytopathol.* 16:117−27

41. Katsuya, K., Green, G. J. 1967. Reproductive potentials of races 15B and 56 of wheat stem rust. *Can. J. Bot.* 45:1077−91

42. Knott, D. R. 1971. Can losses from wheat stem rust be eliminated in North America? *Crop Sci.* 11:97−99

43. Knutson, K. W. 1962. Studies on the nature of field resistance of the potato to late blight. *Am. Potato J.* 39: 152−61

44. Kochman, J. K. Brown, J. F. 1975. Host and environmental effects on post-penetration development of *Puccinia graminis avenae* and *P. coronata avenae. Ann. Appl. Biol.* 81:33−41

45. Lapwood, D. H. 1961. Potato haulm resistance to *Phytophthora infestans.* II. Lesion production and sporulation. *Ann. Appl. Biol.* 49:316−30

46. Lapwood, D. H. 1961. Laboratory assessments of the susceptibility of potato haulm to blight *(Phytophthora infestans). Eur. Potato J.* 4:117−28

47. Lapwood, D. H. 1963. Potato haulm resistance to *Phytophthora infestans.* IV. Laboratory and field estimates compared, and further field analysis. *Ann. Appl. Biol.* 51:17−28

48. Lapwood, D. H., McKee, R. K. 1966. Dose-response relationship for infection of potato leaves by zoospores of *Phytophthora infestans. Trans. Br. Mycol. Soc.* 49:679−86

49. Leonard, K. J. 1969. Factors affecting rates of stem rust increase in mixed plantings of susceptible and resistant oat varieties. *Phytopathology* 59: 1845−50

50. Leonard, K. J. 1969. Selection in heterogeneous populations of *Puccinia graminis* f. sp. *avenae. Phytopathology* 59:1851−57

51. Littlefield, L. J. 1969. Flax rust resistance induced by prior inoculation with an avirulent race of *Melampsora lini. Phytopathology* 59:1323−28

52. Loegering, W. Q. 1951. Survival of races of wheat stem rust in mixtures. *Phytopathology* 41:55−65

53. Wu Lung-Chi, Stahmann, M. A. 1975. Chromatographic estimation of fungal mass in plant materials. *Phytopathology* 65:1032-34

54. Lupton, F. G. H., Johnson, R. 1970.

Breeding for mature plant resistance to yellow rust in wheat. *Ann. Appl. Biol.* 66:137–43

55. Martens, J. W. 1973. Competitive ability of oat stem rust races in mixtures. *Can. J. Bot.* 51:2233–36

56. Mehta, Y. R., Zadoks, J. C. 1970. Uredospore production and sporulation period of *Puccinia recondita* f.sp *triticina* on primary leaves of wheat. *Neth. J. Plant Pathol.* 76:267–76

57. Mehta, Y. R., Zadoks, J. C. 1971. Note on the efficiency of a miniaturized cyclone spore collector. *Neth. J. Plant Pathol.* 77:60–63

58. Müller, K. O., Haigh, J. C. 1953. Nature of 'field resistance' of the potato to *Phytophthora infestans*. *Nature* 171:781–83

59. Nelson, R. R., Tung, G. 1973. Influence of some climatic factors on sporulation by an isolate of race T of *Helminthosporium maydis* on a susceptible male-sterile corn hybrid. *Plant Dis. Reptr.* 57:304–7

60. Nelson, R. R., MacKenzie, D. R., Scheifele, G. L. 1970. Interaction of genes for pathogenicity and virulence in *Trichometasphaeria turcica* with different numbers of genes for vertical resistance in *Zea mays*. *Phytopathology* 60:1250–54

61. Ogle, H. J., Brown, J. F. 1970. Relative ability of two strains of *Puccinia graminis tritici* to survive when mixed. *Ann. Appl. Biol.* 66:273–79

62. Ogle, H. J., Brown, J. F. 1971. Some factors affecting the relative ability of two strains of *Puccinia graminis tritici* to survive when mixed. *Ann. Appl. Biol.* 67:157–68

63. Parlevliet, J. E. 1975. Partial resistance of barley to leaf rust, *Puccinia hordei*. I. Effect of cultivar and development stage on latent period. *Euphytica* 24: 21–27

64. Parlevliet, J. E., van Ommeren, A. 1975. Partial resistance of barley to leaf rust, *Puccinia hordei*. II. Relationship between field trials, micro plot tests and latent period. *Euphytica* 24:293–303

65. Paxman, G. J. 1963. Variation in *Phytophthora infestans*. *Eur. Potato J.* 6:14–23

66. Priestley, R. H., Doling, D. A. 1972. A technique for measuring the spore production of yellow rust (*Puccinia striiformis*) on wheat varieties. *Proc. Eur. Mediterr. Cereal Rusts Conf. Prague, 1972* 1:219–23

67. Robinson, R. A. 1969. Disease resistance terminology. *Rev. Appl. Mycol.* 48:593–606

68. Rodenhiser, H. A., Holton, C. S. 1953. Differential survival and natural hybridization in mixed spore populations of *Tilletia caries* and *T. foetida*. *Phytopathology* 43:558–60

69. Schafer, J. F. 1971. Tolerance to plant disease. *Ann. Rev. Phytopathol.* 9: 235–52

70. Schaper, S. 1951. Die Bedeutung der Incubationszeit für die Züchtung Krautfäuleresistenter Kartoffelsorten. *Z. Pflanzenzuecht.* 30:292–99

71. Schwarzbach, E. 1974. Untersuchung der Mehltauanfälligkeit von Gerstensorten durch Messung der Sporenproduktion befallener Blätter. *Z. Pflanzenzuecht.* 73:311–15

72. Shaner, G. 1973. Estimation of conidia production by individual pustules of *Erysiphe graminis* f.sp. *tritici*. *Phytopathology* 63:847–50

73. Shaner, G. 1973. Evaluation of slow-mildewing resistance of Knox wheat in the field. *Phytopathology* 63:867–72

74. Shaner, G. 1973. Reduced infectability and inoculum production as factors of slow mildewing in Knox wheat. *Phytopathology* 63:1307–11

75. Sharp, E. L. 1973. Additive genes for resistance to stripe rust. *Second Int. Plant Pathol. Congr. Minneapolis, 1973* (Abstr. 0487)

76. Simkin, M. B., Wheeler, B. E. J. 1974. The development of *Puccinia hordei* on barley cv. Zephyr. *Ann. Appl. Biol.* 78:225–35

77. Stakman, E. C., Stewart, D. M., Loegering, W. Q. 1962. Identification of physiologic races of *Puccinia graminis* var. *tritici*. *US Dep. Agric. Res. Bull. E 617.* 53 pp.

78. Talboys, P. W., Garrett, C. M. E., Ainsworth, G. C., Pegg, G. F., Wallace, E. R. 1973. A guide to the use of terms in plant pathology. *Commonw. Mycol. Inst. Phytopathol. Pap. 17.* 55 pp.

79. Thurston, H. D. 1961. The relative survival ability of races of *Phytophthora infestans* in mixtures. *Phytopathology* 51:748–55

80. Thurston, H. D. 1971. Relationship of general resistance: late blight of potato. *Phytopathology* 61:620–26

81. Torres, E., Browning, J. A. 1968. The yield of uredospores per unit of sporulating area as a possible measure of tolerance of oats to crown rust. *Phy-*

topathology 58:1070 (Abstr.)

82. Toxopeus, H. J. 1958. Note on the variation in expression of hypersensitivity to *Phytophthora infestans* in potato seedlings. *Euphytica* 7:38−40

83. Tsai, S. D., Erwin, D. C. 1975. A method of quantifying numbers of microsclerotia of *Verticillium albo-atrum* in cotton plant tissue and in pure culture. *Phytopathology* 65:1027−28

84. Turkensteen, L. J. 1973. Partial resistance of tomatoes against *Phytophthora infestans,* the late blight fungus. *Agric. Res. Rept. 810 PUDOC Wageningen, 1973.* 91 pp.

85. Turner, M. T., Hart, K. 1975. Field spore production of *Helminthosporium turcicum* on *Zea mays* with and without monogenic resistance. *Phytopathology* 65:735−36

86. Umaerus, V. 1970. Studies on field resistance to *Phytophthora infestans.* 5. Mechanisms of resistance and application to potato breeding. *Z. Pflanzenzuecht.* 63:1−23

87. Vanderplank, J. E. 1963. *Plant Diseases: Epidemics and Control.* New York & London: Academic. 349 pp.

88. Vanderplank, J. E. 1968. *Disease Resistance in Plants.* New York & London: Academic. 206 pp.

89. Vanderplank, J. E. 1971. Stability of resistance to *Phytophthora infestans* in cultivars without R genes. *Potato Res. J.* 14:263−70

90. van der Zaag, D. E. 1959. Some observations on breeding for resistance to *Phytophthora infestans. Eur. Potato J.* 2:278−86

91. Ward, S. V., Manners, J. G. 1974. Environmental effects on quantity and viability of conidia produced by *Erysiphe graminis. Trans. Br. Mycol. Soc.* 62:119−28

92. Watson, I. A. 1942. The development of physiologic races of *Puccinia graminis tritici* singly and in association with others. *Proc. Linn. Soc. NSW* 67:294−312

93. Williams, R. J., Owen, H. 1975. Susceptibility of barley cultivars to leaf blotch and aggressiveness of *Rhynchosporium secalis* races. *Trans. Br. Mycol. Soc.* 65:109−14

94. Wolfe, M. S., Schwarzbach, E. 1975. The use of virulence analysis in cereal mildews. *Phytopathol. Z.* 82:297−307

95. Wolfe, M. S., Barrett, J. A., Shattock, R. C., Shaw, D. S., Whitbread, R. 1976. Phenotype-phenotype analysis: field application of the gene-for-gene hypothesis in host-pathogen relations. *Ann. Appl. Biol.* 82:369−74

96. Yarwood, C. E. 1956. Cross protection with two rust fungi. *Phytopathology* 46:540−44

97. Yarwood, C. E. 1961. Uredospore production by *Uromyces phaseoli. Phytopathology* 51:22-27

98. Zadoks, J. C. 1972. Methodology of epidemiological research. *Ann. Rev. Phytopathol.* 10:253−76

MICROBIAL COLONIZATION OF PLANT ROOTS

•3635

G. D. Bowen and A. D. Rovira
CSIRO Division of Soils, Glen Osmond, South Australia, Australia

Research in microbial growth and interactions on roots is likely to increase in the future as the need for integration of plant pathology and soil microbiology is recognized. On the one hand, some diseases are suppressed by soil microorganisms (4) and on the other hand, the soil microflora can also suppress beneficial root symbionts; for example, the establishment of a selected mycorrhizal fungus is easier in sterile than in nonsterile soil (21, 143). Also, poor responses to *Rhizobium* inoculation may result from competition from less effective indigenous strains and antagonists in soil. Interest in controlling the rhizosphere comes also from the effects of certain rhizosphere microorganisms themselves on plant growth; many organisms in the general soil microflora may be subclinical pathogens reducing root and root hair growth or acting directly on the plant (8, 19). Inoculation of many plant species with certain soil bacteria sometimes increases yield, advances flowering, or increases internode extension (27, 93, 117).

After a long period devoted largely to isolation of organisms from roots, there is new impetus into finding mechanisms of colonization of the root; also, experimental approaches to the population dynamics of microorganisms on roots are emerging. In this review we summarize the present information and suggest possible future research. Good data are sparse and conclusions have necessarily often been made on the basis of very few experiments.

DISTRIBUTION OF ORGANISMS ON ROOTS

Light Microscope Studies

Until recently, studies on plants grown in solution or sand culture suggested that almost the entire surface of roots in soil was covered by microorganisms, forming a continuous compartment between soil and root. However, this is usually not so except for such associations as sheathing mycorrhizae (78, 86). After measuring microbial occupation of several representative microscope fields on roots of *Pinus radiata* grown in soil, Bowen & Theodorou (21) concluded that great variability occurred in the microbial cover of roots but that the percentage of surface covered was usually less than 10% with root segments that were three weeks old. Often the cover of the root by microorganisms was extremely low, especially in segments 1−2 days old, but even on root segments that were 90 days old the cover was only 37%. By taking microscope fields at random, Rovira et al (127) found that bacteria covered 4−10% of the root surface of 8 plant species grown for 12 weeks in a fertile grassland soil. Little loss of organisms occurred during washing and staining the roots. Variability was large, and 10 fields chosen at random gave an LSD of 5.2% necessary for significant differences at the 5% probability level between any two species; however, by counting 60 fields, 7.7% cover on *Lolium perenne* was significantly different from 6.3% cover on *Plantago lanceolata*. With a random line intercept method (98) the length of fungal hyphae/mm^2 was calculated at 12.1 and 14.3 mm/mm^2 on the two species respectively. The number of bacteria estimated by direct microscopy was approximately 10 times that obtained from dilution-plating techniques to laboratory media. The reasons for this discrepancy (frequently observed with counts of soil bacteria) may be that not all bacteria will grow on the laboratory medium, that all cells on the root may not be viable, or that colonies detected in dilution plates may frequently arise from groups of cells not dispersed during the dilution procedure. Trolldenier (148) stained roots of clover with acridine orange to distinguish between live and dead bacteria. The use of such vital stains is a technique with considerable potential for soil microbiologists and plant pathologists.

The low surface occupancy by microorganisms leads to two questions: 1. Are there relatively few points of inoculation in soil followed by little spread over the surface? 2. Are some parts of the surface more conducive to microbial growth? Preferential growth of bacteria occurs along the longitudinal junctions of epidermal cells (116), which is often observed with fungi also. Pattern analysis, devised for plant distribution studies (56, 57), was used by Newman & H. J. Bowen (99) to demonstrate that rhizoplane bacteria are not randomly distributed but tend to be aggregated. Nonrandomness was shown at the microcolony scale and over larger general areas unrelated to age of the root segment. Newman & Bowen pointed out the usefulness of pattern analyses to compare the effects of different soils with the same species, different species in the same soil, or differences between different parts of the same root system.

The sources of nonrandomness of organisms on roots are now being identified. In recent studies (18) seedlings of *P. radiata* were grown in sterile soil for 14 days, washed lightly, dried with paper tissue, replanted to a moist nonsterile sandy soil, and the colonization of new growth, the original apical centimeter, the 4−5 cm segment, and the 9−10 cm segment from the original tip were examined daily for 5 days. For the first 2 days colonization at all positions was slight, with a few microcolonies, up to 16−20 cells, occurring on the root surface and occasionally at the junction of epidermal cells. There was an apparently high association between microcolonies and pieces of organic matter, suggesting that colonized organic fragments are a major source of inoculum. High multiplication had occurred in many of the epidermal cell junctions (and to a lesser extent on other parts) by the third and fourth day suggesting that the junctions may be preferred sites of growth and migration along the root. Colonization of sloughed root cells was heavy and often arose from organic matter. In these and other studies, in association with R. C. Foster and J. E. Mitchell (unpublished), heavy microbial growth occurred on some individual cells on the apparently healthy root while neighboring cells were almost devoid of bacteria; electron microscope studies (45) show such growth to be frequently accompanied by senescence of those cells. Foster (44) observed that some epidermal cells of *Sinapis alba* L. growing in solution culture ceased to grow and degenerated while neighboring cells grew normally. Degenerate cells may be a natural phenomenon but particular bacteria (subclinical pathogens?) may also cause or hasten senescence by increasing the leakiness of the cells, for example, 12−18% of photosynthetically fixed carbon may be released from wheat and barley plants growing in nonsterile soil compared with 5−10% in sterile soil (10).

The paucity of microbial cover on the roots, especially in the first day (see above), offers little protection against pathogens that move to the root and infect rapidly, for example, *Phytophthora cinnamomi,* and biological control of such organisms must occur by attack on the propagule in the soil, for example, inhibition of sporangium production (22). More protection by rhizosphere microorganisms can be envisaged when pathogens develop more slowly and spread along the same routes as those occupied by the colonizing soil microflora, especially if these colonizing microorganisms are antagonistic to the pathogen. At first the relative paucity of root colonization seems surprising considering the many millions of bacteria per cubic centimeter of soil; however, bacteria tend to be in microcolonies (69) and in a sand dune soil only about 0.04% of the surface area of the soil particles may be colonized by microorganisms (53). Almost all microorganisms are attached to surfaces rather than free in soil solution (53). When sterile roots of *P. radiata* are firmly pressed into soil, only some 100 organisms/mm² root are reisolated immediately (20); this approximates the theoretical calculations on the numbers of organisms a root is likely to touch or come to within a few microns as it grows through soil.

The hypothesis that organic debris is a major source of inoculum for the root is consistent with the observation (53) that in a sand dune soil, organic matter particles provided less than 15% of the available solid surfaces but carried between 57 and 64% of the soil bacteria. Particulate organic matter often has the majority of the pseudomonad population (128) and these grow well in the rhizosphere. Plant debris may also act as inoculants of plant pathogens; for example, *Gaeumannomyces graminis* and other pathogens survive on infected crop residues. In crop monoculture the sequence of living root − dead root − living root approximates a tightly coupled enrichment culture system leading to evolution of a microflora compatible with a new plant species within a few seasons, for example, development of soils suppressive to *G. graminis* on wheat (4).

Preferential growth and migration along cell junctions may correspond with these being sites of greater exudation, and containing a more continuous and substantial volume of moisture and mucilage. The greater growth along cell junctions has been measured by counting the bacteria in quadrats on the root surface at high magnification (84); on *Eucalyptus calophylla,* quadrats with cell junctions had 3.9 ($\pm$ 0.65, SE) bacteria, and adjoining quadrats had 0.07 ($\pm$0.07). Although fungi and actinomycetes tend also to grow along cell junctions, their ability to translocate nutrients enables them to grow more easily in sites of low nutrition.

Frequent occurrence of bacterial colonies with fungal hyphae (145, 157) may be due to their both occupying energy-rich localities, to bacterial growth being enhanced by exudates from the hyphae, to an effect of the hyphae on increasing leakage from the root, or simply to migration of bacteria along the water films at the hypha–root cell interface.

Microscopy of the rhizoplane and its microflora has been limited and most studies have used relatively high soil moisture, so there is a need to extend studies to other conditions. Problems in quantification can be large because of high variability, but many statistical methods can be transposed from plant ecology. Although the logistics of a quantitative approach in a large, broad study are overwhelming, they are quite feasible for testing specific hypotheses such as the association of plant debris and the first microcolonies, routes of migration, preferential growth at cell junctions, and the occurrence and importance of senescing cells. Furthermore, there are a number of light microscope techniques of great potential but little used by rhizosphere ecologists: 1. The fluorescent antibody technique is a powerful tool for following growth and distribution of specific organisms in natural soils (130). Such techniques have been used for several bacteria and fungi in soil (130) and colonization of roots by *G. graminis* (67), *P. cinnamoni* (85), *Verticillium dahliae* (43), ectomycorrhizal fungi (131), and rhizobia (147). 2. Labeling inoculum with a fluorescent brightener allows it to be followed (1, 157). 3. Distinctive morphological features of bacteria can be used in experimental studies with mixtures of organisms in sterile soil but are of limited use in natural soils. 4. Microautoradiography of incorporated tritiated thymidine could be used to distinguish between growing and dormant cells (25, 146). 5. Histochemical and autoradiographic techniques have great potential in the study of ecophysiology of microorganisms on the root. Brock (25) has discussed a number of aspects of studying growth rates of microorganisms in nature.

Electron Microscope Studies

Transmission electron microscopy (TEM) of the soil-root interface (46, 47) has revealed much more on the spatial relationships of microorganisms, soil, and roots than light microscopy. A prominent feature of TEM studies is the mucilage or mucigel (68) which surrounds the root and in which microorganisms develop (38, 48, 50, 54). The mucigel is 1–10 μm wide and is thicker at the zone of elongation of roots and at the junctions of epidermal cells (45, 102). Although principally polysaccharide (70), it contains sufficient phenolic materials to react with lead to reveal granular and fibrillar structures (54, 59). The mucigel could provide contact between soil and roots for diffusion of nutrients and may protect younger parts of roots from desiccation during temporary wilting periods. It may also give some protection from microorganisms. Mucigel excreted through the root cap serves as a lubricant to roots growing through soil (101). The mucigel is produced by the plant under axenic conditions, but Greaves & Darbyshire (54) demonstrated it to be thicker in nonsterile roots; much of this increased thickness may have been due to bacterial capsular material often seen in TEM (50) and scanning electron microscopy (SEM) (30). Bacteria penetrate the mucigel, creating electron-transparent zones 1−2 μm wide around them, but the mucigel is not usually completely digested and may be continuously produced by the root; SEM of wheat roots (125) has shown disappearance of the mucigel upon infection with *G. graminis* and prolific growth of bacteria on the root surface. TEM studies of colonization of surfaces of nonbiological materials have shown attachment of bacteria (45, 87) and cysts of *Phytophthora palmivora* (134) to surfaces by microfibrils; the frequent polar attachment between bacteria and root seen in TEM suggest that a similar phenomenon occurs with roots. TEM of ectomycorrhizae (80) shows bacterial growth in a mucilaginous layer around and in the layers of the fungus sheath, and this has been confirmed by SEM (83).

TEM studies of roots of subterranean clover (50) showed a layer of bacteria 10−20 μm deep between the root surface and the clay particles with some 10 different bacterial types distinguishable on cell morphology, structure of the cell walls, and cell contents. Many were in microcolonies of 5−12 cells with quite distinct boundaries between the mucilage within the colonies and between the colonies. Counts of the different morphological types showed a zonation of the organisms with certain types dominating the rhizoplane and others dominating the outer rhizosphere. In the presence of microorganisms many of the surface cells of roots were either dead or had lost cell wall structure (54). This breakdown of root cells has also been demonstrated for sand dune grasses (102) and wheat at the flowering stage (48, 49). TEM is a powerful tool to study microbial interactions on roots; for example, structures resembling parasites of bacteria such as *Bdellovibrio* and phage in association with certain bacterial types have been observed (50). SEM has also shown that when biological control of *G. graminis* occurs, bacterial colonization of the hyphae is accompanied by the appearance of pits and holes in the hyphae with complete lysis following (125).

TEM and SEM have much to offer in studies of microbial ecology on roots; the understanding of microbial ecophysiology will be enhanced by further use of histochemical, autoradiographic, and electron microprobe analysis techniques.

MOVEMENT OF ORGANISMS TO THE ROOT

Volatile and dissolved exudates from seeds and roots can affect microbial growth for relatively large distances; for example, chlamydospores of *Fusarium* may germinate 10 mm from seeds (141) and inhibition of rhizobia can occur from seed exudates (14). The distances individual compounds diffuse are governed by soil conditions such as porosity and moisture, and by their absorption and metabolism by microorganisms. The principles of diffusion and convection of solutes toward roots in soil (103) apply to diffusion of exudates away from plant roots. As exudates diffuse away from roots there could be a tendency for them to be swept back to the root by convective movement of water. Exudates from uncolonized recent root growth may diffuse into soil, but a developing rhizoplane population could reduce such movement considerably so that there would be little effect beyond the immediate vicinity of the root after 3–4 days.

Although bacteria can grow at soil matric suctions of 5 to 50 bars (58), bacterial mobility is quite limited below 0.5 to 0.1 bar suction (approximately field capacity) as they need a continuous film of water for movement. Discontinuities of water films can occur at lower suctions than those required to empty soil pore neck diameters of bacterial width or bacterial radii of gyration, and Hamdi (60) found little migration of *Rhizobium trifolii* below 0.035, 0.20, and 0.40 bar suction in coarse sand, fine sand, and silt loam respectively. In sterilized silt loam supplemented with growth medium, the bacteria moved 20, 9, and 5 mm at 0.03, 0.06, and 0.20 bar suction respectively in 3 days. The physical and nutritional nature of the substrate is important, for Wong (159) found that after 24 hr at 0.15 bar suction, *Pseudomonas fluorescens* moved 43 mm in a sterile grit amended with a growth medium, 11 mm in sterile amended soil, and 5 mm in sterilized unamended soil; however, even in sterilized soil itself there is a high growth of pseudomonads (110). At 0.15 bar suction, *B. subtilis* and *Azotobacter vinelandii* moved 3 mm and 2 mm respectively in sterilized unamended soil. These distances are large in relation to the rhizosphere but it is difficult to relate the findings to natural soil for various reasons; viz high levels of inocula were used, high levels of bacterial nutrients were usually added, the use of sterilized soil eliminated competition from other soil organisms. The numbers of bacteria often decline drastically with distance from the nutrient source (60), and it is hard to envisage any appreciable influence on the bacterial composition of the rhizoplane from organisms other than those almost touching the root, except possibly where soil moisture is extremely high. Movement of actinomycete spores in soil is limited even with free water, largely because of adsorption (129). A further probable restriction on movement of bacteria to the root is the steep moisture gradient which can occur close to the root during the day with high transpiration rates (104).

Zoospores of *Pythium* and *Phytophthora* need a continuous film of water for movement; soil pores approximating zoospores in size, 5–10 μm, drain at 0.2 to 0.3 bar suction and are filled during or immediately after rain. Passive movement of zoospores in moving water may be more important than movement by flagella (58). Zoospores are probably the main infecting agents of *Phytophthora* in soil (72) and although movement of *Phytophthora* zoospores to roots is the subject of many papers, the work is usually in solution culture and there is little quantitative study on the distances over which zoospores move to roots in soil of different matric potentials. There is at least one report of zoospores in wet soils not conforming to the expected accumulation sites along roots (65).

Hyphae can grow across soil gaps that have a high humidity but there are few quantitative data on hyphal growth through soil. There are some studies on growth through soil tubes amended with nutrients, growth from an energy-rich agar plug or along Rossi-Cholodny slides, but extrapolation to real soil situations is difficult. Henis & Ben-Yephet (63) found that propagules of *Rhizoctonia solani* placed up to 50 mm from bean roots could cause some disease, some 6% of the test plants being diseased compared with 75% when propagules were placed at 25 mm. Sets of 32 propagules placed 20 mm from seedlings caused severe disease but 1–2 propagules did not. In view of the importance of such studies to disease forecasting, it is surprising that more have not been performed either by direct placement of the propagule (63) or by attachment of it to an inert base at specified distances from the root (12). Some guide to the distances of fungal growth to the roots is given by data on the germination of propagules away from the root, assuming that the germination stimuli also provide sufficient and sustained growth requirements. Chlamydospores of *Fusarium solani* f. sp. *solani* germinated 10 mm from bean seeds at 0.05 bar water suction and 8 mm from seeds at 0.10 bar suction after 24 hr at 20°C (141). At 12 hr, *Pythium ultimum* sporangia had germinated 8 mm from the seed at 0.05 bar suction and 4 mm at 0.1 bar suction. A 7 mm spermatosphere effect on *F. solani* f. sp. *pisi* has been shown with peas in soil (133). Stanghellini & Hancock (141) found germination of *P. ultimum* sporangia 0–2 mm from the seed within 1.5 hr, and 2–4 mm from the seed at 6 hr. Because germ tubes grew at 130–300 μm/hr at 20°C (0.05 and 0.10 bar suction) and approximately 100 μm/hr at 12°C, such organisms can obviously quickly effect a foothold on the young root surface. The data above are somewhat ideal with seed exudates providing a rich food source—the energy sources may be much less luxuriant in exudates from roots.

Some factors in fungal growth through soil are (*a*) energy either from root exudates or from reserves in the propagule; saprophytic growth of *R. solani* is closely related to the size of the propagule (63); (*b*) environment, such as soil temperature; (*c*) soil factors such as water, oxygen, and carbon dioxide (58); (*d*) soil compaction; up to 90% reduction in growth of mycelial strands of *Rhizopogon luteolus* occurs on increasing soil density from 1.2 g/cm^3 to 1.6 g/cm^3 (136); such effects may be due to oxygen depletion and/or carbon dioxide increase; and (*e*) microbial activity; for example, *Fomes annosus* and *Coniophora puteosa* may not grow through nonsterile soil (66, 107). *Pythium aphanidermatum* and *Phytophthora erythroseptica* make little or no growth in nonsterile soil in the absence of a rich food base (80, 152).

There is a need for quantitative migration studies in nonsterile unamended soils, using tracer methods such as immunofluorescence.

Fungus Populations and Root Colonization

Baker & Cook (4) related the hazardous numbers of plant pathogen propagules in soil to their ability to grow intensively on the root and overcome host resistance. However, as the pathogen must first get to the root, we suggest the hazardous number of propagules per gram of soil is also closely related to the size of the propagule and its previous nutrition, this being an indication of its reserves for growth through soil to the root. This is still substantially in agreement with the three groupings of Baker & Cook. Henis & Ben-Yephet (63) noted that propagules of *R. solani* of 500−1000 μm diameter produced colonies of 1.3 cm diameter on glass slides but that propagules of 150−250 μm diameter gave colonies of 0.52 cm diameter.

Hazardous counts in a soil cannot be divorced from the rooting intensity of the plant. The length of root per cubic centimeter of soil in the surface 10 cm may vary from 25−50 cm for grass species, to 20−30 cm for herbs, to 2−4 cm for woody plants (11). For *Lolium multiflorum* the half distance between roots has been calculated at 1.5 mm at 2 cm depth (11), but for some tree species it may be several times this (16). In contrast to organisms with large reserves in the propagule it is apparent that organisms with small propagules, and thus unable to grow through soil for large distances, need to produce many more propagules for some to come within root colonization distances. Species with low rooting density may well avoid serious infection at populations likely to give many infection points on a more intensively rooting species. If the differing distances over which a spore type can grow to the root can be regarded as analogous to the differences between ion species in their diffusivities through soil, one might expect the number of infection points (or colonization points) on roots (with a low inoculum density) to follow curves similar to those below in Figure 1 [after Barley (11) for ions of high, moderate, and low mobility]. We suggest that models based on those developed for ion uptake by roots in soil (11), including effects of non-uniform distribution of roots (6), will be useful in studying epidemiology of root infections. Such models will need to accommodate factors such as aggregates of propagules rather than their uniform distribution; they will be somewhat different from earlier simplified models of root infection (5).

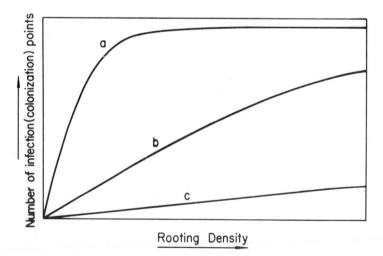

Figure 1   Hypothesized infection (or colonization) points with varying root densities and low (constant) inoculum density, by propagules growing (*a*) over long distances to the root, (*b*) moderate distance, and (*c*) very small distances.

GROWTH RATES OF BACTERIA ON ROOTS

Although growth on surfaces is probably the most common situation in microbial ecology, dynamics of such growth have been little studied. Growth dynamics on root surfaces are so very different from these in rich, stirred laboratory media or even from growth on agar that there is little validity in the extrapolation from studies in laboratory media to growth in soil or on root surfaces. However, a model of bacterial growth on agar surfaces, in which growth occurs only at the edges of the colony (108) may be appropriate to colonies on roots. When isolating from roots, the number of cells actually growing at the time of isolation rather than being dormant is not known. However, growth rates can be obtained experimentally by counting for particular microorganisms at time intervals. This enables a calculation of "generation time," which we suggest as the appropriate measure in most rhizosphere studies as it allows comparison of growth rates between different substrates, e.g. between different species, between different parts of a root, and between root surface and soil, thus overcoming difficulties in extrapolating from one type of surface to another.

Bowen & Rovira (20) examined growth rates of bacteria at three parts of the roots of *P. radiata* seedlings by raising the plants in sterile soil, washing them, and transferring them to nonsterile soil so that all parts of the root, regardless of age, encountered soil microorganisms at the same time. Dilution plate counts of "total" bacteria show generation times of 7.5 hr in the apical centimeter, 9.1 hr in the midroot, and 6.6 hr in the basal portion of the root (9−10 cm from the tip) over the first 2 days. After this logarithmic growth, multiplication was slower, approaching a plateau at approximately 10^{4-5} bacteria/mm^2 of root surface at 4 days. We postulate that this stationary phase is reached when exudation balances maintenance requirements of the microcolonies and that the population will then increase slowly as cells migrate to neighboring parts of the surface not yet colonized. Following the growth of organisms on different parts of roots may be a more direct and relevant method of assessing total energy sources coming from the root than indirect methods such as analyses of root exudates for a range of compounds. We suggest that the characteristics of the root that support microbial growth can be defined by (*a*) an "intensity" factor, i.e. the suitability of the substrate for growth, e.g. exudate, sloughed cells etc., indicated by the generation times of the bacteria, and (*b*) a "capacity" factor which is the total supply of energy-yielding materials from roots, which determines the total biomass on the root.

Table 1 shows the following: 1. The long generation times of organisms on the root (this is much longer than generation times for bacteria in well-aerated, stirred laboratory media); these data agree with calculated generation times of *Rhizobium japonicum* of 9–12 hr on seeds and roots (15). 2. The marked stimulation of growth in the rhizosphere compared with soil (the rhizosphere effect) and the very much faster growth of pseudomonads than of bacilli. Thus, inoculation of roots with *Pseudomonas* spp. is more likely to lead to modification of the rhizosphere microflora than if *Bacillus* spp. are used. 3. The large proliferation of pseudomonads following soil sterilization and a smaller stimulation of bacilli. Sterilized soils are useful for defining gross principles in a highly simplified biological situation, but it should be remembered that sterilization changes the soil physically and chemically and while some organisms, e.g. pseudomonads, are stimulated (110), others are inhibited, e.g. nitrifiers (123) in sterilized soil. In non-sterilized soils quantification of direct microscopy methods using tracers such as fluorescence techniques may be tedious and the study of growth rates can be facilitated by the use of selective media. Such media are often not selective at the strain or species level, and subsequent identification of strains may be carried out on colony characters, reactions on media, or serological reactions (42, 151).

Table 1   Generation times of microorganisms on roots of *Pinus radiata* and in corresponding unplanted soil

Sample	Total viable bacteria[a]	*Pseudomonas* spp.	*Bacillus* spp.
Untreated soil			
Root[b]	7.2[c]	5.2	39
Unplanted soil	14.5	77	>100
Irradiation sterilized-reinoculated soil			
Root	4.8	4.8	25
Unplanted soil	8.0	6.6	26

[a]Dilution plate counts (110). Although the *Bacillus* counts were on spores, subsidiary experiments counting vegetative cells confirm this data.
[b]Counts on 1 cm segments 1–2 cm from root apex.
[c]Generation time in hours.

MIGRATION OF BACTERIA ALONG ROOTS

Rapid migration and occupancy of new root growth would be important in establishing introduced organisms in the rhizosphere. However, the data on migration along roots in natural soils are sparse and conflicting. In a soil conducive to *Azotobacter* growth, inoculation of wheat seed with this organism led to considerably higher populations of *Azotobacter* over the whole root system at six weeks, greater than 20 cm migration (26). In soils not conducive to growth of *Azotobacter,* migration along the root has been poor (26, 112). Soil physical and chemical environments conducive to microbial growth are not the only factors involved, for Ridge (111) was unable to increase the levels of *P. fluorescens* in the rhizosphere of wheat for more than 2-3 cm from the inoculation point in a natural soil while inoculation into sterile soil resulted in spread of the *Pseudomonas* over the whole root. This emphasises the importance of microbial interactions.

Migration rates of *P. fluorescens* along roots of *P. radiata* under varying soil moistures have been studied experimentally in sterile soil by inoculating sterile roots 2.5 cm from the apex and plate counting along the root after two days (18). Growth and migration were rapid at 0.2 bar suction, the cells reaching the apex in two days, but above 0.2 bar growth and migration were severely restricted; this agrees with Griffin (58) on movement of pseudomonads through sandy soil. Such studies in sterilized soil show a potential for migration to keep up with root growth under some conditions.

A number of methods of dispersal of bacteria along roots have been suggested (20): 1. Motility in a film of water on the root, perhaps especially in the cell junctions. 2. Convection in films of water (7) on the root surface. There is no experimental evidence for this. The localization of microcolonies and their embedment in mucigel (30, 124) suggest that such dispersal on roots may be minor if it exists. 3. Being carried along on hyphae or on the root tip and elongating cells. Wong (159) showed that there is a meniscus of water of sufficient thickness between hyphae and surfaces for bacterial migration. Some spread of *B. subtilis* occurred along killed hyphae on a membrane filter buried in grit up to 0.15 bar suction but only up to 0.05 bar with some live hyphae. Wong suggested that migration along mycelial strands would be greater than along individual hyphae because of the water films in the strands.

GROWTH OF FUNGI ALONG ROOTS

While fungi show a preference for growth along cell junctions, the translocation ability of many fungi allows them to grow over areas where exudation is low. This translocation property also allows them some spatial escape from antagonistic bacteria. Four approaches to the study of fungal growth rates along roots have been made: 1. Measurement of growth from a point of inoculation in natural soil; for example, growth rates of *G. graminis* have been recorded from almost zero to 7 mm/day on susceptible species depending on the host, isolate, and soil conditions (100). 2. Taylor & Parkinson (142) measured growth of fungi along broad bean roots from a nonsterile soil into sterile perlite of 3 mm/day, much slower than that of the root which was 9 mm/day, and they concluded that lateral colonization of roots from soil was more significant for young root surfaces than migration of existing fungi along the root. This simple technique, or variations of it, has potential for screening effects of environmental conditions on microbial growth along roots. 3. Robinson & Lucas (114) inoculated *Agropyron repens* and *Agrostis stolonifera* with ^{32}P-labeled hyphae of *G. graminis* and by autoradiography measured a hyphal growth rate of 0.7–0.8 mm/day along the root. Such a tracer technique is applicable only to translocating fungi and assumes no loss of the tracer from the hyphae (which Robinson and Lucas checked). A similar study by Ridge using bacteria (111) failed because of isotope loss and isotope dilution with bacterial growth. Where applicable, however, such a method may be easier to use than fluorescence techniques. 4. Colonization of the rhizoplane of *P. radiata,* growing in sterile soil, by mycorrhizal fungi has been studied by inoculating roots with fungi on washed discs of agar and measuring the length of roots colonized after four weeks, before infection occurred. Bundles of glass fibers were inoculated to provide an inert control (148). The importance of various factors is seen by the following results of lengths of root colonized by *Rhizopogon luteolus* (21): (*a*) Temperature; 27.3 mm at 25°C vs. 4.1 mm at 16°C, (*b*) Moisture: 26.0 mm at 50% field capacity vs 7.0 mm at 25% field capacity, (*c*) Reaction: 26.5 mm at pH 5 vs 0 at pH 8, and (*d*) Soil microorganisms: 22.8 mm with no bacteria vs 19.6 mm with *Bacillus* sp. vs 10.4 mm with *P. fluorescens* isolate A vs 1.2 mm with *P. fluorescens* isolate B. Maximum rates observed were 1.5 mm/day when roots were growing at some 3-4 mm/day. Growth of mycelial strands from mature mycorrhizae along older parts of the roots can be 2-3 times that of individual hyphae (144). The studies above were made to distinguish the effects of various factors on rhizosphere colonization from those on infection processes, a distinction infrequently made in root infection studies. Although all mycorrhizal fungi grew well at 16°C in laboratory media, some failed to colonize roots at this temperature, which is higher than soil temperatures often encountered during the growing season. Such data emphasize the danger of extrapolating from studies on laboratory media to growth in the rhizosphere.

Some rates of root growth per day under conditions of good soil moisture and nutrient status are broad bean, 9 mm (142); small seeded grasses and legumes, 0.5-2 cm, and wheat and barley, 2-7 cm (34); apple, 5 mm (77); *P. radiata,* 3 mm (10°C) to 12 mm (25°C) (17). Thus roots will usually outgrow fungi; hence the importance of lateral colonization by fungi from soil, especially because spores of fungi such as *Pencillium* spp. and *Helminthosporium* spp. can germinate in 1-10 hr when fungistasis is relieved. As root growth slows or ceases, fungal growth can catch up to the root tip. Growth rates of mycorrhizal fungi in and around roots probably play a major part in replacement of some mycorrhizal fungi by other types (55).

Soil physical and chemical conditions have been shown to influence the ectotrophic growth of root pathogens. Garrett (52) considers that the mechanisms responsible for the increased growth of *G. graminis* along roots of wheat seedlings with improved soil aeration and increasing soil pH involve the level of undissociated carbon dioxide in the rhizosphere. Most studies on microbial growth rates along roots have been on young plants; there is scope for study on older roots, and in woody species this could be studied in situ by using methods such as those developed by Smith (138) for root exudate studies from roots of mature trees. Similarly, methods such as split root techniques with roots of the one plant in different soil treatments hold promise for studying interactions between soil conditions and colonization. Not only may this be experimentally advantageous, but it may also start to mimic natural soil heterogeneity.

MICROBIAL INTERACTIONS

Plant pathologists are oriented toward interactions of pathogens and soil microorganisms but soil microbiologists are also concerned with microbial interactions that result in inhibition of symbiotic organisms (such as mycorrhizal fungi and rhizobia) which can be just as disastrous economically as a plant disease. Selection of symbionts for their ability to establish and persist in soil is as important as selecting for effectiveness. Management practices and introduction of antagonistic organisms to control root pathogens should not act against symbionts; for example, stunting of plant growth in fumigated soil has been associated with the elimination of vesicular-arbuscular mycorrhizae (71).

Stimulation of Microbial Growth

Symbiosis between microorganisms in soil and on the root undoubtedly occurs (119) and a mixture of species in microcolonies may act in concert, but these matters have been little studied. A number of records of stimulation of one partner (synergism) occurs, for example, the stimulation of zoospore formation of *P. cinnamomi* by soil organisms (3) and the enhancement of hyphal growth through soils by groups of propagules of *Rhizoctonia solani* (63). Bacteria are stimulated in the fungal mantle of mycorrhizae (86). The length of roots of *Pinus radiata* colonized by the mycorrhizal fungus *Corticium bicolor* was increased from 15.8 mm to 27.2 mm in the presence of a *Bacillus* sp. (144). Similarly, adding a bacteriophagous soil amoeba has stimulated growth of *Pseudomonas* sp. on peas in axenic solution (37). Stimulation can arise directly, or indirectly by an organism increasing root exudation, for example, the stimulation of some root pathogens in association with nematodes. Magyarosy & Hancock (82) found that infection of squash with squash mosaic virus influenced exudation and the laimosphere microflora which in turn gave protection against *Fusarium* stem rot. Beute & Lockwood (13) found that infection of peas with bean yellow mosaic virus increased root rot caused by *Fusarium;* greater exudation from roots of virus-infected plants occurred due to the virus infection increasing cell permeability.

Effects of one microorganism on another also occur in successions on roots; for example, secondary colonizers may live on metabolites from primary colonizers or their decomposition products.

Suppression of Microbial Growth

Although the terms *competition* and *antagonism* are freely used by microbial ecologists, the dividing line between these two phenomena is often not clear. Competition for common substrates is undoubtedly important in microbial growth in soil and around the root. One expression of this competition is the use of diffusible exudates by organisms growing at the root surfaces, thus preventing movement of these exudates further into the soil; another may be the decreased ability of zoospores of *P. cinnamomi* to germinate at the surface of the ectomycorrhiza mantle (90). To control a deleterious organism by competitive means, one may need to consider inoculation with a wide range of fast-growing organisms to swamp out the possible niches the deleterious organism could fill. Successful introduction of desired noninfective organisms would be assisted by introduction of an elite food base (via exudates) which that organism but few others could utilize.

Another possible method of control by competitive means is to use an avirulent organism that is closely related to the pathogen in question, so that it occupies the same ecological niches. Such an approach has been used successfully in the field for crown gall (97) and in laboratory experiments with *Fusarium oxysporum* (76, 94) and *G. graminis* (39, 135, 158). Competition may be involved in the control of *F. oxysporum* on Douglas fir seedlings by *Laccaria laccata* (140).

The isolation from soil of *Bdellovibrio* which parasitizes *Rhizobium* (105) and TEM observations in the rhizosphere of clover showing *Bdellovibrio*-like cells in association with bacteria similar in size and structure to *Rhizobium* sp. (50) suggest that parasitism occurs in the rhizospheres of field-grown plants. De la Cruz & Hubbell (40) have recorded control of *Macrophomina phaseolina* by a hyperparasite (a basidiomycete), and *Trichoderma viride* has been found to parasitize *Poria weirii* (96). SEM studies on *G. graminis* in suppressive soils frequently show heavy colonization of hyphae by bacteria (125) but it is difficult to know whether this is parasitism or luxuriant growth of particular organisms on fungal cells killed by antibiosis.

Antagonism can occur by antibiosis or indirectly by the change of physical and chemical microenvironments by the antagonist. Examples of the former in soil are the attack on resting propagules in soil, decreasing their ability to grow effectively through soil after germination, lysis of fungus germlings (4), and the inhibition of zoospore formation of *P. cinnamomi* by some organisms in suppressive soils (22). The inhibition of fungal growth in the rhizosphere by pseudomonads, for example, *G. graminis* (35), *P. cinnamomi* (23, 83), and *R. luteolus* (21), appears to be due largely to antibiosis rather than competition, as some closely related pseudomonads growing equally well in the rhizosphere do not show inhibition. The production of antibiotics by pseudomonads in laboratory media is well known; however, inhibition of test fungi in laboratory media depends very much on the medium and there is a poor correlation between inhibition on agar and in the rhizosphere (24, 83, 122). A similar situation occurs with actinomycete antibiosis toward *P. cinnamomi* (83). Inhibition of *P. cinnamomi* attack on mycorrhizae formed by *Leucopaxillus cerealis* var. *piceina* is due to antibiotics (91) and probably volatile terpenes (74).

Antagonisms to rhizobia in the field are well recognized (62). Cass-Smith & Holland (31) overcame problems in establishing clover in newly cleared land by fumigating the soil, thus removing indigenous organisms which prevented nodulation by *Rhizobium*. Within a species, e.g. *R. trifolii*, there are strains that are superior in their ability to migrate and colonize roots; this property has been referred to as *incursiveness* (61) or *saprophytic competence* (32).

Knowledge of the growth of symbiotic organisms in the rhizosphere has been used to facilitate their introduction via alternate crops. Diatloff (41) ensured high levels of rhizobia for soybeans by inoculation of the wheat crop preceding soybean. However, marked differences occur between rhizobia in their growth in the rhizosphere; for example, *R. trifolii* colonized the rhizosphere of *Trifolium, Medicago,* and *Poa* while *R. meliloti* colonized only *Medicago* (113). Thus it would be easier to introduce *R. trifolii* by alternate crops than *R. meliloti*. Tuzimura & Watanabe (149) showed that *R. trifolii* reached higher numbers in the rhizospheres of legumes than in the rhizospheres of nonlegumes but that in these the populations still exceeded those in unplanted soil. The principle of using alternate crops to build up inocula of desirable organisms may have great, but as yet unexplored, potential for introducing organisms for biological control of diseases. The reduction in disease caused by *P. cinnamomi* in forests (83) and orchards (55) following underplanting with legumes may be due to stimulation of antagonists to the pathogen.

A number of cases exist where the antagonism of one organism for another can be almost completely nullified by a third organism which has little effect on the other two alone (144, 154); this may be due to the second organism absorbing antibiotics formed by the aggressive organism or modifying a deleterious chemical or physical microenvironment. Such phenomena indicate the precarious balance of microbial interactions and suggest that control of the rhizosphere must work in the context of biological balance (4).

Atkinson, Neal & Larson (2) cite an instance in which resistance of wheat to *Cochliobolus sativus* could not be conferred solely by stimulation of antagonists in the rhizosphere. They showed that the rhizosphere population differed between host genotypes—resistant varieties supported lower populations of bacteria than did susceptible varieties—and that substitution of a single resistance chromosome into an otherwise susceptible variety changed the rhizosphere microflora (2). This demonstration that the numbers and types of rhizosphere microorganisms are under host genetic control makes it possible to contemplate plant breeding programs in which the rhizosphere population is manipulated to alter the susceptibility of the plant to root disease.

There is little doubt that the change of mycorrhizal types in forest soils with age (75) is due largely to change in the overall microbial balance. In a similar way that ammonium fertilization of wheat enhances resistance to *G. graminis* by encouraging suites of antagonistic microorganisms (137), it may well be that the difficulty in establishing some types of endomycorrhizal fungi in acid soils in the absence of lime (95) is due largely to microbial antagonism. Conversely the domination of mycorrhizal formation by *Pisolithus tinctorius* on pine species in inhospitable sites, e.g. mine wastes, in contrast to its low incidence in more fertile soils, may be due to sensitivity to other organisms present in fertile soils but absent from mine wastes.

ROOT EXUDATES

In view of the comprehensive reviews on root exudates (21, 118, 121, 126, 132, 139) our comments are confined to recent studies with direct relevance to microbial colonization of roots in soil. Exudation of organic materials from cereal roots is doubled when the roots grow in a solid medium rather than in solution (9); thus many exudate studies with solution-grown plants have little quantitative relevance to plants growing in soil. Exudation increased with greater mechanical resistance, which may be a factor in increased root rot of beans by *Fusarium* in compacted soil (28).

Wheat and barley labeled with ^{14}C-carbon dioxide release into sandy soil some 8 and 15% of their total plant carbon under sterile and nonsterile conditions respectively (10); this is 5—20 times that reported for solution-grown plants. Much of the difference between sterile and nonsterile roots was due to carbon dioxide considered to be of microbial origin. Contrary to earlier studies on exudates from plants grown in solution, Martin (88) found that over 60% of the labeled, water-soluble material from nonsterile roots has a molecular weight in excess of 1000; in more recent studies Martin (89) has found less high molecular weight material from sterile roots in soil, indicating that these compounds originate from both the plants and microorganisms. The proliferation of bacteria on wheat roots infected with *G. graminis* (125) is evidence of increased leakage of organic compounds following infection, which also occurs in infected leaves (29). Regardless of increased release of assimilate by lysis and colonization of cortical cells (49, 54), many noninfecting microorganisms produce auxins or gibberellins known to affect membrane permeability (64, 160) which could increase the food supply in the immediate area of microbial colonies without cell lysis.

Reid (109) showed that exudation was increased with water suctions above 11.9 bar at which fungi but not bacteria can grow. Thus, water stress would modify the balance of bacteria·and fungi in the rhizosphere; this must be taken into account in our attempts at biological control of fungal root pathogens with bacteria.

The greater diffusion of volatile compounds through soil could make volatile root exudates (73, 74, 79) significant in the ability of roots to stimulate, attract, or inhibit microorganisms. A two- to eightfold increase in the volatile terpenes from mycorrhizal compared with uninfected roots (73) and the inhibition of *P. cinnamomi* and *Fomes annosus* by terpenes (74) suggest their involvement in the resistance of mycorrhizae to root pathogens. A more common volatile compound from roots, however, is carbon dioxide, and the concentration gradient, from roots into soil may affect the growth of microorganisms (52, 58).

Pearson & Parkinson (106) showed that the root tip was a major site of exudation of ninhydrin reacting material but later studies using ^{14}C-labeled assimilate show that diffusible exudate is released along the full length of roots while nondiffusible carbon compounds come from the root tip region (120). More precise methods are necessary to measure exudation at the cellular level; this precision would help us to understand the dynamics of root colonization. Two such techniques are (*a*) microautoradiography of roots of plants assimilating ^{14}C-carbon dioxide and (*b*) direct observation of colonization patterns on roots inoculated uniformly with bacteria (18). Inoculation of roots of *Helianthus annuus* with nutritional mutants of *Neurospora crassa* showed that threonine and asparagine came from the root tip and that leucine, valine, phenylalanine, and glutamic acid came from the root hair zone (51). Until now most root exudate studies have concentrated on analysis for readily assayed nutrients rather than using microbial growth as a more direct and sensitive assay; the emphasis upon the former approach has been criticized (21).

POPULATION BIOLOGY CONCEPTS

We are working toward concepts of how roots are colonized and how the biology of the rhizosphere can be influenced. Modification of root exudates is one strategy, and agronomic manipulation is another, for example, ammonium vs nitrate fertilization (137). In the definition of hypotheses (i.e. models) on how microorganisms colonize and interact on roots it should be remembered that models are tools, not proof, and that it behooves the modeler to test his model, not assume its truth. Most existing mechanistic models in ecology and eco-physiology are simplistic and have shortcomings in reality, especially as they are often deterministic rather than stochastic. However, conceptual and simple mathematical models could be extremely useful even now in microbial ecology for their heuristic value, defining some of the complexities of a system, and in helping to give a ranking to the relative importance of microbial and en-vironmental properties in establishing organisms in the rhizosphere. Their predictive value for detailed composition of the rhizosphere would be low. This is to be expected as microorganisms on roots behave more as a collection of spatial-ly separate communities, determined largely by the organisms the root touches or approaches, rather than as one large interacting community. The simple data we present suggest that it is possible to superimpose noninfecting rhizosphere or-ganisms upon an existing microflora, especially with fast-growing organisms and under high soil moisture conditions; obviously inoculated rhizobia and mycorrhizal fungi establish in the rhizosphere before they infect.

Rapid migration of organisms to newly formed root surfaces would aid their establishment. High migration rates are probably closely linked to high growth rates; high growth rate is a characteristic of general biological importance in primary colonization (81). As more members of other species are present for ini-tial colonization, individual species tend to be more restricted to particular habitats, that is, they go from their "fundamental" niche to their "realized" niche (150). Late colonizers on the root probably establish by growing on products from senescing cells and their associated organisms, on substrates not used by primary colonizers, on exudates from sites not already colonized, or by locally in-creasing the permeability of the plant cell; these hypotheses, however, await ex-perimental evaluation.

Microbial interactions have a large bearing on the establishment or control of an organism in the rhizosphere. Microbial ecologists have been slow to ap-preciate the usefulness of population biology concepts developed mostly in other fields (156). Some of the biological properties in other systems are inapplicable or only partly applicable to microorganisms on roots, for example, searching for food, complicated life cycles, several trophic levels in a food web. Also we cannot identify microorganisms and define variation in the microenvironment as easily as zoologists and botanists recognize different animals, plants, and their environ-ment. However, techniques in rhizosphere microbiology are sufficiently advanced to allow examination of population principles with mixtures of morphologically recognizable microorganisms. There is a need to understand principles of colonization in addition to selecting organisms on an ad hoc basis for their vigor and persistence under a range of conditions. Many statistical methods of analyses are available; for example, competition can be analyzed by replacement series studies, and diallel analysis (156) already developed in other fields.

Species interactions are central to consideration of the population biology of the rhizosphere. Microbiologists tend to regard organisms that coexist on the one root as occupying different niches, but these different organisms may be occupying the same niche, viz sites with identical properties, but at different parts of the root. It is not helpful to merely conclude that when two species coexist they must be occupying different niches; indeed two normally mutually exclusive predators can coexist in the presence of a prey (36) and this may have microbial parallels. It is more helpful to ask the extent to which species can overlap in their niches and still coexist. Theoretical treatments (92) suggest that in a strictly unvarying environment, stability of an ecosystem sets no limit to the degree of overlap, short of complete congruence, but in a randomly fluctuating environment (stochastic), stability demands a minimum distance between species in the resource spectrum. Contrary to the conventional wisdom that complexity begets stability, May (92) suggested that high complexity is a fragile thing permitted by environmental steadiness. We conclude that small physical, chemical, or biological changes may markedly affect species abundance in the rhizosphere. Roughgarden (115) produced a competition model indicating a community with many tightly packed species; those with narrow resource utilization ($\equiv$ specialists) are less susceptible to invasion than one of fewer more generalistic species and this is consistent with the views of Cody (33) on the use of resources by specialists and generalists and their coexistence. Other studies have suggested that fluctuations of a species increase as the proportion of the environment that it can utilize increases and that the stability of a species decreases as the number of species attacking it increases (153). *If* this applies to microorganisms on roots, the most appropriate strategy of inoculating to combat a "generalist" species would be to use a number of species not competing with each other and occupying different niches, but with each species competing with or attacking the pathogen.

These are possibilities which must be studied experimentally. We believe the time has arrived for the application of such basic population studies to rhizosphere microbiology.

Literature Cited

1. Anderson, J. R., Slinger, J. M. 1975. Europium chelate and fluorescent brightener staining of soil propagules and their photomicrographic counting. I. Methods. II. Efficiency. *Soil Biol. Biochem.* 7:205−15
2. Atkinson, T. G., Neal, J. L., Larson, R. I. 1975. Genetic control of the rhizosphere of wheat. ln *Biology and Control of Soil-Borne Plant Pathogens,* ed. G. W. Bruehl, 116−22. St. Paul, Minn.: Am. Phytopathol. Soc. 216 pp.
3. Ayers, W. A., Zentmyer, G. A. 1971. Effect of soil solution and two soil pseudomonads on sporangium production by *Phytophthora cinnamomi. Phytopathology* 61:1188–93
4. Baker, K. F., Cook, R. J. 1974. *Biological Control of Plant Pathogens.* San Francisco: Freeman. 435 pp.
5. Baker, R. 1968. Mechanisms of biological control of soil-borne pathogens. *Ann. Rev. Phytopathol.* 6:263−94
6. Baldwin, J. P., Tinker, P. B., Nye, P. H. 1972. Uptake of solutes by multiple root systems from soil. II. The theoretical effects of rooting density and pattern on uptake of nutrients from soil. *Plant Soil* 36:693−708
7. Bandoni, R. J., Koske, R. E. 1974. Monolayers and microbial dispersal. *Science* 183:1079−80
8. Barber, D. A., Bowen, G. D., Rovira, A. D. 1976. Effects of microorganisms on the absorption and distribution of phosphate in barley. *Aust. J. Plant Physiol.* In press
9. Barber, D. A., Gunn, K. B. 1974. The effect of mechanical forces on the exudation of organic substances by the roots of cereal plants grown under sterile conditions. *New Phytol.* 73:39−45
10. Barber, D. A., Martin, J. K. 1976. The release of organic substances by cereal roots into soil. *New Phytol.* 76:69−80
11. Barley, K. P. 1970. The configuration of the root system in relation to nutrient uptake. *Adv. Agron.* 22:159−201
12. Barton, R. 1957. Germination of zoospores of *Pythium mamillatum* in response to exudates from living seedlings. *Nature* 180:613–14
13. Beute, M. K., Lockwood, J. L. 1968. Mechanism of increased root rot in virus-infected peas. *Phytopathology* 58:1643−51
14. Bowen, G. D. 1961. The toxicity of legume seed diffusates toward rhizobia and other bacteria. *Plant Soil* 15:155−65
15. Bowen, G. D. 1961. *Some aspects of the symbiotic relationships between Centrosema pubescens and strains of Rhizobium Benth.* MSc thesis. Univ. Queensland, Brisbane. 148 pp.
16. Bowen, G. D. 1968. Phosphate uptake by mycorrhizas and uninfected roots of *Pinus radiata* in relation to root distribution. *Trans. 9th. Int. Congr. Soil Sci., Adelaide* 2:219−28
17. Bowen, G. D. 1970. Effects of soil temperature on root growth and on phosphate uptake along *Pinus radiata* roots. *Aust. J. Soil Res.* 8:31−42
18. Bowen, G. D. Unpublished
19. Bowen, G. D., Rovira, A. D. 1968. The influence of micro-organisms on growth and metabolism of plant roots. In *Root Growth,* ed. W. J. Whittington, 170−201. London: Butterworths. 450 pp.
20. Bowen, G. D., Rovira, A. D. 1973. Are modelling approaches useful in rhizosphere biology? *Bull. Ecol. Res. Commun. Stockholm* 17:443−50
21. Bowen, G. D., Theodorou, C. 1973. Growth of ectomycorrhizal fungi around seeds and roots. In *Ectomycorrhizae: Their Ecology and Physiology,* ed. G. C. Marks, T. T. Kozlowski, 107−50. New York: Academic. 444 pp.
22. Broadbent, P., Baker, K. F. 1974. Association of bacteria with sporangium formation and breakdown of sporangia in *Phytophthora* spp. *Aust. J. Agric. Res.* 25:139−45
23. Broadbent, P., Baker, K. F. 1974. Behaviour of *Phytophthora cinnamomi* in soils suppressive and conducive to root rot. *Aust. J. Agric. Res.* 25:121−37
24. Broadbent, P., Baker, K. F., Waterworth, Y. 1971. Bacteria and actinomycetes antagonistic to fungal root pathogens in Australian soils. *Aust. J. Biol. Sci.* 24:925−44
25. Brock, T. D. 1971. Microbial growth rates in nature. *Bacteriol. Rev.* 35:39−58
26. Brown, M. E., Burlingham, S. K., Jackson, R. M. 1962. Studies on *Azotobacter* species in soil. II. Populations of *Azotobacter* in the rhizosphere and effects of artificial inoculation. *Plant Soil* 17:320−32
27. Brown, M. E., Jackson, R. M., Burlingham, S. K. 1967. Growth and

effects of bacteria introduced into soil. In *The Ecology of Soil Bacteria,* ed. T. R. C. Gray, D. Parkinson, 531– 51. Liverpool: Liverpool Univ. Press. 681 pp.

28. Burke, D. W., Holmes, L. D., Barker, A. W. 1972. Distribution of *Fusarium solani* f. sp. *phaseoli* and bean roots in relation to tillage and soil compaction. *Phytopathology* 62:550–54

29. Burkowicz, A., Goodman, R. M. 1969. Permeability alterations induced in apple leaves by virulent and avirulent strains of *Erwinia amylovora. Phytopathology* 59:314–18

30. Campbell, R., Rovira, A. D. 1973. The study of the rhizosphere by scanning electron microscopy. *Soil Biol. Biochem.* 5:747–52

31. Cass-Smith, W. P., Holland, A. A. 1958. The effect of soil fungicides and fumigants on the growth of subterranean clover on new light land. *J. Dep. Agric. West. Aust.* 7:225–31

32. Chatel, D. L., Greenwood, R. M., Parker, C. A. 1968. Saprophytic competence as an important character in the selection of *Rhizobium* for inoculation. *Trans. 9th. Int. Congr. Soil Sci.* 2:65–73

33. Cody, M. L. 1974. Optimization in ecology. *Science* 183:1156–64

34. Cohen, V., Tadmor, N. H. 1969. Effects of temperature on the elongation of seedling roots of some grasses and legumes. *Crop Sci.* 9:189–92

35. Cook, R. J., Rovira, A. D. 1976. The role of bacteria in the biological control of *Gaeumannomyces graminis* by suppressive soils. *Soil Biol. Biochem.* 8:In press

36. Cramer, N. F., May, R. M. 1972. Interspecific competition, predation and species diversity: a comment. *J. Theor. Biol.* 34:289–93

37. Darbyshire, J. F., Greaves, M. P. 1971. The invasion of pea roots, *Pisum sativum* L., by soil microorganisms, *Acanthamoeba palestinensis* (Reich) and *Pseudomonas* sp. *Soil Biol. Biochem.* 3:151–55

38. Dart, P. J., Mercer, F. V. 1964. The legume rhizosphere. *Arch. Microbiol.* 47:344–78

39. Deacon, J. W. 1976. Biological control of the take-all fungus, *Gaeumannomyces graminis,* by *Phialophora radicicola* and similar fungi. *Soil Biol. Biochem.* 8:In press

40. De la Cruz, R. E., Hubbell, D. H. 1975. Biological control of the charcoal root rot fungus *Macrophomina*

phaseolina on slash pine seedlings by a hyperparasite. *Soil Biol. Biochem.* 7:25–30

41. Diatloff, A. 1969. The introduction of *Rhizobium japonicum* to soil by seed inoculation of non-host legumes and cereals. *Aust. J. Exp. Agric. Anim. Husb.* 9:357–60

42. Dudman, W. F., Brockwell, J. 1968. Ecological studies of root-nodule bacteria introduced into field environments. I. A survey of field performance of clover inoculants by gel immune diffusion serology. *Aust. J. Agric. Res.* 19:739–47

43. Fitzell, R. D., Evans, G. Unpublished

44. Foster, R. C. 1962. *Cell wall structure and growth.* PhD thesis. Univ. Leeds, Leeds. 152 pp.

45. Foster, R. C. Unpublished

46. Foster, R. C., Marks, G. C. 1966. The fine structure of mycorrhizas of *Pinus radiata* D. Don. *Aust. J. Biol. Sci.* 19:1027–38

47. Foster, R. C., Marks. G. C. 1967. Observations on the mycorrhizas of forest trees. II. The rhizosphere of *Pinus radiata* D. Don. *Aust. J. Biol. Sci.* 20:915–26

48. Foster, R. C., Rovira, A. D. 1973. The rhizosphere of wheat roots studied by electron microscopy of ultra-thin sections. *Bull. Ecol. Res. Commun. Stockholm* 17:93–102

49. Foster, R. C., Rovira, A. D. 1976. Ultrastructure of wheat rhizosphere. *New Phytol.* 76:343–52

50. Foster, R. C., Rovira, A. D. 1976. The ultrastructure of the rhizosphere of *Trifolium subterraneum* L. *Can. J. Microbiol.* In press

51. Frenzel, B. 1960. Zur ätiologie der anreicherung von aminosäuren und amiden im wurzelraum of *Helianthus annus* L. Ein beitrag zur klärung der probleme der rhizosphäre. *Planta* 55:169–207

52. Garrett, S. D. 1970. *Pathogenic Root-Infecting Fungi.* Cambridge: Cambridge Univ. Press. 282 pp.

53. Gray, T. R. G., Baxby, P., Hill, I. R., Goodfellow, M. 1967. Direct observation of bacteria in soil. In *The Ecology of Soil Bacteria,* ed. T. R. G. Gray, D. Parkinson, 171–97. Liverpool: Liverpool Univ. Press. 681 pp.

54. Greaves, M. P., Darbyshire, J. F. 1972. The ultrastructure of the mucilaginous layer on plant roots. *Soil Biol. Biochem.* 4:443–49

55. Greenhalgh, F. C. Unpublished

56. Greig-Smith, P. 1961. Data on pattern

within plant communities. I. The analysis of pattern. *J. Ecol.* 49: 695−702

57. Greig-Smith, P. 1964. *Quantitative Plant Ecology.* London: Butterworths. 256 pp. 2nd ed.

58. Griffin, D. M. 1972. *Ecology of Soil Fungi.* London: Chapman & Hall. 187 pp.

59. Guckert, A., Breisch, H., Reisinger, O. 1975. Interface sol-racine. I. Etude au microscope electronique des relations mucigel-argile-microorganismes. *Soil Biol. Biochem.* 7:241−50

60. Hamdi, Y. A. 1971. Soil-water tension and the movement of rhizobia. *Soil Biol. Biochem.* 3:121−26

61. Harris, J. R. 1954. Rhizosphere relationships of subterranean clover. I. Interactions between strains of *Rhizobium trifolii. Aust. J. Agric. Res.* 5:247−70

62. Hely, F. W., Bergersen, F. J., Brockwell, J. 1957. Microbial antagonism in the rhizosphere as a factor in the failure of inoculation of subterranean clover. *Aust. J. Agric. Res.* 8:24−44

63. Henis, Y., Ben-Yephet, Y. 1970. Effect of propagule size of *Rhizoctonia solani* on saprophytic growth, infectivity, and virulence on bean seedlings. *Phytopathology* 60:1351−56

64. Higinbotham, N. 1968. Cell electrotropential and ion transport in higher plants. In *Transport and Distribution of Matter in Cells of Higher Plants,* ed. K. Mothes et al, 167-77. Berlin: Akad-Verlag. 215 pp.

65. Ho, H. H. 1969. Notes on the behaviour of *Phytophthora megasperma* var *sojae* in soil. *Mycologia* 61:835−38

66. Hodges, C. S. 1969. Modes of infection and spread of *Fomes annosus. Ann. Rev. Phytopathol.* 7:247-66

67. Holland, A. A., Fulcher, R. G. 1971. A study of wheat roots infected with *Ophiobolus graminis* (Sacc.) using fluorescence and electron microscopy. *Aust. J. Biol. Sci.* 24:819−23

68. Jenny, H., Grossenbacher, K. 1963. Root-soil boundary zones as seen in the electron microscope. *Proc. Soil. Sci. Soc. Am.* 27:273−77

69. Jones, D., Griffiths, E. 1964. The use of thin soil sections for the study of soil micro-organisms. *Plant Soil* 20:232−40

70. Juniper, B. E., Roberts, R. M. 1966. Polysaccharide synthesis and the fine structure of root cells. *J. R. Microsc.*

Soc. 85:63−72

71. Kleinschmidt, G. D., Gerdemann, J. W. 1972. Stunting of citrus seedlings in fumigated nursery soils related to the absence of endomycorrhizae. *Phytopathology* 62:1447−53

72. Kliejunas, J. T., Ko, W. H. 1974. Effect of motility of *Phytophthora palmivora* zoospores on disease severity in papaya seedlings and substrate colonization in soil. *Phytopathology* 64:426−28

73. Krupa, S., Fries, N. 1971. Studies on ectomycorrhizae of pine. I. Production of volatile organic compounds. *Can. J. Bot.* 49:1425−31

74. Krupa, S., Nylund, J. E. 1972. Studies on ectomycorrhizae of pine. III. Growth inhibition of two root pathogenic fungi by volatile organic constituents of ectomycorrhizal root systems of *Pinus sylvestris* L. *Eur. J. For. Pathol.* 2:88−94

75. Lamb, R. J. 1974. *Autoecology of ectomycorrhizal fungi of Pinus.* PhD thesis. Univ. New England, Armidale, NSW. 146 pp.

76. Langton, F. A. 1969. Interactions of the tomato with two formae speciales of *Fusarium oxysporum. Ann. Appl. Biol.* 62:413−27

77. Last, F. T. 1971. The role of the host in the epidemiology of some nonfoliar pathogens. *Ann. Rev. Phytopathol.* 9:341−62

78. Lewis, D. H. 1973. Concepts in fungal nutrition and the origin of biotrophy. *Biol. Rev. Cambridge Philos. Soc.* 48: 261−78

79. Linderman, R. G., Gilbert, R. G. 1975. Influence of volatiles of plant origin on soil-borne plant pathogens. See Ref. 2, 90−99

80. Luna, L. V., Hine, R. B. 1964. Factors influencing saprophytic growth of *Pythium aphamidermatum* in soil. *Phytopathology* 54:955−59

81. MacArthur, R. H., Wilson, E. O. 1967. *Theory of Island Biogeography.* Princeton: Princeton Univ. Press. 203 pp.

82. Magyarosy, A. C., Hancock, J. G. 1974. Association of virus-induced changes in laimosphere microflora and hypocotyl exudation with protection to *Fusarium* stem rot. *Phytopathology* 64:994−1000

83. Malajczuk, N., Unpublished

84. Malajczuk, N., Bowen, G. D. Unpublished

85. Malajczuk, N., McComb, A. J., Parker, C. A. 1975. An immuno-

fluorescence technique for detecting *Phytophthora cinnamomi* Rands. *Aust. J. Bot.* 23:289–309

86. Marks, G. C., Foster, R. C. 1973. Structure, morphogenesis, and ultrastructure of ectomycorrhizae. In *Ectomycorrhizae—Their Ecology and Physiology,* ed. G C. Marks, T. T. Kozlowski, 1–41. New York: Academic. 444 pp.

87. Marshall, K. C., Cruikshank, R. H. 1973. Cell surface hydrophobicity and the orientation of certain bacteria at interfaces. *Arch. Microbiol.* 91:29–40

88. Martin, J. K. 1975. ^{14}C-labelled material leached from the rhizosphere of plants supplied continuously with $^{14}CO_2$. *Soil Biol. Biochem.* 7:395–99

89. Martin, J. K. Unpublished.

90. Marx, D. H. 1969. Ectomycorrhizae as biological deterrents to pathogenic root infections. In *Mycorrhizae,* ed. E. Hacskaylo, 81–96. Washington DC: GPO. 255 pp.

91. Marx, D. H., Davey, C. B. 1969. The influence of ectotrophic mycorrhizal fungi on the resistance of pine roots to pathogenic infections. III. Resistance of aseptically formed mycorrhizae to infection by *Phytophthora cinnamomi.* *Phytopathology* 59:549–58

92. May, R. M. 1974. On the theory of niche overlap. *Theor. Popul. Biol.* 5: 297–332

93. Merriman, P. R., Price, R. D., Kollmorgen, J. F., Piggott, T., Ridge, E. H. 1971. Effect of seed inoculation with *Bacillus subtilis* and *Streptomyces griseus* on the growth of cereals and carrots. *Aust. J. Agric. Res.* 25:219–26

94. Meyer, J. A., Maraite, A. 1971. Multiple infection and symptom mitigation in vascular wilt diseases. *Trans. Br. Mycol. Soc.* 57:371–77

95. Mosse, B., Hayman, D. S., Arnold, D. J. 1973. Plant growth responses to vesicular-arbuscular mycorrhiza. V. Phosphate uptake by three plant species from P-deficient soils labelled with ^{32}P. *New Phytol.* 72:809–15

96. Nelson, E. E. 1964. Some probable relationships of soil fungi and zone lines to survival of *Poria weirii* in buried wood blocks. *Phytopathology* 54:120–21

97. New, P. B., Kerr, A. 1972. Biological control of crown gall: field measurements and glasshouse experiments. *J. Appl. Bacteriol.* 35:279–87

98. Newman, E. I. 1966. A method of estimating the total length of root in a sample. *J. Appl. Ecol.* 3:139–45

99. Newman, E. I., Bowen, H. J. 1974. Patterns of distribution of bacteria on root surfaces. *Soil Biol. Biochem.* 6:205–9

100. Nilsson, H. E. 1969. Studies of root and foot rot disease of cereals and grasses. I. On resistance to *Ophiobolus graminis* Sacc. *Ann. Agric. Coll. Sweden* 35:275–807

101. Oades, M. The mucilaginous layer of roots: a review. In preparation

102. Old, K. M., Nicholson, T. H. 1975. Electron microscopical studies of the microflora of roots of sand dune grasses. *New Phytol.* 74:51–58

103. Olsen, S. R., Kemper, W. D. 1968. Movement of nutrients to plant roots. *Adv. Agron.* 20:91–151

104. Papendick, R. I., Campbell, G. S. 1975. Water potential in the rhizosphere and plant and methods of measurement and experimental control. See Ref. 2, 39–49

105. Parker, C. A., Grove, P. L. 1970. *Bdellovibrio bacteriovorus* parasitizing *Rhizobium* in Western Australia. *J. Appl. Bacteriol.* 33:253–55

106. Pearson, R., Parkinson, D. 1961. The sites of excretion of ninhydrin-positive substances by broad bean seedlings. *Plant Soil* 13:391–96

107. Pentland, G. D. 1967. The effect of soil moisture on the growth and spread of *Coniophora puteana* under laboratory conditions. *Can. J. Bot.* 45:1899–1906

108. Pirt, S. J. 1967. A kinetic study of the mode of growth of surface colonies of bacteria and fungi. *J. Gen. Microbiol.* 47:181–97

109. Reid, C. P. P. 1974. Assimilation, distribution, and root exudation of ^{14}C by ponderosa pine seedlings under induced water stress. *Plant Physiol.* 54: 44–49

110. Ridge, E. H. 1976. Studies of soil fumigation. II. Effects on bacteria. *Soil Biol. Biochem.* 8:In press

111. Ridge, E. H. Unpublished

112. Ridge, E. H., Rovira, A. D. 1968. Microbial inoculation of wheat. *Trans. 9th Int. Congr. Soil Sci. Adelaide* 3:473–81

113. Robinson, A. C. 1967. The influence of host on soil and rhizosphere populations of clover and lucerne root nodule bacteria in the field. *J. Aust. Inst. Agric. Sci.* 33:207–9

114. Robinson, R. K., Lucas, R. L. 1963. The use of isotopically labelled mycelium to investigate the host range and rate of spread of *Ophiobolus graminis.*

New Phytol. 62:50−52

115. Roughgarden, J. 1974. Species packing and the competition function with illustrations from coral reef fish. *Theor. Pop. Biol.* 5:163−86

116. Rovira, A. D. 1956. A study of the development of the root surface microflora during the initial stages of plant growth. *J. Appl. Bacteriol.* 19:72−79   •

117. Rovira, A. D. 1965. The effect of *Azotobacter, Bacillus* and *Clostridium* on the growth of wheat. In *Plant Microbe Relations,* ed. J. Macura, V. Vancura, 193−200. Prague: Czech. Acad. Sci. 333 pp.

118. Rovira, A. D. 1965. Plant root exudates and their influence upon soil microorganisms. In *Ecology of Soilborne Plant Pathogens—Prelude to Biological Control,* ed. K. F. Baker, W. C. Snyder, 170−86. Berkeley: Univ. of Calif. Press. 571 pp.

119. Rovira, A. D. 1965. Interactions between plant roots and soil microorganisms. *Ann. Rev. Microbiol.* 19:241−66

120. Rovira, A. D. 1969. Diffusion of carbon compounds away from wheat roots. *Aust. J. Biol. Sci.* 22:1287−90

121. Rovira, A. D. 1969. Plant root exudates. *Bot. Rev.* 35:35−57

122. Rovira, A. D. Unpublished

123. Rovira, A. D., Bowen, G. D. 1969. The use of radiation sterilized soil to study nitrogen nutrition of wheat. *Aust. J. Soil Res.* 2:57−65

124. Rovira, A. D., Campbell, R. 1974. Scanning electron microscopy of micro-organisms on the roots of wheat. *Microbiol. Ecol.* 1:15−23

125. Rovira, A. D., Campbell, R. 1975. A scanning electron microscope study of interactions between micro-organisms and *Gaeumannomyces graminis* (Syn. *Ophiobolus graminis*) on wheat roots. *Micro. Ecol.* 3:177−85

126. Rovira, A. D., Davey, C. B. 1974. Biology of the rhizosphere. In *The Plant Root and its Environment,* ed. E. W. Carson, 153−204. Charlottesville, Va: Univ. Press of Virginia. 691 pp.

127. Rovira, A. D., Newman, E. I., Bowen, H. J., Campbell, R. 1974. Quantitative assessment of the rhizoplane microflora by direct microscopy. *Soil Biol. Biochem.* 6:211−16

128. Rovira, A. D., Sands, D. C. 1971. Fluorescent pseudomonads—a residual component in the soil microflora? *J. Appl. Bacteriol.* 34:253−59

129. Ruddick, S. M., Williams, S. T. 1972.

Studies on the ecology of actinomycetes in soil. V. Some factors influencing the dispersal and adsorption of spores in soil. *Soil Biol. Biochem.* 4:93−103

130. Schmidt, E. L. 1973. Fluorescent antibody techniques for the study of microbial ecology. *Bull. Ecol. Res. Commun. Stockholm* 17:67−76

131. Schmidt, E. L., Biersbrock, J. A., Bohlool, B. B., Marx, D. H. 1974. Study of mycorrhizae by means of fluorescent antibody. *Can. J. Microbiol.* 20:137−39

132. Schroth, M. N., Hildebrand, D. C. 1964. Influence of plant exudates on root-infecting fungi. *Ann. Rev. Phytopathol.* 2:101−32

133. Short, G. E., Lacy, M. L. 1974. Germination of *Fusarium solani* f. sp. *pisi* chlamydospores in the spermosphere of pea. *Phytopathology* 64:558−62

134. Sing, V. O., Bartnicki-Garcia, S. 1975. Adhesion of *Phytophthora palmivora* zoospores: electron microscopy of cell attachment and cyst wall fibril formation. *J. Cell Sci.* 18:123−32

135. Sivasithamparam, K. 1975. *Phialophora* and *Phialophora*-like fungi occurring in the root region of wheat. *Aust. J. Bot.* 23:193−212

136. Skinner, M. F., Bowen, G. D. 1974. The penetration of soil by mycelial strands of ectomycorrhizal fungi. *Soil Biol. Biochem.* 6:57−61

137. Smiley, R. W., Cook, R. J. 1973. Relationship between take-all of wheat and rhizosphere pH in soils fertilized with ammonium *vs.* nitrate-nitrogen. *Phytopathology* 63:882−90

138. Smith, W. H. 1970. Root exudates of seedling and mature sugar maple. *Phytopathology* 60:701−3

139. Smith, W. H. 1973. Tree root exudates and forest soil ecosystems: exudate chemistry, biological significance and alteration by stress. In *The Belowground Ecosystem: A Synthesis of Plant Associated Processes,* ed. J. K. Marshall, Vol. 2. Stroudsburg: Dowden, Hutchinson & Ross. In press

140. Stack, R. W., Sinclair, W. A. 1975. Protection of douglas-fir seedlings against *Fusarium* root rot by a mycorrhizal fungus in the absence of mycorrhiza formation. *Phytopathology* 65:468−72

141. Stanghellini, M. E., Hancock, J. G. 1971. Radial extent of the bean spermosphere and its relation to the behavior of *Pythium ultimum*. *Phytopathology* 61:165−68

142. Taylor, G. S., Parkinson, D. 1961. The growth of saprophytic fungi on root surfaces. *Plant Soil* 15:261—67
143. Theodorou, C. 1971. Introduction of mycorrhizal fungi into soil by spore inoculation of seed. *Aust. For.* 35: 23—26
144. Theodorou, C. Unpublished
145. Theodorou, C., Bowen, G. D. 1969. The influence of pH and nitrate on mycorrhizal relations of *Pinus radiata* D. Don. *Aust. J. Bot.* 17:59—67
146. Thomas, D. R., Richardson, J. A., Dicker, R. J. 1974. The incorporation of tritiated thymidine into DNA as a measure of the activity of soil microorganisms. *Soil Biol. Biochem.* 6: 293—96
147. Trinick, M. J. 1970. *The ecology of Rhizobium—interactions between Rhizobium strains and other soil micro-organisms.* PhD thesis, Univ. of Western Australia, Perth. 353 pp.
148. Trolldenier, G. 1971. *Bodenbiologie: Der Boden-Organismen im Haushalt der Natur.* Stuttgart: Kosmos-studienbücher. 152 pp.
149. Tuzimura, K., Watanabe, I. 1962. The effect of rhizosphere of various plants on the growth of *Rhizobium.* Ecological studies of root nodule bacteria (Part 3). *Soil Sci. Plant Nutr. Tokyo* 8(4):13—17
150. Vandermeer, J. H. 1972. Niche theory. *Ann. Rev. Ecol. Syst.* 3:107—32
151. Vincent, J. M., Waters, L. M. 1953. The influence of host on competition amongst clover root-nodule bacteria. *J. Gen. Microbiol.* 9:357—70
152. Vujicic, R., Park, D. 1964. Behaviour of *Phythophthora erythroseptica* in soil. *Trans. Br. Mycol. Soc.* 47:455—58
153. Watt, K. E. F. 1961. Community stability and the strategy of biological control. *Can. Entomol.* 97:887—95
154. Wells, T. R., Kreutzer, W. A., Lindsey, D. L. 1972. Colonization of gnotobiotically grown peanuts by *Aspergillus flavus* and selected interacting fungi. *Phytopathology* 62:1238—42
155. Wilcox, H. E. 1968. Morphological studies of the roots of red pine, *Pinus resinosa.* II. Fungal colonization of roots and the development of mycorrhizae. *Am. J. Bot.* 55:688—700
156. Williamson, M. 1972. *The Analysis of Biological Populations.* London: Arnold. 180 pp.
157. Wilson, H. M. 1966. Conidia of *Botrytis cinerea:* Labeling by fluorescent vital staining. *Science* 151: 212—13
158. Wong, P. T. W. 1975. Cross-protection against the wheat and oat take-all fungi by *Gaeumannomyces graminis* var. *graminis. Soil Biol. Biochem.* 7:189—94
159. Wong, T. M. 1972. *Effect of soil water on bacterial movement and streptomycete-fungal antagonism.* PhD thesis. Univ. of Sydney, Sydney. 167 pp.
160. Wood, A., Paleg, L. G. 1972. The influence of gibberellic acid on the permeability of model membrane systems. *Plant Physiol.* 50:103—8

PAPILLAE AND RELATED ♦3636
WOUND PLUGS OF PLANT CELLS

James R. Aist
Department of Plant Pathology, Cornell University, Ithaca, New York 14853

INTRODUCTION

Over 100 years ago deBary (29) discovered localized apparent wall thickenings on the inner surfaces of plant walls at sites of penetration by fungi. Shortly thereafter (30, 31) he described these structures as cell wall ingrowths and noted a correlation between their occurrence and the failure of further fungal development. By the turn of the century these wall-like depositions were thought to be paramural, i.e. located between the plasmalemma and cell wall (106) (Figure 2). The depositions were later shown to be extremely common, and although not ubiquitous (5, 47, 54, 67, 120), they apparently accompanied every host wall penetration in certain parasite-host pairings (3, 8, 37, 47, 54, 71, 96, 107).

If intracellular fungal development is slight or absent at an encounter site (potential or actual penetration site) the depositions are usually hemispherical (3, 46, 54, 71, 96, 107), but if considerable development has occurred there, they may conform to some extent to the intracellular parasite structures around which they form and thus encase them (Figure 1) (2, 17, 30, 31, 37, 46, 53, 71, 77, 80). Because these various forms of wall-like depositions commonly have similar substructure, composition, origin, and mode of formation (Figure 1), I refer to them all as *papillae* (106) [or callosity, lignituber, callus (4)]. The term *papillae* is often used to refer to hemispherical depositions only, with other terms being applied to morphological variants (15). However, viewing them all as papillae emphasizes the concept that they are all products of similar processes. The term *sheath* was rather consistently used in the old literature (97) to refer to papillae, but it has quite a different meaning in current literature (15). Several articles (4, 15, 20, 38, 110, 119) have briefly reviewed some aspects of papillae and should be consulted for additional information and viewpoints.

Attack by pathogenic agents other than fungi may result in papilla formation or related responses. Ingestion of plant cell contents by nematodes is sometimes preceded by rapid aggregation of cell cytoplasm at the feeding site (122, 123), a

145

cytoplasmic response that usually accompanies papilla deposition. Papilla-like structures have been reported in plant cells affected by viruses (7, 23, 66) and a nematode (98).

Plant cell responses to fungal intrusion have much in common with certain other responses to physical and chemical wounding and to hypertonicity (93). Of the available reviews (12, 13, 18, 76) of plant wound responses, only one (18) discusses responses related to papilla formation, and the coverage is incomplete.

This review summarizes current knowledge of papillae and related wound plugs and evaluates the evidence concerning alleged functions of these structures. Emphasis is given to papillae incited by fungal attack. Wound plugs are discussed first, however, in order to lend perspective to the subsequent discussion of papillae.

WOUND PLUGS

Incitement

Plant cells have long been known to react to various types of wounds by reorientation of protoplasm toward the wound and concomitant deposition of new materials along cell walls bordering the damaged or moribund cells (87, 99, 111). Such responses have been experimentally incited by cutting tissues with a sharp blade (25, 87, 99, 111, 125), by applying pressure to tissues (125) or to individual cells (114) by means of microneedles, by inserting microneedles into tissues (26, 84, 87, 94) or individual cells (9, 62, 88, 89, 114, 116), or by topical application of heat (87) or chemicals (50, 78, 87, 92, 99, 118). In nature, wounds produced by adventitious root formation incite polarization of cytoplasm in adjacent cortical cells (94).

Formation

Observations of in vivo responses of plant cells to wounding were made by Tangl in 1884 (111). Other authors (9, 88, 89, 99, 114, 116) have since greatly extended our knowledge of the sequential events involved. The first response is flowing of cytoplasm toward the focal point of the injury. In the case of a tissue wound, cytoplasm of cells bordering the wound aggregates against the wall closest to the trauma, and cells several cell rows away may react (87, 99, 111). The nuclei of these cells are usually displaced in a like manner, a response also seen as a result of parasitic attack by certain fungi (6, 24, 49, 91, 94, 113). These cellular responses, termed *traumatotaxis* (99), sometimes become obvious in one to two hours (87, 125). In individual cells reacting to microneedle puncture, cytoplasm aggregates at the wound within a few minutes after insertion (88, 89, 114, 116). Regardless of whether the injury is to a single cell or to a tissue, the cytoplasmic aggregates are accumulations of seething granular cytoplasm (111), channeled to a target area of the cell by reorientation of cyclosis. After several hours or days they disperse.

The depositions of new materials into and onto wall regions bordering the injury is coincidental with cytoplasmic aggregation (25, 89, 94, 114, 116). This deposition may appear as a general wall thickening (87, 111) or as localized, sub-cellular plugs (25, 88, 89, 94, 111, 114, 116). Stratification of wound plugs has been observed in some cases (25). Of the accounts of wound plug formation in vivo (88, 89, 114, 116), Walker's (116) is the best illustrated. Plug formation usually begins a few minutes after the injury (89, 114, 116), but it has, on occa-sion, been reported to begin even sooner (25, 26, 116).

Nature

The composition of wound plugs has been little studied, but callose is probably a common constituent (25, 89). When viewed with bright field optics they appear highly refractive (89, 111, 116). At first the plugs are soft, but with passage of time they become hard enough to withstand turgor pressure (89, 116). When a micro-electrode tip becomes covered with a plug, the measured potential difference across the plasmalemma drops dramatically (114, 116), perhaps because the plas-malemma has re-formed around the plug, isolating the electrode tip from the cytoplasm.

Functions

A likely function of the cytoplasmic aggregate would be to deposit materials into and onto cell walls affected by the injury. This suggestion is supported by the observation that the cytoplasmic aggregates occur before and during the deposi-tion process (88, 116). The depositions themselves have historically been con-sidered to function in plugging or sealing the wound. One consequence of this plugging would be to prevent the expulsion, by turgor pressure, of cytoplasm (89, 116). Wheeler (118) suggested that such structures restrict the movement of molecules and thus prevent excessive loss of ions from damaged cells or move-ment of toxic materials into them.

PAPILLAE

Incitement

Smith (106) already had considerable insight into papilla incitement in 1900: "There are not data at hand to determine whether the penetrating tube, by means of some chemical substance, excites the cell protoplasm to unusual activity in the production of cellulose over the region of penetration, or whether microscopical-ly small needles would cause such a production mechanically." We now know that microneedles cause papilla-like wound plugs, and it has accordingly been proposed (20, 79) that papillae are incited by the mere physical intrusion of fungi into plant cells.

There is considerable evidence that indicates that a chemical stimulus is also sufficient to incite papillae in the apparent absence of physical stress. Plant cells bordering those under attack respond by both directed cytoplasmic aggregation (Figure 1) (6, 21, 94) and papilla formation (40, 104, 119; H. C. Hoch, unpub-lished). Papilla formation in advance of host wall penetration would also suggest chemical incitement. Recent work by Aist & Israel (2) provided evidence that papillae sometimes precede host wall penetration by *Olpidium* and *Erysiphe*. They monitored the in vivo development of penetration tubes and pegs and determined the time of their appearance relative to that of papillae.

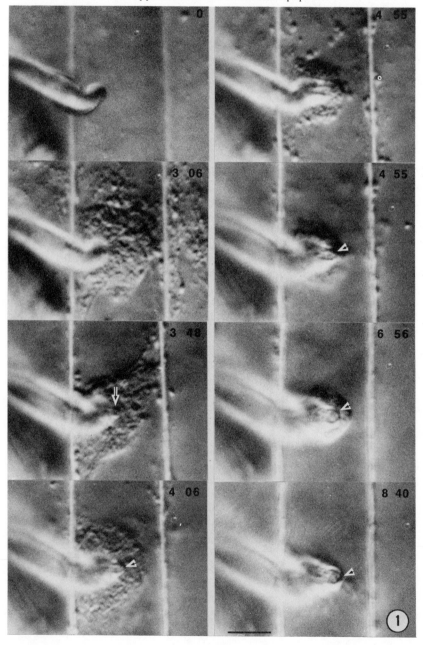

Most ultrastructural reports claiming that papillae are formed before physical penetration of the host wall begins (34, 36, 40, 46, 51, 82, 107, 112) failed to provide proof of this by means of near-median sections through the incipient bore-hole. This left open the possibility that penetration pegs already present at such sites incited the papillae by physical intrusion. Recent work (H. C. Hoch, unpublished) in which serial thin sections through encounter sites were analyzed is exceptional in this regard. Hoch showed that certain fungal hosts of *Pythium acanthicum* produce large papillae prior to the onset of host wall penetration. The work by Sargent et al (100) is also excellent, but physical stress (pressure) exerted by the parasite may well have provided the stimulus for early papilla formation in their material, since host walls were conspicuously indented under appressoria. One other observation that indicates that chemical stimuli can incite papilla formation is the occurrence of papillae at points of contact between intercellular hyphae and plant cell walls in rust-infected plants (56, 104). Young (124), in an effort to test the hypothesis, applied culture filtrates of fungal pathogens to host tissues. Although papillae were not incited in that study, more recent work (50, 78, 92) has shown that certain toxic metabolites of pathogenic fungi will incite formation of papilla-like structures when applied to plant tissues. The evidence clearly indicates that chemical stimuli are sufficient to incite papillae.

Formation

It was suggested in the last century (17, 73) that papillae represent original cell wall material which swells or partly solubilizes during intrusion by fungi. This idea must now be abandoned in view of the bulk of ultrastructural studies that show that the original walls are not so affected, as well as numerous other investigations that reveal that papillae differ chemically from the original walls.

The earliest reports of cytoplasmic aggregation in response to fungal penetration are probably those of Nemec (86) and Guttenberg (49). Since then, cytoplasmic aggregates have been encountered in studies of numerous and varied parasite-host combinations (22, 67, 71, 75, 82, 100, 108). Some observations of in

---

Figure 1 Time-lapse interference contrast photomicrographs of in vivo interactions during an attempt at penetration of a barley coleoptile epidermal cell by a single appressorium of *Erysiphe graminis hordei.* The material was prepared and observed according to Aist & Israel (2). The numbers in the upper right corner of each frame correspond to hours and minutes after the first frame was taken (at 10' 20" postinoculation). At 3' 06" a large cytoplasmic aggregate had formed below the appressorial hook, and a similar response was seen momentarily in the adjacent cell to the right. The penetration peg (*arrows*) was first seen at 3' 48", at which time papilla deposition was not apparent. By 4' 06" a collar-like papilla was present, as evidenced by an apparent thickening of the lateral walls of the penetration peg. Two different focal planes are shown at 4' 55". The first shows a very thick collar-like papilla along the lateral walls of the now greatly elongated penetration peg. The second shows the tip of the growing peg; it was free of both the papilla and the cytoplasmic aggregate. At 6' 56" the cytoplasmic aggregate had penetrated deeply into the cell so as to surround the apex of the penetration peg, which had just begun to enlarge to form the haustorial central body. Development did not proceed further, and the outcome (8' 40") was an encased haustorial initial. By this time the entire cytoplasmic aggregate had dispersed. Calibration = 10 μm ($\times$ 1125).

vivo papilla formation (2, 21, 71, 103) have shown that a conspicuous cytoplasmic aggregate is present before and during papilla deposition. The first description of such a cytoplasmic aggregate in vivo was probably that of Kusano (71). The description by Bushnell & Bergquist (21) is more complete, and is reminiscent of earlier descriptions of aggregates incited by wounding. Sometimes papilla deposition occurs in the absence of a cytoplasmic aggregate (3); conversely, some cytoplasmic aggregates are apparently not accompanied by a papilla (21, 67).

The question of whether papillae are deposited by the host protoplast, as suggested by Smith (106), or by the fungus, is paramount. Insofar as papillae are related to wound plugs, their incitement by microneedles shows that they can be entirely of plant cell origin. Similarly, papillae often form in cells that are not in direct contact with the fungus (40, 104, 119; H. C. Hoch, unpublished). Papillae formed in cabbage root hairs in response to penetration by *Plasmodiophora brassicae* normally do not appear until after the parasite protoplast has been swept away from the encounter site by cyclosis (3). Sometimes papillae are formed before the host cell wall has been completely penetrated (36, 77, 100, 107), which makes it seem unlikely that the penetrating fungus deposits them. Is there evidence that verifies this apparent depositional activity of the host protoplast? Several ultrastructural studies (3, 51, 75, 77, 107; H. C. Hoch, unpublished) conclude that host secretory vesicles in the region of the encounter site contribute to the formation of papillae; some of these studies (3, 77, 107: H. C. Hoch, unpublished) further show that some components of papillae resemble the contents of such vesicles in texture, structure, and/or straining intensity. The plasmalemma depressions in Figures 4 and 5 provide presumptive evidence for the accompanying membrane flow. Furthermore, a number of published electron micrographs (22, 51, 56, 75, 82, 100, 107, 108) demonstrate that cytoplasmic aggregates contain the common secretory apparatus of plant cells, i.e. Golgi apparatus, endoplasmic reticulum, and (presumably) secretory vesicles. As might be expected, papillae are absent in plasma-poor host cells infected by *Olpidium* (75). The evidence warrants the conclusion that papillae are deposited by the host protoplast.

Pitfalls of interpreting microscopic images of nonliving preparations have led to unwarranted conclusions concerning developmental aspects of papillae and penetration pegs. First, the fact that a papilla, but not a penetration peg, is present at an encounter site does not necessarily show that papillae precede host wall penetration. Such encounter sites may be abortive; that is, penetration might not have occurred there even if the material had not been fixed. The latter is substantiated by the high percentage of abortive encounter sites that were shown to occur, even in compatible parasite-host pairings (2, 21, 57, 107). Second, the occurrence of papillae at sites of actual penetration does not provide evidence that those particular papillae were grown through by the fungus; they could well have formed after host wall penetration, as was shown in some cases (2, 3, 21). Thus, extreme caution should be used when one attempts to describe a living process on the basis of images of fixed cells.

What conclusions, then, can be made concerning the timing of papilla formation? There is no doubt that the times at which papilla formation and host wall penetration are initiated often closely coincide (2, 31, 74, 96, 121). The rapidity and duration of the deposition process vary. Sometimes conspicuous papillae form within a few minutes after stimulatory activity of a fungus (3; H. C. Hoch, unpublished). This agrees with the timing of wound plug formation. Papilla formation sometimes lasts ten to twenty minutes (21; H. C. Hoch, unpublished), but it can occur over a much greater period of time, e. g. when a well-developed penetration peg is being encased (Figure 1).

Nature

When viewed with bright field light optics, papillae often appear highly refractive (30, 31, 71, 80, 106, 121), a feature they share with wound plugs. In addition they sometimes have a natural yellow or brown color (17, 29, 45, 62, 74, 86, 102, 124), autofluoresce (60; J. R. Aist and H. W. Israel, unpublished), or are birefringent (44, 45, 102).

The ultrastructure of papillae has been extensively reviewed (15). One striking feature is their heterogeneity; papillae occurring in different plants, different cells of the same plant, or even different papillae in the same cell may be ultrastructurally diverse. Individual papillae, as seen in ultrathin sections, usually are composed of variously shaped areas of graded texture and staining intensities. They may contain membrane fragments or vesicles and apparently hydrophobic (lipid?) droplets (Figures 2, 3). Finger-like projections (Figure 4) of papillae in cabbage root hairs sometimes invaginate the host cytoplasm, and the plasmalemma may be altered where it is in contact with the papilla (Figure 5). Very often papillae have an overall appearance of a lightly staining or nonstaining matrix (light papillae) in which scattered areas of more intensively stained materials are embedded (10, 22, 39, 57, 75, 81, 90, 105). Other (dark) papillae appear as a nearly homogeneous darkly staining mass (36, 46, 60, 83, 96, 107), which appearance in some cases (83, 107) develops during maturation of light papillae. The vesicles, membrane fragments, and densely staining portions of papillae seem to arise, at least in part, as a result of entrapment of organelles and cytoplasmic enclaves in papillae during papilla deposition, and of the subsequent degeneration of these as the papillae mature (75, 84; H. C. Hoch, unpublished). Papillae, like wound plugs, are sometimes layered (6, 8, 44, 45, 49, 62, 102), and ultrastructural details of the layering are available (82, 85, 96, 107). A significant feature of this stratification is the fact that papillae often are covered by a layer of material that closely resembles the normal host wall in texture and staining intensity and merges with it at the borders of the papillae (3, 52, 57, 63; H. C. Hoch, unpublished).

The chemical nature of papillae has been of interest largely because of the possible implications regarding papilla function. In 1895 Mangin (80) reported the common occurrence of large proportions of callose (a β-1,3-glucan) in papillae. This has since been extended (3, 28, 84, 89, 100, 102, 107) to the point that callose is the most commonly identified chemical constituent in papillae. Most of the recent work has relied heavily on the aniline blue–fluorescence method (41) for identification of callose, but there is reason to believe that this test is insufficient when used alone (41, 42).

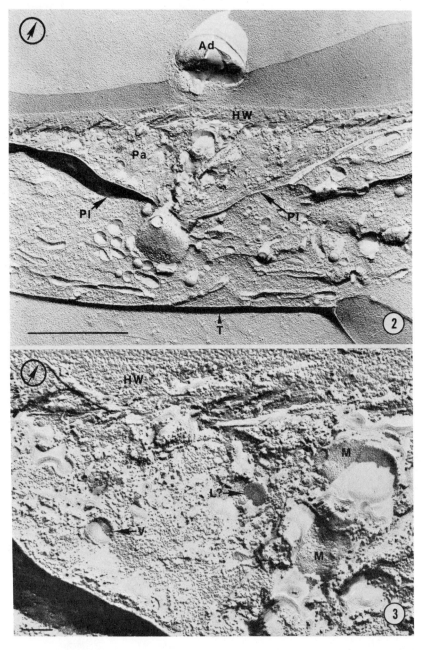

Kusano (70) may have been the first to identify lignin as a major component of papillae. Several other reports (5, 43, 47, 48, 62, 102) of lignified papillae have since appeared, and the work of Sherwood & Vance (102) is particularly significant in that it shows that the lignin is added to an already existing papilla, rather than being deposited with the rest of the papilla from its initiation.

Cellulose has occasionally been identified in papillae (45, 49, 80, 102). The coincidence of a positive test for cellulose and of papilla birefringence (45, 102) strongly supports the notion of a cellulosic component. Furthermore, Sherwood & Vance (102) found that cellulose staining and birefringence were confined to the outer cover of papillae, which is precisely the location of the cell-wall-like papilla layer seen in electron micrographs. Other reported chemical constituents of papillae include protein (32, 35), peroxidase (32, 33), pectin (62), suberin (43), gums (62), and silicon (69). The work of H. C. Hoch (unpublished) is of interest in this regard. Hoch found that papillae in many fungal hosts of *Pythium acanthicum* showed a positive test for polysaccharides, but that they did not contain detectable levels of callose. Subsequent cytochemical work revealed a significant chitin (or chitosan) component in papillae of the fungal host *Phycomyces blakesleeanus*. Thus, although papillae are in many ways distinct from normal cell walls, their chemical makeup may reflect a degree of dependence on normal cell wall physiology.

A synopsis of the chemical components of papillae would not be complete without mention of the many cytochemical tests that have yielded negative results in specific cases. Tests for the following compounds have been negative: cellulose (14, 43, 53, 68, 109), callose (14, 43; H. C. Hoch, unpublished), pectin (14, 43, 53, 107), lignin (68), chitin (14), suberin (102), protein (35, 53), hemicellulose (43), cutin (102), gums (102), and tannins (102). A negative test for callose is sometimes associated with papilla maturation (83, 107), and Sherwood & Vance (102) have suggested that this may be due to lignin interfering with staining (117).

The chemical heterogeneity of papillae agrees well with their structural heterogeneity. In spite of numerous ultrahistochemical procedures now available, there is almost no published work on the ultrastructural localization of components within papillae. This seems the next logical step for this area of investigation.

Regarding the physical nature of papillae, very little information is available other than pure speculation. Bushnell (19) plucked a papilla from a cell wall with a microneedle and surmised that it was hard and insoluble. But what about such things as permeability, porosity, and the relative hardness of the papilla at different times after its initiation? These are all important questions that may have implications with respect to papilla function, but about which we have little

----

Figures 2 and 3   Freeze-etching electron micrographs of cabbage root hairs infected by *Plasmodiophora brassicae,* prepared according to Aist (1). *Figure 2.* A cross-fractured encounter site showing a paramural papilla below the adhesorium. *Figure 3.* An enlargement of part of the papilla in Figure 2, showing membranous, vesicular, and possibly lipoidal components embedded in a finely granular matrix. Abbreviations: Ad = adhesorium, HW = host wall, L? = possible lipid body, M = membrane, Pa = papilla, Pl = host plasmalemma, T = tonoplast, V = vesicle. Arrows at upper left indicate the direction of shadowing. Calibrations: Figure 2 = 1 μm ($\times$ 27,200); Figure 3 = 0.1 μm ($\times$ 85,000).

information. The possibility that callose (58) or papilla-like wall formations (118) restrict the passage of solutes is significant and needs experimental verification.

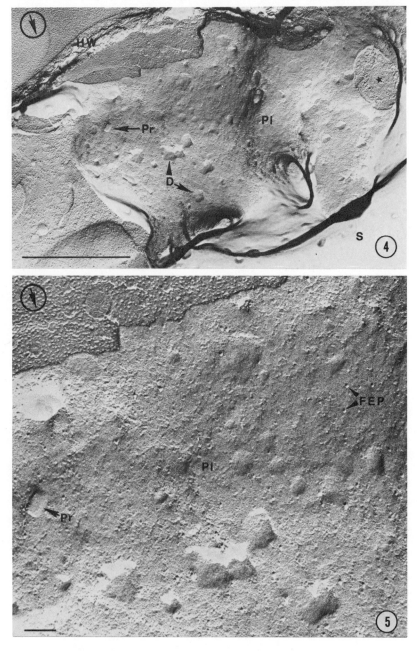

Functions

Smith (106) displayed another remarkable bit of scientific insight in 1900 when he stated, "Just what significance there is in this thickening of the host wall cannot be conclusively determined without experiment." With one exception this advice has been ignored by those who have addressed the question. The necessity for experimentation should become clear during the ensuing discussion.

In 1900, Smith (106), in referring to the damage caused to the host cell wall by the invading fungus, remarked, "There would then be the same reason for this local thickening as there was for the original production of a thick outer wall in the first place." Thus was born the concept of papilla formation as a wound-healing process, a concept that has since gained support (3, 20, 51, 72, 82). Presumably, wound healing in this context has the same meaning as it does in reference to physical and chemical wounds, i. e. to provide a barrier to the expulsion of cytoplasm and/or to prevent excessive loss of diffusates through the affected region of the host plasmalemma. In view of the many similarities between plant cell responses to physical and chemical wounds on the one hand, and papilla formation on the other, it seems reasonable to assume, as a working hypothesis, that papilla formation is one manifestation of a wound-healing process that plant cells in general are capable of undergoing. It therefore becomes easy to explain why papillae form in response to fungal penetration; plant cells are designed to react to a wound in such a manner. Because papillae sometimes do not form during penetration of living cells by fungi (21, 54, 67, 77, 120), the important question is one of "how the deposition processes are turned off" (20).

Another possible function of papillae would be to prevent damage to the host by establishing a permeability barrier that would exclude toxic materials. This idea also seems to have originated with Smith (106) and has been revived recently by others (51, 59, 84, 104, 118). There is very little evidence for this idea, but the observation by H. C. Hoch (unpublished) that the degenerative effects of *P. acanthicum* infection do not readily spread from an infected fungal host cell to a noninfected one if the intervening cross wall is covered with a papilla is suggestive. Similar observations have been made by Heath (54) on cowpea rust. We

---

Figures 4 and 5 Freeze-etching electron micrographs of cabbage root hairs infected by *Plasmodiophora brassicae,* prepared according to Aist (1). *Figure 4.* A convex fracture passing primarily through the host plasmalemma where it borders a papilla. The papilla has several finger-like projections, one of which was fractured across at (*). This plasmalemma region, in contrast to nonpapilla-associated regions (1), has numerous, wide, shallow depressions and a single protrusion of similar size. *Figure 5.* An enlargement of part of Figure 4 showing details of part of the plasmalemma [extra-cytoplasmic fracture face (16)] region associated with the papilla. This region is similar in fine structural detail to plasmalemma regions not associated with papillae (1), except for the wide depressions. The depressions are free of freeze-etch particles and surface irregularities, which suggests they may represent newly added portions of the plasmalemma (95). Abbreviations: D = depression, FEP = freeze-etch particle, HW = host wall, Pl = host plasmalemma, Pr = protrusion, S = shadow. Arrows at upper left indicate the direction of shadowing. Calibrations: Figure 4 = 1 μm (× 29,400); Figure 5 = 0.1 μm (× 85,000).

need to know the permeability characteristics of papillae in order to evaluate this possibility more fully.

Leitgeb (74) suggested in 1881 that papillae may be a necessary prerequisite for successful establishment of the parasitic relationship. Papilla formation would allow the fungus to enter the host cell without causing significant damage. The same idea has recently been used by Held (57), who interpreted the coincidence of sites of successful penetration and papillae in *Rozella*-infected *Allomyces* as evidence that papillae promote penetration. While such a coincidence could be due to a favorable effect of papillae on the ability of the fungus to become established in its host, it is equally possible that it is due to the fact that papillae are only formed in response to (i. e. as a result of) penetration by the parasite. It is significant that the only data Held presented that relate to the timing of papilla formation with respect to that of host wall penetration is his Figure 10, which apparently shows the parasite protoplast caught in the act of passing through the host cell wall where no papilla is present. The interaction is probably not unlike that between *Plasmodiophora brassicae* and cabbage root hairs (3) in which in vivo time course analyses revealed that papillae form after host cell wall penetration, even though general observations had indicated a coincidence between successful penetration and papilla formation. A careful in vivo time course analysis of the sequence of events in the *Rozella-Allomyces* system might clarify the point, but at present there is not sufficient basis for concluding that papillae promote parasite establishment.

In 1900 Smith (106) suggested that papillae, formed in response to penetration, impede the progress of the penetration pegs. This has developed into the most popular theory regarding an active role of papillae in the penetration process (5, 15, 21, 43, 46, 55, 60, 61, 79, 115). The theory is designed to explain a number of instances (5, 17, 30, 31, 47, 59, 65, 71, 101, 115) in which the incidence, morphology, or size of papillae correlates with penetration failure, but not with successful fungal establishment. In many parasite-host pairings only a portion of the parasite units would be impeded by papillae (2, 21, 71, 79, 101). The mechanism(s) whereby papillae might perform such a function would include the establishment of a physical (4, 36, 69, 70, 82, 96, 107, 115) or chemical (11, 28, 52, 97, 107) barrier, or the accumulation of antifungal toxins (36, 60, 115) in papillae. Unfortunately the only evidence that favors this notion is the aforementioned correlation, which alone is not sufficient to warrant a conclusion concerning the role of papillae because other interpretations have not been ruled out.

What alternative interpretations could explain such penetration failures? One obvious possibility is that the fungus is inhibited by a biochemical resistance mechanism unrelated to papilla formation. The theory was very succinctly phrased by Kusano (72) in 1939, and translated by Ito (62):

> The host plants produce the callosities to heal the injuries caused by the causal organisms, but never to check the penetration of them, and whether the fungi grow out of the callosity and continue further development or not is wholly due to a certain substance in cytoplasm. According to the kind of the plant, this unknown substance is different in both quantity and quality and consequently in the plants which give a great

deal of unfavorable action to the entered organisms the invasion appears to be stopped earlier, as if the callosities surved [sic] to check the penetration of the fungi.

That such a mechanism is sometimes operative during or shortly after penetration is illustrated by O. C. Yoder's observations (personal communication) that penetration pegs of *Helminthosporium victoriae* typically cease development in epidermal cells of resistant oat varieties in the absence of any visible host response, including papilla formation, whereas the same fungus inoculated onto susceptible varieties ramifies and colonizes extensively. A second alternative interpretation would be that fungal development fails in the absence of·a susceptibility factor which is induced only in susceptible plants, an idea recently promoted by Daly (27) to explain certain parasite-host interactions. Such a mechanism could account for the failure of *Olpidium* protoplasts to enter penetrated dead host cells in the absence of papillae (65). Finally, the possibility that these penetration failures are due to a reduced penetrating ability (inherent developmental deficiency) of the particular parasite units involved has been discussed (2, 101). Some evidence in favor of this idea is available. In several instances (2, 21, 57, 107) it is obvious that certain parasite units ceased development, prior to host wall penetration, in the absence of papilla formation. The data of Bushnell & Bergquist (21) are particularly instructive in this regard. A significant percentage of the unsuccessful appressoria ceased development after various degrees of penetration activity; some incited no visible host response at all, others incited only the cytoplasmic aggregate, and still others incited both the cytoplasmic aggregate and a papilla. There are not sufficient bases at present for concluding that the unsuccessful appressoria that were associated with papillae aborted for a different reason than did those which were not. The data, taken as a whole, can also be explained by the hypothesis that the unsuccessful appressoria are inherently incapable of producing haustoria.

Were any of these alternative interpretations correct, why would papillae correlate with these penetration failures, but not with successful penetration attempts? One possibility is that such weakened, damaged, or inhibited parasites cause greater trauma than do their successful counterparts. Guttenberg (49) suggested that degenerating haustoria secrete harmful products not secreted by healthy ones, and that this secretion incites papilla formation. He observed that epidermal haustoria which had been torn away from their mycelial connections due to sporulation activity of the white rust fungus soon became encased in papillae. Hirata (59) performed an experiment in which he stripped mycelium of *Erysiphe* from the leaf surface and noted that some of the haustoria thus detached were subsequently encased. Kajiwara & Takahashi (64) treated roots of rust-infected chrysanthemum plants with oxycarboxin, a systemic fungicide that controls the disease. Death of haustoria in the leaves was accompanied by gradual encasement of them by papillae. Lupton (79) observed that when susceptible plants were inoculated with *Erysiphe* under environmental conditions unfavorable for infection, penetration failures, associated with papillae, occurred with much greater frequency than when the conditions favored infection. The effect of the unfavorable conditions could have been to weaken the fungus (thereby preventing it from forming haustoria) and cause it to damage its host. I would like to emphasize that all of these alternative hypotheses to explain the correlation of papillae with penetration failures would view papilla formation as a spurious event (in this case, a general wound response) associated with, but not causally related to, failure of penetration.

If any one of the above hypotheses is sufficient to explain any particular correlation of this type it "cannot be conclusively determined without experiment" (106). The study by Vance & Sherwood (115) is an effort to do that. They were able to prevent protein synthesis, papilla formation, and resistance to various nonpathogens of reed canary grass by pretreating leaf disks with cycloheximide. These results are consistent with the hypothesis that papillae prevent penetration, but they are also consistent with the hypothesis that papillae are spurious and that resistance is not papilla related, since the treatment would be expected to affect any active resistance mechanism, not just papilla formation. Further experimentation is thus needed to elucidate the role of papillae in reed canary grass.

CONCLUSIONS

Plant cells respond to nonlethal wounds by redirecting normal cyclosis so that cytoplasm aggregates at the point or region in the cell nearest the wound. Some kind of wall secretion or apposition (wound plug) ensues and the wound is thus healed. These responses are often rapid and are of common occurrence.

Plant cells under attack by fungal pathogens commonly react in a like manner. That the resulting wall appositions, termed *papillae,* are a response to wounding by the encroaching pathogens appears to be a useful concept. Whether or not they play a role in addition to wound healing has been the subject of considerable speculation. The correlative data which seemingly implicate papillae in resistance to fungal pathogens have several, highly feasible alternative interpretations. Whether any of these interpretations, including a role in disease resistance, are applicable to a given parasite-host interaction remains to be demonstrated by experimentation. Greater knowledge of the nature of papillae would then be important.

Reference to the literature has demonstrated that important information and ideas about papillae were already available by 1925; little real progress has occurred in the last half-century, particularly with regard to papilla function. Our challenge is to utilize the advanced technology and increased knowledge of cell biology to build upon inherited information and concepts.

ACKNOWLEDGMENTS

I am indebted to R. P. Korf, M. A. Waterman, and J. B. Piraino for French translations; to H. W. Israel, M. A. Waterman, and S. J. Aist for helpful suggestions concerning the manuscript; and to K. Mühlethaler, Institute for Cell Biology, Zürich, for use of laboratory and freeze-etch equipment. The preparation of this review was supported in part by a NATO postdoctoral fellowship to the author and by USDA-CSRS Grant #316–15–53 to J. R. Aist and H. W. Israel.

Literature Cited

1. Aist, J. R. 1974. A freeze-etch study of membranes in *Plasmodiophora*-infected and non-infected cabbage root hairs. *Can. J. Bot.* 52:1441–49
2. Aist, J. R., Israel, H. W. 1975. *Induction Mechanism of Biochemical and Cytological Responses*, ed. S. Ouchi. Tokyo: Phytopathol. Soc. Jpn. In press
3. Aist, J. R., Williams, P. H. 1971. The cytology and kinetics of cabbage root hair penetration by *Plasmodiophora brassicae. Can. J. Bot.* 49:2023–34
4. Akai, S. 1959. *Plant Pathology, An Advanced Treatise,* ed. J. G. Horsfall, A. E. Dimond, 1:391–434. New York: Academic. 674 pp.
5. Akai, S. et al 1971. *Morphological and Biochemical Events in Plant-Parasite Interaction,* ed. S. Akai, S. Ouchi, 329–47. Tokyo: Phytopathol. Soc. Jpn. 415 pp.
6. Allen, R. F. 1923. Cytological studies of infection of Baart, Kanred and Mindum wheats by *Puccinia graminis tritici* forms III and XIX. *J. Agric. Res.* 26:571–604
7. Allison, A. V., Shalla, T. A. 1974. The ultrastructure of local lesions induced by potato virus X: a sequence of cytological events in the course of infection. *Phytopathology* 64:784–93
8. Aronescu, A. 1934. *Diplocarpon rosae:* from spore germination to haustorium formation. *Bull. Torrey Bot. Club* 61:291–329
9. Benda, G. T. A. 1959. Nuclear movement in an injured cell. *Protoplasma* 50:410–12
10. Berlin, J. D., Bowen, C. C. 1964. The host-parasite interface of *Albugo candida* on *Raphanus sativus. Am. J. Bot.* 51:445–52
11. Blackwell, E. M. 1953. Haustoria of *Phytophthora infestans* and some other species. *Trans. Br. Mycol. Soc.* 36:138–58
12. Bloch, R. 1941. Wound healing in higher plants. *Bot. Rev.* 7:110–46
13. Bloch, R. 1952. Wound healing in higher plants. II. *Bot. Rev.* 18:655–79
14. Bracker, C. E. 1968. Ultrastructure of the haustorial apparatus of *Erysiphe graminis* and its relationship to the epidermal cell of barley. *Phytopathology* 58:12–30
15. Bracker, C. E., Littlefield, L. J. 1973. *Fungal Pathogenicity and the Plant's Response,* ed. R. J. W. Byrde, C. V. Cutting, 159–318. London: Academic. 499 pp.
16. Branton, D. et al 1975. Freeze-etching nomenclature. *Science* 190:54–56
17. Brefeld, O. 1895. *Untersuchungen aus dem Gesammtgebiete der Mykologie XI. Die Brandpilze II.* Münster: Schoningh. 236 pp.
18. Bünning, E. 1959. *Handbuch der Pflanzenphysiologie,* ed. W. Ruhland, 17(1):119–34. Berlin: Springer. 716 pp.
19. Bushnell, W. R. 1971. See Ref. 5, pp. 229–54
20. Bushnell, W. R. 1972. Physiology of fungal haustoria. *Ann. Rev. Phytopathol.* 10:151–76
21. Bushnell, W. R., Bergquist, S. E. 1975. Aggregation of host cytoplasm and the formation of papillae and haustoria in powdery mildew of barley. *Phytopathology* 65:310–18
22. Chou, C. K. 1970. An electron-microscope study of host penetration and early stages of haustorium formation of *Peronospora parasitica* (Fr.) Tul. on cabbage cotyledons. *Ann. Bot.* 34:189–204
23. Conti, G. G., Vegetti, G., Bassi, M., Favali, M. A. 1972. Some ultrastructural and cytochemical observations on Chinese cabbage leaves infected with cauliflower mosaic virus. *Virology* 47:694–700
24. Contreras, M. R., Boothroyd, C. W. 1975. Histological reactions and effects on position of epidermal nuclei in susceptible and resistant corn inoculated with *Helminthosporium maydis* Race T. *Phytopathology* 65:1075–78
25. Currier, H. B. 1957. Callose substance in plant cells. *Am J. Bot.* 44:478–88
26. Currier, H. B., Shih, C. Y. 1968. Sieve tubes and callose in *Elodea* leaves. *Am. J. Bot.* 55:145–52
27. Daly, J. M. 1972. The use of near-isogenic lines in biochemical studies of the resistance of wheat to stem rust. *Phytopathology* 62:392–400

28. Davison, E. M. 1968. Cytochemistry and ultrastructure of hyphae and haustoria of *Peronospora parasitica* (Pers. ex. Fr.) Fr. *Ann. Bot.* 32:613–21

29. deBary, A. 1863. Recherches sur le developpement de quelques champignons parasites. *Ann. Sci. Nat. Bot. Biol. Veg.* 20:5–148

30. deBary, A. 1867. Ueber den Krebs und die Hexenbesen der Weistanne (*Abies pectinata* D. C.). *Bot. Z.* 25: 257–64

31. deBary, A. 1870. *Eurotium, Erysiphe, Cicinnobolus*, nebst Bemerkungen über die Geschlechtsorgane der Ascomyceten. *Abh. Senckenb. Naturforsch. Ges.* 7:361–455

32. Delon, R. 1974. *Etude ultrastructurale et comparative des relations plante hote/champignon de deux agents pathogenes (Pyrenochaeta lycopersici Gerlach et Schneider, et Colletotrichum coccodes (Wallr.) Hughes) des Racines de Tomates.* PhD thesis. Univ. Nancy I, Nancy. 69 pp.

33. Delon, R. 1974. Localisation d'activites polyphenoloxydasiques et peroxydasiques dans les cellules racinaires de *Lycopersicon esculentum* parasitees par *Pyrenochaeta lycopersici. Phytopathol. Z.* 80:199–208

34. Delon, R., Reisinger, O., Mangenot, F. 1973. Étude aux microscopes photonique et électronique de racines de Tomates var. "Marmande" atteintes de maladie liégeuse. *Ann. Phytopathol.* 5:151–62

35. Edwards, H. H. 1970. A basic staining material associated with the penetration process in resistant and susceptible powdery mildewed barley. *New Phytol.* 69:299–301

36. Edwards, H. H., Allen, P. J. 1970. A fine-structure study of the primary infection process during infection of barley by *Erysiphe graminis* f. sp. *hordei. Phytopathology* 60:1504–9

37. Ehrlich, M. A., Ehrlich, H. G. 1966. Ultrastructure of the hyphae and haustoria of *Phytophthora infestans* and hyphae of *P. parasitica. Can. J. Bot.* 44:1495–1503

38. Ehrlich, M. A., Ehrlich, H. G. 1971. Fine structure of the host-parasite interfaces in mycoparasitism. *Ann. Rev. Phytopathol.* 9:155–84

39. Ehrlich, M. A., Schafer, J. F., Ehrlich, H. G. 1968. Lomasomes in wheat leaves infected by *Puccinia graminis* and *P. recondita. Can. J. Bot.* 46:17–20

40. Endo, R. M., Colt, W. M. 1974.

Anatomy, cytology and physiology of infection by Pythium. *Proc. Am. Phytopathol. Soc.* 1:215–23

41. Eschrich, W., Currier, H. B. 1964. Identification of callose by its diachrome and fluorochrome reactions. *Stain Technol.* 39:303–7

42. Faulkner, G., Kimmins, W. C., Brown, R. G. 1973. The use of fluorochromes for the identification of β $(1 \rightarrow 3)$ glucans. *Can. J. Bot.* 51:1503–4

43. Fellows, H. 1928. Some chemical and morphological phenomena attending infection of the wheat plant by *Ophiobolus graminis. J. Agric. Res.* 37:647–61

44. Fraymouth, J. 1956. Haustoria of the Peronosporales. *Trans. Br. Mycol. Soc.* 39:79–107

45. Fritz, F. 1937. Beitrage zur Pathologie der Zellmembran. *Z. Pflanzenkr.* 47: 532–41

46. Griffiths, D. A. 1971. The development of lignitubers in roots after infection by *Verticillium dahliae* Kleb. *Can. J. Microbiol.* 17:441–44

47. Griffiths, D. A., Lim, W. C. 1964. Mechanical resistance in root hairs to penetration by species of vascular wilt fungi. *Mycopathol. Mycol. Appl.* 24: 103–12

48. Griffiths, D. A., Lim, W. C. 1966. Patterns in host reaction following infection by *Fusarium solani* Mart. (Sacc.). *Mycopathol. Mycol. Appl.* 33:17–27

49. Guttenberg, H. 1905. *Beiträge zur Physiologischen Anatomie der Pilzgallen.* Leipzig: Engelmann. 70 pp.

50. Hanchey, P., Wheeler, H., Luke, H. H. 1968. Pathological changes in ultrastructure: effects of victorin on oat roots. *Am. J. Bot.* 55:53–61

51. Hanchey, P., Wheeler, H. 1971. Pathological changes in ultrastructure: tobacco roots infected with *Phytophthora parasitica* var. *nicotianae. Phytopathology* 61:33–39

52. Hardwick, N. V., Greenwood, A. D., Wood, R. K. S. 1971. The fine structure of the haustorium of *Uromyces appendiculatus* in *Phaseolus vulgaris. Can. J. Bot.* 49:383–90

53. Heath, M. C. 1971. Haustorial sheath formation in cowpea leaves immune to rust infection. *Phytopathology* 61: 383–88

54. Heath, M. C. 1972. Ultrastructure of host and nonhost reactions to cowpea rust. *Phytopathology* 62:27–38

55. Heath, M. C. 1974. Light and electron microscope studies of the interactions of host and non-host plants with cowpea rust—*Uromyces phaseoli* var. *vignae*. *Physiol. Plant Pathol.* 4:403–14

56. Heath, M. C., Heath, I. B. 1971. Ultrastructure of an immune and a susceptible reaction of cowpea leaves to rust infection. *Physiol. Plant Pathol.* 1:277–87

57. Held, A. A. 1972. Host-parasite relations between *Allomyces* and *Rozella*. Parasite penetration depends on growth response of host cell-wall. *Arch. Mikrobiol.* 82:128–39

58. Heslop-Harrison, J. 1966. Cytoplasmic continuities during spore formation in flowering plants. *Endeavor* 25:65–72

59. Hirata, K. 1967. Notes on haustoria, hyphae, and conidia of the powdery mildew fungus of barley *Erysiphe graminis* f. sp. *hordei*. *Mem. Fac. Agric. Niigata Univ.* 6:207–59

60. Holland, A. A., Fulcher, R. G. 1971. A study of wheat roots infected with *Ophiobolus graminis* (Sacc.) using fluorescence and electron microscopy. *Aust. J. Biol. Sci.* 24:819–23

61. Hyde, P. M., Colhoun, J. 1975. Mechanisms of resistance of wheat to *Erysiphe graminis* f. sp. *tritici*. *Phytopathol. Z.* 82:185–206

62. Ito, K. 1949. *Bull. Gov. For. Exp. Stn.* 43:1–126

63. Kajiwara, T. 1973. Ultrastructure of host-parasite interface between downy mildew fungus and spinach. *Shokubutsu Byogai Kenkyu* 8:167–78

64. Kajiwara, T., Takahashi, K. 1975. *Proc. 1st Int. Congr. IAMS*, ed. T. Hasegawa, 2:131–35. Japan: Sci. Counc. Jpn. 675 pp.

65. Karling, J. A. 1948. An *Olpidium* parasite of *Allomyces*. *Am J. Bot.* 35:503–10

66. Kim, K. S., Fulton, J. P. 1973. Plant virus-induced cell wall overgrowth and associated membrane elaboration. *J. Ultrstruct. Res.* 45:328–42

67. Kitazawa, K., Inagaki, H., Tomiyama, K. 1973. Cinephotomicrographic observations on the dynamic responses of protoplasm of a potato plant cell to infection by *Phytophthora infestans*. *Phytopathol. Z.* 76:80–86

68. Kunoh, H. 1972. *Mie Univ. Bull. Fac. Agric.* 44:141–224

69. Kunoh, H., Ishizaki, H. 1974. *Proc. Post Congr. Osaka Symp. Phytopathol.*, Sept. 9–11, pp. 23–26

70. Kusano, S. 1911. *Gastrodia elata* and its symbiotic association with *Armillaria mellea*. *J. Agric. Coll. Tokyo Univ.* 4:1–65

71. Kusano, S. 1936. On the parasitism of *Olpidium*. *Jpn. J. Bot.* 8:155–87

72. Kusano, S. 1939. Relation between the fungus infection and the callosity-formation of the hosts. *Ann. Phytopathol. Soc. Jpn.* 9:111–12

73. Lang, W. H. 1899. The prothallus of *Lycopodium clavatum* L. *Ann. Bot.* 13:279–317

74. Leitgeb, M. H. 1881. *Completoria complens* Lohde, ein in Farnprothallien schmarotzender Pilz. Sitzungsber. *Akad. Wiss. Wien. Math. Naturwiss. Kl.* 84:288–324

75. Lesemann, D. E., Fuchs, W. H. 1970. Die Ultrastruktur des Penetrationsvorganges von *Olpidium brassicae* on Kohlrabi-Wurzeln. *Arch. Mikrobiol.* 71:20–30

76. Lipetz, J. 1970. Wound-healing in higher plants. *Int. Rev. Cytol.* 27:1–28

77. Littlefield, L. J., Bracker, C. E. 1972. Ultrastructural specialization at the host-pathogen interface in rust-infected flax. *Protoplasma* 74:271–305

78. Luke, H. H., Warmke, H. E., Hanchey, P. 1966. Effects of the pathotoxin victorin on ultrastructure of root and leaf tissue of *Avena* species. *Phytopathology* 56:1178–83

79. Lupton, F. G. H. 1956. Resistance mechanisms of species of *Triticum* and *Aegilops* and of amphidiploids between them to *Erysiphe graminis* DC. *Trans. Br. Mycol. Soc.* 39:51–59

80. Mangin, L. 1895. *Bull. Soc. Hist. Nat. Autun.* 8:55–108

81. Mason, D. L. 1973. Host-parasite relations in the *Gibberidea* disease of *Helianthus strumosus*. *Mycologia* 65:1158–70

82. McKeen, W. E., Rimmer, S. R. 1973. Initial penetration process in powdery mildew infection of susceptible barley leaves. *Phytopathology* 63:1049–53

83. Mercer, P. C., Wood, R. K. S., Greenwood, A. D. 1971. *Ecology of Leaf Surface Micro-organisms*, ed. T. F. Preece, C. H. Dickinson, 381–89. London: Academic. 640 pp.

84. Mercer, P. C., Wood, R. K. S., Greenwood, A. D. 1974. Resistance to anthracnose of French bean. *Physiol. Plant Pathol.* 4:291–306

85. Mercer, P. C., Wood, R. K. S., Greenwood, A. D. 1975. Ultrastructure of the parasitism of *Phaseolus*

vulgaris by *Colletotrichum lindemuthi-anum. Physiol. Plant Pathol.* 5:203–14

86. Němec, B. 1904. Über die Mykorrhiza bei *Calypogeia trichomanis. Beih. Bot. Zentralbl.* 16:253–68

87. Nestler, A. 1898. Über die durch wundreiz bewirkten Bewegungser-scheinungen des Zellkernes und des Protoplasmas. *Sitzungsber. K. Akad. Wiss. Math. Naturwiss. Kl. I* 107: 708–30

88. Nichols, S. P. 1925. The effect of wounds upon the rotation of the protoplasm in the internodes of *Nitella. Bull. Torrey Bot. Club* 52: 351–63

89. Nims, R. C., Halliwell, R. S., Ros-berg, D. W. 1967. Wound healing in cultured tobacco cells following mi-croinjection. *Protoplasma* 64:305–14

90. Orcival, J. 1969. Infrastructure des sucoirs et relations hôte-parasite dans des stades ecidiens d'Uredinales. *C.R. Acad. Sci. D.* 269:1973–75

91. Pappelis, A. J., Pappelis, G. A., Kulfinski, F. B. 1974. Nuclear ori-entation in onion epidermal cells in relation to wounding and infection. *Phytopathology* 64:1010–12

92. Park, P. 1974. *Proc. Post Congr. Osaka Symp. Phytopathol.,* Sept. 9–11, pp. 32–35

93. Pearce, R. S., Withers, L. A., Willison, J. H. M. 1974. Bodies of wall-like material ("wall-bodies") produced intracellularly by cultured isolated protoplasts and plasmolysed cells of higher plants. *Protoplasma* 82:223–36

94. Pearson, N. L. 1931. Parasitism of *Gibberella saubinetii* on corn seedlings. *J. Agric. Res.* 43:569–96

95. Pfenninger, K. H., Bunge, R. P. 1974. Freeze-fracturing of nerve growth cones and young fibers. *J. Cell Biol.* 63:180–96

96. Politis, D. J. 1976. Ultrastructure of penetration by *Colletotrichum graminicola* of highly resistant oat leaves. *Physiol. Plant Pathol.* 8:117–22

97. Rice, M. A. 1927. The haustoria of certain rusts and the relation between host and pathogene. *Bull. Torrey Bot. Club* 54:63–153

98. Riggs, R. D., Kim, K. S., Gipson, I. 1973. Ultrastructural changes in Pe-king soybeans infected with *Hetero-dera glycines. Phytopathology* 63:76–84

99. Ritter, G. 1911. Über traumatotaxis und chemotaxis des zellkernes. *Z. Bot.* 3:1–42

100. Sargent, J. A., Tommerup, I. C.,

Ingram, D. S. 1973. The penetration of a susceptible lettuce variety by the downy mildew fungus *Bremia lactucae* Regel. *Physiol. Plant Pathol.* 3:231–39

101. Schnepf, E., Deichgräber, G., Hege-wald, E., Soeder, C.-J. 1971. Elektro-nenmikroskopische Beobachtungen an Parasiten aus *Scenedesmus*- Mas-senkulturen 3. *Chytridium* sp. *Arch. Mikrobiol.* 75:230–45

102. Sherwood, R. T., Vance, C. P. 1976. Histochemistry of papillae formed in reed canarygrass leaves in response to noninfecting pathogenic fungi. *Phytopathology.* 66:503–10

103. Skipp, R. A., Deverall, B. J. 1972. Relationships between fungal growth and host changes visible by light microscopy during infection of bean hypocotyls (*Phaseolus vulgaris*) sus-ceptible and resistant to physiological races of *Colletotrichum lindemuthia-num. Physiol. Plant Pathol.* 2:357–74

104. Skipp, R. A., Harder, D. E., Sam-borski, D. J. 1974. Electron microsco-py studies on infection of resistant (Sr6 gene) and susceptible near-iso-genic wheat lines by *Puccinia graminis* f. sp. *tritici. Can. J. Bot.* 52:2615–20

105. Slusher, R. L., Haas, D. L., Carothers, Z. B., Sinclair, J. B. 1974. Ultrastruc-ture at the host-parasite interface of *Phytophtora megasperma* var. *sojae* in soybean rootlets. *Phytopathology* 64:834–40

106. Smith, G. 1900. The haustoria of the Erysipheae. *Bot. Gaz.* 29:153–84

107. Stanbridge, B., Gay, J. L., Wood, R. K. S. 1971. See Ref. 83, pp. 367–79

108. Stavely, J. R., Pillai, A., Hanson, E. W. 1969. Electron microscopy of the development of *Erysiphe polygoni* in resistant and susceptible *Trifolium pratense. Phytopathology* 59:1688–93

109. Stevens, F. L. 1922. The Helmintho-sporium root-rot of wheat, with obser-vations on the morphology of *Hel-minthosporium* and on the occurrence of saltation in the genus. *Ill. Dep. Registr. Ed. Bull. Div. Nat. Hist. Surv.* 14:77–185

110. Swart, H. J. 1975. Callosities in fungi. *Trans. Br. Mycol. Soc.* 64:511–15

111. Tangl, E. 1884. Zur Lehre von der Continuität des Protoplasmas im Pflanzengewebe. *Sitzungsber. K. Akad. Wiss. Math. Naturwiss. Kl. I* 90:10–38

112. Temmink, J. H. M., Campbell, R. N. 1969. The ultrastructure of *Olpidium brassicae.* III. Infection of host roots. *Can. J. Bot.* 47:421–24

113. Tomiyama, K. 1956. Cell physiology

studies on the resistance of potato plant to *Phytophthora infestans*. IV. On the movements of cytoplasm of the host cell induced by the invasion of *Phytophthora infestans*. *Ann. Phytopathol. Soc. Jpn.* 21:54–62

114. Umrath, K. 1932. Die Bildung von Plasmalemma (Plasmahaut) bei *Nitella mucronata*. *Protoplasma* 16:173–88

115. Vance, C. P., Sherwood, R. T. 1976. Cycloheximide treatments implicate papilla formation in resistance of reed canarygrass to fungi. *Phytopathology*. 66:498–502

116. Walker, N. A. 1955. Microelectrode experiments on *Nitella*. *Aust. J. Biol. Sci.* 8:476–89

117. Wardrop, A. B. 1971. *Lignins: Occurrence, Formation, Structure and Reactions*, ed. K. V. Sarkanen, C. H. Ludwig, 19–41. New York: Wiley. 916 pp.

118. Wheeler, H. 1974. Cell wall and plasmalemma modifications in diseased and injured plant tissues. *Can. J. Bot.* 52:1005–9

119. Wheeler, H. 1975. *Plant Pathogenesis*. Berlin: Springer. 106 pp.

120. Williamson, B., Hadley, G. 1970. Penetration and infection of orchid protocorms by *Thanatephorus cucumeris* and other *Rhizoctonia* isolates. *Phytopathology* 60:1092–96

121. Wolff, R. 1873. Beitrag zur Kenntniss der Ustilagineen. *Z. Bot.* 31:673–77

122. Wyss, U. 1971. Der Mechanismus der Nahrungsaufnahme bei *Trichodorus similis*. *Nematologica* 17:508–18

123. Wyss, U. 1973. Reaktion von Cytoplasma und Zellkern in von *Trichodorus similis* besaugten Wurzelhaaren. Wirt: *Nicotiana tabacum*. *Mitt. Biol. Bundesanst. Land Forstwirtsch. Berlin-Dahlem* 151:305–6

124. Young, P. A. 1926. Penetration phenomena and facultative parasitism in *Alternaria, Diplodia* and other fungi. *Bot. Gaz.* 81:258–79

125. Ziegler, M. 1964. Untersuchungen über Wundreizreaktionen an Pflanzen. *Protoplasma* 44:350–62

THE PLANT DISEASE CLINIC—A THORN IN THE FLESH, OR A CHALLENGING RESPONSIBILITY

♦3637

Robert Aycock
Department of Plant Pathology, North Carolina State University, Raleigh, North Carolina 27607

Whether to operate a plant-disease clinic or not is a question that has perplexed many departments of plant pathology. The obligation to provide both instruction to students in diagnostic procedures, and service to the general public, often conflicts with other demands and goals that departments have set for themselves. This article examines these conflicts, relates some of the inherent problems in *combining* an instructional and a service arm of a department of plant pathology, and reports progress made in recent years in the United States toward fulfilling both commitments as they relate to diagnosis of plant diseases and other plant problems.

DIAGNOSIS

Diagnosis is the art or act of identifying a disease from the signs and symptoms (*Webster's Third New International Dictionary*). Although stated simply, the process is often not so simple—at least in the terms of modern agriculture. The predilection of plant pathologists to identify with physicians is a childish effort to claim status for a profession where none is needed. However, consideration of diagnosis in the medical context is useful. According to Ham (3), a physician is responsible for the care of the patient and for prevention of disease. This implies the need for accurate diagnosis, which depends on a professional evaluation of the person *and his environment,* and application of knowledge from the preclinical and clinical fields to the care of the patient. A physician (who practices the art) remains an *investigator* for life and, by the intellectual process of distinguishing

165

between fact and theory is judging information by a scientific method of thought. Diagnosis, prognosis, and treatment are derived from observations and data of various kinds, each of which requires critical analysis for validity and clinical significance. These sources of data usually include (a) information from the patient, including his activities and life-style as they may relate to his history of health or illness, (b) data obtained by physical examination, and (c) results of laboratory examinations. Diagnosis which is followed by treatment is then based on the evaluation and integration of these data.

In medicine, laboratory examinations may represent only a small portion of the data. The history of discomfort, which only the patient himself can give, is often of paramount importance and is an advantage over veterinarians that physicians are quick to acknowledge. It is an advantage that they also have over plant pathologists.

In plant pathology the patients are sick plants, or perhaps several thousand of them in many acres of plants if we are dealing with commercial agriculture. The grower himself may, in extreme cases, be sick, but professionally the plant pathologist can deal with him only indirectly as he attempts to ameliorate impending economic disaster. The plant diagnostician often plays a role different from that of the physician. The latter bases his diagnosis upon his own personal observations and history of the patient, plus whatever additional information he can obtain from a pathologist in a medical laboratory. The plant pathologist in the laboratory or clinic, on the other hand, is usually expected to make the diagnosis based on plant specimen material submitted to him from fields, forests, orchards, or groves, sometimes hundreds of miles away, and more often than not in poor condition. In the absence of a patient who can talk, therefore, the information accompanying a plant specimen is of paramount importance. Ideally, diagnosis should be made at the site where the plant grows, by someone who is knowledgeable about the host species and requirements for its successful culture and growth. The technical aspects of plant disease diagnosis have been dealt with in an excellent fashion by Streets (11).

Any report from the plant clinician becomes a valuable additional tool in the hands of a capable and experienced observer. The report reveals the presence of a known or previously unreported pathogen, rules out infectious agents, or suggests the possible role of unfavorable environment or imbalance of nutrients. It should not be considered more than a tool. If the experienced observer and the laboratory plant pathologist can be the same individual so much the better—but this is seldom the case in plant-disease clinics and diagnostic laboratories.

Although the principles of diagnosis in plant pathology, medicine, and veterinary science are much the same, the plant pathologists have some unique handicaps. Like his medical colleague, however, the plant diagnostician is an *investigator* in the truest sense of the word. By the intellectual process of distinguishing between observed or known fact and speculation or opinion, he is evaluating information by a scientific method of thought.

Diagnosis and prescription (recommendation) may often be based on expediency or urgent need. Because the diagnostician is often not sustained by a deep understanding of the biochemical and physiological nature of the particular problem there may be a lack of scientific appreciation by his research colleagues. Although the aim of diagnosis is to provide sufficient understanding to preserve or protect host plants, there are for plants, as Jubb & Kennedy have suggested for animals (6), many endemic and epidemic diseases concerning which our understanding is necessarily postponed in the interest of quickly finding a way to avoid disease or exert some control over it.

Both research and diagnosis are, therefore, absolutely necessary, and complementary to each other. It is, however, a paradox that preoccupation with research (10) has diverted the best minds in the profession from consideration of diagnosis per se and from refinement of old diagnostic procedures and discovery of new ones. In our value system, particularly that of the 1950s and 1960s, the education of the diagnostician received short shrift, and the need for professional acumen in this area was largely ignored.

Like many professions, plant pathology has been unable to maintain consistently that balance of learning and experience necessary to the health and longevity of our profession. We have worshipped at the altar of many false gods and climbed upon many bandwagons along the way—and it appears that we will continue to do so. Many of us have been unable or unwilling to differentiate between those things that have the greatest appeal at the moment and those things that will do most toward advancing the discipline.

We claim that our mission, basically, is to help people. Our science, as described by Stakman (10), was born out of need in the fields and granaries—"more than in halls of ivy." Society needed agriculture and agriculture needed plant pathology. The "emphasis on the practical," he said, was nothing to be ashamed of; plant pathology could take pride in the fact that it is really needed and not merely tolerated by society. The simple justification for the formation of the American Phytopathological Society, according to L. R. Jones (5), was to secure greater efficiency in *service*. Service, he said, always meant some distraction from ideals and always entailed some temporary sacrifice, but in "science as in other human pursuits, service in the end is indispensable to the highest development."

THE CLINIC

The word *clinic,* according to *Webster's Third New International Dictionary,* originates from a Greek term that means "medical practice at the sickbed." In modern terms, it means medical instruction, using patients. Webster's definition of first choice is "a session or class of medical instruction in a hospital held at the bedside of patients serving as case studies." The important point is, however, that in every definition of Webster, *instruction* is either stated or implied.

The relative proportion of instruction versus service has been an important consideration by academic departments in the establishment of plant disease clinics. A diagnostic service is one thing; a clinic is something else. The question of how well the two can be combined is at the crux of problems associated with their operation. The operation of a diagnostic service poses problems and certainly has its limitations. However, the availability of a plant pathologist who has curiosity and a sense of obligation will go a long way toward making the operation a success. Establishing a clinic, or place of instruction, even if combined with a diagnostic service, may be much more difficult. Some of the problems and limitations of each type of venture will be detailed in our later consideration of clinics now operating in the United States.

Certain educational benefits aside from direct student training and exposure should accrue in clinic operation. In states that have had operating clinics for a number of years, increased capability in disease diagnosis, particularly with certain crops, has been noted. Submission of only the atypical or hard-to-diagnose specimens has become routine with tobacco extension agents in North Carolina. Similar observations with major crops have been reported from other states, but this situation is by no means universal.

The value of clinics as teaching aids has never been fully appreciated or exploited, for the necessity of rapidly processing many plant-disease specimens of varying quality and the time and patience required for teaching students are not mutually supportive. It is probably true that many graduate advisers attached little importance to a student's experience in diagnostic work during what has been called the fleeting Golden Age of Graduate Education (1) in the 1950s and 1960s, a period when "graduate education became a major industry in its own right."

Students were not encouraged by the example set by many faculty. The excitement of tracking down a causal agent through scientific detective work was not emphasized. This was a treasure that was almost lost as the great mycological age faded away a generation before. There were, as always, some notable exceptions. Some students were not seduced by visions of instant fame in those highly specialized fields which had been sanctified by the major granting agencies. These pursued practical plant pathology for its own sake, but they were not considered among the elite of that period.

Attitudes and values of those years were in conflict, but Dean Pound's address (7) to the American Phytopathological Society in 1965, which pointed eloquently to the justifiable need for more research in basic biology, expressed a view which carried the day. His statement that most of our diseases had been described and that reasonable control measures had been worked out, spoken from the vantage point of unquestioned leadership, had a profound effect upon young people in our profession. The claim that biological research was being carried out increasingly by sophisticated instrumentation and that those researchers who failed to use the great technology available to them would settle into a lower category of eminence (7) reflected the mood of the period. In the shadow of Uncle Sam's largesse, our vision of L. R. Jones's concept of "greater efficiency in service" had become slightly blurred.

It was not that the emphasis on basic biology was wrong—the real sadness was that the leadership in scientific societies and in the profession seldom spoke out, encouraging excellence in applied research and obligation in service aspects of plant pathology. Although Dean Pound in 1968 (8) noted with alarm the growing disparity of funding between basic and production research and emphasized the need for balance in both areas, support and appreciation for all aspects of plant pathology was neither achieved nor maintained. Horsfall later (4) surmised that we were not very "smart outside." He called attention to Sharvelle's plea for relevance made also in 1965. We failed to "carry water on both shoulders." Our value system was corrupted.

Now partly because of employment opportunities, the pendulum has swung back again and students themselves cry for practical experience and apprenticeship training. It is indeed difficult to "manage pests" without being able to recognize them, or the damage they do. Many departments are concerned and are seeking to improve these opportunities, but there is, perhaps, more talk than progress. We as faculty assert our commitment to see that students are trained (by someone, it is not quite clear by whom) for all those important practical tasks which some of us have no intention of doing ourselves. The need for appropriate education in diagnostic methods and philosophy can now be justified and is generally accepted by the profession (even without calling attention to the world food crisis). How to provide the high quality experience in combination with the service aspects of diagnostic clinics is yet to be fully resolved.

WHAT ACADEMIC DEPARTMENTS ARE DOING

Data from a survey made by Ridings (9) in 1969, then a graduate student at North Carolina State University, showed that 25 states operated plant disease clinics. They were established to fulfill several functions: (a) to provide reliable and prompt disease diagnosis to growers and extension agents, (b) to train graduate students and faculty, (c) to instruct agents, (d) to record disease occurrences, (e) to provide insight into new disease situations, and (f) to coordinate and save time in departmental responses to inquiries from the field.

The situation appears to have changed markedly since that time. Data collected by the author recently from questionnaires circulated to all 50 states, showed that, of 48 institutions responding, 43 regularly operate diagnostic clinics or services.[1] Many of these not only do routine microscopic and cultural studies for fungi and bacteria but also nematode assays [35] and serological or other studies for virus determinations [24]. In states that do not operate clinics or ser-

---

[1] Information on clinics and diagnostic services has been gratefully received from the following individuals: L. L. Farrar, Alabama; C. E. Logsdon, Alaska; M. C. McDaniel, Arkansas; A. O. Paulus, California; L. E. Dickens, Colorado; R. B. Carroll, Delaware; M. S. Khan, District of Columbia; R. S. Mullin and T. A. Kucharek, Florida; N. E. McGlohon, Georgia; A. P. Martinez, Hawaii; H. S. Fenwick, Idaho; M. C. Shurtleff, Illinois; D. H. Scott, Indiana; A. H. Epstein, Iowa; W. G. Willis, Kansas; J. R. Hartman, Kentucky; R. B. Carver, Louisiana; G. Kleinschmidt, Maine; L. O. Weaver, Maryland; C. J. Gilgut and W. J. Manning, Massachusetts; F. Laemmlen, Michigan; W. C. Stienstra, Minnesota; W. F. Moore, Mississippi; E. W. Palm and C. H. Baldwin, Jr., Missouri; D. S. Wysong, Nebraska; J. Gallian, Nevada; W. E. MacHardy, New Hampshire; S. H. Davis, New Jersey; E. L. Shannon, New Mexico; A. F. Sherf and C. E. Williamson, New York; R. K. Jones, North Carolina; E. H. Lloyd, Jr., North Dakota; C. W. Ellett, Ohio; R. V.

vices, problems are handled by the appropriate extension and/or research specialist, or by the state's department of agriculture or regulatory agency. In some states, such as California and Florida, the state department, or its equivalent, plays a particularly strong and effective role. The number of specimens handled by university departments vary from 10–20 per year in Alaska to 11,000 in Georgia. The latter includes nematode assays, however. Plant specimens average over 4,000 per year in North Carolina. These include many soil samples accompanying diseased plants which are assayed for nematodes. They do not include soil samples submitted for assays to provide information necessary for predicting the advisability of soil fumigation for succeeding crops.

The sources of specimens sent to most clinics or diagnostic services include small growers, commercial farmers, nurserymen, foresters, flower growers, urban homeowners, extension agents, and commercial consultants of one kind or another. Nearly everyone feels that better service and greater educational opportunities are available when specimens are submitted by extension agents and reports are returned to them. In Georgia and Indiana only those from agents are processed. The percentage of specimens received from agents compared to those obtained from the general public varies considerably, but most institutions report that over half of the material submitted is from agents. Nearly everyone would like to work through agents, although how to legitimately turn away taxpayers, rural or urban, when specimens are brought in personally is a difficult question which most clinics have not satisfactorily resolved.

An increased competence in diagnostic ability among agents has been noted in a number of states as a result of the availability of clinics, particularly with respect to certain major commodities. Agents who change crop assignments are much more likely to send in large numbers of specimens the first year than in succeeding seasons. In North Carolina where tobacco is the largest cash crop, agents assigned to this commodity seldom send in specimens with infectious disease, or if so, only those showing atypical symptoms on resistant varieties. This reflects a longtime association of many of them (since 1951) with the plant disease clinic (2) where suspected problems were checked out and verified.

On the other hand, A. O. Paulus (personal correspondence) of California believes that the operation of "formal" clinics is not justified, at least in his state. Student examination of specimens mailed in would be a waste of time in 50–75% of the cases. Farm advisers are well trained in California, are excellent diagnosticians, and send in only highly unusual problems.

Although many departments expressed a keen disappointment in the progress made by a significantly large number of agents, the educational role of the clinic in agent training was generally recognized, except in a few states where the

Sturgeon, Jr., Oklahoma; I. C. MacSwan, Oregon; J. F. Tammen, Pennsylvania; D. B. Wallace, Rhode Island; F. H. Smith and W. M. Epps, South Carolina; M. A. Newman and C. Hadden, Tennessee; C. W. Horne, R. W. Berry, and J. Amador, Texas; A. R. Gotlieb, Vermont; R. C. Lambe, Virginia; O. C. Maloy, Washington; J. F. Baniecki, West Virginia; G. L. Worf, Wisconsin; G. H. Bridgmon, Wyoming.

agents' role is apparently very limited. The rapid turnover of agents in some states was cited as a serious handicap. Some respondents point out that the necessity of explaining diagnosis and control recommendations to growers is an important educational exercise in itself. The agent also learns how important it is sometimes to say, "I don't know." Other respondents emphasize that workshops, site visits, or field clinics are sometimes more useful than mere examination of limited plant material.

The utilization of graduate students in clinic operations varies greatly. In Delaware, students serve in the clinic if they are interested in extension or urban horticulture; in Georgia, one or two students assist and receive academic credit; in Hawaii the clinic serves as a practicum for graduate students; both Indiana and Illinois utilize half-time graduate students who are in charge of clinic operations, while all students participate; Mississippi utilizes two half-time MS students and a full-time PhD student. Iowa, Louisiana, and New Jersey make use of the clinic or clinic specimens in teaching courses in clinical plant pathology or plant-disease diagnosis. Minnesota uses the clinic in the last two years of the Plant Health Technology program. At North Carolina State University all graduate students participate in clinic service for two hours per day for a period of one week on a rotating basis during late spring and summer. Participation fulfills part of the requirements for a three-hour course entitled "Current Phytopathological Research under Field Conditions" in which students visit field-research activities and problem areas in the state. Graduate student participation is on a voluntary basis at a number of institutions including Kentucky, Wisconsin, and West Virginia, while Texas, Vermont, New Mexico, and Alaska report no participation by students.

Making the clinic experience of maximum mutual benefit to both student and clinician is a problem faced by most institutions. Most clinics, especially in their formative years are understaffed and the clinician is more often than not faced with a deluge of material, which should be diagnosed and reported on as soon as possible. Such pressures leave little opportunity to spend the time with beginning students so necessary in an exciting and profitable teaching experience. The number of students that can be handled is thus necessarily extremely limited. With more advanced students who have had previous experience, the situation is different. They are often able to play a responsible role in assisting in the diagnostic and reporting process. Usually, however, some supplementary effort must be made by clinicians or faculty if students are to receive the experience they need. In North Carolina, specimens are accumulated during the week and displayed on Friday with opportunities for students to work with the material and have informal conferences and discussions with faculty. Further adaptations are planned by this department and by numerous others reporting, as attempts are made to provide greater opportunities for students in diagnostic studies.

Most clinics or services are operated by extension faculty [25], but 15 respondents indicated that both research and extension faculty were responsible. Subprofessional assistance in the form of technicians is provided by 25 institutions. To be most successful, clinics should rely upon research, extension, and teaching faculty. The proximity of resource people in nearly all specialties in the larger departments argues for the advantage of locating such services in university departments rather than with state departments of agriculture or regulatory agencies.

Diseases new (or unknown) to the state or region have been discovered in 40 clinics. Witchweed (*Striga asiatica*), a parasitic seed plant of corn, was discovered in the clinic at North Carolina State University (2) by a graduate student from India who was familiar with related species on other hosts in his homeland. Findings have influenced the direction of research and extension programs, according to 31 respondents. These include programs on the oat cyst nematode and eastern filbert blight in Oregon, sorghum downy mildew on corn, wheat spindle streak virus, and anthracnose of corn in Indiana. Clinic observations have led to research on root rot and to the initiation of intensive research on nematode diseases of woody ornamentals in North Carolina. In Ohio, extension and research programs have been influenced by clinic findings; these include attention to southern corn leaf blight, soybean and wheat seed-quality problems, corn ear rots, and root rots.

Among trends noted during recent years are (*a*) increasing numbers of ornamental plants from urban dwellers, (*b*) more awareness by agents of disease damage, (*c*) increasing numbers of specimens and a greater percentage of difficult-to-diagnose specimens, with (*d*) a concomitant decrease in easily diagnosable diseases.

The clinician also has a problem with specimens that show symptoms of insect damage, herbicide injury, environmental or physiological stress, or nutritional imbalance. Some diagnose only infectious disease and refer other problems to appropriate departments. Others attempt to rule out as many factors as possible, even if an infectious disease is not present. At North Carolina State University the Departments of Entomology and Plant Pathology combined their efforts and established a plant disease and insect clinic in 1970. This merger has provided a greater convenience to agents and permitted more rapid diagnosis of specimens submitted. Preliminary reading on soil pH and soluble salts are also made if specimen examination so indicates. If unusual acidity or conductivity readings are obtained, the agent is notified immediately and further consultation is carried on with the appropriate specialist. Preliminary screening, using these procedures is very helpful and time-saving to those in the field. Illinois will probably be the first state to house in a single unit a coordinated plant-health diagnostic clinic involving personnel from plant pathology, weed science, soil science, entomology, the Natural History Survey, and botany.

A number of clinicians note the important point that it *does not* pay to advertise. Most of those who operate clinics feel that they would be swamped with poorly prepared specimens if wide publicity were given, but nearly all clinics provide free diagnostic services. One exception is the diagnostic service operated at Portageville, Missouri (C. Baldwin, personal correspondence), serving the Bootheel area. A team approach is used here that takes advantage of the expertise of the extension personnel located at this center, including a horticulturist, an agronomist, a weed control specialist, and a plant pathologist. No graduate students are available. A variable fee structure has been set for the services available, e.g. soil analysis, certain minor element detection, nematode assays. Meanwhile, the clinic at the University of Missouri, Columbia, is operated by Dr. Einor Palm in conjunction with a clinical plant pathology course, and combines teaching and service.

The Plant Disease Clinic, established at Pennsylvania State University in 1970, also operated on a fee basis for a number of years. It was eventually returned to the Department of Plant Pathology to be supported by regularly budgeted funds for departmental operation.

Pennsylvania State University now works very closely with the state department of agriculture to establish a statewide system for plant disease detection and early warning. Clinics of the two agencies have been completely integrated through a mutually accessible computer and the use of remote entry computer terminals in the clinics. Each agency has immediate access to the diagnosis and control recommendations by the other. It is hoped that the currently operational Blitecast system for forecasting potato late blight can soon be implemented on a fee basis. Apple scab and powdery mildew will also be included as this program is expanded and perfected. The eventual development of a nationwide computer network for plant disease detection and early warning information through such linkages in each state is strongly supported by the Penn State group.

A number of states now use mobile clinics, including Minnesota, Oklahoma, Virginia, Louisiana, and Illinois. In Oklahoma the mobile unit is part of a multi-crop pest-management program, while in Minnesota heavily populated urban areas are serviced. Mobile clinics are used in Virginia, but it is not possible to provide the same level of support because of limited technical assistance and travel funds. Staffing of mobile clinics in Texas was considered impossible because of the large distances involved.

Mobile clinics, according to most respondents, require too great an investment for the returns expected. It was emphasized (particularly by those who do not have them) that they were chiefly of value as a public-relations tool and that it would not be possible to provide information of the same quality as that available in a departmental clinic or diagnostic service. It appears that mobile clinics, except perhaps in very affluent states, or in very special circumstances, may not play a significant role on a continuing basis. Success in "bringing the mountain to Mahomet" will capture the attention of the public only until the event becomes commonplace. In the context of our current priorities for educational and research needs, the time does not appear ripe for widespread use of mobile clinics.

THE FUTURE

New techniques in diagnosis will undoubtedly be developed in the years ahead. The prospect for rapid and accurate identification of pathogens by procedures perhaps yet undreamed of may be available to our successors. Then as now, however, the successful operation of plant disease clinics will probably rest not so much upon the availability of sophisiticated techniques and hardware, as upon attitude, motivation, and sense of commitment shown by those responsible for diagnosis and prescription. Motivation (and attitude) are conditioned by the system of values in vogue during any particular period.

The art of plant pathology requires much knowledge, common sense, and skill, but those who have practiced it thus far have been considered second-class citizens by their peers. Our value system has not yet placed the emphasis on classroom teaching and extension (another form of teaching) that has been accorded research, and has not awarded deserved recognition to individuals in those categories. Preeminence has nearly always been associated with so-called basic research (bad or good), seldom with teaching, even less with extension. Moreover, attitudinal problems among both junior and senior scientists resulting from the "lushness of support" for basic research during the 1960s was noted by Pound (8). Erosion of institutional loyalties and an unwillingness to sacrifice individuality for larger goals became a serious problem of that period.

During the years since 1965, when Fellow awards were established by the American Phytopathological Society, only one extension plant pathologist has been accorded this honor. Even plant pathologists with nationally recognized educational or disease-control programs may be conspicuously absent from the roll call of our heroes. Our inability to establish a system of rewards (tangible or otherwise) which recognizes that competence, dedication, innovation, and learning are as esteemed in the art as they are in the science of plant pathology harms our profession and has influenced adversely attitudes of many young people entering the profession. Moreover, such attitudes have tempted extension people to neglect their primary educational responsibilities and move almost entirely into the applied-research field in order to gain that esteem and adulation accorded their colleagues in research. The results of such moves in some cases, admittedly, have been highly beneficial, but challenging opportunities to develop strong educational programs (including diagnostic training) have been sacrificed.

In order to ensure that our science and art remain strong, we must show our students the real world and demonstrate an attitude of appreciation toward excellence in all aspects of the art as well as the science. We must cultivate a sense of responsibility to those who pay the bill and reaffirm that our mission *is* service, no less than discovery.

Literature Cited

1. Dresch, S. P. 1974. *An Economic Perspective on the Evolution of Graduate Education.* Natl. Board Grad. Educ. *Tech. Rept. No.* 1:1–76
2. Ellis, D. E. 1955. The plant disease clinic at North Carolina State College. *Down to Earth* 11(3):4–5
3. Ham, T. H. 1951. *A Syllabus of Laboratory Examinations and Clinical Diagnosis.* Cambridge, Mass.: Harvard Univ. Press. 496 pp.
4. Horsfall, J. G. 1969. Relevance: Are we smart outside? *Phytopathol. News* 3(12):5–9
5. Jones, L. R. 1911. The relations of plant pathology to the other branches of botanical science. *Phytopathology* 1:39–44

6. Jubb, K. V. F., Kennedy, P. C. 1970. *Pathology of Domestic Animals,* Vol. 1. New York & London: Academic. 593 pp. 2nd ed.
7. Pound, G. S. 1965. *The Position of Plant Pathology in Current Administrative Changes in Biology and Agriculture.* Presented October 6, 1965, before the Am. Phytopathol. Soc. Miami Beach, Fla. (mimeo., 11 pp.)
8. Pound, G. S. 1969. A midstream view of plant pathology. *Phytopathol. News* (Excerpts from a paper presented to Am. Phytopathol. Soc., Columbus, Ohio. September 5, 1968) 3(2): p. 1, 3–4
9. Ridings, W. H. 1969. *Plant Disease Clinics in the USA. NC State Univ. Dep. Plant Pathol. Semin.* (mimeo.)
10. Stakman, E. C. 1964. Opportunity and obligation in plant pathology. *Ann. Rev. Phytopathol.* 2:1–12
11. Streets, R. B. 1972. *The Diagnosis of Plant Diseases. A Field and Laboratory Manual.* Tucson: Univ. Arizona Press. 234 pp. 2nd ed.

GENETIC CHANGE IN HOST-PARASITE POPULATIONS

Clayton Person
Department of Botany, The University of British Columbia, Vancouver, British Columbia, Canada

J. V. Groth
Department of Plant Pathology, University of Minnesota, St. Paul, Minnesota 55101

O. M. Mylyk
Department of Botany, The University of British Columbia, Vancouver, British Columbia, Canada

INTRODUCTION

Although the total amount of factual genetic information relating to population changes in host-parasite systems is small, the recent literature in phytopathology abounds with interpretation and discussion of events that have taken place or may have taken place at the population level. Accordingly, this review does not catalog the detailed results of a large number of studies. Its purpose is to review, in the light of the theory of population genetics, the mainly speculative literature that is accumulating. Because the existing observational and experimental data pertain almost exclusively to parasitic systems in which host populations have been manipulated by man, this review centers mainly on changes that have occurred, or may occur, in parasite populations as a direct consequence of human intervention.

The process by which pathogen populations change as a direct result of human activity was referred to by Johnson (17) as "man-guided" evolution. The need for accurate information that would have relevance to this interpretation was referred to in an earlier review (27). That need still exists. There is also a pressing need for adoption of a clearly defined set of terms (a problem recently discussed in another connection by Nelson, 26). It is also important to distinguish between organisms that reproduce asexually (either as haploids or diploids) and those that do not. Population genetics has traditionally dealt with those that do not reproduce asexually. Relatively little attention has been directed to organisms such as the rusts, mildews, and other parasites that undergo repeated cycles of asexual division which are only occasionally interrupted by nuclear fusion and meiosis, and in which genetic recombination may occur only occasionally or not at all.

177

THE ABSENCE OF A SEXUAL PHASE

The advantage conferred by sexual reproduction and meiotic recombination to asexually reproducing haploid species has been examined in a number of recent papers. In one of these, Maynard Smith (22) came to the conclusion that for haploid populations that are sufficiently large, sexual reproduction need not alter the rate of evolution. Because of its possible applicability to the mildews and other haploid parasites, it is worth examining the theoretical basis for this conclusion. Maynard Smith's analysis was applied to a large haploid population containing four genotypes, ab, Ab, aB, and AB, (fitnesses 1, $1 - H$, $1 - h$, and $1 - H \cdot 1 - h$, respectively) that were being maintained by recurrent mutation ($a \rightarrow A$ at rate u_A; $b \rightarrow B$ at rate u_B) at two genetic loci. He showed that at equilibrium the four genotypes would be present in proportions described by the relationship $p_{ab} \cdot p_{AB} = p_{Ab} \cdot p_{aB}$. It was because this relationship remained unchanged when genotypes ab, Ab, aB, and AB were assigned altered fitnesses ($1 + K \cdot 1 + k$, $1 + k$, $1 + K$, and 1, respectively) that he concluded that sex would not confer an evolutionary advantage. The interesting point is that, providing the population is sufficiently large, it will already contain both single and double mutants; i.e. recombinants that might otherwise have been produced by nuclear fusion and meiosis. A number of subsequent analyses of this particular problem have been summarized by Maynard Smith (23), who states that the different assessments (arrived at by other authors) of the relative advantage of sex to haploid, asexually reproducing populations were, in effect, predetermined by the different sets of conditions that were assumed by these others to exist. But providing the populations are of sufficient size, haploid asexually reproducing populations will be capable of evolution even in the absence of a sexual phase.

PATHOGEN POPULATIONS

Would the Maynard Smith conclusion apply to an asexually reproducing mildew population? Data are badly needed to answer questions of this kind. Information is needed on mutation rates, on relative fitnesses of "wild" and mutant genotypes, and on population sizes. Here we may take our wild population as one which is flourishing on a susceptible host, and which is producing mutants that have potential virulence on a resistant variety that has not yet been introduced (cf discussion by Day 3, p. 180). Day's estimate of the rate of mutation (10^{-8}) was based on the rate of back mutation in *Neurospora*. It seems to us that an assumed rate of 10^{-7} would have been realistic. [Leijerstam, cited by Wolfe (41), estimated mutation frequencies to be as high as 2000 per locus per hectare per day for certain loci of *Erysiphe graminis* f. sp. *tritici*.]

As with mutation rates, the relative fitnesses of wild and mutant genotypes is also a subject for speculation. If we assume that the host that supports large numbers of parasites is fully susceptible, offering little or no resistance to the parasite (as would be the case if a resistance gene were needed), the difference in fitness could be relatively small. (The discrimination between virulent and avirulent genotypes is normally made by resistant hosts.) And, where the difference in fitness between virulent and avirulent genotypes is small, the expectation is that relatively large numbers of mutant genes would be maintained on susceptible hosts. [The relationship is described by $q = u/s$; with mutant gene a at a 1% reproductive disadvantage ($s_a = 0.01$) and assumed mutation $A \rightarrow a$ at a rate $u_A = 10^{-7}$, the frequency (q_a) of mutant gene a becomes 10^{-5}; where the same set of assumptions is applied to two different loci the frequency of double mutants would be 10^{-10}; and for the lower mutation rate, $u_A = 10^{-8}$, assumed by Day (selection coefficients remaining at 0.01) the frequency of the double mutant becomes 10^{-12}]. For an estimate of population size we can take Slootmaker's estimate of 10^9 lesions per hectare, each producing 10^4 conidia per day (these figures were quoted by Day 3, p. 180), which led to an expected total production of 10^{13} conidia per hectare per day. In 1975 about 3.5×10^6 hectares were sown to barley in the United States. Although it is not certain that this large area would be occupied by a unit population of *Erysiphe,* it is clear that the totals for lesions and daily conidial production for a mildew population could exceed those quoted by Day by several orders of magnitude.

This estimate suggests that, even in the absence of a sexual stage, potentially virulent single-mutant genotypes would almost certainly be maintained on a continuing basis in large asexually reproducing mildew populations. In addition the estimate suggests that double mutants will probably also be maintained. When it is recognized that sexual reproduction and genetic recombination, in fact, do occur, this possibility (in our opinion) is also elevated to the level of near certainty.

These expectations would not be substantially different for diploid or dikaryotic populations (e.g. the rusts) that reproduce vegetatively over long periods of time. Providing the populations are sufficiently large, these populations also would be capable of evolution even in the absence of a sexual stage (C. Person and J. V. Groth, unpublished).

From the foregoing discussion we think it incorrect to assume that sequential changes in virulence have to occur as a direct consequence of sequential mutation in large parasite populations. In our opinion such changes are more reasonably explained in terms of changed conditions of selection which have operated sequentially on preexisting genotypes that are maintained in genetically heterogeneous parasite populations. Thus, for the mildew population described by Day (3, p. 180) it would have been equally reasonable *not* to have assumed that the 10^9 lesions present per hectare all developed from a single initiating infection. In our view the initial inoculum (i.e. the "spore shower") from which lesions have developed to this density should be regarded as a sample of genotypes generated by, and arriving from, an earlier mildew population. Depending on the size of this sample, mutant genotypes could have been present in the initiating inoculum.

If it is accepted that single-mutant genotypes are continuously maintained in large parasite populations, and that many different genetic loci are capable of producing mutants, it would follow that many different kinds of single-mutant genotypes are continuously present. For some of these loci the normal and mutant alleles will have the potential of functioning as alleles for avirulence and virulence, if and when matching R genes are introduced into the host population. For each such locus the frequency of the mutant allele (the allele with potential virulence) will be determined by its fitness (relative to that of the normal allele) and by its rate of production via mutation. These frequencies are expected to vary from locus to locus: for some potential v genes the numbers will be relatively large and for others relatively small. Although these frequencies will be determined quite independently of the matching R gene (which is not yet part of the system) their assessment is made possible only when the presence of virulent genotypes is revealed through use of the matching R gene. For v genes that were maintained on susceptible hosts in unusually high frequencies, the parasite population will be judged to have "preexisting" virulence against the R gene which (we have assumed) has not been previously present. The preexisting virulence in stem rust to specific resistance of oats fits this expectation. Martens et al (21) have reported that this virulence occurs widely in stem rust populations even though the matching R gene has never been incorporated into commercially grown oat varieties. Vanderplank (35, p. 73) discusses a similar example, involving high frequencies in *Phytophthora* of matching virulence on the R_4 gene in areas in which potatoes with this gene have not been grown. In our opinion both these examples show that equilibrial frequencies for certain mutant genotypes can be unexpectedly high. Where this is found to be the case, the possibility that the mutant gene has pleiotropic effects should not be overlooked. For reasons outlined later in this review we would not postulate that stabilizing selection does not operate, (cf Martens et al 21) or that high rates of forward and reverse mutation (cf Vanderplank 35, p. 73) are involved.

DIRECTIONAL SELECTION, STABILIZING SELECTION, AND AGGRESSIVENESS

The foregoing serves as a starting point for discussing directional selection, stabilizing selection, and aggressiveness. The basic assumptions will be that large parasite populations growing on susceptible hosts will be genetically heterogneous, that they will contain preexisting or potential v genes at a number of different loci, that the frequencies of these will vary from locus to locus, and that because of genetic heterogeneity in the population as a whole, the normal and mutant subpopulations in respect to any particular locus need not be genetically homogeneous.

These assumptions, as they would apply to any particular genetic locus are partly illustrated in Figure 1, where the "quadratic check" is separated into two parts, and where the two interactions relevant to susceptible hosts are shown at the left side. With the matching R gene not yet part of the system, the equilibrial frequency of mutant genotype would be described by $q = u/s$. If we now imagine that the matching R gene has been introduced, this preexisting equilibrium will be immediately disturbed. Interaction with resistant hosts is now possible and, as illustrated on the right-hand side of Figure 1, every such interaction will involve a reversal in relative fitnesses of A and a. Also, because A reproduces poorly (or perhaps not at all) in interaction with resistant hosts, the difference in relative fitness on resistant hosts is expected to be much larger than it was on susceptible hosts. On the mixed host population, genotypes A and a will initiate disease reactions on both resistant and susceptible hosts. On resistant hosts genotype a will be favored by directional selection; on susceptible hosts genotype A will be favored by stabilizing selection. These opposing selection pressures are also indicated in Figure 1.

On the mixed host population, and with directional and stabilizing selection both operating, the important parameters will be (a) selection coefficients operating against a and A when these interact with susceptible and resistant hosts, respectively; and (b) the proportions of total disease interactions that take place on each of the two kinds of host. These are indicated in Figure 1, where the

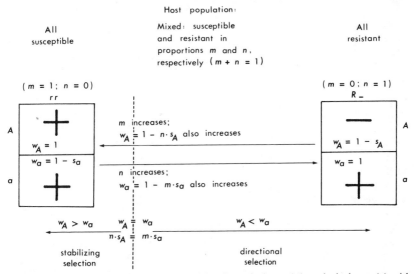

Figure 1   The interactions of parasite alleles for avirulence (A) and virulence (a) with susceptible (rr) hosts are shown at the left, and those with resistant ($R-$) hosts at the right side of the figure. Taken collectively these four interactions represent a quadratic check, in which the plus sign denotes host susceptibility and parasite virulence, and the minus sign denotes host resistance and parasite avirulence. For interactions taking place on susceptible (rr) hosts, alleles A and a are assigned fitness values of $w_A = 1$ and $w_a = 1 - s_a$, respectively; for interactions taking place on resistant ($R-$) hosts, A and a are assigned fitness values of $w_A = 1 - s_A$ and $w_a = 1$, respectively. In making these assignments it is assumed that the reproductivities of A on susceptible hosts and of a on resistant hosts are identical. The figure also shows that on mixed populations in which the proportions of susceptible and resistant hosts are represented by m and n, respectively ($m + n = 1$), there is one set of values of m and n for which fitnesses of A and a would be equal. From this equilibrial position ($n \cdot s_A = m \cdot s_a$) the net effect of an increase in the proportion of resistant hosts (n) would be attributable to directional selection, and the net effect of a decrease in the proportion of resistant hosts would be attributable to stabilizing selection.

simplifying assumption has been made that the fitnesses of A on susceptible hosts and of a on resistant hosts are identical. Setting these at unity, fitness $1 - s_A$ is assigned to genotype A for its interaction with resistant hosts, and fitness $1 - s_a$ is assigned to genotype a for its interaction with susceptible hosts. If we now set the proportions of total disease interactions on susceptible and resistant hosts at m and n, respectively, $(m + n = 1)$, the fitnesses of genotypes A and a when interacting with mixed host populations will be as follows:

$$w_A = m + n(1 - s_A) = 1 - n \cdot s_A, \text{ and}$$
$$w_a = m \cdot (1 - s_a) + n = 1 - m \cdot s_a,$$

from which it can be seen not only that relative fitnesses of A and a are dependent on m and n, but also that these fitnesses will be equal when $n \cdot s_A = m \cdot s_a$. This relationship was described earlier by Leonard (18a), whose analysis was based on measured reproductivities of oat stem rust on mixed host populations. Because the selection coefficient against A on resistant hosts is greater than it is against a on susceptible hosts (i.e. $S_A > S_a$) this equality is expected to hold only when the proportion of total interactions taking place on resistant hosts is small (i.e. when $m > n$). At this position of equality, $(w_A = w_a)$, no change in relative frequencies of A and a is expected. But for this particular position of equality, the stability of relative frequencies is highly sensitive to small changes in relative fitnesses. With a slight increase in the proportion (m) of interactions on susceptible hosts, the net result will be stabilizing selection, and similarly, with a slight decrease in m the net result will be directional selection. A direct relationship is also to be expected between the amount by which m (and therefore n) is changed and the magnitude of the new selection pressure that the change invokes.

Stabilizing selection was defined by Vanderplank (35, p. 60) as the force that keeps virulence genes rare: "This force is stabilizing selection in favor of races of the pathogen with no unnecessary virulence." We have just seen that this force is selection pressure acting against genotype a on susceptible hosts and that the condition of stability toward which stabilizing selection would tend is also the equilibrial condition (described by $q = u/s$) that would be achieved when the opposing pressures of mutation and stabilizing selection are balanced. We have seen in addition to this that these opposing pressures would operate even though the matching R gene was totally absent from the system. In fact the selection pressure operating against a will be maximal when the matching R gene is totally absent from the system.

These relationships raise serious doubts concerning the validity of Vanderplank's classification of R genes according to their strength. He states "We define the strength of a resistance gene in terms of the strength with which stabilizing selection acts against the gene's complementary race" (35, p. 73). This definition does not accommodate the fact that the strength of stabilizing selection is maximal when the matching R gene is totally absent. There is no way that a totally absent R gene can influence or be influenced by stabilizing selection. As for the R gene whose presence is assumed, the definition does not accommodate the fact that the kind of selection which takes place (i.e. whether it is directional or stabilizing) and its intensity are both determined by the frequency of the matching R gene in the host population.

For parasite populations in which directional selection has taken place, and in which a has largely replaced A, total removal of the R gene would initiate stabilizing selection. But even in this special case, because of the relatively small fitness difference separating A and a on susceptible hosts, the relatively slow replacement of a by A would extend over a relatively long time interval. Clearly the continuing high frequencies of matching v genes following removal (complete or otherwise) of R genes cannot be taken as proof positive that stabilizing selection is not taking place. Yet several authors have questioned stabilizing selection on this basis (e.g. 11, 21, 28). Vanderplank did present persuasive evidence for stabilizing selection against matching virulence on gene R_1 when this gene had been removed. However, the role previously played by directional selection in establishing the frequencies of virulent genotypes that existed before stabilizing selection began to operate (and, if we may assume that some R_1 genes were not removed, in opposing stabilizing selection during its operation) was not considered by Vanderplank.

In his recent book Vanderplank formulated a second gene-for-gene hypothesis, as follows: "In host-parasite systems in which there is a gene-for-gene relationship, the quality of a resistance gene in the host determines the fitness of the matching gene to survive *when this virulence is unnecessary;* and reciprocally the fitness of the virulence gene to survive *when it is unnecessary* determines the quality of the matching resistance as judged by the protection it can give to the host" (36, p. 148; italics ours). From the phrases we have italicized it is evident that this definition would apply to situations in which the R gene is not present. Accordingly, the second part of the definition, which relates the level of preexisting virulence to the potential effectiveness of a matching R gene that has not been present, is quite acceptable. However, for reasons outlined earlier, the first part of the definition is not acceptable; the relationship described by $q = u/s$ that would maintain virulence when it is unnecessary would not be influenced in any way by the absent R gene. The causal connection (36, p. 156) that would imply "that the lack of fitness to survive when virulence is unnecessary is no haphazard accident of creation, but is determined by the matching resistance gene" is, in our opinion, undemonstrable.

The Vanderplank proposals concerning nonspecific components of disease interaction (i.e. horizontal resistance, aggressiveness) stand as valuable theoretical contributions to our developing understanding of the genetic complexities of host-parasite interaction. Vanderplank proposed "that races which do not interact differentially with varieties of the host be said to vary in aggressiveness"; that "pathogenicity includes both aggressiveness and virulence"; and that "there is no evidence for a positive correlation between aggressiveness and virulence; as far as is known, greater aggressiveness is not linked with greater virulence" (35, p. 16–17). Robinson extended these proposals by pointing out that the absence of differential interaction would lead to a system of "constant ranking of pathodemes and pathotypes" (31). Although a few examples of oligogenically controlled horizontal resistance were cited by Vanderplank, both he and Robinson expressed the opinion that horizontal resistance would more usually be under polygenic control. And, for resistance that is known to be polygenically determined, it was Robinson's opinion that such resistance would always be manifested as horizontal resistance. This viewpoint, when extended to include the parasite, would lead to the expectation that the polygenically controlled nonspecific component of pathogenicity will always be manifested as aggressiveness. The discussion of aggressiveness that follows is based on this viewpoint.

Selection in parasite populations is expected to operate on both the specific and nonspecific components of fitness. For the nonspecific component (aggressiveness) its expected effect is to increase the mean fitness of the parasite population regardless of the genetic composition of the host population, and regardless of whether directional and stabilizing selection may be taking place simultaneously. However, since we have assumed that many genes each with small effect are likely to be involved, selection for aggressiveness is expected to proceed more slowly than selection for virulence. Evidence for selection operating in this way (on entire genotypes rather than on specific alleles for virulence) was presented by Green (11) who interpreted the "erosion" of effectiveness of resistance genes $Sr7a$ and $Sr11$ in wheat as being due to the action (in the parasite) of modifiers that had accumulated at genetic loci other than those responsible for specific virulence. In our opinion this interpretation is correct, since it properly relates erosion of resistance to nonspecific genetic changes (i.e. to selection for aggressiveness) in the parasite. The specifically adapted variants found within virulent races (16, 18a) also can be regarded as products of selection for aggressiveness. But these opinions are not supported by a substantial body of genetic information. Relatively little is known about the genetic basis of aggressiveness. Based on cultures derived from two teliospores, Emara & Sidhu (7) concluded that variability in aggressiveness of *Ustilago hordei* was polygenically determined. Ebba & Person (5), also working with *U. hordei,* showed that a changed genetic background had a profound effect on the expression of specific virulence and that this effect was expressed in only one of two test environments. Other evidence for quantitative differences in the expression of specific virulence has been obtained in studies with *Puccinia recondita* (4, 13) and with *Puccinia hordei* (2). Some of the problems associated with genetic study of aggressiveness are referred to by Wolfe (41).

Evidence for polygenic determination of horizontal (nonspecific) resistance is more substantial. Studies are being carried out on a larger number of systems (1, 4, 8, 9, 12, 13, 15, 18, 19, 24, 32–34, 37–40) and, in addition to quantitative differences attributable to polygenes, maternal effects have been reported (19, 20). This area of study is also not free of difficulty. Hayes (14) has pointed out that specific resistance will have the appearance of horizontal resistance so long as parasite strains with matching virulence remain undiscovered. Riley (29) has reviewed the evidence that major genes that have been overcome may also contribute to nonspecific resistance. In this interesting paper Riley emphasizes the urgent need for accurate genetic information relating to nonspecific host resistance and to parasite aggressiveness. In our opinion this need would include a search for, and studies of, systems that do in fact show constant ranking.

MANAGEMENT OF DISEASE

Along with virulence and aggressiveness, which have been treated here as specific and nonspecific components of parasite fitness, we must also regard environmental influences as key components of parasite fitness. Even though the parasite may be genetically competent, its reproductive success also will be determined by temperature, humidity, and other factors of the physical environment. These are not normally controllable. By controlling the genetic composition of host populations with which parasites must interact in order to reproduce, however, plant breeders can, and do, manipulate the genetic components of parasite fitness. Seen from this point of view, the primary objective of breeding for disease resistance is to reduce the mean fitness of parasite populations. Extending this view to epidemiology, the objective of plant breeding is to reduce the extent to which parasite fitness contributes to R, the parameter that Vanderplank described as the rate of increase of disease per unit of infectious tissue per unit of time (36, p. 91).

Different approaches to this objective have been reviewed on numerous occasions. They include (a) use of multigenic varieties (several R genes are incorporated into a single cultivar), (b) use of multiline varieties (several R genes are incorporated, singly, into different constituent lines of a composite cultivar), (c) localized usage of R genes, (d) cycling of R genes, (e) use of polygenic resistance, and (f) combinations of the foregoing.

Although use of multigene cultivars would undoubtedly result in a maximal initial loss of parasite fitness, it would also initiate selection favoring the extremely rare genotype with matching virulence. For reasons given earlier in this review we think that extremely rare genotypes may be less rare, and parasite populations larger, than many authors have assumed them to be. The population whose size approximates the inverse of the frequency of a rare genotype is likely to contain it. We do not agree with those who have assumed that selection for the rare matching genotype would be made possible only after a highly improbable sequence of mutational events has occurred (e.g. Day 3, Green 11, Nelson 26, Vanderplank 35); a population large enough to maintain single mutants would be capable of producing recombinants by sexual or parasexual processes. Where the multigene variety is not universally grown, or where the parasite can exist on residual hosts (e.g. wild grasses, 6) the parasite population may be of sufficient size to generate rare matching genotypes.

Compared with a multigene cultivar, a smaller initial reduction in parasite fitness is to be expected when a multiline variety is introduced; also, the initial reduction in size of the parasite population is expected to be smaller. For each of the several R genes contained in the multiline, the interactions with matching A and a alleles will conform with the conditions illustrated in Figure 1. The fitness of each virulence allele will relate directly to n, the proportion of the total host population that contains the matching R gene. When n is made smaller, fitness of the matching v allele decreases. In theory it should be possible (by adding R genes to multiline varieties in suitably adjusted proportions) to eliminate directional selection for matching v genes. This was first proposed by Leonard (18a), who advocated the inclusion of a fully susceptible component in the multiline. But, so long as directional selection has not been eliminated in this way, selection would favor parasite genotypes containing added matching v alleles. Whether the genotype with greatest fitness would in fact contain *all* the required matching v genes cannot be surmised, since this would depend on whether the effects of stabilizing selection are cumulative (i.e. each compatible interaction occurring in the multiline would involve directional selection for only one of the several matching v genes; each such interaction would also involve stabilizing selection for all other v genes which, in the particular interaction, are unnecessary; we would need to know whether there is an added increment of stabilizing selection for each added, but unnecessary, v gene). The immediate reduction in fitness of the parasite population, following introduction of a multiline variety, has been described in terms of reduced infectivity (and reduced "r") by Frey et al (10). The theory of multilines illustrates again that the link that connects epidemiology with population genetics of pathogens is one that connects Vanderplank's term R and the fitness of the pathogen populations.

The localized usage and the cycling of R genes would both impose temporal discontinuities in the process of directional selection for matching virulence. Geographical localization would be most effective against parasites that migrate over long distances; and cycling against those that do not. In both cases the selection would be disruptive; alleles that were favored at one time are selected against at another. In theory the process of stabilizing selection could be switched on and off at will; directional selection could be placed under human control; used R genes could be stored and used again at some later date.

Polygenic resistance is regarded by many as stable resistance. As mentioned by Riley (29) the supporting evidence for this point of view is largely empirical. A theoretical analysis has been carried out (28) which did support the conclusion that, compared with major-gene resistance, polygenic resistance would be relatively much more stable.

Geographic localization and cycling of R genes are strategies that could be used in conjunction with the multigene and multiline approaches. Horizontal resistance could also be made an integral part of any management program. In our opinion, however, it is premature to attempt to identify any "best" approach at this time (cf Nelson, 26). We agree with Riley's (29) opinion that major genes should not be abandoned. The various possible strategies need to be rigorously assessed, and there is too little information for making assessments. Until it is known which of the possible strategies will most extensively prolong the useful life span of a single R gene, no intelligent choice can be made. And, as Robinson has shown (30), we should not expect to discover any single stratagem that will work for all disease problems.

Literature Cited

1. Boukema, I. W., Garretsen, F. 1975. Uniform resistance to *Cladosporium fulvum* Cooke in tomato (*Lycopersicon esculentum* Mill.). 1. Investigations on varieties and progenies of diallel crosses. *Euphytica* 24:99–104
2. Clifford, B. C., Clothier, R. B. 1974. Physiologic specialization, of *Puccinia hordei* on barley hosts with non-hypersensitive resistance. *Trans. Br. Mycol. Soc.* 63(3):421–30
3. Day, P. R. 1974. *Genetics of Host-Parasite Interaction.* San Francisco: Freeman. 238 pp.
4. Dyck, P. L., Samborski, D. J. 1974. Inheritance of virulence in *Puccinia recondita* on alleles at the Lr_2 locus for resistance in wheat. *Can. J. Genet. Cytol.* 16:323–32
5. Ebba, T., Person, C. 1975. Genetic control of virulence in *Ustilago hordei.* IV. Duplicate genes for virulence and genetic and environmental modification of a gene-for-gene relationship. *Can. J. Genet. Cytol.* 17:631–36
6. Eshed, N., Wahl, I. 1975. Role of wild grasses in epidemics of powdery mildew on small grains of Israel. *Phytopathology* 65:57–63
7. Emara, Y. A., Sidhu, G. 1974. Polygenic inheritance of agressiveness in *Ustilago hordei. Heredity* 32:219–24
8. Ferreira, S. A., Webster, R. K. 1975. Genetics of stem rot resistance in rice and virulence in *Sclerotium oryzae. Phytopathology* 65:968–71
9. Ferreira, S. A., Webster, R. K. 1975. The relationship of sporulation, sclerotia production, and growth rate to virulence and fitness of *Sclerotium oryzae. Phytopathology* 65: 972–76
10. Frey, K. J., Browning, J. A., Simons, M. D. 1973. Management of host resistance genes to control diseases. *Z. Pflanzenkr. Pflanzenschutz* 80:160–80
11. Green, G. J. 1975. Virulence changes in *Puccinia graminis* f. sp. *tritici* in Canada. *Can. J. Bot.* 53:1377–86
12. Habgood, R. M. 1974. The inheritance of resistance to *Rhyncosporium secalis* in some European spring barley cultivars. *Ann. Appl. Biol.* 77:191–200
13. Haggag, M. E. A., Samborski, D. J., Dyck, P. L. 1973. Genetics of pathogenicity in three races of leaf rust on four wheat varieties. *Can. J. Genet. Cytol.* 15:73–82
14. Hayes, J. D. 1973. Prospects for controlling cereal disease by breeding for increased levels of resistance. *Ann. Appl. Biol.* 75:140–44

15. Jennings, C. W., Russell, W. A., Guthrie, W. D. 1974. Genetics of resistance in maize to first- and second-brood European corn borer. *Crop Sci.* 14:394–98
16. Johnson, R., Taylor, A. J. 1972. Isolates of *Puccinia striiformis* collected in England from the wheat varieties Maris Beacon and Joss Cambier. *Nature* 238:105–6
17. Johnson, T. 1961. Man-guided evolution in plant rusts. *Science* 133:357–62
18. Leakey, C. L. A. 1973. A note on *Xanthomonas* blight of beans (*Phaseolus vulgaris* L. Savi) and prospects for its control by breeding for tolerance. *Euphytica* 22:132–40
18a. Leonard, K. J. 1969. Selection in heterogeneous populations of *Puccinia graminis* f. sp. *avenae. Phytopathology* 59:1851–57
19. Lim, S. M. 1975. Diallel analysis for reaction of eight corn inbreds to *Helminthosporium maydis* race T. *Phytopathology* 65:10–15
20. Lunsford, J. N., Futrell, M. C., Scott, G. E. 1975. Maternal influence on response of corn to *Fusarium moniliforme. Phytopathology* 65:223–25
21. Martens, J. W., McKenzie, R. I. H., Green, G. J. 1970. Gene-for-gene relationships in the *Avena: Puccinia graminis* host-parasite system in Canada. *Can. J. Bot.* 48:969–75
22. Maynard Smith, J. 1968. Evolution in sexual and asexual populations. *Am. Nat.* 102:469–73
23. Maynard Smith, J. 1974. Recombination and the rate of evolution. *Genetics* 78:299–304
24. Muehlbauer, F. J., Kraft, J. M. 1973. Evidence of heritable resistance to *Fusarium solani* f. sp. *pisi* and *Pythium ultimum* in peas. *Crop Sci.* 13:34–36
25. Nelson, L. R., Scott, G. E. 1973. Diallel analysis of resistance of corn (*Zea mays* L.) to corn stunt. *Crop Sci.* 13:162–64
26. Nelson, R. R. 1975. *Breeding Plants for Disease Resistance.* Univ. Park, Pa.: Penn. State Univ. Press. 401 pp.
27. Person, C. 1968. Genetical adjustment of fungi to their environment. *The Fungi—an Advanced Treatise.* Vol. III, Chap. 15. New York & London: Academic. 738 pp.
28. Person, C., Groth, J. V. 1973. Constant ranking and stability in polygenically based host-parasite systems. Presented at 2nd Int. Congr. Plant Pathol., Minneapolis, Minn.

29. Riley, R. 1973. Genetic changes in hosts and the significance of disease. *Ann. Appl. Biol.* 75:128–32
30. Robinson, R. A. 1971. Vertical resistance. *Rev. Plant. Pathol.* 50:233–39
31. Robinson, R. A. 1973. Horizontal resistance. *Rev. Plant Pathol.* 52:483–501
32. Shepherd, R. L. 1974. Transgressive segregation for root-knot nematode resistance in cotton. *Crop Sci.* 14:872–75
33. Simons, M. 1975. Heritability of field resistance to the oat crown rust fungus. *Phytopathology* 65:324–28
34. Sparnaaij, L. D., Garretsen, F., Bekker, W. 1975. Additive inheritance of resistance to *Phytophthora cryptogea* Pethybridge and Lafferty in *Gerbera jamesonii* Bolus. *Euphytica* 24:551–56
35. Vanderplank, J. E. 1968. *Disease Resistance in Plants.* New York & London: Academic. 206 pp.
36. Vanderplank, J. E. 1975. *Principles of Plant Infection.* New York, San Francisco & London: Academic. 216 pp.
37. Weerapat, P., Sechler, D. T., Poehlman, J. M. 1974. Barley yellow dwarf virus resistance in crosses with Pettis oats. *Crop Sci.* 14:218–20
38. Wilcox, J. R., Laviolette, F. A., Martin, R. J. 1975. Heritability of purple seed stain resistance in soybeans. *Crop Sci.* 15:525–26
39. Wilkins, P. W. 1975. Inheritance of resistance to *Puccinia coronata* Corda and *Rhyncosporium orthosporum* Caldwell in Italian ryegrass. *Euphytica* 24:191–96
40. Wilkins, P. W. 1975. Tolerance to ryegrass mosaic virus and its inheritance. *Ann. Appl. Biol.* 78:187–92
41. Wolfe, M. S. 1973. Changes and diversity in populations of fungal pathogens. *Ann. Appl. Biol.* 75:132–36

VIRUS AND VIRUS-LIKE   ✦3639
DISEASES OF CEREAL CROPS[1]

John T. Slykhuis[2]
Ottawa Research Station, Agriculture Canada, Ottawa, Ontario

This review includes diseases caused by agents recognized or assumed to be viruses or mycoplasma-like organisms that have been found to infect cereals in the field. The cereals included are Gramineae that are grown for edible seed, principally wheat (*Triticum aestivum, T. compactum,* and *T. durum*); barley (*Hordeum vulgare* and *H. distichon*); oats (*Avena sativa* and *A. byzantina*); rye (*Secale cereale*); maize (*Zea mays*); rice (*Oryza sativa*); sorghum (*Sorghum vulgare*); and millets (mainly *Setaria italica, Panicum miliaceum,* and *Eleusine coracana*).

The first disease of cereals found to be caused by an infectious agent, later recognized as a virus, was the dwarf disease of rice in Japan. It was the first plant virus shown to be transmitted by an insect, the first shown to multiply within an insect, and the first to be recognized with an electron microscope in both the host plant and the vector insect (26).

Other early discovered viruses found to infect cereals include the aphid-transmitted sugarcane mosaic virus reported by Brandes in southern USA in 1919 (66), the planthopper-transmitted maize mosaic virus (corn stripe) reported by Kunkel in Hawaii in 1921 (38), the leafhopper-transmitted maize streak virus reported by Storey in Africa in 1925 (8), and the soil-borne wheat mosaic virus (rosette) reported in the USA in 1925 (12). The latter was the first plant virus shown to be carried from season to season in soil.

During the 1930s viruses were recognized as the causes of several more diseases of cereals including rice stripe in Japan (50), wheat streak mosaic in the USA (11), and oat pseudo-rosette in the USSR (96). About seven more were recognized during the 1940s, but virus diseases were not recognized to be common or very significant in most cereal-growing areas of the world until the mid-to-late 1950s. Now virus diseases are known on cereals in most areas of the world where cereals are grown.

---

[1]Contribution number 450, Ottawa Research Station, Agriculture Canada, Ottawa.

[2]Present address: Research Station, Agriculture Canada, Summerland, British Columbia, Canada.

The symptoms and effects of the virus diseases affecting cereals vary widely. The symptoms include various patterns of leaf discoloration, deformations, and stunting. The effects on yield of grain range from insignificant to loss of the entire crop in many fields. Such effects are influenced by the virulence of the virus, the extent and time of infection, the susceptibility of the crop, the simultaneous presence of other viruses and diseases, and the effects of cultural and environmental conditions. Since the cereals are annuals and are not propagated vegetatively, unless a virus is seed borne its severe incidence in any crop will depend on a rapid spread into the crop during a short critical period each crop year.

Despite the wide range of effects produced by the different viruses, their spectra of symptoms and host range overlap so much that few can be identified by host reactions alone. However, the differentiation of most cereal virus diseases has been aided by studies on morphology with the electron microscope, physicochemical properties, and serological relationships. Some virus diseases have been described under different names in different countries. For example, the diseases of rice called *penyaket merah* in Malaya, *yellow orange leaf* in Thailand, and *leaf yellowing* in India all appear to be identical with diseases described earlier under other names including *rice tungro* in the Philippines (50). The barley yellow dwarf virus is recognized as the cause of oat red leaf, oat yellow leaf, cereal yellow dwarf, and probably millet red leaf and other diseases as well as barley yellow dwarf. On the other hand, Rochow (71) pointed out that some of the variants of what is now called *barley yellow dwarf virus* are so different serologically and in interactions in cross protection tests that they probably should be considered distinct viruses. Hence, the problems in differentiating the cereal viruses are not all resolved.

For this discussion the viruses are grouped according to their primary means of spread, and by particle morphology. They are designated by the common names used in *Plant Virus Names,* compiled by the Commonwealth Mycological Institute (54), and *Descriptions of Plant Viruses,* published by the Commonwealth Mycological Institute and the Association of Applied Biologists, England, 1971 to 1974. These publications also list synonyms and cryptograms not included in this review.

SEED-BORNE VIRUSES

Only the following two cereal-infecting viruses have been reported to be carried from generation to generation primarily in the seed.

Barley Stripe Mosaic Virus   Described by McKinney in 1951 (1). Also called Barley false stripe virus, Barley mosaic virus, and Oat stripe mosaic virus. Consists of short, tubular particles 128 × 20 nm. Seed borne in barley and wheat, and mechanically transmissible. Causes mosaic and striping of barley, wheat, oats, rye, maize, sorghum, millets, a number of other grasses, and some dicotyledonous plants. Distributed worldwide with seed.

Barley Mosaic Virus   Described by Dhanraj & Raychaudhuri in 1969 (24). Particles isometric 40 nm diam. Reported to be seed borne and transmitted by aphids (*Rhopalosiphum maidis*) to barley and transmitted mechanically, causing mosaic, chlorosis, and stunting of barley, oats, and wheat in India.

Barley stripe mosaic virus (BSMV) is the only virus affecting cereals that is known to be widely distributed with seed (57). Although its spread from diseased to healthy plants occurs through leaf contact and transfer of pollen, no insect or other vectors are known. It is most readily found in plant introduction nurseries and is most likely to spread in breeders' plots where plants are handled and pollen transferred in crossing experiments. Perhaps cultivars like Glacier and Compana barley were infected at very early stages in the breeding program. In the first years after BSMV was recognized, most seed lots of these varieties contained a trace to 50% infected seed, and infection probably occurred in other seed lots in which it was not detected. In some samples of Glacier, 89% of the seed produced diseased plants, and yields were reduced by 31% (25).

The disease is difficult to eliminate from infected seed lots. Screening out poorly filled kernels has limited effect because seeds of normal size also may be infected (41). The virus is not inactivated while the seed remains viable in long-term storage or by heat treatments (35, 97). Even under the best conditions for symptom expression, some barley plants that do not show symptoms may carry the virus and produce infected seed (56). Serological methods for rapid detection of BSMV in single seeds have been useful in detecting BSMV in seed lots and hence have aided in limiting the occurrence of this disease (36).

SOIL-BORNE VIRUSES

Because wheat mosaic (wheat rosette) was the first serious disease of wheat shown to be caused by a virus and because it was the first plant virus known to be soil borne, soil transmission was thought to be a special trait of the cereal mosaics. For a period, most newly discovered cereal mosaics were suspected to be caused by soil-borne viruses until proven otherwise. Listed below are soil-borne viruses now recognized in cereals.

Transmission Associated with the Fungus, Polymyxa graminis

PARTICLES ROD-SHAPED

Wheat (soil-borne) mosaic virus (WSBMV)   Described by McKinney in 1925 (12). Stiff rod-shaped particles of two lengths 110–160 nm and 300 nm. Causes light green to yellow mosaic or rosetting with extreme stunting and excessive tillering in some varieties. Affects wheat, barley, and rye in USA, Japan, and Italy.

Barley yellow mosaic virus (BYMV)   Described by Miamoto in 1958 (42, 100). Rod-shaped particles 100–600 × 15–20 nm. Causes chlorotic mottle on barley in Japan.

PARTICLES FILAMENTOUS

Oat mosaic virus (OMV)   Described by McKinney in 1946 (16, 37b, 53, 55). Filamentous particles 700 nm long. Causes varying degrees of chlorotic mottling, streaking, or spotting of oats (*Avena sativa* and *A. byzantina*) in eastern USA, Washington State, and England.

Wheat spindle streak mosaic virus (WSSMV)   Described by Slykhuis 1970 (87, 107). Filamentous particles 2000–3000 × 12 nm. Causes light green to yellow spindle-shaped streaks and necrosis of leaves, reduced heading, and slight stunting of winter wheat in Canada (Ontario) and eastern USA.

Vector Not Known, Particles Rod-Shaped

Rice necrosis mosaic virus (RNMV)   Described by Fujii et al in 1967 (50). Particles 275 and 550 × 13–14 nm. Causes necrotic lesions on basal portions of stems and sheaths, elongated yellowish flecks especially on lower leaves, decumbent growth of the whole plant. Found on rice in Japan.

Transmitted by Soil-Inhabiting Nematodes, Particles Isometric

Brome mosaic virus (BMV)   Described by McKinney et al in 1942 (4). Particles 25 nm diam. Reported to be transmitted by nematodes *Xiphinema paraelongatum* and *X. coxi* in Germany. Causes mosaic on barley, wheat, oats, rye, and necrosis of maize. Reported in USA, South Africa, Germany, Finland, and Russia.

Some of the above viruses, WSBMV, OMV, BYMV, and WSSMV, have many characteristics in common. They cause mosaics and other effects on fall-sown or winter-grown but not on spring-sown cereals. Infection of each is associated with a root-infecting fungus, *Polymyxa graminis* (Plasmodiophorales) (12, 53, 100); it takes place during and after the seedling stage, in the fall or early winter depending on the climate. Symptoms develop in the spring.

Although all these viruses are transmissible mechanically by the leaf rub method, transmission is generally more efficient by the airbrush spray technique (98). The viruses are unstable in crude leaf extracts but the infectivity of the viruses persists for several years in soil stored moist or dry (55, 80, 87). However, soil can be rendered noninfectious by heating it to 55°C or higher, or by treating it with certain fungicidal chemicals.

WSBMV (12) has been differentiated into two strains: yellow mosaic and green mosaic. The latter induces a rosetting type of growth in some cultivars. Similar strains serologically related to each other and to the USA strains occur in Japan. An apparently similar virus occurs in Italy, but it has a wider range of graminaceous hosts than reported for the viruses from the USA and Japan.

WSSMV has been found only on wheat. No strains have been differentiated. It consists of particles as long or longer than any other known virus (87, 107), and its temperature requirements appear to be the lowest for any known virus. The optimum temperature for infection from soil is about 15°C, but symptoms do not develop unless the plants are grown predominantly at 10°C (90).

Usually the soil-borne viruses whose transmission is associated with the fungus, *Polymyxma graminis,* affect their host cereals only mildly, but severe damage can result under conditions favoring disease development. For example, there appears to be a widespread latent infection of wheat spindle streak mosaic virus in most soils in Ontario (87). Growing wheat frequently in the same field increases the infectivity of the soil. When infectivity is high, disease development depends on the susceptibility of the crop and on the weather. Most commercial cultivars of winter wheat grown in Ontario are susceptible. Prolonged periods in spring with temperatures between 5° and 15°C favor severe development of symptoms. Rapid onset of summer with temperatures predominantly above 20°C suppresses disease development.

Although the temperature optima may not be as low for WSBMV, OMV, and BYMV as for WSSMV, somewhat similar cool growing conditions appear to favor the development of these diseases. Under such conditions WSBMV caused almost complete crop failure in susceptible varieties in Illinois in some years (46). In 1957, approximately 347,000 acres of wheat were diseased in Kansas causing an estimated loss of about 2,000,000 bushels resulting from reductions in yield of 5 to 12 bushels per acre in many fields (81). Serious crop losses also have been attributable to barley yellow mosaic, oat mosaic, and wheat spindle streak mosaic when cool temperatures favored the diseases.

Even before the cause of wheat rosette was known in Illinois in the early 1920s, some varieties of wheat were found to be resistant, and resistant lines were selected from susceptible varieties (46). Losses in barley and oats also can be minimized by using varieties resistant to the soil-borne viruses affecting them (55, 86). Some varieties of wheat from the Ukraine are highly resistant to WSSMV in Ontario.

Although the viruses persist for long periods in the soil, the diseases undoubtedly could be controlled by long intervals between susceptible crops as is evident with WSSMV in Ontario (87). Also the regular use of poultry and animal manures has beneficial effects. Although soil can be disinfected by heating to 55°C or higher, and by treating with some fungicidal chemicals, these measures are not practical in North America. In Japan, Miyamoto (58) prescribed the use of flame to heat the soil, or drenching the seed furrows with dilutions of pyroligneous acid or wood vinegar, a by-product of charcoal production, to control the soil-borne wheat and barley mosaics.

RNMV, which also consists of rod-shaped particles, is the only soil-borne virus reported to infect rice (50). Since 1959 it has been found in eight prefectures, and was reported infecting 269 ha in Okayama prefecture in 1964. In severely affected areas, 47% of the grain was lost. Symptoms often take 2 months or longer to develop from soil-borne infection. The disease can be controlled by avoiding the reuse of upland seed beds in which the disease has occurred unless the soil has been sterilized with heat or chemicals, or by raising seedlings in wet bed nurseries. Some rice varieties are resistant.

Brome mosaic virus (BMV), first reported in the USA, probably has the widest range of grass hosts of any mechanically transmissible virus, but it does not appear to spread extensively in the USA except with clones of plants and by physical association of plants. However, it has been reported on wheat, oats, and barley in Finland (15), South Africa (104), and Russia (103). It was reported to be transmitted by *Xiphinema* spp. to barley (4). When transmitted mechanically to cereals, BMV can cause severe mosaic and death of some cultivars of maize, but it does not appear to have spread extensively enough to cereals in the field to cause serious crop losses.

BEETLE-TRANSMITTED VIRUSES

The following two viruses have been reported to be transmitted to cereals in the field by chewing insects.

Cocksfoot mottle virus   Described by Serjeant in 1963 (18). Particles isometric 30 nm diam. Transmitted mechanically and in a semipersistent manner by *Lema melanopa* and *L. lichenis*. Causes mottling of wheat as well as *Dactylis glomerata* in England.

Rice yellow mottle virus   Described by Bakker in 1970 (2, 3). Particles isometric 32 nm diam. Transmitted mechanically and in a semipersistent manner by *Sesselia pusilla,* a number of other beetles and grasshoppers. Causes stunting, reduced tillering, crinkling, mottling, yellow streaking, sterility, sometimes death of rice in Kenya.

Cocksfoot mottle virus was the first virus known to be transmissible to cereals by a chewing insect. The virus is widespread in *D. glomerata* seed crops in the east Midlands of England in the third harvest year. Although the beetles *L. melanopa* and *L. lichenis* can transmit the virus efficiently in experiments, animal grazing and cutting implements appear to be more important to spread within each field. The virus can be transmitted readily to wheat, oats, and barley experimentally, but it has been reported in cereals only once, in a wheat crop (7). The absence of the virus in cereals results from inefficient transmission by the vectors in the field. The beetles do not move into cereals until after the cereals have begun to tiller and are no longer susceptible.

RYMV, reported only near Lake Victoria in Kenya, appears to cause serious damage to rice. It is noteworthy in being transmitted by a wide range of chewing insects, i.e. beetles in seven families, and a species of grasshopper.

ERIOPHYID MITE-TRANSMITTED VIRUSES

Since 1952, two mechanically transmissible viruses, and a pathogen with undetermined morphology, have been shown to be transmitted by mites.

Filamentous Particles, Transmissible Mechanically

Wheat streak mosaic virus (WSMV)   Described by McKinney in 1937 (11). Particles 700 × 15 nm. Transmitted by *Aceria tulipae* and *A. tosichella* (99). Causes severe mosaic, stunting, and necrosis resulting in devastating losses of wheat. Also infects barley, maize, oats, rye, millets, and other grasses. Reported in USA, Canada, Jordan, Rumania, Russia, Turkey (14), and Yugoslavia.

Agropyron mosaic virus (AMV)   Described by McKinney in 1937 (89). Particles 717 × 15 nm. Transmitted by *Abacarus hystrix*. Causes mosaic on wheat and *Agropyron repens* in USA, Canada, Finland, and Germany.

Particle Uncertain, Not Transmissible Mechanically

Wheat spot mosaic agent (WSpMA)   Described by Slykhuis in 1956 (85). [Also called Wheat spot chlorosis (61).] Transmission by *Aceria tulipae*. Found associated with WSMV. Causes chlorotic spotting, chlorosis, and necrosis on wheat, barley, maize, oats, rye, annual grasses. Found in Canada (Alberta) and USA (Ohio).

Two of the viruses listed above, WSMV and AMV, have flexuous thread-like particles and a number of characteristics in common. Both are readily transmitted mechanically and cause chlorotic dashes and streaks in the leaves and various degrees of stunting. They have a number of hosts in common, but they are not closely related serologically. WSpMA has not been transmitted mechanically.

WSMV is one of the most spectacular pathogens of wheat. Despite its destructiveness WSMV can be controlled by simple management practices. It has caused total loss of the wheat crop in many winter wheat fields in the Great Plains areas of North America from Oklahoma and Kansas to Montana and Alberta, Canada. It also occurs in Ohio, Michigan, and Ontario. It has also been found in Jordan and a number of countries in Europe (86) including Turkey (14). *Aceria tulipae*, a mite, is the major vector but *A. tosichella* has been described as a new vector of WSMV in Yugoslavia (99). The mites, which are dispersed principally by wind, feed with piercing-sucking mouthparts. They multiply parthenogenetically and can complete the life cycle from egg through two nymphal instars to mature adult in 7–10 days at 24–27°C. The mites acquire the virus after feeding for 15 minutes or longer in the nymphal stages and can carry it through the molts. Although they do not acquire infectivity during the adult stage, all feeding stages can transmit the virus. Usually 30–50% of the mites can transmit the virus when tested singly. High concentrations of WSMV particles have been found in the mid and hind gut of *A. tulipae* (64).

WSMV probably originated in a perennial grass but it can be perpetuated on susceptible annuals. In many areas, such as in southern Alberta, winter wheat crops sown early in the fall are subject to infection from virus-carrying mites that may be present on any immature wheat or other susceptible plants in or adjacent to the field. Both the mites and the virus survive the winter in the wheat plants and then multiply and spread rapidly when the weather turns warm the next spring and summer. Any immature spring or volunteer wheat, or regrowth from hail-damaged crops, as well as other susceptible grasses, can perpetuate the virus and mites after the winter wheat crop matures, and if still present, these plants provide the sources of infection for fall-sown crops. Cultural control of wheat streak mosaic can be achieved easily in most areas simply by eliminating over-summering hosts of the virus and mites before sowing winter wheat.

The WSpMA was discovered when *A. tulipae* was found to be a vector of WSMV in 1952 (85) and has been isolated repeatedly in Alberta. A similar disease called wheat spot chlorosis was found in Ohio (61). Ovoid, double-membrane bodies 0.1–0.2 nm in diam. were reported in plants infected with the latter (10). Like WSMV, the WSpM agent is carried efficiently by all stages of *A. tulipae* except the eggs, and mites retain infectivity up to 12 days on immune hosts. Single mites may simultaneously carry both WSMV and WSpMA. Wheat infected with both agents is more severely stunted and dies more quickly than plants infected with either alone. Since new lines of wheat resistant to WSMV may not be resistant to the WSpMA, it appears highly desirable to develop resistance to the mites (47).

AMV has been found on wheat as well as on *Agropyron repens* in various locations in the USA and Canada, and in several countries in Europe. In Ontario we find it each year in *A. repens* and in wheat growing in close association with diseased *A. repens,* but the extent of infection in wheat is limited, and the mosaic symptoms very mild. The limited spread into wheat is related to the low efficiency of transmission by the vector mite, *Abacarus hystrix.* There are strains of AMV in Ontario that affect wheat severely. If these were to spread rapidly into wheat when the plants are very young, significant yield losses could result.

Ryegrass mosaic virus, also transmitted by *A. hystrix,* can be transmitted to oats and rice, but it has not been reported on cereals in the field (94).

APHID-TRANSMITTED VIRUSES

Many plant viruses are transmitted by aphids, including about 12 that have been found to infect cereals. These may be considered in two distinct groups, *circulative viruses* and *stylet-borne viruses.* The circulative viruses, which are ingested by the aphids and pass from the gut by way of the hemolymph to the salivary glands, are introduced into new plants with saliva while the aphids are feeding. The aphids cannot transmit the virus immediately after acquiring it, but once they become infective they remain so for long periods. None of the circulative viruses are transmissible manually by the leaf rub method. In contrast, the stylet-borne viruses are retained by the vector for only short periods, usually less than 24 hr, but are readily transmitted to plants mechanically. The viruses in each group are either isometric or filamentous.

Circulative Viruses

PARTICLES ISOMETRIC

Barley yellow dwarf virus (BYDV)   Described by Oswald & Houston in 1951 (71). Particles 20–24 nm diam. Transmitted by 14 spp. of grass-feeding aphids including *Machrosiphum avenae, Metopolophium dirhodum, Rhopalosiphum maidis, R. padi,* and *Schizaphis graminum.* Causes chlorosis and stunting of oats, barley, wheat, millet, sorghum, maize, and many other grasses. Distribution worldwide.

Rice giallume virus   Described by Baldacci et al in 1970 (50, 63). Particles 20 nm diam. Transmitted by *R. padi.* Causes yellowing and stunting of rice in Italy.

PARTICLES FILAMENTOUS

Wheat yellow leaf virus   Described by Inouye et al in 1973 (44). Particles flexuous 1850 × 10 nm. Transmitted semipersistently by *R. maidis.* Causes yellowing, reddening, and leaf blight. Infects wheat, barley, oats, rye, and several other grasses. Found in several prefectures in Japan.

PARTICLES UNDETERMINED

Maize leaf fleck virus   Described by Stoner in 1952 (95). Transmitted by *Myzus persicae, R. maidis,* and *R. padi.* Causes yellow to orange spots, tip and marginal burn, and necrosis of leaves of maize and *Phalaris tuberosa* var. *stenoptera.* Found in California.

Stylet-Borne Viruses

PARTICLES ISOMETRIC

Cucumber mosaic virus   Also called southern celery mosaic strain. Described by Wellman in 1934 (30). Particles isometric 30 nm diam. Transmitted by more than 60 spp. of aphids including *Aphis gossypii, M. persicae, R. maidis,* and *R. padi.* Infects many dicots and has been transmitted to sorghum, wheat, and rye. Causes chlorotic stripes, stunting, and necrosis on maize in the USA and Yugoslavia (65, 106).

Cardamom (greater) mosaic streak virus   Described by Raychaudhuri & Chatterjee in 1961 (29, 69). Particles isometric 40 nm diam. Transmitted by *R. maidis* and *Brachycaudus helichrysi* on wheat and greater cardamom (*Amomum subulatum*) in eastern India.

PARTICLES FILAMENTOUS

Sugarcane mosaic virus (SCMV)   Described by Brandes in 1919 (66). Particles 750 nm long. Transmitted by many aphids including *Dactynotus ambrosiae, Hysteronura setariae, R. maidis,* and *S. graminum.* Causes mosaic of sugarcane, maize, sorghum, millet, wheat, barley, rye, rice, and other grasses. Reported in many parts of the world. Different strains are important on different cereals.

Maize dwarf mosaic strain   Described by Williams & Alexander in 1965 (66). Causes mosaic and occasional reddening and dwarfing of maize in the USA.

Strain A infects *Sorghum halepense.* Strain B does not infect *S. halepense* or sugarcane. [Maize mosaic virus, reported in India by Chona & Seth in 1960 (21, 73) appears to be similar to strain B.]

Sorghum red stripe strain   Described by Lovisolo in 1957 (66). Causes mosaic, red striping, and necrosis of sorghum and infects maize and *S. halepense* in Yugoslavia.

PARTICLE UNDETERMINED

Eleusine mosaic virus   Described by Venkatarayan in 1946 (68). Transmitted by *R. maidis, M. persicae,* and five other spp. Causes a mosaic on *Eleusine coracana, E. aegyptii,* and *E. indica* and can infect *Sorghum vulgare, Zea mays,* and *Setaria italica* in India. (Probably a strain of SCMV.)

BYDV is the most common and widely distributed of the viruses recognized on small grains. About 100 species of grasses, but no dicotyledonous plants, have been reported to be susceptible (71). Transmission of the virus depends on many factors including aphid species or clones, virus isolates, temperature, test plant species, age, and condition of the source plant.

Five strains of BYDV have been differentiated on the basis of their vector specificity as follows:

1. RMV—Transmitted regularly by *Rhopalosiphum maidis* but infrequently by *R. padi, Macrosiphum avenae,* and *Schizaphis graminum;* weakly virulent in Coast Black oats.
2. RPV—Transmitted regularly by *R. padi,* erratically by *S. graminum,* and rarely by *R. maidis* and *M. avenae;* weakly virulent in Coast Black oats.
3. MAV—Transmitted regularly by *M. avenae* but rarely by *R. padi, R. maidis,* and *S. graminum;* moderately virulent in Coast Black oats.
4. PAV—Transmitted regularly by *R. padi* and *M. avenae,* erratically by *S. graminum,* and rarely by *R. maidis;* strongly virulent in Coast Black oats.
5. SGV—Transmitted regularly by *S. graminum* but rarely by *M. avenae, R. padi,* or *R. maidis;* weakly virulent in Clintland 64 oats.

Antisera have been prepared against the MAV, PAV, and RPV strains. Serological tests showed a close relationship between MAV and PAV but indicated that RPV is different. This difference, confirmed by cross protection and other tests, indicates that RPV could be considered an unrelated virus.

Losses from BYDV in spring oats and barley in the northern USA and Canada are usually less in early sown crops that are well established before the virus carrying aphids arrive than in late sown crops. In some situations where aphids survive the winter on cereals, as occurs in New Zealand, spraying with systemic organophosphate insecticides to kill the aphids has been effective in reducing secondary spread of BYDV, thereby increasing the yields of wheat and barley (93). Many cereal breeders are selecting for tolerance to the strains of BYDV prevalent in their area. The most effective control appears to be improved tolerance combined with cultural practices to avoid infection at critical stages of crop development.

Neither of the two stylet-borne viruses with isometric particles causes a major widespread disease of cereals. One of these, the southern celery strain of cucumber mosaic virus, was one of the first manually transmissible viruses recognized in cereals. It has been found in maize in Florida since 1919, sometimes causing death of some sweet corn seedlings (55, 94).

Cardamom (greater) mosaic streak, or Cardamom "chirke" has been reported to cause a streak mosaic on some varieties of wheat growing near cardamom in the mountainous regions of eastern India, but other varieties were resistant or tolerant.

Strains of SCMV or related viruses are the only stylet-borne viruses that have been of widespread concern in cereal crops. Some strains of SCMV have been reported in maize and sorghum in many parts of the world where sugarcane, *Sorghum halepense* (Johnson grass), or other hosts harbor them (66). A number of strains, which have been distributed with sugarcane cuttings, cause mosaic on sugarcane and rarely if ever infect *S. halepense.* Sorghum red stripe and maize mosaic in southern Europe are attributed to SCMV perpetuated beyond the range of sugarcane on *S. halepense* and on spontaneous plants of *S. vulgare.*

Maize dwarf mosaic has caused considerable concern in a number of corn belt states since its discovery in Ohio in 1962 (108). It caused chlorotic streaks, stunting, and early death of some varieties of maize. The virus has similar particles, is serologically related, and has aphid vectors and many hosts in common with other strains of SCMV. Its distribution is related to the presence of *S. halepense,* which grows as a perennial weed and harbors the virus. Several additional perennial grasses have also been reported to be susceptible. Seed transmission (0.4%) has been reported in maize. Carry-over to a few plants via seed could provide an initial infection in a new crop from which the virus could be spread by aphids.

LEAFHOPPER- AND PLANTHOPPER-TRANSMITTED VIRUSES

About 25 viruses of cereals are transmitted by leafhoppers in the family Cicadellidae (Deltacephalidae) in the superfamily Cicadelloidea, and by planthoppers in the family Delphacidae in the superfamily Fulgoroidea. Two of them are now characterized as mycoplasma-like organisms with pleomorphic bodies, a third as spiroplasma, and the others as viruses with isometric, bacilliform, or flexuous, thread-like particles. However, the morphology of the causal agent is uncertain for some diseases including rice grassy stunt, oat pseudo-rosette, and rice hoja blanca; two types of particles have been associated with all of these diseases.

The viruses and mycoplasma are transmitted by either leafhoppers or planthoppers. They have many similarities in relationships to their hosts and their respective vectors. None is transmissible to plants by leaf rub methods, and none is seed or soil borne or transmitted by other arthropods. Some have several vectors in the same family of insects, but I know of no virus or mycoplasma infecting cereals that is transmitted both by a leafhopper and a planthopper.

Leafhopper (Cicadellidae)-Transmitted Viruses

LEAFHOPPER-TRANSMITTED MYCOPLASMA-LIKE ORGANISMS

Aster yellows   Described by Kunkel in 1926. Transmitted to cereals by *Macrosteles fascifrons, Endria inimica,* and *Elymana virescens* in USA and Canada (5, 19, 20) and by *M. laevis* in Europe (59). Causes chlorotic blotches, sterility, and stunting on barley, wheat, oats, rye, and other grasses and infects many dicots in North America and Europe.

Maize stunt   Described by Alstatt in 1945 (23, 32). A spiroplasma transmitted by *Dalbulus maidis, D. elimatus,* and *Graminella nigrifrons.* Causes chlorotic to bronze stripes, stunting and bushy growth on maize, teosinte, and *Tripsacum* sp. Found in southwestern USA, Mexico, Central America, Cuba, and Puerto Rico.

Rice yellow dwarf   Described by Iida and Shinkai in 1950 (50). Transmitted by *Nephotettix cincticeps, N. impicticeps,* and *N. apicalis.* Causes yellowing, stunting, excess tillering, and reduced panicle development in rice. Also infects *Alopecurus aequalis, Glyceria acutiflora,* and *Oryza cubensis.* Found in Japan, Taiwan, Malaysia, Philippines, and many other countries of Asia.

LEAFHOPPER-TRANSMITTED VIRUSES WITH ISOMETRIC PARTICLES

Maize chlorotic dwarf virus   Described by Rosenkranz in 1969 (9, 57a, 62, 72). Particles 26–31 nm diam. Transmitted in a semipersistent manner by *Graminella nigrifrons* (lost ability to transmit in 48 hr), and infrequently by *Deltocephalus sonorus.* Causes stunting, chlorotic blotches on leaf blades, reddish purple discoloration of tips and margins of leaves of sweet corn. Found in maize and Johnson grass in Ohio, Kentucky, Illinois, and some southern states.

Maize streak virus   Described by Storey in 1925 (8). Particles 20 nm diam. Usually in pairs 30 × 20 nm. Transmitted by *Cicadulina mbila, C. storeyi, C. bipunctella zeae, C. latens,* and *C. parazeae.* Causes chlorotic spots and streaks on maize, wheat, oats, millet, and *Eleusine* in Africa and Southeast Asia.

Maize wallaby ear virus   Described by Schindler in 1942 (34). Particles 85 nm diam. Transmitted by *Cicadulina bipunctata* and *Nesoclutha pallida.* Causes swellings in secondary veins on undersurface of the top leaves, dwarfing, abnormally deep green color, and reduction in cob and grain growth of maize in Australia.

Oat blue dwarf virus   Described by Goto & Moore in 1952 (6). Particles 28–30 nm diam. Transmissible by 25–30% of natural population of *Macrosteles fascifrons.* In oats causes bluish-green foilage, enations on leaf and stem veins, and sterile florets. Also infects barley, many other grasses and dicots. Occurs in North America from Kansas to Manitoba. Similar virus transmitted by *M. laevis* common in Czechoslovakia (102).

Rice dwarf virus   Described by Takata on 1895 and by Fukushi in 1931 and 1934 (40, 50). Particles 70 nm diam. Transmitted by *Nephotettix cincticeps, N. apicalis,* and *Inazuma dorsalis.* Multiplies in the vectors and is transmitted transovarially. Causes stunting and fine chlorotic specks on foliage of rice (*Oryza sativa*), wheat, oats, barley, rye, millet, and *Echinochloa crus-galli* var. *oryzicola* in Japan and Korea.

Rice orange leaf virus   Described by Ou in 1963 (50). Particles 15 nm diam. Transmitted by 7–14% of *Recilia dorsalis.* Causes golden yellow to deep bright orange leaves, in-rolling and drying out of leaves, death of plants. Reported in the Philippines, Thailand, Ceylon, Malaysia, and India.

Rice tungro virus   Described by Rivera & Ou in 1965 (28, 50). (Penyaket merah, Yellow-orange leaf, Leaf yellowing.) Particles 30–33 nm diam. Transmitted in a semipersistent manner by *Nephotettix impicticeps, N. apicalis,* and *Recitia (Inazuma) dorsalis.* Causes various symptoms on rice depending on the variety, including stunting, mottling, and yellowing. Also infects some other grasses. Reported in the Philippines, Malaysia, Indonesia, Pakistan,Thailand, and India. Considered the most important disease of rice in Southeast Asia.

Wheat striate virus (eastern)   Described by Nagaich & Sinha in 1974 (60). Particles 40 nm diam. Transmitted by *C. mbila.* Found on wheat and barley near Simla, India. (This virus has many characteristics like those of maize streak virus except that the particles are larger.)

LEAFHOPPER-TRANSMITTED VIRUSES WITH BACILLIFORM OR BULLET-SHAPED PARTICLES

Rice transitory yellowing   Described by Chiu, Lo, Pi & Chen in 1965 (76). Particles 129 × 96 nm in dip preparations and 193 × 94 nm in ultrathin sections. Transmitted in a persistent manner by *Nephotettix apicalis, N. cincticeps,* and *N. impicticeps.* Causes severe yellowing, stunting, and moderate yield losses on the second rice crop in central and southern Taiwan.

Wheat striate mosaic virus (American)   Described by Slykhuis in 1953 (84). Particles 200–250 × 75 nm. Transmitted in a persistent manner by *Endria inimica* and *Elymana virescens.* Causes fine line streaks, chlorosis, necrosis, and stunting, principally of wheat (*Triticum aestivum* and *T. durum*). Also infects barley, oats, maize, and some other grasses. Reported in South and North Dakota and Nebraska, USA, and in Manitoba and Saskatchewan, Canada.

Wheat (winter) mosaic virus   Described by Zazhurilo & Sitnikova in 1939 (70, 74, 75, 109). Particles 260 × 20 nm. Transmitted by *Psammotettix striatus* including transovarial, and by *M. laevis.* Causes chlorotic spots and stripes and dwarfing of wheat. Also infected millet, rye, and grasses in the Ukraine.

LEAFHOPPER-TRANSMITTED VIRUSES WITH UNCERTAIN PARTICLE MORPHOLOGY

Wheat striate mosaic virus (Australian)   Described by Grylls in 1963 (33). Transmitted by *Nesoclutha obscura.* Causes fine yellow to gray broken streaks, and dwarfing of wheat (*Triticum* spp.), oats, barley, maize, and a number of other grasses in Australia not found to be of great economic importance.

Planthopper (Delphacidae)-Transmitted Viruses

PLANTHOPPER-TRANSMITTED MYCOPLASMA

Rice grassy stunt   Mycoplasma-like bodies observed in diseased plants; isometric particles reported in vector (see below).

PLANTHOPPER-TRANSMITTED VIRUSES WITH ISOMETRIC PARTICLES

Maize rough dwarf virus   Described by Biraghi in 1949 (51). Particles 70 nm diam. Transmitted by *Laodelphax (Delphacodes) striatellus* in which it multiples and by *Javesella (Delphacodes) pellucida* and *Sogatella vibix.* In maize causes severe dwarfing, leaf enations, and poorly developed ears. In wheat, oats, barley, and other grasses causes excessive tillering and dwarfing. Occurs in Europe and Israel. Cereal tillering disease in Sweden seems closely related (49).

Oat sterile dwarf virus   Described by Prusa in 1958 (67). (Oat sterility, oat stunting, oat dwarf tillering disease, base tillering disease.) Particles 73 nm diam (13). Transmitted by *Javesella (Calligypona) pellucida* (0.2% transovarial transmission), *C. discolor,* and *Dicranotropis hamata* (101). Causes stunting, excess tillers and sterility of oats, wheat, maize, and other grasses in Czechoslovakia, Sweden, Finland, and Britain (17).

Rice black streaked dwarf virus   Described by Kurebayashi & Shinkai in 1952 (77). Particles 75–80 nm diam. Transmitted by *Laodelphax striatellus, Unkanodes sapporonus,* and *Ribautodelphax albifascia.* Causes severe stunting, darkening of leaves, twisting of distal portions of young leaves, white waxy swellings along the veins on the underside of leaf blades, sheaths and culms later becoming dark brown forming black streaked tumors on rice. Also affects maize, wheat, oats, barley, and rye. Reported in Japan.

Rice grassy stunt virus   Described by Rivera, Ou & Iida in 1966 (50). Also called Rice rosette. Particles 70 nm diam in vector. (Mycoplasma-like bodies have been reported in diseased plants.) Transmitted by *Nilaparvata lugens.* Causes stunting, excessive tillering; leaves short, narrow, pale green to yellow, often with dark brown spots or blotches on rice. Reported in Philippines, India, and Malaysia.

Rice stripe virus   Described by Kuribayashi in 1931 (50). Particles 25–35 nm diam. Transmitted by *L. striatellus, U. sapporonus,* and *R. albifascia.* Causes twisting, narrowing, elongation, and drooping of young leaves; yellowish green to white stripes; reduces development of rice grain. Can also infect barley, oats, wheat, rye, millet, maize, and many other grasses. Reported in Japan, Korea, and Taiwan.

Wheat streak virus (African)   Also called African cereal streak virus. Described by Harder & Bakker in 1973 (37). Particles 24 nm diam. Transmitted by *Toya catilina*. Causes phloem necrosis resulting in streaking of wheat, oats, barley, rye, triticale, rice, and other grasses in East Africa.

PLANTHOPPER-TRANSMITTED VIRUSES WITH BACILLIFORM PARTICLES

Maize mosaic virus   Described by Kunkel in 1921 (38). Particles bullet-shaped 242 × 48 ɩ.m in plant sections, 225 × 90 nm in leaf dip preparations. Transmitted by 5% of natural population of *Peregrinus maidis* in which it multiplies. Causes yellow spots and stripes in leaves, and stunting of maize and sorghum. Reported in Cuba, Hawaii, Mauritius, Puerto Rico, Surinam, Tanzania, Trinidad, Venezuela, and probably India.

Oat pseudo-rosette virus   Described by Sukov & Vovk in 1938 (94, 96). Also called Oat "zakuklivanie" ("pupation disease") virus, Siberian oat mosaic virus; northern cereal mosaic virus as described by Ito & Fukushi in 1944 in Japan (45) may also be a synonym. Two types of particles described; bacilliform or bullet-shaped, 60 nm diam, and flexuous rods 500–600 × 40 nm. Transmitted by *L. striatellus, U. sapporonus,* and *R. albifascia.* Causes light green stripes and spots on leaves, reddish-brownish discoloration, excess tillers, stunting, and reduced spikelets in oats. Also infects maize, barley, rye, millets, rice, wheat, and other grasses. Reported in Siberia and Japan.

Barley yellow striate mosaic virus   Described by Conti in 1969 (22). Particles 330 × 60 nm. Transmitted by *L. striatellus.* Causes fine chlorotic streaking of wheat in Italy. Also infects barley, oats, sweet corn, and some grasses. (Appears similar to oat pseudo-rosette.)

PLANTHOPPER-TRANSMITTED VIRUSES WITH FLEXUOUS, THREAD-LIKE PARTICLES

Rice hoja blanca virus   Described by Malaguti et al in 1956 (27, 50). Particles thread-like (78). [Isometric particles 42 nm diam have also been reported (39).] Transmitted by *Sogata orizicola* (also transmitted transovarially) and by *S. cubanus.* Causes white striping, mottling, stunting, and sterility of rice. Can infect wheat, barley, oats, rye, *Echinochloa colonum,* and other grasses. Reported in Cuba, Japan, Surinam, USA, and Venezuela.

PLANTHOPPER-TRANSMITTED VIRUSES WITH UNCERTAIN PARTICLE MORPHOLOGY

Wheat striate mosaic virus (European)   Described by Slykhuis & Watson in 1958 (92, 94). Also called oat striate and red disease (48). Transmitted by *Javesella (Delphacodes, Calligypona) pellucida* (virus passes through eggs to nymphs) and

J. dubia. Causes fine chlorotic streaks, necrosis, and stunting of wheat, oats, barley, rye, and some other grasses in England, Czechoslovakia, Finland, Rumania, Spain, and Sweden.

Most of the viruses and the mycoplasma transmitted by leafhoppers and planthoppers are persistent or circulative in their vectors. The insects require several minutes or hours of feeding on diseased plants to acquire the pathogen, but they may not be able to transmit it for several hours or days, until after the ingested virus or mycoplasma has passed from the gut to the salivary glands by way of the hemolymph. Infection of plants occurs by means of saliva while the insect is feeding. Some viruses and mycoplasma have been shown to multiply in the insect as well as in plants. The insects can be made infective by injecting extracts of infective insects or diseased plants into their abdomen (26).

Transmission through the eggs to the progeny of the vectors has been demonstrated for the leafhopper-transmitted rice dwarf (26) and winter wheat mosaic (74) viruses, and for the planthopper-transmitted rice stripe (26), European wheat striate mosaic (92), rice hoja blanca (27), and maize rough dwarf (51) viruses. Transovarial transmission can be important in assuring that the virus will be perpetuated with the vector insects even if virus-susceptible plants are not available for long periods, such as over winter, but some of the viruses are pathogenic to the insects. Maize rough dwarf virus and European wheat striate mosaic virus reduce the number of progeny produced by infected females (51, 105).

Two of the leafhopper-transmitted viruses, rice tungro and maize chlorotic dwarf, have semipersistent relationships with their vectors; that is, the vectors can transmit the viruses immediately after feeding on diseased plants, but they lose most of their infectivity within one day, and all of it within five days (50, 72).

Although some diseases have been reported only on one host genus, e.g. maize wallaby ear, maize chlorotic dwarf, rice orange leaf, rice transitory yellowing, and rice grassy stunt, the others are known to occur on, or can infect experimentally several cereals and other grasses. Aster yellows mycoplasma and oat blue dwarf virus appear to be the only leafhopper-transmitted pathogens of cereals that also readily infect dicotyledonous plants.

Each of the leafhopper- and planthopper-transmitted viruses of cereals has been found only in limited geographic or climatic regions and none is distributed throughout the geographic range of any cereal crop. The ranges of these viruses are limited by the range of the vectors as well as by the range of susceptible hosts.

Antisera developed for a number of the leafhopper- and planthopper-transmitted viruses should be useful in determining presence or absence of relationships between viruses in different countries. For example, rice black streaked dwarf and maize rough dwarf viruses are closely related serologically (52). R. C. Sinha (unpublished) did not detect a reaction between Australian wheat striate mosaic virus from leaves of diseased wheat and an antiserum prepared for North American wheat striate mosaic virus (83).

There have been many reports of leafhopper- and planthopper-transmitted viruses on cereals over widespread areas, but there have been no reports of widespread devastating epidemics. Severe losses have occurred only locally where

there were abundant sources of susceptible plants on which the virus and its vector could multiply, and the crop in question was highly susceptible and planted at
such a time as to expose the young plants to heavy infestations of infective insects
from nearby sources.

Symptoms of some of the diseases caused by mycoplasma have been suppressed by treating the plants with tetracyclines, but there are no chemical treatments suitable for practical control of cereal-infecting mycoplasma or viruses
(50).

There are varieties of cereals with immunity or high degrees of resistance to
most of the known leafhopper- and planthopper-transmitted viruses (50, 94).

MECHANICALLY TRANSMITTED VIRUSES
WITH NO KNOWN VECTOR

Panicum mosaic virus   Described by Still in 1957 (79, 82). Causes yellow-green
mosaic, mottling, chlorosis, and stunting of *Panicum miliaceum, Setaria italica,*
and some other grasses including the native perennial *Panicum virgatum* in Kansas.

Hordeum mosaic virus   Described by Slykhuis & Bell in 1966 (91). Particles
filamentous 700 × 15 nm. Causes light green to yellow mottling of barley, wheat,
oats, rye, *Hordeum jubatum,* and *Agropyron trachycaulum* in Alberta, Canada.

Oat necrotic mottle virus   Described by Gill in 1967 (31). Particles flexuous.
Causes light green to necrotic mottling, stunting, and yield loss of oats. Also infects *Poa pratensis* and some other grasses in Manitoba, Canada.

Wheat yellow mosaic   Described by Inouye in 1969 (43). Filamentous particles
275–300 and 575–600 × 13–14 nm. Causes yellow mosaic of wheat in Japan.

Poa semilatent virus   Described by Slykhuis in 1971 (88). Particles short, stiff
rods 160 × 30 nm. Causes chlorosis and blight of wheat; mottling, stunting, and
necrosis of oats, barley, and maize; transient light green mottling of *Poa palustris;*
light green mottling of *Agropyron trachycaulum* and other grasses. Found in
Alberta, Canada.

Maize chlorotic mottle   Described by Hebert & Castillo in 1973 (37a). Particles
isometric to 30 nm diam. Does not react with antisera to several known viruses
with similar particles. Causes chlorotic mottling, stunting, malformation of ears,
and yield reduction of maize in. Peru. Also infects *Euchlaena mexicana, Panicum
virgatum, Sorghum halepense, S. vulgare, Setaria glauca, Bromus arvensis,* and
Triticum aestivum.

It is assumed that the mechanically transmitted viruses with no known vectors
have vectors like those of other cereal viruses with similar morphologies. Some
may merely be unidentified variants of known viruses. Although each can cause
severe effects on some cereals, none appears to have caused sufficiently important

recurrent damage to create a sense of urgency about determining the natural means of transmission.

GENERAL CONSIDERATIONS

Virus diseases are recognized on cereal crops on all continents and probably occur in all cereal-growing areas. Although some cereal-infecting viruses have been discovered recently, the total number does not appear to have increased over the last 10 years because some diseases with different names in different countries are now known to be caused by the same virus.

Some viruses have worldwide distribution, others are limited in range. Barley stripe mosaic virus, which is efficiently seed borne, can occur in all areas where barley is grown from infected seed. Barley yellow dwarf virus appears to be distributed throughout the range of cereal crop production because it infects many graminaceous species and has many widely distributed aphid vectors. Most other viruses are more limited in distribution because of limited distribution of harboring hosts or of vectors.

The cereals are annuals and are not propagated vegetatively; hence, unless a virus is efficiently seed borne, like barley stripe mosaic virus, the number of diseased plants in each crop each year depends on the rapidity of virus spread during susceptible phases of crop development each crop season. Some viruses, including wheat streak mosaic virus, barley yellow dwarf virus, and oat sterile dwarf virus, have frequently caused devastating damage in situations where a heavy buildup of the respective viruliferous vectors was present when the susceptible crops were very young.

High incidence of infection does not necessarily mean devastating losses. Some seed lots of barley with nearly all seeds carrying BSMV may still produce large yields. The soil-borne mosaic viruses of wheat, oats, and barley frequently infect all plants in patches or entire fields in which the soil is infectious, but the presence or absence of severe loss in yield will depend on the duration of temperatures favoring disease development as well as on the susceptibility of the crop variety.

Generally the most promising approach to practical control of the cereal viruses appears to be the avoidance of buildup of the viruses and vectors on carry-over hosts, the timing of planting to avoid exposure of the young crop to periods of heavy virus spread, and the use of immune, resistant, or tolerant varieties.

There are a number of viruses not mentioned in this review that can be transmitted to cereals but have not been found on cereals in the field. Some of these could cause losses that have not been recognized.

Note added in proof   Cereal enanismo, described by Gibler in 1957 (28a), is caused by an agent transmitted by the leafhopper *Cicadulina pastusae* in Colombia and Equador. *Wheat dwarf,* described by Vacke in 1961 (100a, 102a), is caused by an agent transmitted by the leafhopper *Psammotettix alienus* in Czechoslovakia, Rumania, Sweden, and the Ukraine.

Literature Cited

1. Atabekov, J. G., Novikov, V. K. 1971. Barley stripe mosaic virus. *Descr. Plant Viruses, No. 68,* Commonw. Mycol. Inst., Surrey, England
2. Bakker, W. 1970. Rice yellow mottle, a mechanically transmissible virus disease of rice in Kenya. *Neth. J. Plant Pathol.* 76:53–63
3. Bakker, W. 1974. Characterization and ecological aspects of rice yellow mottle virus in Kenya. *Wageningen, Neth. Agric. Res. Rept. No. 829.* 152 pp.
4. Bancroft, J. B. 1970. Brome mosaic virus. *Descr. Plant Viruses, No. 3,* Commonw. Mycol Inst., Surrey, England
5. Banttari, E. E., Moore, M. B. 1960. Transmission of aster yellows virus to barley. *Plant Dis. Reptr.* 44:154
6. Banttari, E. E., Zeyen, R. J. 1973. Oat blue dwarf virus. *Descr. Plant Viruses, No. 123,* Commonw. Mycol. Inst., Surrey, England
7. Benigno, D. A., A'Brook, J. 1972. Infection of cereals by cocksfoot mottle and *Phleum* mottle viruses. *Ann. Appl. Biol.* 72:43–52
8. Bock, K. R. 1974. Maize streak virus. *Descr. Plant Viruses, No. 133,* Commonw. Mycol. Inst., Surrey, England
9. Bradfute, O. E., Louie, R., Knoke, J. K. 1972. Isometric viruslike particles in maize with stunt symptoms. *Phytopathology* 62:748
10. Bradfute, O. E., Whitmoyer, R. E., Nault, L. R. 1970. Ultrastructure of plant leaf tissue infected with mite-borne viral-like pathogens, *28th Ann. Proc. Electron Microsc. Soc. Am.*
11. Brakke, M. K. 1971. Wheat streak mosaic virus. *Descr. Plant Viruses, No. 48,* Commonw. Mycol. Inst., Surrey, England
12. Brakke, M. K. 1971. Soil-borne wheat mosaic virus. *Descr. Plant Viruses, No. 77,* Commonw. Mycol. Inst., Surrey, England
13. Brčák, J., Králick, O., Vacke, J. 1972. Virus origin of the oat sterile dwarf disease. *Biol. Plant.* 14:302–4
14. Bremer, K. 1971. Wheat streak mosaic virus in Turkey. *Phytopathol. Mediterr. 10:280–82*
15. Bremer, K. 1973. The brome grass mosaic virus as a cause of cereal disease in Finland. *Ann. Agric. Fenn.* 12:207–14
16. Bruehl, G. W., Damsteegt, V. D. 1961. Soil-borne mosaic of fall-seeded oats in western Washington. *Plant Dis. Re-*

ptr. 45:884–88
17. Catherall, P. L. 1970. Oat sterile dwarf virus. *Plant Pathol.* 19:75–78
18. Catherall, P. L. 1970. Cocksfoot mottle virus. *Descr. Plant Viruses, No. 23,* Commonw. Mycol. Inst., Surrey, England
19. Chiykowski, L. N. 1963. *Endria inimica* (Say), a new leafhopper vector of a celery-infecting strain of aster-yellows virus in barley and wheat. *Can. J. Bot.* 41:669–72
20. Chiykowski, L. N., Sinha, R. C. 1969. Comparative efficiency of transmission of aster yellows by *Elymana virescens* and *Macrosteles fascifrons* and the relative concentration of the causal agent in the vectors. *J. Econ. Entomol.* 62:883–86
21. Chona, B. L., Seth, M. L. 1960. A mosaic disease of maize (*Zea mays* L.) in India. *Indian J. Agric. Sci.* 30:26–32
22. Conti, M. 1969. Investigations on a bullet-shaped virus of cereals isolated in Italy from planthoppers. *Phytopathol. Z.* 66:275–79
23. Davis, R. E. 1974. Spiroplasma in corn stunt-infected individuals of the vector leafhopper *Dalbulus maidis.* *Plant Dis. Reptr.* 58:1109–12
24. Dhanraj, K. S., Raychaudhuri, S. P. 1969. A note on barley mosaic in India. *Plant Dis. Reptr.* 53:766–67
25. Eslick, R. F. 1953. Yield reductions in Glacier barley associated with a virus infection. *Plant Dis. Reptr.* 37:290–91
26. Fukushi, T. 1969. Relationships between propagative rice viruses and their vectors. In *Viruses, Vectors, and Vegetation,* ed. K. Maramorosch, 279–301. New York: Interscience, 666 pp.
27. Galvez, G. E. 1968. Transmission studies of the hoja blanca virus with highly active, virus-free colonies of *Sogatodes oryzicola.* *Phytopathology* 58:818–21
28. Galvez, G. E. 1971. Rice tungro virus. *Descr. Plant Viruses, No. 67,* Commonw. Mycol. Inst., Surrey, England
28a. Galvez, G. E., Thurston, H. D., Bravo, G. 1963. Leafhopper transmission of Enanismo of small grains. *Phytopathology* 53:106–8
29. Ganguly, B., Raychaudhuri, S. P. 1971. Purification of wheat streak mosaic virus. *Phytopathol. Z.* 70:11–14
30. Gibbs, A. J., Harrison, B. D. 1970. Cucumber mosaic virus. *Descr. Plant Viruses, No. 1,* Commonw. Mycol. Inst., Surrey, England

31. Gill, C. C. 1974. Inclusions and wall deposits in cells of plants infected with oat necrotic mottle virus. *Can. J. Bot.* 52:621–26

32. Granados, R. R., Maramorosch, K., Everett, T., Pirone, T. P. 1966. Transmission of corn stunt virus by a new leafhopper vector, *Graminella nigrifrons* (Forbes). *Contrib. Boyce Thompson Inst.* 23:275–80

33. Grylls, N. E. 1963. A striate mosaic virus disease of grasses and cereals in Australia, transmitted by the cicadellid, *Nesoclutha obscura. Aust. J. Agric. Res.* 14:143–53

34. Grylls, N. E. 1975. Leafhopper transmission of a virus causing maize wallaby ear disease. *Ann. Appl. Biol.* 79:283–96

35. Hagborg, W. A. F. 1955. Heat tolerance of host and parasite in false stripe of barley. *Proc. Can. Phytopathol. Soc.* 23:15–16

36. Hamilton, R. I. 1965. An embryo test for detecting seed-borne barley stripe mosaic virus in barley. *Phytopathology* 55:798–99

37. Harder, D. E., Bakker, W. 1973. African cereal streak, a new disease of cereals in East Africa. *Phytopathology* 63:1407–11

37a. Hebert, T. T., Castillo, J. 1973. A new virus of maize in Peru. In *2nd Int. Congr. Plant Pathol., Minneapolis, Minn.* (Abstr. 0072)

37b. Hebert, T. T., Panizo, C. H. 1975. Oat mosaic virus. *Descr. Plant Viruses, No. 145,* Commonw. Mycol. Inst., Surrey, England

38. Herold, F. 1972. Maize mosaic virus. *Descr. Plant Viruses, No. 94,* Commonw. Mycol. Inst., Surrey, England

39. Herold, F., Trujillo, G., Munz, K. 1968. Virus like particles related to hoja blanca disease of rice. *Phytopathology* 58:546–47

40. Iida, T. T., Shinkai, A., Kimura, I. 1972. Rice dwarf virus. *Descr. Plant Viruses, No. 102,* Commonw. Mycol. Inst., Surrey, England

41. Inouye, T. 1962. Studies on barley stripe mosaic in Japan. *Ber. Ohara Inst.* 11:413–96

42. Inouye, T., Saito, Y. 1975. Barley yellow mosaic virus. *Descr. Plant Viruses, No. 143,* Commonw. Mycol. Inst., Surrey, England

43. Inouye, T. 1969. Filamentous particles as the causal agent of yellow mosaic disease of wheat. *Nogaku Kenkyu* 53:61–68

44. Inouye, T., Mitsuhata, K., Heta, H.,

Hiura, U. 1973. A new virus of wheat, barley and several other plants in Gramineae, wheat yellow leaf virus. *Nogaku Kenkyu* 55:1–16

45. Ito, S., Fukushi, T. 1944. Studies on northern cereal mosaic. *J. Sapporo Soc. Agric. For.* 36(3):62–89, 36(4):65–88

46. Koehler, B., Bever, W. M., Bonnett, O. T. 1952. Soil-borne wheat mosaic. *Bull. Ill. Agric. Exp. Stn.* 556:567–99

47. Larson, R. I., Atkinson, T. G. 1972. Isolation of an *Agropyron elongatum* chromosome conferring resistance to the wheat curl mite on a *Triticum-Agropyron* hybrid. *Can. J. Genet. Cytol.* 14:731–32

48. Lindsten, K. 1970. Investigations on the spread and control of oat sterile dwarf. *Natl. Swed. Inst. Plant Prot. Contrib.* 14:134, 407–46

49. Lindsten, K., Gerhardson, B., Pettersson, J. 1973. Cereal tillering disease in Sweden and some comparisons with oat sterile dwarf and maize rough dwarf. *Natl. Swed. Inst. Plant Prot. Contrib.* 15:151, 375–97

50. Ling, K. C. 1972. *Rice Virus Diseases.* Los Baños, Philippines: Int. Rice Res. Inst. 142 pp.

51. Lovisolo, O. 1971. Maize rough dwarf virus. *Descr. Plant Viruses, No. 72,* Commonw. Mycol. Inst., Surrey, England

52. Luisoni, E., Lovisolo, O., Kitagawa, Y., Shikata, E. 1973. Serological relationship between maize rough dwarf virus and rice black-streaked dwarf virus. *Virology* 52:281–83

53. MacFarlane, I., Jenkins, J. E. E., Melville, S. C. 1968. A soil-borne virus of winter oats. *Plant Pathol.* 17:167–70

54. Martyn, E. B., ed. 1968. *Plant Virus Names. Phytopathol. Pap. No. 9.* Kew, England: Commonw. Mycol. Inst. 204 pp.

55. McKinney, H. H. 1953. *Virus Diseases of Cereal Crops. Yearb. Dep. Agric., Plant Diseases,* pp. 350–60

56. McKinney, H. H. 1954. Culture methods for detecting seed-borne virus in Glacier barley seedlings. *Plant Dis. Reptr.* 38:152–62

57. McKinney, H. H., Greeley, L. W. 1965. Biological characteristics of barley stripe mosaic virus strains and their evolution. *US Dep. Agric. Tech. Bull. 1324.* 84 pp.

57a. Milbrath, G. M., Tosic, M. 1974. Occurrence of maize chlorotic dwarf virus in Illinois. *Plant Dis. Reptr.* 58:420

58. Miyamoto, Y. 1961. Studies on soil-borne cereal mosaics. VII. Controlling of soil-borne cereal mosaics with special reference to the effect of soil treatment with pyroligneous acid (wood vinegar). *Ann. Phytopathol. Soc. Jpn.* 26:90–97

59. Murtomaa, A. 1966. Aster yellows-type virus infecting grasses in Finland. *Ann. Agric. Fenn.* 5:324–33

60. Nagaich, B. B., Sinha, R. C. 1974. Eastern wheat striate: a new viral disease. *Plant Dis. Reptr.* 58:968–70

61. Nault, L. R., Styer, W. E. 1970. Transmission of an eriophyid-borne wheat pathogen by *Aceria tulipae.* *Phytopathology* 60:1616–18

62. Nault, L. R., Styer, W. E., Knoke, J. K., Pitre, H. N. 1973. Semipersistent transmission of leafhopper-borne maize chlorotic dwarf virus. *J. Econ. Entomol.* 66:1271–73

63. Osler, R., Amici, A., Nelli, G. 1974. Transmission of rice 'giallume' by an aphid, *Rhopalosiphum padi.* *Rev. Patol. Veg.* IV 10:5–17

64. Paliwal, Y. C., Slykhuis, J. T. 1967. Localization of wheat streak mosaic virus in the alimentary canal of its vector *Aceria tulipae* Keifer. *Virology* 32:344–53

65. Panjan, M. 1966. About some manifestations of mosaic of corn in Yugoslavia. *Rev. Roum. Biol. Ser. Bot.* 11:159–62

66. Pirone, T. P. 1972. Sugarcane mosaic virus. *Descr. Plant Viruses, No. 88,* Commonw. Mycol. Inst., Surrey, England

67. Prusa, V. 1958. Die Sterile Verzwergung des Hafers in der Tschechoslowakischen Republik. *Phytopathol. Z.* 33:99–107

68. Rao, D. G., Varma, P. M., Capoor, S. P. 1965. Studies on mosaic diseases of *Eleusine* in the Deccan. *Indian Phytopathol.* 18:139–50

69. Raychaudhuri, S. P., Chatterjee, S. N. 1961. Chirke a new virus threat to cardamom. *Indian Farming* 9:11–12

70. Razvyazkina, G. M., Polyakova, G. P. 1967. Electron microscope study of wheat mosaic virus transmitted by the cicada *Psammotettix striatus* L. Transl. from *Dokl. Akad. Nauk SSSR* 174:1435–36

71. Rochow, W. F. 1970. Barley yellow dwarf virus. *Descr. Plant Viruses, No. 32,* Commonw. Mycol. Inst., Surrey, England

72. Rosenkranz, E. 1969. A new leafhopper-transmissible corn stunt disease agent in Ohio. *Phytopathology* 59:1344–46

73. Seth, M. L., Raychaudhuri, S. P. 1967. Further studies on mosaic disease of maize (*Zea mays* L.). *Biol. Plant.* 9:372–76

74. Shaskol'skaya, N. D. 1962. Transovarial transmission of winter wheat mosaic virus by the leafhopper *Psammotettix striatus* L. *Zool. Zh.* 41:717–20 (in Russian, English summary)

75. Sherban, O. D., Onishchenko, A. M. 1972. Winter wheat mosaic and its vectors in Moldavia. *Mikrobiol. Zh. Kiev* 34:757–62

76. Shikata, E. 1972. Rice transitory yellowing virus. *Descriptions of Plant Viruses, No. 100,* Commonw. Mycol. Inst., Surrey, England

77. Shikata, E. 1974. Rice black-streaked dwarf virus. *Descriptions of Plant Viruses, No. 135,* Commonw. Mycol. Inst., Surrey, England

78. Shikata, E., Gálvez, G. E. 1969. Fine flexuous threadlike particles in cells of plants and insect hosts infected with rice hoja blanca virus. *Virology* 39:635–41

79. Sill, W. H. Jr. 1957. *Panicum* mosaic, a new virus disease of *Panicum virgatum* and related grasses. *Phytopathology* 47:31

80. Sill, W. H. Jr. 1958. A comparison of some characteristics of soil-borne wheat mosaic viruses in the Great Plains and elsewhere. *Plant Dis. Reptr.* 42:912–24

81. Sill, W. H. Jr., King, C. L. 1958. The 1957 soil-borne wheat mosaic epiphytotic in Kansas. *Plant Dis. Reptr.* 42:513–16

82. Sill, W. H. Jr., Talens, L. T. 1962. New hosts and characteristics of the *Panicum* mosaic virus. *Plant Dis. Reptr.* 46:780–83

83. Sinha, R. C. 1968. Serological detection of wheat striate mosaic virus in extracts of wheat plants and vector leafhoppers. *Phytopathology* 58:452–55

84. Sinha, R. C., Bekki, R. M. 1972. American wheat striate mosaic virus. *Descr. Plant Viruses, No. 99,* Commonw. Mycol. Inst., Surrey, England

85. Slykhuis, J. T. 1956. Wheat spot mosaic caused by a mite-transmitted virus associated with wheat streak mosaic. *Phytopathology* 46:682–87

86. Slykhuis, J. T. 1967. Virus diseases of cereals. *Rev. Appl. Mycol.* 46:401–29

87. Slykhuis, J. T. 1970. Factors determining the development of wheat

spindle streak mosaic caused by a soilborne virus in Ontario. *Phytopathology* 60:319-31

88. Slykhuis, J. T. 1971. Poa semilatent virus from native grasses. *Phytopathology* 62:508-13

89. Slykhuis, J. T. 1973. Agropyron mosaic virus. *Descr. Plant Viruses, No. 118,* Commonw. Mycol. Inst., Surrey, England

90. Slykhuis, J. T. 1974. Differentiation of transmission and incubation temperatures for wheat spindle streak mosaic virus. *Phytopathology* 64:554-57

91. Slykhuis, J. T., Bell, W. 1966. Differentiation of *Agropyron* mosaic, wheat streak mosaic and a hitherto unrecognized *Hordeum* mosaic virus in Canada *Can. J. Bot.* 44:1191-1208

92. Slykhuis, J. T., Watson, M. A. 1958. Striate mosaic of cereals in Europe and its transmission by *Delphacodes pellucida* (Fab.). *Ann. Appl. Biol.* 46: 542-53

93. Smith, H. C. 1963. Control of barley yellow dwarf virus in cereals, *N.Z. J. Agric. Res.* 6:229-44

94. Smith, K. M. 1972. *A Textbook of Plant Virus Diseases.* Longman. Edinburgh: Longman Group J. Div. 684 pp. 3rd ed.

95. Stoner, W. N. 1952. Leaf fleck, an aphid-borne persistent virus disease of maize. *Phytopathology* 42:683-89

96. Sukhov, K. S., Vovk, A. M. 1938. Mosaic of cultivated cereals and how it is communicated in nature. *C. R. Dokl. Acad. Sci. URSS* 20:745-48

97. Timian, G. 1965. Heat treatments fail to inactivate barley stripe mosaic virus in seed. *Plant Dis. Reptr.* 49:696-97

98. Toler, R. W., Hebert, T. T. 1962. Factors affecting transmission of oat mosaic virus by inoculation with infective juice. *Phytopathology* 52:30

99. Tošić, M. 1971. Virus diseases of wheat in Serbia. I. Isolation and determination of wheat streak mosaic virus and brome mosaic virus. *Phytopathol. Z.* 70:145-62

100. Toyama, A., Kusaba, T. 1970. Transmission of soil-borne barley yellow mosaic virus. II. *Polymyxa graminis* Led. as vector. *Ann Phytopathol. Soc. Jpn.* 36:223-29

100a.Vacke, J. 1961. Wheat dwarf virus disease. *Biol. Plant.* 3:228-33

101. Vacke, J. 1966. Study of transovarial passage of the oat sterile dwarf virus. *Biologia Plant.* 8:127-30

102. Vacke, J. 1970. Some new findings on oat dwarf transmitted by the leafhopper *Macrosteles laevis* Rib. *Vyzk. Ustavo Rostl. Praha-Ruzyne* 16:21-30

102a.Vacke, J. 1972. Host plant range and symptoms of wheat dwarf virus. *Výzkumných Ústavo Rostlinné Výroby Praha-Ruzyně* 17:151-62

103. Vlasov, Yu. I., Artemyeva, N., Larina E. 1965. Virus diseases of the Graminae. *Zashch. Rast. Vredit. Bolez.* 8:43-44

104. Von Wechmar, M. B. 1965. Virus diseases endanger our grain. *Farming S. Afr.* 41:22-25

105. Watson, M. A., Sinha, R. C. 1959. Studies on the transmission of European wheat striate mosaic virus by *Delphacodes pellucida* Fabricius. *Virology* 8:139-63

106. Wellman, F. L. 1934. Infection of *Zea mays* and various other Gramineae by the celery virus in Florida. *Phytopathology* 24:1035-37

107. Wiese, M. V., Hooper, G. R. 1971. Soil transmission and electron microscopy of wheat spindle streak mosaic. *Phytopathology* 61:331-32

108. Williams, L. E., Alexander, L. J. 1965. Maize dwarf mosaic, a new corn disease. *Phytopathology* 55:802-4

109. Zazhurilo, V. K., Sitnikova, G. M. 1939. Mosaic of winter wheat. *C. R. Dokl. Acad. Sci. URSS* 25:798-801

ALIEN GERM PLASM AS A SOURCE OF RESISTANCE TO DISEASE

♦3640

D. R. Knott and J. Dvořák
Crop Science Department, University of Saskatchewan, Saskatoon, Saskatchewan, Canada

INTRODUCTION

In the breeding of crop plants man is constantly looking for new sources of germ plasm. Cultivated species often lack genes required by the plant breeder, particularly genes for disease resistance. In many genera, and especially in those that comprise a polyploid complex, species related to the cultivated species provide an important reservoir of genes for use by the breeder. More rarely, related genera have been used successfully as a source of desired genes.

The current trend towards genetic uniformity and the loss of the old "land" cultivars in many crops is resulting in an erosion of genetic variability. Furthermore, in a number of crops the known genes for disease resistance are being used up as they are released in cultivars and then overcome by new races of a pathogen. Thus it is likely that wild species will become increasingly important sources of germ plasm in the breeding of many crops (142).

A survey of the literature reveals many examples of the successful transfer of disease resistance from one species to another. It also reveals the wide range of problems that can arise in interspecific hybridization and interspecific transfer of genes. Progress is often slow and difficult. In this review, although some of the successes are described, the emphasis is on the problems and methods of overcoming them. Because of the huge volume of literature available, the authors have had to restrict themselves to those papers that appeared to be most useful in meeting the objectives of the review.

ALIEN GERM PLASM

Alien germ plasm could be defined as germ plasm from a species related to a crop species. However, that definition is too broad. Much of the early taxonomic work was based solely on differences in plant morphology rather than on cytogenetic

211

and evolutionary evidence. Since major morphological differences can be simply inherited, two supposed species may differ basically in alleles at only one or a few loci. For example, in the older classifications hexaploid wheats were separated into five species, *Triticum aestivum, T. compactum, T. spelta, T. sphaerococcum,* and *T. vavilovii* which differ from one another by alleles at only five loci (88). Hybridization among these species is no different from crosses among cultivars, and their germ plasms cannot be realistically described as alien. Similarly, in tomatoes *Lycopersicon pimpinellifolium* has frequently been used as a source of germ plasm in breeding *L. esculentum.* Hybrids between the species show a high degree of fertility and some taxonomists regard the parents as one species (71).

In this paper germ plasm is considered as alien only if the two species involved are sufficiently distinct that they show some degree of genetic isolation.

At present, genes can be transferred only between related species although further development of techniques such as somatic cell hybridization, transduction, and DNA transformation may change the picture in the future. Many procedures can be used in making gene transfers and the one that is appropriate for a particular situation depends on the relationship between the two species involved. An understanding of the cytotaxonomic relationships, cytogenetic structure, and evolutionary history of the two species is essential in determining the appropriate procedure and in making a successful transfer.

The interspecific transfer of a gene for disease resistance requires the transfer of chromatin from a chromosome of a donor species to a chromosome of a recipient. Ideally the alien chromosome segment should be as small as possible, particularly if the two species show extensive genetic and chromosomal differentiation. Otherwise, the substitution of an alien chromosome segment for a chromosome segment of the recipient species may result in undesirable duplications, deletions, and linkages of genes. The effect is particularly large in diploids but is often reduced in polyploids where the effect of single chromosomes is smaller because of the duplication of genetic material in two or more different genomes. The removal of deleterious genes linked to a desired gene is possible if the chromosomes involved are homologous and crossing-over occurs. However, it becomes increasingly difficult as the degree of differentiation increases and the chromosomes involved are only homoeologous and not homologous.

Crop species may be divided into four broad groups on the basis of their interrelationships with their close relatives and, therefore, the ease with which interspecific gene transfers can be made.

I.   Diploid crops whose relatives are mostly diploid.
   A. Species groups that are in an early stage of evolutionary differentiation. The species in such a group usually cross readily and show little or no chromosomal differentiation. Consequently, pairing in hybrids is good and chiasma frequency is high, e.g. *Lycopersicon, Pyrus,* and *Malus.*
   B. Species groups that are in an advanced stage of evolutionary differentiation. Crosses within such a group are more difficult to make and chromosomes of different species are homoeologous but not homologous.

Pairing and chiasma frequencies are often considerably reduced. The ease of gene transfer varies greatly and special techniques are often necessary, e.g. *Beta* and *Ribes.*

II. Polyploid crops whose relatives form a polyploid complex.
  A. Polyploid complexes in which a diploidizing system to restrict homoeologous pairing has not evolved. This situation is found most frequently in polyploid species that do not require meiotic regularity and high fertility for their reproduction and, therefore, is often associated with vegetative reproduction. In most interspecific hybrids the chromosomes pair, crossing-over takes place, and gene transfer is reasonably straightforward, e.g. *Solanum.*
  B. Polyploid complexes in which a diploidizing system has evolved. A genetic diploidizing system that prevents pairing between homoeologous chromosomes but allows homologues to pair normally has evolved in a number of polyploid species that reproduce only by seeds so that high fertility is essential. In interspecific crosses chromosome pairing varies greatly from one hybrid to another. The strength of the diploidizing system probably differs in different genera. For example, it is highly restrictive in *Triticum* and *Avena.* However, if a diploidizing system is present in *Nicotiana* as suggested by Kimber (58) it must be much less restrictive since crossing-over between homoeologous chromosomes frequently occurs without any manipulation of the system.

In this review, specific examples of the techniques that have been used to transfer genes in each of the four groups of species are described. For genera in which interspecific gene transfer has not yet been used, the review should provide a useful guide in the selection of an appropriate procedure based on the cytotaxonomic and cytogenetic features of the species.

SOURCES OF DISEASE RESISTANCE

Any sizable program using alien germ plasm in breeding for disease resistance must begin with a survey of the resistance present in species related to the crop. For most important crops, collections of wild relatives along with some information on their disease resistance are available. However, more detailed testing of various species collections to the particular disease, and often specific races of the disease, must be carried out. For example, 61 tomato workers collaborated to test the resistance of 144 accessions of wild species of *Lycopersicon* and suspected hybrids to 15 bacterial, fungal, and virus diseases and to nematodes (1). Many sources of resistance were found. In tobacco Sievert (130) tested about 1000 lines of *Nicotiana tabacum* and 60 *Nicotiana* species to anthracnose (*Colletotrichum destructivum*) in an attempt to find resistance. The *N. tabacum* lines were all susceptible but nine other species carried resistance. As Sievert points out it is important to remember that not all accessions of a species will necessarily carry resistance to a disease and there may even be segregation within a single accession.

ALIEN TRANSFERS

Transfers Between Diploids with Little Chromosomal Differentiation

LYCOPERSICON   The genus *Lycopersicon* contains six or seven diploid species that are divided into two subgenera largely on the basis of fruit color. The red-fruited subgenus *Eulycopersicon* includes two species, *L. esculentum* and *L. pimpinellifolium.* As mentioned earlier, genes can be transferred so readily between the two species that their transfer can hardly be considered a use of alien germ plasm. The green-fruited subgenus *Eriopersicon* includes a much more diverse group of four or five species. The diploid chromosome number for all *Lycopersicon* species is 24 and interspecific hybrids normally show 12 pairs of chromosomes at meiosis (101).

When *L. esculentum* is crossed with species in the subgenus *Eriopersicon,* the crosses are successful only if *esculentum* is the female parent (101). Similarly, the F_1 hybrids can be used only as male parents in backcrosses to *esculentum* and as female parents in backcrosses to the wild parent. This is contrary to the situation commonly observed in other genera where interspecific hybrids frequently can be used only as female parents because of pollen sterility. Interspecific crosses in *Lycopersicon* are complicated by self-incompatibility which is common to most members of the subgenus *Eriopersicon* and is transmitted as a dominant character.

Of the species in the *Eriopersicon* subgenus, *L. hirsutum* probably hybridizes most readily with *esculentum* (101). It has been used as a source of resistance to *Septoria lycopersici* (72), *Corynebacterium michiganense* (67), and tobacco mosaic virus (94). Kuriyama et al (67) reported considerable differences in the fertility of F_1 hybrids involving different *hirsutum* parents. Hybrids with the var. *glabratum* could be both selfed and backcrossed to *esculentum* while another hybrid could not. Bud pollination increased both the fruit set and seed production.

The two species, *L. chilense* and *L. peruvianum,* appear to be the most promising sources of disease resistance in tomatoes. Unfortunately, they are also difficult to cross with *esculentum.* While *peruvianum* can be crossed with *esculentum,* the embryos often abort. Thus a number of authors resorted to embryo culture to obtain hybrids. Kuriyama et al (67) reported that in crosses between *esculentum* and *peruvianum* the F_1 and F_2 plants were self-incompatible but interfertile and that the F_1 plants could be used as males in backcrosses to *esculentum.* When *peruvianum* was used as a source of resistance to nematodes (*Heterodera marioni*), McFarlane et al (82) were able to obtain hybrid plants using embryo culture. The F_1 plants could not be crossed with *esculentum* but did set some selfed seed. Selection over several generations resulted in highly resistant derivatives which, however, were highly incompatible with *esculentum* and were eventually dropped (34) in favor of material produced by Smith (132) and Watts

(145). In this material, resistance proved to be linked to undesirable fruiting characteristics. The linkage was eventually broken by Gilbert & McGuire (40) and the resistance has been utilized in commercial cultivars in a number of countries (69).

Several of the wild species have been used in breeding for tobacco mosaic virus (TMV) resistance in tomato (94). Multiple crosses involving *peruvianum, pimpinellifolium, hirsutum,* and *chilense* were used in a major program of breeding for disease resistance in Hawaii (35). Two genes for resistance to TMV (Tm_1 and Tm_2) were identified and both were found to be linked to undesirable characters. In particular, Tm_2 proved to be very tightly linked to a gene for a netted virescent character which caused stunting and yellowing of the leaves. The linkage has apparently not been broken. Khush et al (57) found that the virescent locus is located in a heterochromatic segment and this may account for the lack of crossing-over between it and Tm_2.

The interspecific transfer of genes in *Lycopersicon* proved to be surprisingly difficult even though all species have the same chromosome number and the chromosomes pair well in hybrids. Thus chromosome homology is not a problem, but gene-cytoplasm interactions, genetic incompatibility, sterility, and undesirable linkages are.

Transfers Between More Differentiated Diploids

RIBES   The genus *Ribes* provides an example of species that have the same chromosome number but where interspecific gene transfers are often difficult, in part because of reduced chromosome pairing (54, 91). The cultivated forms, red currants, black currants, and gooseberries, are all diploids with $2n = 16$ but they belong to different sections of the genus. Hybrids between species within a section show fairly regular meiosis and are usually fertile while intersectional hybrids are difficult to obtain, show reduced chromosome pairing, and are usually sterile. Interspecific hybridization has been used extensively in breeding in *Ribes* and in many cases gene transfers are fairly simple.

The ornamental *Ribes* species in the section *Calobotrya* carry resistance to many insects and pests that attack currants and gooseberries (53). Unfortunately, hybrids between these species and the cultivated species frequently show semilethal dwarfing and male sterility. Nevertheless, resistance to mildew (*Sphaerotheca mors-uvae*) has been transferred from *R. sanguineum* and *R. glutinosum* to the black currant, *R. nigrum,* by backcrossing. A hybrid in which *sanguineum* was used as the female parent proved to be male sterile while the reciprocal was male fertile.

With the more difficult intersectional hybrids, hybrid plants are treated with colchicine to produce fertile amphidiploids and these are then backcrossed to either the diploid or a colchicine-produced autotetraploid form of the cultivated species (54, 91). Backcrossing to the diploid results in a more rapid elimination of the donor genome and recovery of the cultivated genotype. However, the first backcross seedlings are triploids containing two genomes from the recurrent parent and one from the donor and tend to be sterile. A backcross to the tetraploid results in tetraploids with three genomes from the recurrent parent and one from the donor. The tetraploids are more fertile and may provide more opportunity for recombination between the recurrent and donor genomes.

Knight et al (62) used this procedure to transfer resistance to the gall mite (*Cecidophyopsis ribis*) from the gooseberry, *R. grossularia,* to black currant, *R. nigrum,* and made backcrosses to both the diploid and tetraploid forms of *nigrum.* In the second backcross to the diploid, 2% of the plants were resistant while in the tetraploid 50% were resistant. However, since resistance was successfully transferred to the diploid, work at the tetraploid level was dropped.

BETA   The genus *Beta* comprises mostly diploid species, and frequent attempts have been made to use wild species of the section *Patellares* as sources of resistance to diseases and pests of sugar beet (*Beta vulgaris*), particularly to the sugar beet nematode (*Heterodera schachtii*) (114). Most attempts failed because of lack of viability of the hybrids, largely because the F_1 seedlings fail to develop root systems. To overcome this problem F_1 seedlings were grafted onto sugar beets to enable them to survive (21, 50). Nevertheless, the transfer of nematode resistance did not succeed because of poor fertility in the hybrids and a failure to recover resistance in backcrosses. Savitsky (114) finally succeeded in transferring nematode resistance from *B. procumbens* ($2n = 18$) to *B. vulgaris* ($2n = 18$) by means of a cross between $4n$ *B. vulgaris* and $2n$ *B. procumbens.* The F_1 hybrids were grafted onto sugar beets and were grown for several years among diploid sugar beet pollinators. In this way adequate amounts of seed were obtained. In the triploid hybrids the nine *procumbens* chromosomes rarely paired with the *vulgaris* chromosomes. After a backcross to diploid *vulgaris,* four nematode-resistant plants carrying nine pairs of *vulgaris* chromosomes and one added *procumbens* chromosome were obtained. The *procumbens* chromosome paired only occasionally to form a trivalent. From 8834 progeny of plants with the added chromosome, two diploid, resistant plants were obtained in which the resistance had been transferred to a *vulgaris* chromosome. In the addition line, a gene for early bolting was linked to the gene for resistance but the genes were separated when the crossover with a *vulgaris* chromosome occurred.

Transfers from Species in a Polyploid Complex to a Polyploid Crop That Lacks a Genetic Diploidizing System

SOLANUM   The genus *Solanum* is a very large and complex one comprising approximately 2000 species of which about 100 are tuber bearing and belong to the section *Tuberarium* (23, 25, 43). The tuber bearing *Solanums* fall into a polyploid series having a basic chromosome number of 12 and containing forms with $2n = 24, 36, 48, 60,$ and 72. The genomes in the *Solanum* species appear to be relatively similar and genome evolution has not progressed far (42, 43, 48, 80, 96, 97). Because the cultivated potato is almost entirely vegetatively reproduced, it can withstand considerable chromosomal and meiotic irregularities. Although the common cultivated potato is tetraploid, diploids, triploids, and occasionally pentaploids are grown in South America. Since they share a common genetic system, Dodds (26) includes them all in the species *Solanum tuberosum.*

The wild species of *Solanum* have been used very extensively in breeding for resistance to every type of potato disease (33).

After its introduction into Europe, the potato became a major food crop in a number of countries. The appearance of late blight (*Phytophthora infestans*) in Europe about 1845 caused serious problems including the great famine in Ireland. With the discovery that the wild hexaploid, *Solanum demissum* (syn. *S. utile*) ($2n = 72$), carried a high degree of resistance to late blight, breeding for resistance began in a number of countries (6, 8, 100, 108, 110, 112, 113, 115). Reddick et al (100) reported that *demissum* would cross as a pollen parent with a few cultivars of *tuberosum*. The reciprocal cross is difficult to make because many cultivars of *tuberosum* do not produce viable pollen. The F_1 hybrids are pentaploids and are largely self-sterile but backcrosses using *tuberosum* pollen can be made. Sidorov (129) emphasized the importance of backcrossing the F_1 plants to *tuberosum* in order to recover desirable types. Various authors have studied meiosis in the F_1 hybrids and found frequent univalents at metaphase (43). Even after three and four backcrosses, Schnell (117) reported that plants had chromosome numbers similar to the F_1 hybrids, showed irregular meiosis, and retained undesirable characters from *demissum*. Nevertheless, several of the genes for resistance were transferred from *demissum* to *tuberosum* and used in a number of breeding lines and commercial cultivars such as Muller's W-types and Aquila (89). Ross (108) reported that 70% of the cultivars in Germany in 1966 carried *demissum* genes. Toxopeus (136) found that backcrosses of *demissum* to *tuberosum* tended to be more productive than crosses among *tuberosum* cultivars.

S. stoloniferum ($2n = 48$) (syn. *S. ajascoense* and *S. antipovichii*) has also been used in breeding for resistance to late blight (111). *Stoloniferum* and *demissum* appear to have two genomes in common and carry similar genes for resistance to late blight (83).

The early promise of the work with late blight resistance from *demissum* was shattered with the discovery that the hypersensitive type of resistance was rapidly overcome by physiologic races of the pathogen (6, 70, 116). In recent years potato breeders have been more interested in nonspecific resistance because it has been shown to be more permanent (90). However, since nonspecific resistance is generally polygenic, its transfer from alien species is more difficult. Nevertheless, Ross (108) reported that the "field" resistance of *stoloniferum* had been transferred to *S. tuberosum* by backcrossing.

Virus diseases are particularly important in the cultivated potato, and extensive work has been done to transfer resistance from wild species to *tuberosum*. The best resistance to virus X was found in *S. acaule* ($2n = 48$) (109, 134). The crossability of *acaule* with *tuberosum* is improved if its chromosome number is first doubled (140). Its resistance has been successfully transferred to commercial cultivars.

Ross & Baerecke (109) and Stelzner (134) found that *S. stoloniferum* carried good resistance to viruses A and Y and behaved like *acaule* in crosses with *tuberosum*. Resistance to viruses A and Y has been transferred to *tuberosum* by backcrossing.

Cockerham (20) studied the inheritance of resistance to potato viruses X and Y within the wild species. Two of six genes controlling resistance to virus X are associated in one linkage group with two or possibly three loci for resistance to virus Y. There is also evidence for allelic relationships of genes in different species.

Two species of nematodes (*Heterodera rostochiensis* and *H. pallida*) are important parasites on potatoes (36). Several species of *Solanum* have resistance and two in particular, *S. vernei* and *S. multidissectum,* have been used in breeding work (29, 30, 36, 108). Both are diploid species ($2n = 24$) that are closely related to *tuberosum,* and no major problems were encountered in crossing. Resistance from *vernei* appeared to be polygenic, was difficult to recover, and was associated with poor cooking quality (36). Combinations of genes from several sources are necessary to give resistance to all races of the nematodes.

The cultivated potato is a tetraploid, while about 70% of its wild relatives are diploid and others range up to hexaploid. The level of ploidy of the parents influences the ease and success of interspecific hybridization. It is possible to produce *tuberosum* derivatives that parallel the range of ploidy in the wild species. Hougas & Peloquin (47) produced diploid lines ($2n = 24$) of *tuberosum* and successfully crossed them with 24 diploid species. Inheritance should be simpler at this chromosome level (disomic rather than tetrasomic) and gene transfers may be easier. Other authors have done the reverse, doubled the chromosome number of a diploid species and then crossed it with *tuberosum.* For example, Livermore & Johnstone (74) found with *S. chacoense* that doubling the chromosome number greatly increased the crossability with *S. tuberosum.*

Most of the interspecific transfers in *Solanum* have involved species that cross fairly readily and show a reasonable degree of chromosome pairing. However, a number of other species carry desirable genes for disease and insect resistance but will not cross with *tuberosum.* Dionne (24) found that *S. acaule* would cross with these species and with *tuberosum.* It could, therefore, be used as a bridge by crossing a species with *acaule* and then crossing the hybrid with *tuberosum.*

Transfers from Species in a Polyploid Complex to a Polyploid Crop That Carries a Genetic Diploidizing System

By far the largest number of interspecific transfers and some of the most complex and fascinating ones have been made in polyploid complexes where diploidizing genetic systems have evolved. Several of the procedures used have been described as chromosome engineering.

NICOTIANA   The genus *Nicotiana* contains a large number of species with a wide range of chromosome numbers (from $2n = 18$ to $2n = 48$) that do not fall into a simple polyploid series (131). Kimber (58) has suggested that a diploidizing system is present in the cultivated tobacco, *Nicotiana tabacum* ($2n = 48$). Interspecific crosses have been used extensively and successfully in breeding (11, 19, 79, 137).

Clausen & Cameron (16) noted that the five diploid species that are most closely related to cultivated tobacco cross readily with it. The chromosomes of the diploids pair with those of *tabacum,* the hybrids show some degree of fertility in backcrosses, and the transfer of genes presents no unique problems. However, these species do not carry high levels of resistance to any of the major diseases of tobacco.

Mann et al (79) divided the *Nicotiana* species into three groups, a first group containing the 5 diploids that are most closely related to *N. tabacum*, a second group of 25 species that can be crossed with *tabacum* with varying degrees of difficulty, and a third group of 30 species that can be crossed with *tabacum* only with considerable difficulty if at all. The species of the second group are the most interesting since a number' of them have been used successfully in breeding disease-resistant tobacco cultivars. In general, their hybrids with *tabacum* show little chromosome pairing and are nearly, if not completely, sterile.

One of the first procedures that was used to overcome problems in the production and fertility of hybrids involved the use of sesquidiploids, that is, hybrids that are diploid for the chromosomes of one species (usually *tabacum*) but haploid for the other. Sesquidiploids can be produced by doubling the chromosome number of *tabacum* and crossing the tetraploid to the alien species. Alternatively if hybrid plants can be obtained from crosses involving the diploids, then the chromosomes of the F_1 plants can be doubled and a backcross made to *tabacum*. The sesquidiploids are usually sufficiently fertile to be either selfed or backcrossed to *tabacum*.

Chaplin & Mann (14) studied the crossability of tetraploid *tabacum* with 24 *Nicotiana* species. In some cases the tetraploid could be used as either the male or the female, but in other cases it could be used successfully either only as the male or only as the female. In general, crossability relationships were the same as for the diploid, but in a few crosses hybrid plants were more vigorous or more fertile.

One of the first successful programs in *Nicotiana* was the transfer of resistance to tobacco mosaic virus from *Nicotiana glutinosa* by Holmes (45). A fertile alloploid, *N. glutinosa* × *N. tabacum*, called *N. digluta*, had been developed by Clausen & Goodspeed (17). It was crossed and backcrossed several times to the cultivar Samsoun, and a homozygous resistant line, similar to Samsoun, was obtained. Gerstel (39) showed that Holmes' Samsoun had a *glutinosa* chromosome substituted for a *tabacum* chromosome. By repeated backcrossing to burley tobacco varieties, Valleau (137) obtained lines in which a segment of the *glutinosa* chromosome carrying resistance had been transferred to a *tabacum* chromosome, presumably by the occasional pairing and crossing-over of homoeologous chromosomes. The resistance derived from *glutinosa* was reported by several authors to be linked to undesirable characters (15). Litton et al (73) registered four cultivars produced from a *glutinosa* derivative (itself a backcross) that had been backcrossed six times to each of four recurrent parents. All four cultivars bloomed three days earlier than their recurrent parent and three of the four were not quite equal to their recurrent parent in yield and dollar value.

Chaplin (13) crossed tetraploid *N. tabacum* with *N. plumbaginifolia* ($2n = 20$) in order to transfer resistance to black shank (*Phytophthora parasitica* var. *nicotianae*). The hybrids were self-sterile but set seed readily when pollinated with *tabacum* pollen. After one further backcross a resistant line was obtained in which resistance had been transferred to a *tabacum* chromosome, probably by homoeologous crossing-over. Apple (3) made a similar transfer of black shank resistance from *plumbaginifolia* and found that two types of resistance were segregating (4). High resistance against race 0 was due to a single, dominant gene while a more moderate resistance to races 0 and 1 was not simply inherited. Goins & Apple (41) transferred the resistance to race 0 to the cultivar Hicks by backcrossing five times. Even after this many backcrosses the resistance was found to have deleterious effects on various agronomic and chemical characters.

Valleau et al (138) transferred black shank resistance from *N. longiflora* (2n = 20) to *N. tabacum*. The cross between the species can be made readily if *tabacum* is used as a female. Hybrid seeds were treated with colchicine, and the resulting plants were backcrossed to *tabacum*. ⬦

Nicotiana longiflora and *N. plumbaginifolia* are closely related. Collins et al (22) crossed the two species and also crossed the *tabacum* lines carrying black shank resistance from the two species. In each case no susceptible segregates were obtained. Thus, not only is the locus for resistance identical in each species but it was transferred to the same *tabacum* chromosome in the two separate programs. Undoubtedly, crossing-over occurred between homoeologus chromosomes.

Wernsman et al (146) used monosomic analysis to attempt to locate the chromosome(s) carrying black shank resistance in NC2326, a cultivar having resistance from *plumbaginifolia,* and L8, a breeding line with resistance from *longiflora.* No positive chromosome identification could be made for NC2326 but in L8 the resistance appeared to be on chromosome V.

An unusual situation involving male sterility in derivatives of *N. tabacum* × *N. plumbaginifolia* hybrids was analyzed by Cameron & Moav (12). The chromosome carrying the gene for black shank resistance (*B5*) was found to carry a pollen killer gene (*K1*) and the two were difficult to separate. In pollen quartets in which *K1* is segregating, pollen carrying the normal complement of *tabacum* chromosomes aborts. Presumably Chaplin (13) and Apple (3) had been successful in breaking this linkage by homoeologous crossing-over.

Nicotiana longiflora has also been used as a source of resistance to wildfire (*Pseudomonas tabaci*). Since hybrids between *tabacum* and *longiflora* die at an early stage, Clayton (18) crossed tetraploid forms of both. Although most seedlings died, a few grew weakly and flowered but proved to be completely sterile. After 18 months three plants produced large galls of callus tissue near the ground surface and their tops died back. New, vigorous shoots developed from the galls. One resistant plant was successfully backcrossed to *tabacum,* and after several generations of selection a homozygous resistant breeding stock was obtained. It also proved to be resistant to blackfire (*Pseudomonas angulata*).

As noted by Mann et al (79) a number of *Nicotiana* species either cannot be crossed with *N. tabacum* or the hybrids are sterile. A bridging cross may be necessary in order to transfer genes from the alien species to *tabacum. Nicotiana repanda* (2n = 48) has resistance to many tobacco diseases, including the rootknot nematode (*Meloidogyne javanica*) but it is generally thought to be impossible to cross it with *N. tabacum.* Burk (9) and Schweppenhauser (118) used *N. sylvestris* (2n = 24) as a bridging species in attempts to transfer nematode resistance from *repanda* to *tabacum.* Burk crossed *repanda* with *sylvestris* and backcrossed twice to *sylvestris.* Eight resistant backcross plants were crossed to

N. tabacum but only three produced seed. Schweppenhauser crossed the amphidiploid, *N. repanda* × *N. sylvestris,* with *N. longiflora.* The F_1 plants were selfed and resistant F_2 plants were crossed and backcrossed twice to *tabacum.* Resistant plants with an added alien chromosome were obtained. As yet a successful transfer of the resistance of *repanda* to a commercial tobacco cultivar has not been made (133).

Stavely et al (133) crossed tetraploid *tabacum* with the amphidiploid, *N. repanda* × *N. sylvestris,* in an attempt to transfer resistance to *Alternaria alternata, Cercospora nicotianae,* and *Meloidogyne javanica.* The sterile F_1 plants were treated with colchicine to double the chromosome number and recover fertility, and were then backcrossed twice to *tabacum.* Resistance to *Cercospora* and *Meloidogyne* but not to *Alternaria* was successfully transferred. The resistant lines almost certainly involve alien chromosome additions, and further work will be needed before they can be used in breeding commercial cultivars. Resistance to *Alternaria* may be complex in inheritance and, therefore, not easily transferred.

Some of the difficulties in interspecific transfers in *Nicotiana* undoubtedly are caused by chromosomal-cytoplasmic interactions (131). Thus, when black root rot (*Thielavia basicola*) resistance was transferred from *debneyi* to *tabacum* in crosses in which *tabacum* was always used as the male parent, male sterility was a major problem. By the third backcross the resistant lines were completely male-sterile as a result of having the *tabacum* chromosomes in *debneyi* cytoplasm. The problem was corrected by using *tabacum* as a female parent so that the *tabacum* chromosomes were now in *tabacum* cytoplasm (19).

Embryo culture is commonly used in interspecific crosses, although not in *Nicotiana,* presumably because of the small size of the seeds. However, special germination conditions have been successfully used. Burk (10) found that seeds of the amphidiploid, *N. sylvestris* × *N. tomentosiformis,* would not germinate under normal greenhouse conditions but did not germinate when placed on soil in clear plastic conditions which were then sealed to maintain a high level of humidity.

BRASSICA   The brassicas form a polyploid complex that may contain a diploid-izing system although there is little evidence available as yet. Resistance to the clubroot (*Plasmodiophora brassicae*) has been transferred from the turnip (*Brassica campestris* $2n = 20$, AA) to the Swede turnip (*B. napus* $2n = 36$, AACC). Lammerink (68) successfully made the transfer by crossing the two species and backcrossing resistant F_2 plants to *napus.* Johnston (51) used colchicine to double the F_1 hybrid to produce an amphidiploid (AAAACC) which can be maintained indefinitely and is more easily backcrossed to *napus.* Resistant second backcross progeny were virtually identical with the recurrent parent.

GOSSYPIUM   The genus *Gossypium* includes about 31 diploid ($2n = 26$) and 4 tetraploid ($2n = 52$) species (95). The cultivated cottons belong to two diploid species, *G. herbaceum* (A_1A_1) and *G. arboreum* (A_2A_2), which have been grown for a very long time in Africa and Asia, and two tetraploid species, *G. hirsutum* and *G. barbadense* (AADD), which originated in the Americas and now provide most of the world's cotton (87). Two of the tetraploid species, *G. barbadense* and *G. tomentosum* cross readily with *G. hirsutum* and the F_1 plants are vigorous and fertile and show normal chromosome pairing. However, F_2 progenies contain many weak, dwarf, chlorotic, and inviable plants. Stephens (135) noted that there was strong selection in favor of the parental types and against intermediate types. In backcrosses he showed that there was a considerable selective elimination of genes from the donor parent. Nevertheless, a number of characters have been transferred to *hirsutum*. For example, the two recessive genes for nectariless, a characteristic that reduces the incidence of boll rot and some insects, were transferred from *tomentosum* to *hirsutum* by Meyer & Meyer (86). However, when Shepherd (128) attempted to transfer resistance to the root-knot nematode (*Meloidogyne incognita acrita*) from a wild *barbadense* to *hirsutum* he was able to recover only part of the resistance. Transfers have also been made in the opposite direction, such as the transfer of three genes for resistance to blackarm (*Xanthomonas malvacereum*) from *hirsutum* to *barbadense* by Knight (59, 61).

The tetraploid *Gossypiums* carry an A genome whose chromosomes pair with the chromosomes in some of the African and Asian diploids and a D genome whose chromosomes pair with chromosomes of some of the wild American diploids. A number of transfers of genes from diploids to tetraploids have been made although the hybrids are more difficult to produce than are hybrids between tetraploids. Knight (60, 61) transferred two genes for blackarm resistance from *arboreum* (AA) to *barbadense* (AADD) by first doubling the diploid and then crossing to *barbadense*. The hybrids (AAAD) showed only slight fertility, but sizable backcross progenies were obtained by growing large F_1 populations and pollinating all flowers each day with *barbadense* pollen. The transfer of the gene B_{6m} was interesting since it has only a modifying effect rather than a direct effect on resistance. First backcross plants were crossed with a strain of *barbadense* homozygous for B_2. Some progenies segregated for much greater resistance than is conditioned by B_2 alone, thus indicating the presence of the modifier, B_{6m}. Meyer (85) crossed *G. armourianum* (D_2D_2) with *hirsutum*, produced the amphidiploid, and backcrossed it to *hirsutum* in order to transfer the D_2 smoothness character that provides resistance to boll weevils, boll worms, and leaf worms. Other crosses are more difficult to make. In some cases hybrids have been produced by first crossing two diploids, doubling the chromosome number, and then crossing the amphidiploid with *hirsutum*.

The recent experience with the corn leaf blight (*Helminthosporium maydis*) epidemic in the United States has shown that disease susceptibility can depend on a particular type of cytoplasm (46). To prepare for such a possibility, cytoplasm from seven *Gossypium* species has been transferred into *G. hirsutum* (87).

TRITICUM   The genus *Triticum* and its close relative *Aegilops* (which some authors include under *Triticum*) provide many examples of the transfer of disease resistance from an alien species to a cultivated crop. The *Triticum-Aegilops* polyploid complex is composed of about 30 species containing various combinations of perhaps 6 basic genomes ($n = 7$) in diploids, tetraploids, and hexaploids (7, 78, 88). The two major cultivated species, *Triticum durum* and *T. aestivum*, cross with many of the other species in the complex and with rye, *Secale cereale*, and certain *Agropyrons*.

The common bread wheat, *T. aestivum*, is a hexaploid ($2n = 42$). Because it has considerable duplication and triplication of genetic material it can withstand many gross chromosomal changes. Sears (120) showed that the *aestivum* chromosomes fall into seven groups of three homoeologous chromosomes. There is increasing evidence that the chromosomes of the relatives of wheat fall into the same homoeologous groups.

With the discovery of physiologic races of stem rust (*Puccinia graminis tritici*) and leaf rust (*Puccinia recondita*) of wheat, it was noted that the tetraploid *Triticum* species carried more resistance to rusts and other diseases than did the hexaploid bread wheat. Thus some of the early work with interspecific crosses involved hybrids of the tetraploid emmers (*T. dicoccum*) and durums (*T. durum*) with bread wheat.

Hayes et al (44) studied crosses of emmer and durum wheats with bread wheat, including a cross between Iumillo durum and Marquis which gave rise to Marquillo, a parent of the very important cultivar Thatcher. Despite the difference in chromosome numbers the crosses were easily made although the hybrid seed was shriveled. However, the F_1 plants showed considerable sterility. Combining rust resistance with desirable bread wheat characters proved to be difficult.

McFadden (81) crossed the highly disease-resistant Yaroslav emmer with Marquis. Again the two main problems were sterility in the F_1 plants and linkage of undesirable emmer characters with the disease resistance. To overcome these problems McFadden resorted to several generations of mass selection including the removal of shrunken seeds. Eventually two selections, Hope and H44–24, were obtained that combined resistance to several diseases and have been used extensively in wheat breeding. Although they were given names, neither Hope nor Marquillo ever became important commercially. Probably at least one backcross to the bread wheat parent should have been used in order to recover the characteristics required. However, this would have made the recovery of resistance more difficult, particularly since resistance to several diseases and, therefore, a number of genes, was being transferred simultaneously.

The Australian cultivars Gabo and Timstein, which were developed from the cross, Gaza durum × Bobbin, and are identical in rust resistance (143), provide an interesting example that is typical of a number of others in wheat. In crosses with Chinese Spring, Sears & Rodenhiser (126) reported that Timstein carried two dominant, complementary linked genes for stem rust resistance. However, in crosses of Timstein with Thatcher, Newthatch, and Mida, Koo & Ausemus (66) found only one gene. Similarly in studies in Australia, Watson & Waterhouse (144) found only one gene in Timstein. The cultivars Gabo, Yalta, and Charter, which have the same resistance to race 126 of stem rust as Timstein, were each crossed with Eureka and Mentana by Luig (76). The three crosses with Eureka all gave ratios of 3 resistant plants: 1 susceptible while the crosses with Mentana gave ratios ranging from 1.0:1 to 2.4:1. Luig proposed that there was only one gene but that differential gametic transmission was involved. Later Loegering & Sears (75) concluded that *Sr11*, the gene for stem rust resistance in Timstein, is linked with a pollen killer locus, *ki*. When Timstein is crossed with cultivars such as Chinese Spring which carries *Ki*, *ki* pollen produced on the *Ki ki* plants aborts. However, in a series of crosses involving Gabo, Luig (77) obtained variable results that suggested a more complex genetic system than just a single pollen killer locus.

T. *timopheevi* probably has more disease resistance than any of the other *Triticum* species. Consequently, there has been considerable interest in using it as a source of resistance. Rather similar results to those with Gabo and Timstein were obtained with C.I. 12633, a derivative from a cross between *timopheevi* and *aestivum* which was backcrossed to the *aestivum* parent. Allard & Shands (2) concluded that the stem rust resistance of C.I. 12633 was due to duplicate dominant genes linked with approximately 15% recombination. Nyquist (92) located the supposed genes on chromosome XIII (2B). However, when Nyquist (93) crossed C.I. 12633 with five susceptible cultivars he obtained a range from 7.1 to 26.6% of susceptible plants. He concluded that rust resistance was controlled by one gene and that differential fertilization was at least partly responsible for the different ratios.

In recent years a number of authors have emphasized the value of the diploid ancestors of wheat as sources of genetic variability which may be tapped with relatively simple cytogenetic techniques (55, 151).

Kerber & Dyck (55, 56) found good sources of leaf and stem rust resistance among the diploid ancestors. They transferred the stem rust resistance of T. *monococcum* ($2n = 14$) first to *durum* ($2n = 28$) by backcrossing and then from *durum* to *aestivum* ($2n = 42$), again by backcrossing (56). The resistance was due to a single dominant gene which, however, showed reduced transmission in some crosses at both the diploid and hexaploid level. Unfortunately, the degree of resistance decreased with increasing levels of ploidy although it was still valuable. Gerechter-Amitai et al (38) transferred stem rust resistance from T. *boeoticum* ($2n = 14$) to *durum* by a similar procedure.

A more complex procedure was used by Kerber & Dyck (55) and Dyck & Kerber (32) to transfer two genes for leaf rust resistance to T. *aestivum* from *Aegilops squarrosa* ($2n = 14$), the donor of the D genome in wheat. In each case a synthetic tetraploid (AABB) extracted from the cultivar Canthatch (AABBDD) was crossed with a strain of *squarrosa* (DD) and colchicine was used to produce an amphidiploid (AABBDD). This amphidiploid was then crossed with *aestivum* (AABBDD). Both types of leaf rust resistance from *squarrosa* were due to single dominant or partially dominant genes and could be readily transferred to cultivars of *aestivum*.

The transfer of disease resistance from alien species is most difficult when the chromosomes involved are not sufficiently similar that crossing-over can occur naturally. To overcome this problem some very complicated and elegant procedures have been developed in wheat (65, 107, 122).

Sears (119) developed the first procedure that used radiation to transfer disease resistance from a chromosome of one species to a chromosome of another. *Aegilops umbellulata* ($2n = 14$) has good resistance to leaf rust, but the cross between it and *aestivum* does not produce viable seeds. Sears, therefore, crossed *A. umbellulata* with *T. dicoccoides* ($2n = 28$) and produced the amphidiploid which could be crossed with *aestivum*. In the latter hybrid the *umbellulata* chromosomes did not pair with *aestivum* chromosomes. The F_1 plants were partially male fertile, and two backcrosses were made to *T. aestivum* cv Chinese Spring with selection for leaf rust resistance. A resistant plant having an added *umbellulata* isochromosome was obtained. Its progeny were radiated with X rays prior to meiosis and then used as pollen parents in crosses with Chinese Spring. Pollen carrying the *umbellulata* isochromosome functioned only rarely and was, therefore, screened out. However, wheat-*umbellulata* translocations were often less deleterious and many of the rust-resistant progeny carried translocations. Only one of the translocations proved to be transmitted entirely normally through the gametes. In this line, later named Transfer, the translocation involved wheat chromosome 6B (121). Athwal & Kimber (5) showed that the *umbellulata* chromosome is homoeologous with group 6 chromosomes in wheat. Transfer has been used widely in wheat breeding and is one of the parents of the soft red winter wheat, Riley 67. Knott (65) made nine backcrosses to add the leaf rust resistance of Transfer to the cultivar Thatcher. Although the backcross lines had 1% higher protein content than Thatcher, they were very inferior in baking quality.

Knott (63) used Sears' procedure to translocate stem rust resistance from an *Agropyron elongatum* ($2n = 70$) chromosome to a wheat chromosome. In this case there is no problem making the cross but the *elongatum* chromosome did not pair with a wheat chromosome in either addition or substitution lines. One of the translocations was transmitted normally through the gametes and has been used as a source of stem rust resistance in the Australian cultivars Eagle and Kite. Knott (64) and Johnson (49) showed that the *Agropyron* chromosome was homoeologous with wheat group 6 and the translocation that was transmitted normally through the gametes proved to involve chromosome 6A.

Sharma & Knott (127) used a similar procedure to transfer leaf rust resistance to wheat from Agrus, a substitution line in which an *Agropyron* chromosome had replaced wheat chromosome 7D. Again one translocation that was transmitted normally through the gametes was obtained and eventually named Agatha. In this case also the translocation involved the homoeologous wheat chromosome, 7D. However, the resistance in Agatha proved to be linked to an undesirable yellow flour color, although it is being used in wheat breeding.

A chromosome carrying leaf, stem, and stripe rust (*Puccinia striiformis*) resistance in *Agropyron intermedium* was transferred to wheat by Wienhues (147). Radiation was used to translocate the leaf rust resistance to a wheat chromosome (148), but in general the lines carrying translocations gave reduced yields (149).

It is interesting to note that although Wienhues and Sharma & Knott worked with different *Agropyron* species, the leaf rust resistance proved in each case to be carried on an *Agropyron* chromosome homoeologous with wheat group 7. In fact, Wienhues (149) obtained a spontaneous translocation between the *Agropyron* chromosome and wheat chromosome 7A.

The radiation method has two major problems: first, it requires a large amount of cytological work and, second, most of the translocations obtained involve nonhomoeologous chromosomes and are deleterious. Only transfers between homoeologous chromosomes appear not to be deleterious.

To overcome these problems Driscoll & Jensen (27) and Sharma & Knott (127) developed procedures that involved detecting nondeleterious translocations by looking for normal 3:1 ratios in segregating families following radiation. Driscoll & Jensen (28) successfully used their procedure to produce a wheat-rye translocation stock, Transec, carrying both leaf rust and powdery mildew (*Erysiphe graminis*) resistance from rye.

The discovery by Riley (102) and Sears & Okamoto (125) of the genetic control of homoeologous pairing in wheat has largely eliminated the use of the radiation procedure except in cases where the chromosomes of the alien species are so dissimilar to those of wheat that homoeologous pairing cannot occur. A gene on chromosome 5B was found to be largely responsible for the suppression of homoeologous pairing in hexaploid wheat and in interspecific hybrids involving it (106). This discovery made possible the transfer of genes between *Triticum* and many of its relatives including *Aegilops* and a number of *Agropyrons.*

Two types of procedure have been developed to make use of crossing-over between homoeologous chromosomes of wheat and an alien species. One involves the use of a cross or crosses that result in a hybrid deficient for chromosome 5B, for example, a cross between monosomic 5B and the alien species. As an alternative, Wall et al (139) and Sears (124) have produced mutants of the gene on chromosome 5B and these permit homoeologous pairing in interspecific hybrids. The second procedure involves the use of a genotype of *Aegilops speltoides* ($2n = 14$) which suppresses the effect of chromosome 5B (107). For example, *A. speltoides* can be crossed with an addition line carrying an alien chromosome.

One of the first successful uses of the 5B mechanism involved the transfer of resistance to stripe rust from *Aegilops comosa* (104, 105). *Aegilops comosa* was crossed and backcrossed to *T. aestivum* to produce a disomic addition line for the *comosa* chromosome carrying stripe rust resistance. In the addition line the *comosa* chromosome does not pair with any wheat chromosome although it will substitute for chromosomes 2A, 2B, and 2D. The addition line was crossed with *A. speltoides,* and the 29-chromosome hybrids were backcrossed using *aestivum* as the pollen parent. After three backcrosses a resistant 42-chromosome line was obtained and designated Compair. In Compair the resistance proved to be on wheat chromosome 2D showing that homoeologous pairing and crossing-over had involved 2D and the *comosa* chromosome.

As mentioned earlier, when *Agropyron* leaf rust resistance was transferred to wheat in Agatha (127), the resistance was associated with a yellow flour color. Sears (123) used homoeologous pairing to produce some new transfers of the *Agropyron* leaf rust resistance from the original parent, Agrus, in which the *Agropyron* chromosome was substituted for wheat chromosome 7D. By a series of crosses, plants were obtained that were nullisomic 5B, monosomic 7D, and monosomic for the *Agropyron* chromosome (7el). In the absence of 5B, homoeologous pairing could take place between 7D and 7el. The plants were pollinated with *aestivum* pollen. Of 74 resistant progeny that were cytologically analyzed 21 presumed crossovers were recovered. Four of the transfers that showed the best transmission through male gametes were tested for yellow pigment in the seed by J. Dvořák (unpublished). Unfortunately, they all carried the undesirable gene for yellow color. This may indicate that the gene is distal to the gene for resistance and any transfer resulting from a single crossover between the centromere and the gene for resistance will automatically carry the second gene.

Since the leaf rust resistance of Agatha is potentially valuable, J. Dvořák (unpublished) used another procedure to break the linkage between leaf rust resistance and yellow flour color. A substitution line carrying *A. elongatum* chromosome 7E which is homoeologous with chromosome group 7 in wheat and lacks the gene for yellow pigment, was crossed with the translocation line, Agatha. In the hybrids, 7E and the translocation chromosome are each present singly and pairing occurs between 7E and the homoeologous *Agropyron* segment of the translocation chromosome (31). Thus, crossing-over may occur between the gene for resistance and the gene for yellow pigment.

Although a number of alien addition and substitutution lines have been produced in wheat, very few have proven to have any potential as commercial cultivars. Alien chromosomes can generally only be substituted for a homoeologous wheat chromosome. Such substitutions are generally stable and occasionally show some promise. A number of European wheats carrying mildew and stripe rust resistance from rye have been shown to be either substitutions of rye chromosome 1R for wheat chromosome 1B or spontaneous 1B/1R translocations (84, 150).

Sears (122) demonstrated that in wheat plants that are monosomic for two chromosomes, both monosomes may misdivide and rejoin to give a chromosome combining an arm from each. Such a procedure could provide a useful way of transferring an alien chromosome arm to a wheat chromosome. Since wheat and rye chromosomes show little homoeologous pairing, it is likely that the spontaneous wheat-rye translocations mentioned above occurred by the joining of alien chromosome arms.

The interest in the commercial production of hybrid wheat in recent years has stimulated research on the effects of genome-cytoplasm interactions. The immediate interest was in cytoplasmic male sterility. However, Washington & Maan (141) found that when they transferred the genomes of two bread wheats and one durum wheat into different cytoplasms, the expression of genes for resistance to leaf rust was altered.

AVENA   The genus *Avena* has many similarities to *Triticum*. It comprises about 20 species containing various combinations of four basic genomes of seven chromosomes each (99). The species range from diploids to hexaploids with the major cultivated species being hexaploid. There is evidence for genetic control of homoeologous pairing in the hexaploid (37), and the genetic control is suppressed by the genotype of a diploid species, *Avena longiglumis* (98). Although much less work has been done on interspecific gene transfers in oats than in wheat, many of the same problems have arisen and similar procedures have been used to overcome them (99).

DISCUSSION

In this final section the problems encountered in alien transfers and the methods used to overcome them are reviewed.

The major obstacle in an alien transfer may come at the first step, making the cross. When the two species being crossed are only very distantly related, fertilization either fails to occur or embryo development stops at an early stage (e.g. crosses between wheat and barley). Immunosuppressive drugs are being tested to see if they will overcome the problem. While initially some favorable results were reported it appears now that the reports were overly optimistic. Whenever a problem is encountered in making a cross, a range of genotypes of both parents should be tried in the hope that one combination may be more compatible than another. For example, Chinese Spring wheat has been shown to carry two genes that make it cross more readily with rye than other genotypes do (103). In a number of cases intervarietal or even interspecific hybrids have been found to be better parents than more homogeneous material. Reciprocal differences are not uncommon, so crosses should be tried in both directions. In some interspecific crosses failure results from slow pollen tube growth and has been overcome by either bud pollination or stylar grafts.

Since difficulties in making crosses between species are often due to differences in levels of ploidy, various procedures involving changes in the chromosome number have been tried. For example, the chromosome number of a wild diploid species may be doubled so that it will cross with a tetraploid cultivated species.

When none of these procedures works, the use of a third species to act as a bridge between two species has been successful in a number of crosses. This third species must be compatible with both of the other two. Thus genes from a diploid *Triticum* can be transferred first to a tetraploid and then to the hexaploid. In some cases two species are crossed, colchicine or nitrous oxide is used to produce an amphidiploid, and the latter is then crossed with the commercial species.

Interspecific crosses probably fail most frequently after fertilization has occurred. Particularly when differences in chromosome numbers are involved, the embryo may abort because of a failure of endosperm development. Treatment with growth hormones such as gibberellin has been reported to be useful in some cases. If the embryo is able to develop normally for a period before breakdown begins, it is often possible to use embryo culture to obtain a hybrid plant. Some cytogeneticists use embryo culture routinely, even for hybrids where the embryo survives but endosperm development is poor and seed viability low.

Even when seedlings are obtained, often the battle is not over. Hybrid plants may die at any stage before reproduction or they may be sterile. Frequently not much can be done except search for other hybrid combinations that may be more viable. However, in several cases it is possible to rear otherwise inviable hybrids by grafting them onto one of the parents. Colchicine treatment of hybrids to produce amphidiploids has been commonly used to increase vigor and particularly fertility. The amphidiploids can then usually be backcrossed to the cultivated species although this is not always true.

The sterility of interspecific hybrids can result from genetic or cytoplasmic factors or from genome-cytoplasm interactions. Plants of an interspecific hybrid may be self-sterile because of genes for incompatibility but fertile when intercrossed. One genotype in a particular cytoplasm may cause sterility while another may result in fertility. Similarly, backcrossing that is always in one direction can result in the genome of one species being present in the cytoplasm of another and cause sterility. Reversing the direction of the cross will put the genome in its own cytoplasm and restore fertility.

Once successful crosses and backcrosses have been produced, problems still arise frequently. The introduction of alien germ plasm into the finely balanced genetic system of a cultivated species causes a major disturbance. Often one or more sizable segments of alien chromatin are involved, and because they are homoeologous but not homologous to the corresponding segments of the recipient species, crossing-over may not occur regularly. Thus it is not surprising that difficulties in breaking linkages between resistance and deleterious characters are often reported. Usually they can be broken with enough backcrosses or sufficiently large populations, but special cytogenetic procedures may be necessary in some cases. The introduced genetic material may be deleterious in one genetic background but not in another so various genetic combinations should be tested. The expression of resistance is often reduced when genes are transferred to a new species, but this also can vary depending on the genetic background.

Most cultivated crops are the result of many years of breeding and combine a multitude of desirable characteristics. Following an interspecific cross, the recovery of all of the characteristics is not easy and may require repeated backcrossing and selecting.

Perhaps this review can be concluded by noting one of the more spectacular recent developments in plant hybridization, the intergeneric fusion of plant protoplasts. There have been some highly optimistic reports on the potential of cell fusion in the production of wide hybrids. Fusion of cells from species as diverse as legumes and grasses have been obtained and the hybrid cells have divided (52). However, the production of plants from these hybrid cells is probably a long way off and perhaps impossible. Even if plants are produced, all of the problems of sterility and gene transfer will still be present in more extreme form than in the hybrids that can now be produced.

For the foreseeable future breeders will have to use more mundane procedures. The wild relatives of cultivated crops are a tremendously valuable reservoir of germ plasm. They have been used with considerable success already but much useful and rewarding work remains to be done. In the future more emphasis should probably be placed on breeding for general or nonspecific resistance although this will no doubt be more difficult than breeding for specific resistance.

Literature Cited

1. Alexander, L. J., Hoover, M. M. 1955. Disease resistance in the wild species of tomato: Report of the National Screening Committee. *Ohio Agric. Exp. Stn. Bull. 752.* 76 pp.
2. Allard, R. W., Shands, R. G. 1954. Inheritance of resistance to stem rust and powdery mildew in cytologically stable spring wheats derived from *Triticum timopheevi. Phytopathology* 44:266–74
3. Apple, J. L. 1962. Transfer of resistance to black shank (*Phytophthora parasitica* var. *nicotianae*) from *Nicotiana plumbaginifolia* to *N. tabacum. Phytopathology* 52:1 (Abstr.)
4. Apple, J. L. 1967. Occurrence of race 1 of *Phytophthora parasitica* var. *nicotianae* in North Carolina and its implications in breeding for disease resistance. *Tob. Sci.* 11:79–83
5. Athwal, R. S., Kimber, G. 1972. The pairing of an alien chromosome with homoeologous chromosomes of wheat. *Can. J. Genet. Cytol.* 14:325–33
6. Black, W. 1952. XVII. Inheritance of resistance to blight (*Phytophthora infestans*) in potatoes: Inter-relationships of genes and strains. *Proc. R. Soc. Edinburgh Sect. B* 64:312–52
7. Bowden, W. M. 1959. The taxonomy and nomenclature of the wheats, barleys and ryes and their wild relatives. *Can. J. Bot.* 37:657–84
8. Bukasov, S. M. 1936. The problems of potato breeding. *Am. Potato J.* 13:235–52
9. Burk, L. G. 1967. An interspecific bridge-cross. *Nicotiana repanda* through *N. sylvestris* to *N. tabacum. J. Hered.* 58:215–18
10. Burk, L. G. 1973. Partial self-fertility in a theoretical amphiploid progenitor of *N. tabacum. J. Hered.* 64:348–50
11. Burk, L. G., Heggestad, H. E. 1966. The genus *Nicotiana:* A source of resistance to diseases of cultivated tobacco. *Econ. Bot.* 20:76–88
12. Cameron, D. R., Moav, R. 1957. Inheritance in *Nicotiana tabacum.* XXVII. Pollen killer, an alien genetic locus inducing abortion of microspores not carrying it. *Genetics* 42:326–35
13. Chaplin, J. F. 1962. Transfer of black shank resistance from *Nicotiana plumbaginifolia* to flue-cured *N. tabacum. Tob. Sci.* 6:184–87
14. Chaplin, J. F., Mann, T. J. 1961. Interspecific hybridization, gene transfer and chromosomal substitution in *Nicotiana. N. C. Agric. Exp. Stn. Tech. Bull. 145.* 31 pp.
15. Chaplin, J. F., Mann, T. J., Apple, J. L. 1961. Some effects of the *Nicotiana glutinosa* type of mosaic resistance on agronomic characters of flue-cured tobacco. *Tob. Sci.* 5:80–83
16. Clausen, R. E., Cameron, D. R. 1957. Inheritance in *Nicotiana tabacum.* XXVIII. The cytogenetics of introgression. *Proc. Natl. Acad. Sci. USA* 43:908–13
17. Clausen, R. E., Goodspeed, T. H. 1925. Interspecific hybridization in *Nicotiana.* II: A tetraploid *Glutinosa-tabacum* hybrid, an experimental verification of Winge's hypothesis. *Genetics* 10:278–84
18. Clayton, E. E. 1947. A wildfire resistant tobacco. *J. Hered.* 38:35–40
19. Clayton, E. E. 1958. The genetics and breeding progress in tobacco during the last 50 years. *Agron. J.* 50:352–56
20. Cockerham, G. 1970. Genetical studies on resistance to potato viruses X and Y. *Heredity* 25:309–48
21. Coe, G. E. 1954. A grafting technique enabling an unthrifty interspecific hybrid of *Beta* to survive. *Proc. Am. Soc. Sugar Beet Technol.* 8:157–60
22. Collins, G. B., Legg, P. D., Litton, C. C., Stokes, G. W. 1971. Locus homology in two species of *Nicotiana, Nicotiana longiflora* Cav. and *N. plumbaginifolia* Viv. *J. Hered.* 62:288–90
23. Correll, D. S. 1962. *The Potato and Its Wild Relatives.* Renner, Texas: Texas Res. Found. 606 pp.
24. Dionne, L. A. 1963. Studies on the use of *Solanum acaule* as a bridge between *Solanum tuberosum* and species in the series *Bulbocastana, Cardiophylla* and *Pinnatisecta. Euphytica* 12:263–69
25. Dodds, K. S. 1962. Classification of cultivated potatoes. See Ref. 23, pp. 517–39
26. Dodds, K. S. 1966. The evolution of the cultivated potato. *Endeavor* 25:83–88
27. Driscoll, C. J., Jensen, N. F. 1963. A genetic method for detecting induced intergeneric translocations. *Genetics* 48:459–68
28. Driscoll, C. J., Jensen, N. F. 1965. Release of a wheat-rye translocation stock involving leaf rust and powdery mildew resistances. *Crop Sci.* 5:279–80

29. Dunnett, J. M. 1960. The role of *Solanum vernei* Bitt. et Wittm. in breeding for resistance to potato root eelworm (*Heterodera rostochiensis* Woll.) *Scott. Plant Breed. Stn. Rept.,* pp. 39–44

30. Dunnett, J. M. 1961. Inheritance of resistance to potato root ellworm in a breeding line stemming from *Solanum multidissectum* Hawkes. *Scott. Plant Breed. Stn. Rept.,* pp. 39–46

31. Dvorak, J. 1975. Meiotic pairing between single chromosomes of diploid *Agropyron elongatum* and decaploid *A. elongatum* in *Triticum aestivum. Can. J. Genet. Cytol.* 17:329–36

32. Dyck, P. L., Kerber, E. R. 1970. Inheritance in hexaploid wheat of adult-plant leaf rust resistance derived from *Aegilops squarrosa. Can. J. Genet. Cytol.* 12:175–80

33. Frandsen, N. O., Baerecke, M.-L., Ross, H., Troka, M., Rudorf, W. 1958. Kartoffel. IV. Grundlagen und Methoden der Züchtung A. Resistenzeigenschaften und ihre vererbung. In *Handbuch der Pflanzenzüchtung,* ed. H. Kappert, W. Rudorf, 3:71–135. Berlin: Parey. 2nd ed.

34. Frazier, W. A., Dennett, R. K. 1949. Isolation of *Lycopersicon esculentum* type tomato lines essentially homozygous resistant to root knot. *Proc. Am. Soc. Hortic. Sci.* 54:225–34

35. Frazier, W. A., Kikuta, K., McFarlane, J. S., Hendrix, J. W. 1946. Tomato improvement in Hawaii. *Proc. Am. Soc. Hortic. Sci.* 47:277–84

36. Fuller, J. M., Howard, H. W. 1974. Breeding for resistance to white potato cyst-nematode, *Heterodera pallida. Ann. Appl. Biol.* 77:121–28

37. Gauthier, F. M., McGinnis, R. C. 1968. The meiotic behavior of a nullihaploid plant in *Avena sativa* L. *Can. J. Genet. Cytol.* 10:186–89

38. Gerechter-Amitai, Z. K., Wahl, I., Vardi, A., Zohary, D. 1971. Transfer of stem rust seedling resistance from wild diploid einkorn to tetraploid *Durum* wheat by means of a triploid hybrid bridge. *Euphytica* 20:281–85

39. Gerstel, D. U. 1945. Inheritance in *Nicotiana tabacum.* XIX. Identification of the *Tabacum* chromosome replaced by one from *N. glutinosa* in mosaic-resistant Holmes Samsoun tobacco. *Genetics* 30:448–54

40. Gilbert, J. C., McGuire, D. C. 1956. Inheritance of resistance to severe root knot from *Meloidogyne incognita* in commercial type tomatoes. *Proc. Am. Soc. Hortic. Sci.* 68:437–42

41. Goins, R. B., Apple, J. L. 1971. Inheritance and phenotypic expression of a dominant factor for black shank resistance from *Nicotiana plumbaginifolia* in a *Nicotiana tabacum* milieu. *Tob. Sci.* 15:7–11

42. Hawkes, J. G. 1958. Significance of wild species and primitive forms for potato breeding. *Euphytica* 7:257–70

43. Hawkes, J. G. 1958. Kartoffel. I. Taxonomy, cytology and crossability. See Ref. 33, pp. 1–43

44. Hayes, H. K., Parker, J. H., Kurtzweil, C. 1920. Genetics of rust resistance in crosses of varieties of *Triticum vulgare* with varieties of *T. durum* and *T. dicoccum. J. Agric. Res.* 19:523–42

45. Holmes, F. O. 1938. Inheritance of resistance to tobacco-mosaic disease in tobacco. *Phytopathology* 28:553–61

46. Hooker, A. L. 1972. Southern leaf blight of corn—Present status and future prospects. *J. Environ. Qual.* 1:244–49

47. Hougas, R. W., Peloquin, S. J. 1962. Exploitation of *Solanum* germ plasm. See Ref. 23, pp. 21–24

48. Howard, H. W. 1970. *Genetics of the Potato: Solanum tuberosum.* New York: Springer. 126 pp.

49. Johnson, R. 1966. The substitution of a chromosome from *Agropyron elongatum* for chromosomes of hexaploid wheat. *Can. J. Genet. Cytol.* 8:279–92

50. Johnson, R. T. 1956. A grafting method to increase survival of seedlings of interspecific hybrids within the genus, *Beta. J. Am. Soc. Sugar Beet Technol.* 9:25–31

51. Johnston, T. D. 1974. Transfer of disease resistance from *Brassica campestris* L. to rape (*B. napus* L.). *Euphytica* 23:681–83

52. Kao, K. N., Constabel, F., Michayluk, M. R., Gamborg, O. L. 1974. Plant protoplast fusion and growth of intergeneric hybrid cells. *Planta* 120:215–27

53. Keep, E. 1973. *Ribes sanguinem* and related species as donors in currant and gooseberry breeding. *J. Yugoslav Pomol.* 25–26:3–7

54. Keep, E. 1975. Currants and gooseberries. In *Advances in Fruit Breeding,* ed. J. Janick, J. N. Moore, 197–268. West Lafayette, Ind.: Purdue Univ. Press. 623 pp.

55. Kerber, E. R., Dyck, P. L. 1969. Inheritance in hexaploid wheat of leaf rust resistance and other characters derived from *Aegilops squarrosa*. *Can. J. Genet. Cytol.* 11:639–47

56. Kerber, E. R., Dyck, P. L. 1973. Inheritance of stem rust resistance transferred from diploid wheat (*Triticum monococcum*) to tetraploid and hexaploid wheat and chromosome location of the gene involved. *Can. J. Genet. Cytol.* 15:397–409

57. Khush, G. S., Rick, C. M., Robinson, R. W. 1964. Genetic activity in a heterochromatic chromosome segment of the tomato. *Science* 145:1432–34

58. Kimber, G. 1961. Basis of the diploidlike meiotic behaviour of polyploid cotton. *Nature* 191:98–100

59. Knight, R. L. 1946. Breeding cotton resistant to blackarm disease (*Bact. malvacearum*). Part I: Introductory. Part II: Breeding methods. *Emp. J. Exp. Agric.* 14:153–74

60. Knight, R. L. 1953. The genetics of blackarm resistance. IX. The gene B_{6m} from *Gossypium arboreum*. *J. Genet.* 51:270–75

61. Knight, R. L. 1954. Cotton breeding in the Sudan. Part I and II. Egyptian cotton. *Emp. J. Exp. Agric.* 22:68–92

62. Knight, R. L., Keep, E., Briggs, J. B., Parker, J. H. 1974. Transference of resistance to black-currant gall mite, *Cecidophyopsis ribis*, from gooseberry to black currant. *Ann. Appl. Biol.* 76:123–30

63. Knott, D. R. 1961. The inheritance of rust resistance. VI. The transfer of stem rust resistance from *Agropyron elongatum* to common wheat. *Can. J. Plant Sci.* 41:109–23

64. Knott, D. R. 1964. The effect on wheat of an *Agropyron* chromosome carrying rust resistance. *Can. J. Genet. Cytol.* 6:500–7

65. Knott, D. R. 1971. The transfer of genes for disease resistance from alien species to wheat by induced translocations. In *Mutation Breeding for Disease Resistance*, 67–77. Vienna: Int. At. Energy Agency. 249 pp.

66. Koo, F. K. S., Ausemus, E. R. 1951. Inheritance of reaction to stem rust in crosses of Timstein with Thatcher, Newthatch, and Mida. *Agron. J.* 43:194–201

67. Kuriyama, T., Kuniyasu, K., Mochizuki, H. 1971. Studies on the breeding of disease-resistant tomato by interspecific hybridization. *Bull. Hortic.*

Res. Stn., Jpn., Ser. B, No. 11, 33–60 (In Japanese, Engl. summ.)

68. Lammerink, J. 1970. Inter-specific transfer of clubroot resistance from *Brassica campestris* L. to *B. napus* L. *N. Z. J. Agric. Res.* 13:105–10

69. Laterrot, H. 1973. Selection de varietes de tomate resistantes aux *Meloidogyne*. *OEPP/EPPO Bull.* 3:89–92

70. Lehmann, H., Von 1938. Geschichte und Ergebnisse der Versuche zur Züchtung krautfäulewiderstandsfähiger Kartoffelin. *Dev. Züchter* 10:72–80

71. Lesley, J. W. 1948. Plant breeding methods and current problems in developing improved varieties of tomatoes. *Econ. Bot.* 2:100–10

72. Lincoln, R. E., Cummins, G. B. 1949. *Septoria* blight resistance in the tomato. *Phytopathology* 39:647–55

73. Litton, C. C., Collins, G. B., Legg, P. D., Everette, G. A., Masterson, J. B. 1972. Registration of Mosaic Resistant Little Crittenden, Black Mammoth, Madole, and Little Wood tobaccos. *Crop Sci.* 12:397

74. Livermore, J. R., Johnstone, F. E. Jr. 1940. The effect of chromosome doubling on the crossability of *Solanum chacoense*, *S. jamesii*, and *S. bulbocastanum* with *S. tuberosum*. *Am. Potato J.* 17:170–73

75. Loegering, W. Q., Sears, E. R. 1963. Distorted inheritance of stem-rust resistance of Timstein wheat caused by a pollen-killing gene. *Can. J. Genet. Cytol.* 5:65–72

76. Luig, N. H. 1960. Differential transmission of gametes in wheat. *Nature* 185:636–37

77. Luig, N. H. 1964. Heterogeneity in segregation data from wheat crosses. *Nature* 204:260–61

78. Mac Key, J. 1968. Relationships in the Triticinae. *Proc. 3rd Int. Wheat Genet. Symp.*, pp. 39–50

79. Mann, T. J., Gerstel, D. U., Apple, J. L. 1963. The role of interspecific hybridization in tobacco disease control. *Proc. III World Tobacco Sci. Congr.*, Salisbury, S. Rhodesia, pp. 201–7

80. Marks, G. E, 1955. Cytogenetic studies in tuberous *Solanum* species. I. Genomic differentiation in the group Demissa. *J. Genet.* 53:262–69

81. McFadden, E. S. 1930. A successful transfer of emmer characters to *Vulgare* wheat. *J. Am. Soc. Agron.* 22:1020–34

82. McFarlane, J. S., Hartzler, E., Fra-

zier, W. A. 1946. Breeding tomatoes for nematode resistance and for high vitamin C content in Hawaii. *Proc. Am. Soc. Hortic. Sci.* 47:262–70

83. McKee, R. K. 1962. Identification of R-genes in *Solanum stoloniferum*. *Euphytica* 11:42–46

84. Mettin, D., Blüthner, W. D., Schlegel, G. 1973. Additional evidence on spontaneous 1B/1R wheat-rye substitutions and translocations. *Proc. 4th Int. Wheat Genet. Symp.*, pp. 179–84

85. Meyer, J. R. 1957. Origin and inheritance of D₂ smoothness in Upland cotton. *J. Hered.* 48:249–50

86. Meyer, J. R., Meyer, V. G. 1961. Origin and inheritance of nectariless cotton. *Crop Sci.* 1:167–69

87. Meyer, V. G. 1974. Interspecific cotton breeding. *Econ. Bot.* 28:56–60

88. Morris R., Sears, E. R. 1967. The cytogenetics of wheat and its relatives. In *Wheat and Wheat Improvement*, ed. K. S. Quisenberry, L. P. Reitz, 19–87. Madison, Wis.: Am. Soc. Agron.

89. Muller, K. O. 1951. Über die Herkuft der W-Sorten, ihre Entwicklungsgeschichte und ihre bisherige Nutzung in der praktischen Kartoffelzüchtung. *Z. Pflanzenzuecht.* 29:366–87

90. Niederhauser, J. S. 1961. Genetic studies of *Phytophthora infestans* and *Solanum* species in relation to late blight resistance. In *Recent Advances in Botany*, 1:491–97. Toronto, Ontario: Univ. Toronto Press. 951 pp. 2 vols.

91. Nilsson, F. 1973. *Ribes* breeding by species crossing. *J. Yugoslav Pomol.* 25–26:37–44

92. Nyquist, W. E. 1957. Monosomic analysis of stem rust resistance of a common wheat strain derived from *Triticum timopheevi*. *Agron. J.* 49:222–23

93. Nyquist, W. E. 1962. Differential fertilization in the inheritance of stem rust resistance in hybrids involving a common wheat strain derived from *Triticum timopheevi*. *Genetics* 47:1109–24

94. Pelham, J. 1966. Resistance in tomato to tobacco mosaic virus. *Euphytica* 15:258–67

95. Phillips, L. L. 1974. Cotton (*Gossypium*). In *Handbook of Genetics*, ed. R. C. King, 2:111–33. New York: Plenum

96. Propach, H. 1938. Cytogenetische Untersuchungen in der Gattung *Solanum*, Sect. *Tuberarium*. IV. Tetraploide und sesquidiploide Artbastarde. *Z. Indukt. Abstamm. Vererbungsl.* 74:376–87

97. Propach, H. 1940. Cytogenetische Untersuchungen in der Gattung *Solanum*, Sect. *Tuberarium*. V. Diploid Artbastarde. *Z. Indukt. Abstamm, Vererbungsl.* 78:115–28

98. Rajhathy, T., Thomas, H. 1972. Genetic control of chromosome pairing in hexaploid oats. *Nature New Biol.* 239:217–19

99. Rajhathy, T., Thomas, H. 1974. Cytogenetics of oats. *Genet. Soc. Can. Misc. Publ. No. 2.* 90 pp.

100. Reddick, D., Crosier, W. F., Mills, W. R. 1931. Blight immune potato hybrids. *Proc. Ann. Meet. Potato Assoc. Am.* 18:60–64

101. Rick, C. M., Butler, L. 1956. Cytogenetics of the tomato. *Adv. Genet.* 8:267–382.

102. Riley, R. 1958. Chromosome pairing and haploids in wheat. *Proc. X Int. Congr. Genet.* 2:234–35

103. Riley, R., Chapman, V. 1967. The inheritance in wheat of crossability with rye. *Genet. Res.* 9:259–67

104. Riley, R., Chapman, V., Johnson, R. 1968. Introduction of yellow rust resistance of *Aegilops comosa* into wheat by genetically induced homoeologous recombination. *Nature* 217:383–84

105. Riley, R., Chapman, V., Johnson, R. 1968. The incorporation of alien disease resistance in wheat by genetic interference with the regulation of meiotic chromosome synapsis. *Genet. Res.* 12:199–219

106. Riley, R., Chapman, V., Kimber, G. 1959. Genetic control of chromosome pairing in intergeneric hybrids with wheat. *Nature* 183:1244–46

107. Riley, R., Kimber G. 1966. The transfer of alien genetic variation to wheat. *Rept. Plant Breed. Inst., Cambridge 1964–1965*, pp. 6–36

108. Ross, H. 1966. The use of wild *Solanum* species in German potato breeding of the past and today. *Am. Potato J.* 43:63–80

109. Ross, H., Baerecke, M. L. 1950. III. Selection for resistance to mosaic virus (diseases) in wild species and in hybrids of wild species of potatoes. *Am. Potato J.* 27:275–84

110. Rudorf, W. 1950. IV. Methods and results of breeding resistant strains of potatoes. *Am. Potato J.* 27:332–39

111. Rudorf, W., Schaper, P. 1951. Grundlagen und Ergebnisse der Züchtung krautfäuleresistenter Kartoffelsorten. *Z. Pflanzenzuecht.* 30:29–88

112. Rudorf, W., Schaper, P., Ross, H., Baerecke, M-L., Torka, M. 1950. I.

The breeding of resistant varieties of potatoes. *Am Potato J.* 27:222–35

113. Salaman, R. N. 1931. Recents progres dans la creation de varietes de pommes de terre resistant au mildiou *"Phytophthora infestans."* Deuxieme *Congr. Int. Pathol. Comp.* 2:435

114. Savitsky, H. 1975. Hybridization between *Beta vulgaris* and *B. procumbens* and transmission of nematode (*Heterodera schachtii*) resistance to sugarbeet. *Can. J. Genet. Cytol.* 17:197–209

115. Schick, R. von. 1932. Über das Varhalten von *Solanum demissum, Solanum tuberosum* und ihren Bastarden gegenüber verscheidenen Herkünften von *Phytophthora infestans. Der Züchter* 4:233–37

116. Schick, R., von, Schaper, P. 1936. Das Verhalten von verschiedenen Formen von *Solanum demissum* gegenüber 4 verchiedenen Linien der *Phytophthora infestans. Der Züchter* 8:65–70

117. Schnell, L. O. 1948. A study of meiosis in the microsporocytes of interspecific hybrids of *Solanum demissum* × *Solanum tuberosum* carried through four backcrosses. *J. Agric. Res.* 76:185–212

118. Schweppenhauser, M. A. 1968. Recent advances in breeding tobacco resistant to *Meloidogyne javanica. Coresta Inf. Bull.* 1:9–20

119. Sears, E. R. 1956. The transfer of leaf-rust resistance from *Aegilops umbellulata* to wheat. *Brookhaven Symp. Biol.* 9:1–22

120. Sears, E. R. 1958. The aneuploids of common wheat. *Proc. 1st Int. Wheat Genet. Symp.,* pp. 221–28

121. Sears, E. R. 1966. Chromosome mapping with the aid of telocentrics. *Proc. 2nd Int. Wheat Genet. Symp.,* pp. 370–81

122. Sears, E. R. 1972. Chromosome engineering in wheat. *Stadler Genet. Symp.* 4:23–38

123. Sears, E. R. 1973. *Agropyron*-wheat transfers induced by homoeologous pairing. *Proc. 4th Int. Wheat Genet. Symp.,* pp. 191–99

124. Sears, E. R. 1975. An induced homoeologous-pairing mutant in *Triticum aestivum. Genetics* 80:S74

125. Sears, E. R., Okamoto, M. 1958. Intergenomic chromosome relationships in hexaploid wheat. *Proc. X Int. Congr. Genet.* 2:258–59

126. Sears, E. R., Rodenhiser, H. A. 1948. Nullisomic analysis of stem-rust resistance in *Triticum vulgare* var. Timstein. *Genetics* 33:123–24

127. Sharma, D., Knott, D. R. 1966. The transfer of leaf-rust resistance from *Agropyron* to *Triticum* by irradiation. *Can. J. Genet. Cytol.* 8:137–43

128. Shepherd, R. L. 1974. Breeding root-knot-resistant *Gossypium hirsutum* L. using a resistant wild *G. barbadense* L. *Crop Sci.* 14:687–91

129. Sidorov, F. F. 1937. Züchtung *Phytophthora* widerstandfähiger Kartoffelsorten. *Phytopathology* 27:211–41

130. Sievert, R. C. 1972. Resistance to anthracnose in the genus *Nicotiana. Tob. Sci.* 16:32–39

131. Smith, H. H. 1968. Recent cytogenetic studies in the genus *Nicotiana. Adv. Genet.* 14:1–54

132. Smith, P. G. 1944. Embryo culture of a tomato species hybrid. *Proc. Am. Soc. Hortic. Sci.* 44:413–16

133. Stavely, J. R., Pittarelli, G. W., Burk, L. G. 1973. *Nicotiana repanda* as a potential source for disease resistance in *N. tabacum. J. Hered.* 64:265–71

134. Stelzner, G. 1951. Virusresistenz der Wildkartoffeldin. *Z. Pflanzenzuecht.* 29:135–58

135. Stephens, S. G. 1949. The cytogenetics of speciation in *Gossypium.* I. Selective elimination of the donor parent genotype in interspecific backcrosses. *Genetics* 34:627–37

136. Toxopeus, H. J. 1952. Over de mogelijke betekenis van *Solanum demissum* voor de veredeling gericht op verhoging van de knolopbrengst. *Euphytica* 1:133–39

137. Valleau, W. D. 1952. Breeding tobacco for disease resistance. *Econ. Bot.* 6:69–102

138. Valleau, W. D., Stokes, G. W., Johnson, E. M. 1960. Nine years' experience with the *Nicotiana longiflora* factor for resistance to *Phytophthora parasitica* var. *nicotianae* in the control of black shank. *Tob. Sci.* 4:92–94

139. Wall, A. M., Riley, R., Chapman, V. 1971. Wheat mutants permitting homoeologous meiotic chromosome pairing. *Genet. Res.* 18:311–28

140. Wangenheim, K.-H, F., von. 1955. Zur Ursache der Kreuzungsschwierigkeiten zwischen *Solanum tuberosum* L. and *S. acaule* Bitt. bzw. *S. stoloniferum* Schlechtd. et. Bouche. *Z. Pflanzenzuecht.* 34:7–48

141. Washington, W. J., Maan, S. S. 1974. Disease reaction of wheat with alien cytoplasms. *Crop Sci.* 14:903–5

142. Watson, I. A. 1970. The utilization of wild species in the breeding of cultivated crops resistant to plant pathogens. In *Genetic Resources in Plants— Their Exploration and Conservation,* ed. O. H. Frankel, E. Bennett, IBP Handb. No. 11, pp. 441–57. Oxford: Blackwell. 554 pp.

143. Watson, I. A., Stewart, D. M. 1956. A comparison of the rust reaction of wheat varieties Gabo, Timstein, and Lee. *Agron. J.* 48:514–16

144. Watson, I. A., Waterhouse, W. L. 1949. Australian rust studies. VII. Some recent observations on wheat stem rust in Australia. *Proc. Linn. Soc. NSW* 74:113–31

145. Watts, V. M. 1947. The use of *Lycopersicon peruvianum* as a source of nematode resistance in tomatoes. *Proc. Am. Soc. Hortic. Sci.* 49:233–34

146. Wernsman, E. A., Matzinger, D. F., Powell, N. T. 1974. Genetic investigations of intraspecific and interspecific

147. Wienhues, A. 1966. Transfer of rust resistance of *Agropyron* to wheat by addition, substitution and translocation. *Proc. 2nd Int. Wheat Genet. Symp., Hereditas Suppl.* 2:328–41

148. Wienhues, A. 1967. Die Übertragung der Rostresistenz aus *Agropyrum intermedium* in den Weizen durch Translokation. *Der Züchter* 37:345–52

149. Wienhues, A. 1973. Translocations between wheat chromosomes and an *Agropyron* chromosome conditioning rust resistance. *Proc. 4th Int. Wheat Genet. Symp.,* pp. 201–7

150. Zeller, F. J. 1973. 1B/1R wheat rye chromosome substitutions and translocations. *Proc. 4th Int. Wheat Genet. Symp.,* pp. 209–21

151. Zohary, D., Harlan, J. R., Vardi, A. 1969. The wild diploid progenitors of wheat and their breeding value. *Euphytica* 18:58–65

sources of black shank resistance in tobacco. *Tob. Sci.* 18:15–18

Copyright © 1976 by Annual Reviews Inc.
All rights reserved

FOSSIL FUNGI   ♦3641

K. A. Pirozynski

Palaeontology Division, National Museums of Canada, Ottawa, Canada

INTRODUCTION

Paleomycology is not a new science. Sternberg's descriptions, in 1820, of the first fungi, among the smallest of fossils, precede the naming of some of the largest—the first dinosaurs. Admittedly his *Algacites* and *Carpolites* were not intended for fungi, but they contained descriptions of what was later claimed to be a carpophore and rhizomorphs of Basidiomycota. The credit for the first conscious attempts to record fossil fungi must go to Eichwald and to Lindley & Hutton who, in 1830 and 1833, described the polypores *Daedalea volhynica* and *Polyporus bowmanii*. The identity of the former is obscure; the latter was soon identified as a fish scale. In 1836 Göppert described indeterminate specks on fronds of a Carboniferous fern as *Excipulites neesii*, setting a precedent for unfounded conclusions that, continued in the work of others, brought the number of similar fossil genera to 20 by the mid-nineteenth century. They were given both fossil and modern generic names, and matched with living pyrenomycetes, discomycetes, and coelomycetes.

With the exception of hyphomycetes in amber, which Berkeley (8) studied with a compound microscope, the others—of which the vast majority were leaf-spotting fungi—were described as seen by oculo nudo or with a hand lens. It was not until the latter part of the century that microscopic examination became the rule rather than the exception. European bituminous coals and amber yielded mycelia and reproductive bodies of the lower "phycomycetous" fungi (24, 46); and perithecia accompanied by identifiable ascospores, and dispersed conidia were found in Tertiary lignites (5, 26, 47). The century ended with Seward's excellent review (52), and with the less critical but nevertheless invaluable compilations of Meschinelli, first in Saccardo's *Sylloge Fungorum* and then in the comprehensive *Iconographia* (41) in which 359 accepted fossil species were treated in 54 genera.

After the initial burst of paleomycological activity which left so impressive a record, the first 50 years of this century were relatively uneventful. A few taxonomists, reviewers (43), and a group of workers intrigued by the ancient demonstration of mycotrophism, however, kept the science alive. The 1950s mark a revival of interest in fossil fungi, due mainly to two developments: the birth of palynology with associated elaboration of techniques for isolation of pollen and spores, and an intensified exploration for fossil fuels which opened up new paleohabitats. However, attempts to interpret dispersed fungal spores proved difficult—because mycologists seldom describe spores with the precision characteristic of a palynologist's analysis of the pollen and spores of plants—and led to the frequent dismissal of fossil fungi as being of little value in phylogenetic and stratigraphic considerations. It is true that fragments of mycelia or fructifications rarely provide satisfactory clues to their ordinal, or even their phyletic affiliations, and in modern classifications spores play an increasingly subordinate role to other, much more ephemeral characters. Nevertheless, studies of assemblages, and an interpretation of individual fossil fragments made in the light of contemporary biological and geological developments, promise to provide a stronger basis for speculation on the origin and phylogeny of fungi and their role in the evolution of present-day biota.

FOSSIL RECORD

Proterozoic: Dawn of Oomycota

The concept of Precambrian life is relatively new, but it included fungi from its inception: coenocytic hypha-like filaments associated with stromatolitic blue-green algae. Some, assigned to *Eomycetopsis,* were found in the Middle and Upper Precambrian of most continents; others, as partners of a presumed lichen-like association, are held responsible for the extraction and deposition of South African gold some 2.3–2.7 billion years ago (28). Although the fungal affinity of these filaments is far from proven, there is no reason, judging by later events, to question the existence of Oomycotan water molds in the early Proterozoic, or to dispute their place among the first eukaryotes. The Late Precambrian "ascus-like microfossil" (50) is a perfect match for textbook illustrations of intercalary oogonia of modern Saprolegniaceae.

Early Phanerozoic: Age of Marine Oomycota and Chytridiomycota

The occurrence of saprolegniaceous water molds in ancient marine environments has been on record since the 1850s. A relative of the modern *Leptolegnia marina* may have been responsible for tubular borings in shells and scales of arthropods and fish since Early Cambrian (58). *Ordovicimyces,* showing both Oomycotan and algal affinities, grew in Ordovician bryozoa. *Palaeachlya silurica* allegedly parasitized Silurian and Devonian corals, though its frequent and specific association might perhaps indicate a mutualistic rather than a parasitic relationship.

The progress of early life may have depended not so much on an organism's physical fitness to survive as on its ability to cooperate. The endozoic members of Oomycota and Chytridiomycota—the latter discovered in fragments of calcareous animals ranging in age from Cambrian to recent (65)—are practically indistinguishable from their living descendants (17). This elicited comments from even the earliest investigators, who considered the generic, if not specific longevity "without a parallel in the organic world" as being due to their "lowly organization and simple structure" (31). Similar genetic conservatism was claimed much later for some Cyanophyta, whose usefulness as index fossils was, consequently, questioned (49). Both groups, with eons of evolution in a stable marine environment behind them, appear to indicate that the older the group and the less sophisticated its biology, the longer its life span. Endozoic marine water molds and chytrids do not disappear in the early Phanerozoic, to reappear, apparently unchanged, today. Examples similar to both the ancient and the modern, and described in *Palaeachyla, Phycomycites,* or *Propythium,* have been discovered in corals, bryozoans, etc from the Carboniferous, Jurassic, and Tertiary.

Devonian: Terrestrial Symbiotic Oomycota

Among the ancient inhabitants of calcareous marine metazoa were forms (17) that appeared in the Early Devonian, in some of the earliest vascular plants. Describing these vesicular, coenocytic, endophytic hyphae as *Palaeomyces gordonii* and *P. asteroxylii,* Kidston & Lang noticed their similarity to the modern endotrophs and suspected a symbiotic association—a view that has since acquired new evidence and supporters (9, 14). Pirozynski & Malloch (44) went a step further by postulating the mycotrophic origin of vascular plants, which they consider to be a highly evolved alga/fungus partnership. Given the new role, the fungus becomes less of a paradox: obligately endophytic, yet ubiquitous in host range and geographic distribution, unchanged morphologically despite profound climatic and floristic changes—one of the oldest organisms known that "undoubtedly occupies a unique archaeological niche which entitles it to be termed a living fossil" (63). Today, performing its half a billion-year-old function, it still supports the vast majority of living plants, including our crops. Unfortunately it is rarely recognized that crop plants that "cannot provide their own fertilizer" are not the unsolved problem of millenia of agricultural practice, but their result—the result of perennial disregard of the sustaining function of the less conspicuous fungal component.

There are no explicit records of Devonian terrestrial Chytridiomycota, but that may be because they inhabited waterlogged soils, perhaps even before the establishment of the first terrestrial flora.

The antiquity of today's land-based Oomycota and Chytridiomycota seems clear, both from their present critical dependence on free water, indicative of an aquatic origin, and their genetic conservatism as demonstrated by their fossil record. The major change in their long existence was the transition from a marine to brackish and, eventually, a freshwater habitat. But this must have happened long ago, before separation of the present-day continents, for they have ubiquitous geographical distribution in spite of their present intolerance of salinity and their inefficient dispersal mechanisms.

Carboniferous: Endomycorrhizae; Parasitic Oomycota and Chytridiomycota; Appearance of Zygomycota and Basidiomycota

The discovery of *Mycorrhizonium,* a psilopsid rhizome harboring the familiar endophyte (61) stimulated other investigators to demonstrate the presence of fungi, sometimes in the form of endomycorrhizae, in lepidodendrons, calamites, ferns, and cordaites (1, 14, 37). The symbiotic nature of much of the Carboniferous flora seemed to be established, and a wish was expressed that the fossil record of the fungal component "may continue to reveal significant fragments of their [plants'] ancestral development" (3). Others, however, saw the fungi in the role of parasites or saprophytes (4), or were prevented from accepting the fungi as mycorrhizal because of their caulicolous habit. But the death of the photosynthetic partner need not prevent a fungus from persisting in the decomposing tissues, and as far as their systemic habit is concerned we should remember the modern *Psilotum,* in which the symbiotic endophyte occurs both in the gametophyte and the sporophyte.

The fungi involved in these early associations were given a variety of names which, like names given to their living descendants, reflect the vagueness of specific limits of the vesicular-arbuscular mycorrhizal endophytes. It was proposed that *Peronosporites antiquarius* and an unnamed species of Cash & Hick be classified with *Palaeomyces gracilis* (52) and also with *Protomyces protogenes* and *Rhizophagites* spp., common in younger geological formations and identical with modern endomycorrhizal fungi (14, 63). Not all of these were necessarily symbiotic. They may have been parasites—whose presence was suspected in the Early Devonian—perhaps related to the Carboniferous *Peronosporoides,* a Cretaceous *Peronospora*-like "fungus a" (23), or the Miocene *Peronosporites* and *Pythites* (42). Some may even represent Chytridiomycota not unlike *Grilletia* or *Oochytrium* discovered in seeds and wood of Carboniferous plants (39, 46).

Claims made in the late 1800s of discoveries of Zygomycota in Paleozoic coals are inconclusive. *Mucor combrensis* in a lepidodendron megaspore was based on an impression of a colony of radiating filaments somewhat resembling mycelium of modern Mucoraceae in culture. *Zygosporites,* originally referred to freshwater algae, contained stalked vesicles which indicate a fungus (52). *Sporocarpon* is similar. Williamson's original preparations of the latter were reinvestigated and described as "representing the first record of a Palaeozoic septate fungus" (29). *Sporocarpon* and the segregate genera *Mycocarpon* and *Dubiocarpon* were subsequently found in North America (20) and, while their fungal nature was confirmed, their phyletic affiliation remains unknown. They are hollow bodies with walls made up of one or more layers of cells either (*a*) irregular, or (*b*) in radial files, (*c*) with some cells modified as spines, and are reminiscent of (*a*) the zygospores of modern *Mortierella,* or the sporangium of, (*b*) *Syncephalis-Syncephalastrum,* or (*c*) *Spinalia.* Admittedly the fossil specimens are considerably larger, but what was not oversized in the Carboniferous?

More positive evidence for coal-age Zygomycota is provided by the presumed ascomycotan *Protoascon missouriensis* which was recently referred to as a water mold (4), but which I consider to be a zygospore of an *Absidia* not unlike *A. glauca.* Its substrate, a corroded megaspore, indicates the saprophytic habit which today characterizes the phylum and denies the postulated zygomycotan ancestry of the universal symbiotic endophyte, currently classified in the mucoraceous Endogonaceae. The present-day worldwide distribution of Zygomycota, and their only partial adaptation to xerophytic conditions, betray conservatism, and point to an ancient origin, possibly from conjugating protists.

Many examples of Carboniferous Basidiomycota have shared the fate of the historic *Polyporus bowmanii*. Lesquereux's *Polyporites* was matched with zonate concretions and the same may be true of *Pseudopolyporus*. *Dactyloporus* is now considered a product of an "animal agency," *Rhizomorpha sigillariae* closely resembles insect galleries in wood, and *Incolaria* shows little evidence of being a fungus (30, 52). Likewise, the Cretaceous *Polyporites brownii* and *P. stevensonii* fell victim to interpretations as a syringopore coral and the dental plate of a lung fish, respectively (12).

However, the discovery in the wood of a Carboniferous fern of the dikaryotic, clamp-bearing mycelium of *Palaeancistrus* (21) makes subsequent, if not preceding, records of Basidiomycota more credible. Equally convincing are petrified carpophores of *Phellenites digustoi* on wood of Jurassic *Araucaria*, and of several species of Tertiary *Daedalites, Lenzites,* and *Polyporites* described between 1857 and 1900 (41). More recently, well-preserved specimens of what appears to be *Fomes fomentarius* were found in European Miocene (55), and it was also identified, together with *Ganoderma lucidum* and *G. applanatum,* in the Alaskan Pleistocene (18). *Fomes idahoensis,* closely resembling modern *Fomitopsis pinicola,* has been repeatedly collected in the Idaho Pliocene (2, 13). Other groups are represented by *Hydnum argillae* and a beautifully preserved *Geaster florissantensis* from the Miocene, a modern *Bovista plumbea* from the Alaskan Pleistocene (18) and, possibly, by *Palambages* (56), a *Burgoa*-like bulbil from the Cretaceous.

While Polyporales appear to be well established in the Mesozoic and their presence in the Carboniferous indicated, the other major group of homobasidiomycetes, the Agaricales, are conspicuous by their absence from the fossil record. This is not unexpected if one considers the unsuitability of fleshy carpophores for fossilization, and the inability of most basidiospores to withstand acetolysis (27). It is true that two species of *Agaricites* were described from the Miocene, but the names given to hyphae preserved in wood in the one case, and a mushroom-shaped fossil in the other, do not indicate affinities. The agarics may not be as old as the polypores but we should nevertheless look for their pioneers in the Mesozoic. To James (32) the proof of their former existence was "in the remains of insects that live upon fleshy fungi to exclusion of all other substances." To me, the Mesozoic origin of Agaricales is indicated by their present-day involvement in secondary ectotrophic symbioses, for these may have led to the evolution of the Pinaceae, whose success in conquering boreal Laurasia was not paralleled in the already separated Gondwana (D. Malloch, personal communication).

Early claims of finds of fossil Uredinales, ranging in age from Cretaceous to Early Tertiary, are inconclusive: they were based on streaks or concentric markings that superficially resembled sori of modern rust fungi. The earliest example, a teliospore-like propagule of *Teleutospora millotii* in an unlikely situation inside a lepidodendron megaspore, "cannot be accepted as sufficient evidence of the existence of a Palaeozoic *Puccinia*" (52). More recent claims (7) of the occurrence in Tertiary lignites of spores of Uredinales (and Ustilaginales), though exaggerated—judging by the accompanying illustrations—are nevertheless acceptable in view of other contemporary finds. These include teliospores of *Milesia, Puccinia, Ravenelia, Triphragmium, Uromyces,* and *Xenodochus* in Miocene deposits of North America and India (10, 45, 64). Rusts are credited with the most primitive life cycle among the Basidiomycota and, because the most primitive forms of today occur on ferns and conifers, it has been claimed that they coevolved with their hosts in the Carboniferous, if not the Devonian (33). Rust spores are among the most resistant of propagules, and their absence from earlier sediments is inconsistent with these speculations. Inexplicably, however, they can also be missing from very recent strata (64).

Microscopic organic remains associated with refractive bodies in bituminous coals are called scleronites. Opinions regarding their origin are divided. The Europeans consider them to be remains of fungi. Beneš (6) claimed that among these, preserved as real microfossils, not only fragments of hyphae, spores, and sclerotia can be identified, but also asci, conidia, and ascogenous hyphae. Unfortunately the verbal account is not supported by the accompanying graphic evidence. On the other hand, the English-speaking investigators consider the Paleozoic scleronites to largely represent fusinized resins with little evidence of the physical presence of fungi (59).

Cretaceous: Diversification of Ascomycota

Paleozoic records of Ascomycota, even those in recent literature, are inconclusive (60). The same applies to the Late Triassic *Birsiomyces* which, though assigned to the loculoascomycetes, leaves much to the imagination. However, by the Cretaceous, a diversified mycoflora of Ascomycota and their imperfect states is demonstrated, a mycoflora that must have originated earlier and proliferated with its angiosperm substrates. Superficial leaf fungi belonging to three extant, predominantly tropical families ranged in distribution from Canada and eastern Asia (34, 54) to Argentina and Australia (23, 40). Other groups of leaf-dwellers are well represented by *Pleosporites* and, less reliably, by *Petrosphaeria* (57).

Dispersed spores from Early and Late Cretaceous of Colorado and Alberta include "fungal spore sp. A" (19), a *Dictyosporium* type of conidium whose fragments were recorded subsequently from younger, widely separated localities as *Ctenosporites* (25, 36) and *Pluricellaesporites glomeratus* (56), which closely resembles the modern Javanese palmicolous *Pithomyces pulvinatus. Helicoma*-type conidia (*Involutisporonites*), ascospores similar to those of *Zopfia* (*Dyadosporites*), toruloid mycelium, apparently of sooty molds, and fragments of hyphae with characteristically collapsed, perforated septa complete the assemblage.

The Mesozoic origin of Ascomycota is indicated by their present-day marginal involvement in advanced mycotrophic symbioses, and their often restricted geographic distribution despite specialized adaptations for aerial dispersal. The subordinate role of fungal decay of coal-forming organisms (48) may have been due to the absence of Ascomycota in the Paleozoic. Their late appearance in the fossil record contradicts the existing hypotheses regarding their evolution. If they arose from the red algae, a view currently in vogue, the stepping stone may have been a lichen: not an "ascophyte" devoid of the algal component (15) but an orthodox lichen of the kind so conspicuous in equatorial rain forests of today.

Cenozoic: Age of Ascomycota

The Cretaceous families as well as new groups of superficial leaf Ascomycota appear early in the Tertiary. Some can be confidently matched with living representatives on account of their distinctive mycelium or fructifications, and an often excellent state of preservation. The records of foliicolous fungi were compiled by Dilcher (22), who demonstrated the presence in Eocene Tennessee of essentially modern *Asterina, Asterolibertia (Asterina nodosaria), Patouillardiella, Meliola,* and *Euthalopycnidium (Trichopeltinites).* Further studies led to the transfer of his *Shortensis memorabilis,* first into the modern *Manginula* (35), and then to *Vizella* (51), but I consider it conspecific with *V. oleariae* despite 50 million years of evolution on a different continent and readaptation to a host which evolved subsequently. His *Pelicothallos villosus* appears to be the green alga *Cephaleuros virescens,* and his *Sporidesmium henryense* is a good match for *Hansfordiella asterinarum,* which is still associated with *Asterina* as its parasite.

Among other groups, the powdery mildews *Uncinulites* and *Erysiphites* (42) are Ascomycota, the latter perhaps even Erysiphaceae, but the evidence is insufficient for accurate disposition. What has been identified as *Microsphaera, Phyllactinia,* and *Uncinula* (38) turns out to be dinophytan cysts. Setose and lobed bodies assigned to two modern genera of Chytridiomycota (11) resemble developing fructifications and juvenile conidiophores of certain leaf-litter hyphomycetes, or appressoria of, for example, *Gaeumannomyces.* Perithecia and pycnidia of unspecialized saprophytes such as *Chaetomium* and *Chaetomella* (42), and less distinctive ostiolate or astomous fruit bodies (38), have also been found.

Apothecia of discomycetes and perithecia of wood-inhabiting stromatic Sphaeriales were already recognized from Cenozoic remains in the first half of the nineteenth century. These finds are now known to correspond to modern *Hypoxylon, Rosellinia,* and *Trematosphaeria* (55, 62). The lignicolous *Cryptocolax clarnensis,* with both the cleistothecia and conidia preserved, not only matches but antedates the description of the living *Xylogone sphaerospora.*

Amber has long been the source of excellently preserved hyphomycetes: *Paecilomyces* (8), *Cladosporium, Gonatobotrys, Torula* (16), and others whose proper identity could no doubt be established by reexamination of the original specimens.

However, the majority of Paleogene Ascomycota are represented by dispersed spores, especially the more resistant, highly melanized ascospores and conidia. The practice of earlier workers, who correlated isolated spores with extant genera, is rarely continued by modern micropaleontologists who often have only a marginal interest in, and knowledge of, fungi. It is frequently assumed that fungi recovered from ancient sediments can no longer relate to living mycoflora—a misapprehension that, nevertheless, absolves the investigator from seeking a mycologist's opinion. Fungal spores find their way into "Incertae sedis" or, if recognized, end up in new form-genera or are sorted into morphographic categories (53). Most often, however, their presence is ignored. This regrettable attitude deprives future paleomycologists of valuable data because Tertiary assemblages rarely contain spores that cannot, eventually, be assigned to modern genera. In a recent paper devoted to fungal spores (25), for example, one can recognize in *Striadisporites* the ascospores of *Gelasinospora,* in *Granatisporites* the conidia of the *Brachysporium-Endophragmia-Bactrodesmium* complex, and *Pesavis simplex* is practically indistinguishable from the conidium of *Ceratosporella bicornis.*

CONCLUSION

The fossil record of fungi can be summed up by a sentence written by Gardner in 1886: "The fungi are destitute of chlorophyll and hence, or owing to their parasitic and saprophytic habits, any further development in them seems to have been arrested." Because of this very conservatism which rarely makes them significant as index fossils, they provide a link with the past—as living witnesses of ancient environments. But this potential as interpreters of past environments will not be realized until a coordinated study of fossil and modern fungi is undertaken.

Literature Cited

1. Agashe, S. N., Tilak, S. T. 1970. Occurrence of fungal elements in the bark of arborescent calamite roots from the American Carboniferous. *Bull. Torrey Bot. Club* 97:216–18
2. Andrews, H. N. 1948. A note on *Fomes idahoensis* Brown. *Ann. Mo. Bot. Garden* 35:207
3. Andrews, H. N., Lenz, L. W. 1943. A mycorrhizome from the Carboniferous of Illinois. *Bull. Torrey Bot. Club* 70:120–25
4. Baxter, R. W. 1975. Fossil fungi from American Pennsylvanian coal balls. *Paleontol. Contrib. Univ. Kans.* 77:1–6
5. Beck, R. 1882. Das Oligocän von Mittweida mit besonderer Berücksichtigung seiner Flora. *Z. Dtsch. Geol. Ges.* 34:735–70
6. Beneš, K. 1969. Paleomykologie uhelných sloji. *Sb. Věd. Pr. Vys. Šk. Báňske Ostrave Rada Horn.-Geol.* 15:79–99
7. Beneš, K., Kraussova, J. 1965. Paleomycological investigation of the Tertiary coals of some basins in Czechoslovakia. *Sbor. Geol. Věd. Paleontol.* 6:149–68
8. Berkeley, M. J. 1848. On three species of mould detected by Dr. Thomas in the amber of East Prussia. *Ann. & Mag. Nat. Hist. Ser. 2* 2:380–83
9. Boullard, B., Lemoigne, Y. 1971. Les champignons endophytes du *"Rhynia Gwynne-vaughanii"* K. et L. Étude morphologique et déductions sur leur biologie. *Botaniste* 54:49–89
10. Bradley, W. H. 1931. Origin and microfossils of the oil shale of the Green River Formation of Colorado and Utah. *US Geol. Surv. Prof. Pap. 168.* 58 pp.
11. Bradley, W. H. 1967. Two aquatic fungi (Chytridiales) of Eocene age from the Green River Formation of Wyoming. *Am. J. Bot.* 54:577–82
12. Brown, R. W. 1938. Two fossils misidentified as shelf-fungi. *J. Wash. Acad. Sci.* 28:130–31
13. Buchwald, N. F. 1970. *Fomes idahoensis* Brown. A fossil polypore fungus from the Late Tertiary of Idaho, U.S.A. *Friesia* 9:339–40
14. Butler, E. J. 1939. The occurrences and systematic position of the vesicular-arbuscular type of mycorrhizal fungi. *Trans. Br. Mycol. Soc.* 22:274–301
15. Cain, R. F. 1972. Evolution of the fungi. *Mycologia* 64:1–14

16. Caspary, R. 1907. Die Flora des Bernsteins und anderer fossiler Harze des ostpreussischen Tertiärs. *Abh. K. Preuss. Geol. Landesamtes N.S.* 4:1–181

17. Cavaliere, A. R., Alberte, R. S. 1970. Fungi in animal shell fragments. *J. Elisha Mitchell Sci. Soc.* 86:203–6

18. Chaney, R. W., Mason, H. L. 1936. A Pleistocene flora from Fairbanks, Alaska. *Novit. Am. Mus. Nat. Hist.* 887:1–17

19. Clarke, R. T. 1965. Fungal spores from Vermejo Formation coal beds (upper Cretaceous) of central Colorado. *Mt. Geol.* 2:85–93

20. Davis, B., Leisman, G. A. 1962. Further observations on *Sporocarpon* and allied genera. *Bull. Torrey Bot. Club* 89:97–109

21. Dennis, R. L. 1970. A middle Pennsylvanian basidiomycete mycelium with clamp connections. *Mycologia* 62:578–84

22. Dilcher, D. L. 1965. Epiphyllous fungi from Eocene deposits in western Tennessee, U.S.A. *Palaeontographica B* 116:1–54

23. Douglas, J. G. 1973. The Mesozoic floras of Victoria. Part 3. *Mem. Geol. Surv. Victoria* 29:10–23

24. Duncan, P. M. 1876. On some unicellular algae parasitic within Silurian and Tertiary corals, with a notice of their presence in *Calceola sandalina* and other fossils. *Q. J. Geol. Soc. London* 32:205–11

25. Elsik, W. C., Jansonius, J. 1974. New genera of Paleogene fungal spores. *Can. J. Bot.* 52:953–58

26. Felix, J. 1894. Studien über fossile Pilze. *Z. Dtsch. Geol. Ges.* 46:269–80

27. Graham, A. 1962. The role of fungal spores in palynology. *J. Paleontol.* 36:60–68

28. Hallbauer, D. K., van Warmelo, K. T. 1974. Fossilized plants in thucholite from Precambrian rocks of the Witwatersrand, South Africa. *Precambrian Res.* 1:199–212

29. Hutchinson, S. A. 1955. A review of the genus *Sporocarpon* Williamson. *Ann. Bot. N.S.* 19:425–35

30. James, J. F. 1885. Remarks on a supposed fossil fungus from the coal measures. *J. Cin. Soc. Nat. Hist.* 8:157–59

31. James, J. F. 1893. Notes on fossil fungi. *J. Myc.* 7:268–73

32. James, J. F. 1893. Fossil fungi. *J. Cin. Soc. Nat. Hist.* 16:94–98

33. Kevan, P. G., Chaloner, W. G., Savile, D. B. O. 1975. Interrelationships of early terrestrial arthropods and plants. *Palaeontology* 18:291–417

34. Krassilov, V. A. 1967. [The early Cretaceous flora of the southern Primorye and its stratigraphic significance.] *Moscow Far East Geological Institute.* 364 pp.

35. Lange, R. T. 1969. Recent and fossil epiphyllous fungi of the *Manginula-Shortensis* group. *Aust. J. Bot.* 17:565–74

36. Lange, R. T., Smith, P. H. 1975. *Ctenosporites* and other Paleogene fungal spores. *Can. J. Bot.* 53:1156–57

37. Lignier, M. O. 1906. *Radiculites reticulatus,* radicelle fossile de Séguoinée. *Bull. Soc. Bot. France* 53:193–201

38. Macko, S. 1957. Lower Miocene pollen from the valley of Kłodnica near Gliwice (Upper Silesia). *Trav. Soc. Sci. Wrocław, B* 88:1–14

39. Magnus, P. 1903. Ein von F. W. Oliver nachgewiesener fossiler parasitischer Pilz. *Ber. Dtsch. Bot. Ges.* 21:248–50

40. Martinez, A. 1968. Microthyriales (Fungi, Ascomycetes) fosiles del Cretacico Inferior de la Provincia de Santa Cruz, Argentina. *Ameghiniana* 5:257–63

41. Meschinelli, L. 1902. Fungorum fossilium omnium hucusque cognitorum Iconographia. *Venice: J. Galla.* 144 pp.

42. Pampaloni, L. 1902. I resti organici nel disodile di Melilli in Sicilia. *Palaeontogr. Ital.* 8:121–30

43. Pia, J. 1927. Thallophyten. In *Handbuch der Paläobotanik,* ed. M. Hirmer, 1:31–136. Münich & Berlin: Oldenbourg

44. Pirozynski, K. A., Malloch, D. W. 1975. The origin of land plants: a matter of mycotrophism. *BioSystems* 6:153–64

45. Ramanujan, C. G. K., Ramachar, P. 1963. Sporae dispersae of the rust fungi (Uredinales) from the Miocene lignite of South India. *Curr. Sci.* 32:271–72

46. Renault, B. 1896. Études des gîtes minéraux de la France. Basin houiller et permien d'Autun et d'Épinac. *Flore Fossile,* 4(Part 2):421–47. Paris: Min. Trav. Publ.

47. Richon, C. 1885. Notice sur quelques Sphériacées nouvelles. *Bull. Soc. Bot. France* 32:viii–xii

48. Schopf, J. M. 1952. Was decay important in origin of coal? *J. Sediment. Petrol.* 22:61–69

49. Schopf, J. W. 1968. Microflora of the Bitter Springs Formation, Late Precambrian, central Australia. *J. Paleontol.* 42:651–88

50. Schopf, J. W., Barghoorn, E. S. 1969. Microorganisms from the Late Precambrian of South Australia. *J. Paleontol.* 43:111–18

51. Selkirk, D. R. 1972. Fossil *Manginula*-like fungi and their classification. *Proc. Linn. Soc. NSW* 97:141–49

52. Seward, A. C. 1898. *Fossil Plants,* 1:205–22. Cambridge: Cambridge Univ. Press

53. Sheffy, M. V., Dilcher, D. L. 1971. Morphology and taxonomy of fungal spores. *Palaeontographica B* 133:34–51

54. Singh, C. 1971. Lower Cretaceous microfloras of the Peace River area, northwestern Alberta. *Bull. Res. Counc. Alberta* 28:1–310

55. Skirgiełło, A. 1961. Flora kopalniana Turowa koło Bogatyni II. *Pr. Muz. Ziemi* 4:5–12

56. Srivastava, S. K. 1968. Fungal elements from the Edmonton Formation (Maestrichtian), Alberta, Canada. *Can. J. Bot.* 46:1115–18

57. Stopes, M. C. 1913. The Cretaceous flora. *Catalogue of the Mesozoic plants in the British Museum (Natural History).* 1:267–81

58. Taylor, B. J. 1971. Thallophyte borings in phosphatic fossils from the Lower Cretaceous of south-east Alexander Island, Antarctica. *Palaeontology* 14:294–302

59. Taylor, G. H., Cook, A. C. 1962. Sclerotinite in coal—its petrology and classification. *Geol. Mag.* 99:41–52

60. Tiffney, B. H., Barghoorn, E. S. 1974. The fossil record of the fungi. *Occas. Pap. Farlow Herb. Cryptogam. Bot. Harv.* 7:1–42

61. Weiss, F. E. 1904. A mycorrhiza from the lower Coal measures. *Ann. Bot. London* 18:255–65

62. Willis, J. H., Gill, E. D. 1965. Fossil fungus (*Hypoxylon*) from Tertiary brown coal, Yallourn, Victoria, Australia. *Proc. R. Soc. Victoria* 78:115–17

63. Wolf, F. A. 1969. Nonpetrified fungi in late Pleistocene sediment from eastern North Carolina. *J. Elisha Mitchell Sci. Soc.* 85:41–44

64. Wolf, F. A. 1969. A rust and an alga in Eocene sediment from western Kentucky. *J. Elisha Mitchell Sci. Soc.* 85:57–58

65. Zebrowski, G. 1936. New genera of Cladochytriaceae. *Ann. Mo. Bot. Gard.* 23:553–64

Copyright © 1976 by Annual Reviews Inc.
All rights reserved

MANAGEMENT OF FOOD RESOURCES BY FUNGAL COLONISTS OF CULTIVATED SOILS

♦3642

G. W. Bruehl

Department of Plant Pathology, Washington State University, Pullman, Washington 99163

Preparation of "Systems and Mechanisms of Residue Possession by Pioneer Fungal Colonists" (19) convinced me that food as a factor in the survival and evolution of fungi is not adequately appreciated. This, in spite of the statement by Garrett (38) that the most common cause of death in a microorganism is starvation, and by Clark (25) that food is the major factor limiting microbial development within soil. This paper stresses aspects of food utilization, conservation, and expenditure, with little emphasis upon food assimilation, which is largely accomplished through parasitism.

The greatest portion of arable land is occupied by herbaceous annuals with many fine rootlets and relatively transient structures. Several fungi have achieved prominence by their success as parasites of these plants. For the most part, *Pythium* and *Fusarium* spp. are examined in an effort to determine attributes essential to success—success as evident in breadth of distribution and prevalence. Damage to the host is incidental.

In the first paper (19) active versus passive possession of food, competition for substrate, staling products, antibiotics, and the critical advantage of plant parasites over saprophytes in being able to colonize living tissues in advance of competitors were stressed. This paper is based on the premise that making money is only part of becoming rich. It must be wisely invested also.

COLONIZATION

Colonization, as used in this paper, refers to occupancy of land, not to direct colonization of host tissues or residues. *Fusarium oxysporum* f. sp. *lini* did not occur in abundance in the virgin prairie and plains soils of northcentral North America. It followed the cultivation of flax and increased rapidly in most fields. It was a successful colonist. *Fusarium solani* f. sp. *phaseoli* was not abundant in the virgin soils of the Columbia River Basin in Washington and Oregon; it followed

247

cultivation of beans and multiplied rapidly in most fields. Fungal colonists are like weeds. Weeds are particularly benefited by the activities of man (10). Production of most annuals (flax, beans, corn, etc) usually involves annual tillage and recurrent disruption of the flora. A cultivated field may be rid of growing plants all at once. Seeds of a single species may be planted, followed by synchronous germination, growth, and maturation. A successful fungal colonist exploits these opportunities and survives interim periods with no food source. Chinn (23) reported 984 propagules of *Dendryphion nanum* per gram of soil after rape (*Brassica* sp.), 354 per gram after cereal, and 86 per gram after fallow in a Saskatchewan rotation—rape, cereal, fallow, rape. Only six propagules per gram were found in fields that had not produced rape. This fungus went from feast to famine, declining rapidly in the absence of a host. It lacks some attributes of more successful colonists.

Successful colonists are preadapted (33) to the main factors of "new" environments. *Fusarium solani* f. sp. *phaseoli* chlamydospores carried on bean seed were already well adapted for survival in fields of beans when beans were first sown in the virgin Washington soil. In general, host tissues are a special, relatively constant envrionment. No major evolutionary changes are required. The major preadaptation of parasites is to the host.

Successful colonists need great independence; theoretically, single propagules may initiate a new population. Independence requires an adequate food reserve and no need for a mating partner. Most widespread soilborne plant pathogens are either asexual or homothallic. Ability to act alone is important in colonizing new fields and in advancing within a field.

The ability to benefit from tillage is characteristic of most colonists. *Aphanomyces, Pythium, Phytophthora, Fusarium, Rhizoctonia,* and *Verticillium* occur at some time in soil in small, discrete units. Tillage distributes the propagules more uniformly within the three-dimensional matrix and increases encounters with seeds, seedlings, or rootlets. Menzies (60) stressed the significance of tissue fragmentation to dispersal of *Verticillium albo-atrum* and Ashworth et al (8) stressed that microsclerotia free in the soil, not those bound in host tissue, constitute the effective inoculum.

Some attributes of fungi that have colonized so many cultivated lands are discussed in greater detail in this review.

THE PROPAGULE

Form and Physiology

It is unlikely that fungal morphology or physiology has been altered appreciably by agriculture. Conditions attendant with cultivation favor only a few among the multitude of fungi in the soil. *Pythium* from native grasses is like *Pythium* from corn. Agriculture was selective more than formative.

Garrett (39), in his enunciation of the inoculum potential concept, stressed the need for a critical amount of energy to establish an infection or to compete for colonization of a substrate. The inoculum potentials of various propagules differ. The microconidia of vascular fusaria are adequate to establish these fungi in xylem vessels. The xylem fluids themselves are a suitable medium for growth (48). Single zoospores support penetration of delicate host cells (31, 62, 77, 88, 89) as do basidiospores (30, 56, 88). These propagules are one cent pieces. While basidiospores and conidia of *Fomes annosus* may infect a freshly cut tree stump, this fungus invades woody roots from established mycelium in other tree roots or stumps by means of root grafts. In one case, pennies were efficient; in the other, gold bullion of infected roots was required.

Small sclerotia (11, p. 160) appear to be less efficient as primary propagules than spores when the target is a delicate structure. The same quantity of material in smaller packages has a distributional advantage, especially when the propagules are uniformly mixed in tilled layers of soil. The total mass of 3000 chlamydospores 10 μm in diameter is equivalent to that of 100 oospores 30 μm in diameter. These populations are also about equivalent in infective power. If one tenth of a sclerotium (1000 μm in diameter) is equivalent in infecting power to 3000 chlamydospores or 100 oospores per gram of soil, the sclerotium is one thirty-third as effective on a per volume basis.

Estimates such as these are highly speculative because of the variations in estimates of inoculum abundance and effectiveness. Under optimum conditions, as few as two *Pythium aphanidermatum* oospores per gram of soil are sufficient to blight alfalfa seedlings (80) while 100% infection of rye seedlings did not occur until the soil contained 150 oospores of *Pythium myriotylum* per gram (65). In spite of difficulties in contrasting the efficiencies of propagules on a mass basis, it is probable that the smallest effective unit is the most efficient, especially when the fungus has a wide host range or when food is present at reasonable intervals.

Most propagules of soilborne plant pathogens are sensitive to fungistasis, remaining dormant even though temperature and moisture conditions are favorable for their germination. Fungistasis usually is overcome when carbohydrates and nitrogenous substances in the soil solution exceed a certain threshold concentration. Sensitivity to fungistasis is a major adaptation for survival in that it restricts germination to situations in which the possibilities of obtaining a substrate are good.

Most colonists respond similarly to a wide variety of soluble or volatile substances that are secreted by plants. Sclerotia of *Sclerotium cepivorum* may remain dormant in soil for years but germinate promptly in response to volatile stimulants from the roots of *Allium* spp. (49). *Sclerotium cepivorum* is too specialized or limited in host range to rank as a major colonist of cultivated soils. In contrast, chlamydospores, oospores, and small sclerotia (36) of the most widespread colonists respond to sugars and amino acids from many plants. Some degree of specialization is evident within microsclerotia of *Verticillium albo-atrum;* however, Schreiber & Green (73) found that more of them germinated in the presence of tomato exudates than in exudates of several nonhosts.

Small resting propagules (chlamydospores, sporangia, oospores) usually require greater concentrations of sugars and amino acids to counteract fungistasis than do microsclerotia or sclerotia. They have greater surface:volume ratios and absorb more food more quickly per unit of mass than large propagules (83). Spores requiring exogenous food for germination may increase in dry weight before germ tubes emerge (4). Larger propagules have greater reserves of food to sustain hyphal growth for greater distances through soil, so they theoretically should be able to respond to lower concentrations of exudate with a reasonable chance for success. Small sclerotia may respond to foods (9, 36, 47) but large sclerotia that produce spores (such as *Claviceps purpurea*) as infectious propagules germinate independently of exogenous food (39).

Ko & Chan (50) reported that sporangia of *Phytophthora palmivora* that germinated directly were more effective as infectious propagules than when germination was indirect. They calculated that formation of the usual 16 zoospores per sporangium represented an investment in membranes, flagella, and locomotion that was not recovered. If direct germination is preferred in a nutrient-enriched situations (82), and if zoospore formation is preferred in poorer situations (as in *P. aphanidermatum*), then the zoospore is produced at greater distance from potential food, and a price is exacted for the distributive, seeking activity.

In addition to fungistasis, which limits germination to promising situations, chemotropism and chemotaxis conserve energy. Growth from or movement by a small propagule should be toward the food source, not random and not away from it.

Sclerotium rolfsii sclerotia germinate and its hyphae colonize dead leaflets or debris before they attack healthy peanut stems. *S. rolfsii* is not a true parasite in that it kills tissues in advance of penetration. Boyle (18) termed it a necrophyte. *Botrytis* depends upon dead flower petals, pollen, or other food source prior to penetration (14). Apparently, the propagules of these fungi do not contain reserves sufficient to sustain infection, or these fungi lack the ability to parasitize living cells. Sclerotia of *Whetzelinia sclerotiorum* produce apothecia that provide ascospores as primary inoculum. Exogenous food was required for ascosporic infection of plants in the prebloom stage, either from dead or dying blossoms or wounds. Abawi & Grogan (1), however, did not obtain infection when sclerotia alone were the inoculum. If sclerotia were in contact with dead tissue, the dead tissue was colonized and the fungus then attacked bean seedlings. Invoking the inoculum potential concept seems illogical. It is more likely that mycelium arising directly from these sclerotia is not able to achieve parasitism directly. The amount of energy available in sclerotia is far greater than that in oospores of chlamydospores of other fungi or of its own ascospores. Some other explanation is required.

Garrett (40) hypothesized that some fungi have developed their relationships to such an extent that exudates are not just signals for germination but they also supply an essential part of the energy for penetration and establishment. The fungus has come to "rely upon provision of carbon and nitrogen nutrients at a concentration normally found in plant exudates. . . . in the course of evolution of some fungal spores, their endowment of endogenous nutrients has become adjusted to take advantage of a reliable supplementation by exogenous nutrients in plant exudates."

Pugh et al (72) stressed the importance of anthers as a food base for *Fusarium graminearum* prior to penetration of the wheat spike, and Dickson thought that pollen in the fluids behind leaf sheaths contributed to stalk rot of corn. Undoubtedly a food base strengthens the attack, but Tu (87) and Schroeder (74) found that infection of wheat spikes by macroconidia of *F. graminearum* occurred in situations lacking caught anthers or pollen grains. If a macroconidium has sufficient energy to sustain infection of wheat glumes, it should have sufficient energy to sustain infection of wheat roots. Garrett's (40) hypothesis supports the theme of this entire paper, that food economy is the central factor in fungal evolution, but he may have gone too far.

Oospores of some *Pythium* species have a degree of constitutive dormancy so that only part of them germinate, even when supplied with proper conditions and exogenous food (80).

In summary, exogenous foods may overcome fungistasis (signal the presence of food), guide the organism to the host either by chemotropism or chemotaxis, and may form an important addition to the endogenous energy reserves. This latter is particularly important when germination is stimulated but a suitable substrate is not found, and the organism is capable of forming a replacement chlamydospore (75) or sporangium (82).

Genetic Characteristics

The success of a pathogen is determined not only by the nature and quantity of food under its control, but by its genetic capabilities as well. Pathologists extol the virtues of variation and lament the instability of pathogens. We are unduly biased by the appearance of new races and by variations we observe in culture. We may have messages from nature. Most soil fungi are either asexual or homothallic. Burnett (21) observed that single sexual types of heterothallic fungi were repeatedly found in soil, that zygotes are usually abundant only among homothallic species, and that sexual reproduction in the sense of outbreeding is not common. The strong forces favoring nonrecombination among fungi are commonly overlooked (53). Constancy rather than variation is generally favored by selection among soilborne pathogens.

If a bird drops one propagule under favorable conditions, it should reproduce. If sexual reproduction were essential, and if the fungus were heterothallic, a second propagule would have to be dropped close by. I know of no widespread colonist of agricultural soils that depend upon heterothallic reproduction. Several soilborne fungi that are either asexual, homothallic, or heterothallic with a probably unimportant sexual state are listed in Table 1. Tillage, the three-dimensional disturbance of soil (12), and the limited growth possible from individual small propagules place a premium upon sexual independence.

Asexual or homothallic reproduction confers advantages other than independence. Stebbins (in 53) observed that genetic systems of rapidly reproducing organisms favor fitness over genetic flexibility. Fitness implies close adaptation to a particular niche, and in the case of parasites of herbaceous plants, this implies conditions of short duration. If fitness is favored, the habitat must be relatively constant. How demanding are conditions for a microconidium of

Table 1   The role of the sexual state in several soilborne fungi

Sexual stage unknown

Cephalosporium gramineum, C. gregatum
Cercosporella herpotrichoides
Fusarium culmorum, F. oxysporum
Macrophomina phaseolina
Periconia circinata
Phymatotrichum omnivorum
Sclerotium rolfsii
Thielaviopsis basicola
Verticillium albo-atrum

Sexual stage known but unimportant in nature

Cochliobolus sativus
Phytophthora cinnamomi
Nectria haematococca (= *Fusarium solani*) (foot rot forms)

Sexual stage important in dissemination

Gaeumannomyces graminis
Gibberella zeae (= *Fusarium graminearum*)
Micronectriella nivalis (= *Fusarium nivale*)
Nectria haematococca (= *Fusarium solani*) (tree canker forms)
Thanatephorus cucumeris (= *Rhizoctonia solani*)

Sexual stage important as a survival structure, homothallic

Aphanomyces cochlioides, A. euteiches, A. raphani
Phytophthora cactorum, P. fragariae, P. megasperma
Pythium aphanidermatum, P. graminicola, P. ultimum

Fusarium oxysporum f. sp. *lini* within a xylem vessel of a susceptible flax plant? How variable is it between the second and third layers of cortical cells of young pea epicotyls? It is my belief that close fitness is favored by the relative simplicity of life within a suscept. The host supplies a buffered, relatively stable environment.

Haploid, asexual organisms are suited to close, relatively precise selection. It is probable that few genes are redundant so that single mutations are often expressed. If a given gene mutates once in every 1×10^6 spores, if soil contains 1000 chlamydospores per gram, and if one hectare contains 1×10^6 kilograms of infested top soil, the population in one hectare would provide 1×10^6 mutants at that one locus. When *Periconia circinata* devastated Milo sorghum in California, one resistant plant was found in each 4 ha of land (2). One kilogram of soil containing 1000 propagules per gram should be the equivalent in mutants to the 4 ha of sorghum. By asexual propagation, haploids maintain most individuals in close fitness and provide an element of variation with little expense. Deleterious mutants are eliminated with little loss to the population as a whole. Advantageous mutants would be "pure" and stable at inception, ready for increase and perpetuation.

Of what use is the perfect stage? Burnett (21) suggests that in many cases the sexual stage is important for dispersal rather than for recombination. Among the important cereal fusaria of the USA, the one in the humid central and eastern areas of the country with an epidemiologically important perfect state is *Gibberella zeae*. Ascosporic inoculum is important in head blight (scab) of wheat and barley and in ear and stalk rot of corn. Airborne ascospores are more effectively dispersed than water-splashed conidia from refuse on the ground. In relatively arid south-central Washington *Fusarium graminearum* ($=G.$ *zeae*) exists as a soilborne fungus without a perfect stage. Among the most widespread soil fusaria of the world are species with no known sexual stage (*Fusarium oxysporum, Fusarium culmorum*), or whose sexual stage is known but plays little or no known role in nature (*Fusarium solani*) (58).

Students of fusaria proclaim the great variation within a species, and yet if fresh isolates of *Fusarium culmorum* are obtained directly from diseased wheat in central Washington, they are amazingly similar. Many pathologists have recorded the recurrence of certain wild-type cultures (27, 54, 63). In nature, intense selection eliminates most variants (63), maintaining fitness among pathogenic fusaria.

The form-genus *Helminthosporium* contains many pathogens of cereals and grasses, most of which primarily attack aboveground structures. Barley has three major *Helminthosporium* pathogens. In net blotch (=*Drechslera teres*), the perfect stage, *Pyrenophora teres,* is common and important in humid regions. Spot blotch is caused by *Bipolaris sorokiniana* (=*H. sativum*). This fungus has a wide host range, is the most prevalent soilborne member of the group, and its perfect stage (*Cochliobolus sativus*) is known only in culture (84). It is heterothallic and the two mating types are widely distributed, yet the sexual stage plays no known role. *Drechslera graminea* (= *H. gramineum*) causes a systemic disease of barley; its only important means of survival is as mycelium in infected kernels. Two of these three fungi have essentially eliminated the sexual stage and these two have solved the problem of dispersal by persistence in soil, wide host range, or by a highly effective seedborne stage. The perfect stage of these two was expendable.

Mather (57) asked why the sexual stage persists. He offered two possibilities: Rare beneficial sexual recombinations may have sufficient merit to warrant preservation, or sexual reproduction may produce the only resting spore available within the life cycle. The latter reason probably accounts for the significance of oospores among species of *Aphanomyces, Phytophthora,* and *Pythium* that lack long-lived chlamydospores. Most species dependent upon oospores are homothallic, lending support to the independence of propagules and to the closeness-of-fit hypotheses.

The closeness-of-fit idea is tenuous, however, among homothallic *Phycomycetes* because of variation found among single zoospore cultures (45, 46), and because several generations of inbreeding were required to stabilize some single oospore field isolates of *Phytophthora megasperma* var. *sojae* (55). If suitable genetic markers can be discovered, maybe someone will determine the degree of nuclear exchange occurring among thalli of homothallic *Phycomycetes*. Unlimited mating as well as independence is possible with homothallism.

Let us now digress from the soil. Three rusts attack wheat in Washington. The perfect stage is important only in *Puccinia graminis*. Unlike *P. striiformis* and *P. recondita, P. graminis* is unable to overwinter here on living wheat leaves. Flax rust (*Melampsora lini*) overwinters as teliospores in North Dakota and every first generation of this autoecious rust is sexual. Has annual use of the sexual stage made flax rust more difficult to control by resistance than rusts lacking a sexual stage? The sexual stage may be essential for overwintering, but is it necessary for the production of new races?

The above presentation is an attempt to stimulate examination of the relative importance of genetic stability versus variability in survival of soilborne plant pathogens.

Phenotypic Plasticity

The preceding section stressed the advantage of close selection. This narrow viewpoint should not obscure the remarkable versatility of most soilborne plant pathogens as reflected in phenotypic plasticity, that is, the ability to vary physiologically and structurally within a genotype. Allard (3) admired the phenotypic plasticity of the wild oat. Under favorable conditions it produces many seeds; under great adversity it manages to produce a few seeds or even only one. Thus, one individual differs markedly in response to environmental conditions.

Compare the plasticity of the wild oat plant with that of *Fusarium oxysporum* f. sp. *lini*. A chlamydospore of this fungus germinates in response to exudates from a susceptible flax rootlet. A germ tube emerges, it penetrates the root, and a distributive and absorptive mycelium develops within the host. Conidiophores within a xylem vessel produce microconidia adapted to transport within the xylem vessel. Eventually, if moist, warm conditions prevail, the fungus, by growth, reaches the outer surface of the stem. A sporodochium forms and stimulated by light, macroconidia are produced external to the host. The macroconidia are dispersed by rain. In soils favoring survival, the food within all the cells of the macroconidium is concentrated into one or two cells, a thick wall forms, and the protoplasm of the relatively short-lived macroconidium is now within one or two chlamydospores that are adapted to survival in soil.

The formation of thick, tough hyphae in diseased host tissues by *Fusarium moniliforme* (68) and by *Rhizoctonia solani* (16) are examples of effective but less dramatic morphologic changes with altered physiologic attributes.

A chlamydospore of a *Fusarium* can germinate and convert back to a chlamydospore if adverse conditions develop. Hyphae and macroconidia can convert to chlamydospores. Even microconidia of *Fusarium oxysporum* f. sp. *pisi* form a chlamydospore under adversity (51). Thus, some fungi have a marked ability to retreat into a resting structure. Stanghellini & Hancock (82) observed replacement sporangia and retraction into a parent sporangium in *Pythium ultimum,* but replacement oospores in *Pythium aphanidermatum* were not observed (81). In higher fungi, spore forms tend to form in a regular succession. The ability of some fusaria to retreat into chlamydospores at any point, or to have a reversible developmental cycle, has great survival value.

Oospores, the only survival structure of *Pythium aphanidermatum* in soil (22), germinate by means of germ tubes when exogenous food is present in soil and by means of zoospores when germination occurs in surface soil water. Stanghellini & Burr (81) concluded that near a food source, direct germination is most important, but that in surface waters, zoospores disseminate the fungus during irrigations or hard rains. If food is readily available in the soil solution, the oospore germinates directly. If food is not readily available and the water is well aerated, however, the oospore germinates indirectly by means of zoospores. In a similar way, chlamydospores of *Phytophthora cinnamomi* normally produce germ tubes when food is present, but in low nutrient conditions some sporangia are formed (64). When food levels are minimal, and potential food is at a distance, the motile (searching) stage is favored; when food is close, direct germination is favored.

Another aspect of phenotypic plasticity is illustrated by *Fusarium culmorum* and by *Typhula idahoensis.* A chlamydospore of *F. culmorum* in the soil may germinate in response to a false signal but the protoplasm may form a replacement chlamydospore and save itself. The propagule was expended, the investment was a failure, but the capital was salvaged. In contrast, a sclerotium of *Typhula idahoensis* germinates in the fall, snow falls, hyphae develops over a wheat leaf, but warm rain falls and melts the snow, arresting the development of snow mold. *Typhula* has no retreat mechanism with which to salvage its investment. The spent sclerotium and its hyphae die without replacement. *Typhula* is thus relatively weak in phenotypic plasticity.

Plasticity, encompassing both physiology and morphology, is so common that effort is required to appreciate how remarkable it really is.

SELECTION

A principle of ecology is that two species cannot coexist long in the same niche. If this is true, few soilborne pathogens occupy the same niche because many coexist in our fields. These parasites differ sufficiently in response to physical (temperature, moisture, pH, etc) or to host factors so as to avoid or minimize intense direct competition.

Competition, Radiate Evolution

The genus *Fusarium* contains examples of ways in which distinguishable but similar entities coexist, and they are probably examples of "radiate" evolution. The wilt forms of *Fusarium oxysporum* probably had a common origin, but became physiologically distinct as they were selected into diverse units, primarily by host factors. A soil can contain several f. sp. of *F. oxysporum* with little evidence of effective competition among them.

Fusarium solani also escaped from intense competition by radiate evolution. If a common progenitor gave rise to the various f. spp., they have been separated so long that they no longer interbreed (59) even when properly mated in the laboratory. An experiment of Kraft & Burke (52) demonstrated that populations of *F. solani* f. spp. *phaseoli* and *pisi* can coexist for extended periods of time in the same field. If beans are grown repeatedly, *pisi* is not eliminated; if peas are grown repeatedly, *phaseoli* is not eliminated. Both forms are able to use root exudates of the nonsuscept. *F. solani* f. sp. *pisi* invaded beans to a limited extent, even though it did not damage them and its presence did not reduce the ability of *phaseoli* to attack beans.

Baker (10, p. 172) states that general, all-purpose genotypes are good only for short-term spread and increase; they don't seem suited for incorporation into permanent vegetation. The parasites cited above are examples of this statement. Whether saprophytic clones are general, all-purpose genotypes is not known. Maybe they too are physiologically adapted to different substrates. Competition apparently selected specialists from ancestral generalists.

Another facet of competition is illustrated by animals; each individual usually requires a certain space or territory, and a certain minimal food supply for normal development and reproduction (17, p. 42). If too many larvae of certain nematode species attack an organ, the sex ratio is altered so that few females are differentiated, enabling them to mature and produce viable eggs (86). So far as I know, multiple infection of a root by *Pythium* does not reduce the size of oospores. Some internal regulating system operates to produce units of functional size.

Selection in Soil

The soil exerts a selective force even if the fungus exists in soil only as resting propagules. *Cochliobolus sativus (Helminthosporium sativum)* is a multiclonal species pathogenic both to foliage and below-ground plant parts. Conidia of different clones differ in germinability (24); some germinate readily with little fungistasis and others require larger amounts of food to stimulate germination. Some transform into chlamydospore-like structures (61). Success in soil requires a mechanism preventing germination in the absence of a potential substrate. Success as a foliage pathogen requires little or no dormancy and little or no exogenous food response for rapid leaf-to-leaf spread. It seems mutually exclusive for a conidium of a given genetic constitution to be adapted to both survival in soil and to rapid leaf-to-leaf propagation. *C. sativus* probably depends upon interclonal variation to achieve importance both as a soilborne and as a foliage pathogen.

Chlamydospores of *Fusarium oxysporum* f. spp. *batatas, cubense,* and *lycopersici* require larger amounts of exogenous foods to germinate in soil than do chlamydospores of saprophytic but otherwise similar populations of *F. ox-*

ysporum (76). Chlamydospores of *F. solani* f. sp. *phaseoli* are longer lived in soil than those of the common California clone of f. sp. *cucurbitae* (66).

Pionnotal types arise frequently in culture in certain fusaria, yet they are seldom recovered from nature. Miller (63) and Lin (54) found pionnotal types to be pathogenic in sterile soil but not in natural soil. These variants apparently are capable of parasitic life but fail to survive, even for short periods, in natural soil. If pionnotal mutants occur in nature as frequently as in culture, they must be eliminated in soil.

Cephalosporium gramineum, a vascular parasite of winter wheat and many grasses, is of limited distribution and does not qualify as a widespread fungal colonist. It lacks effective resting structures but survives primarily as active mycelium (20) in undecomposed, infested host debris. The soil microflora exerts selection pressure upon *C. gramineum* because all isolates obtained from nature produce an antifungal antibiotic that prolongs its survival in straw.

These examples, along with *Rhizoctonia solani*, make it clear that the genetic endowment of a fungus must allow for survival in soil, that parasitism alone is not enough.

THE PARASITE

Host Range

How can *Pythium ultimum* penetrate living, vital tissues of so many plants? Why isn't a hypersensitive response elicited in many of them? Is it possible that phenols and phytoalexins do not develop in very young tissues? Most studies of hypersensitivity have involved above-ground plant parts, but Paxton (71) recently reviewed phytoalexin formation below ground. Phytoalexins and phenolic compounds have been reported: from roots of *Pisum sativum, Trifolium pratense, Daucus carota, Ipomoea batatas, Gossypium hirsutum, Persea borbonia* which is a relative of *P. americana* (the cultivated avocado); in hypocotyls of *Phaseolus vulgaris, Vigna sinensis, Glycine max,* and *Carthamus tinctorius;* and in tubers of *Solanum tuberosum.* If we assume that phytoalexins and wound-response phenols are as common below ground as above ground, "primitive" parasites are subtle in that early stages of penetration of many plants are not followed by a quick resistance reaction.

Pythium spp. and most *Rhizoctonia solani* isolates have wide host ranges. In a book on diseases of cereals and grasses, Sprague (78) laboriously gave the host ranges of all pathogens, including *Pythium arrhenomanes.* What grass is not subject to attack by *P. arrhenomanes?* A nonhost list might be more useful, if known, for some pathogens within certain plant families.

Vascular fusaria are quite host specific in that they produce disease in relatively few plants, but many formae speciales have wide host ranges in that they invade and perpetuate themselves in many nonhosts (7). The formae speciales of *Fusarium solani* are quite host specific and they increase to damaging populations only with repeated culture of hosts, but once a population is established, the population can be maintained by nonhosts. *F. solani* f. sp. *phaseoli* became serious in the virgin fields of the Columbia Basin of Washington and Oregon only because beans were grown without rotation in the first years the land was tilled. Some fusaria are secondary invaders of lesions on a plant not normally a host. Nash & Snyder (67) found *Fusarium oxysporum* f. sp. *vasinfectum* in lesions caused by *F. solani* f. sp. *phaseoli* on beans.

Fusaria are poorly represented under evergreen forests, and Toussoun (85) attributes their scarcity largely to the leachates of pine needles. The latter stimulate the germination of chlamydospores (43) without replacement.

A wide host range is characteristic of the best fungal colonists. They may be specialized as to what plants are damaged, but they are able to obtain food from many plants. Lack of host specialization is akin to a weed that grows in many soils.

Synchronization

Fungal propagules must germinate at the proper time to be most effective: fungal developments must be synchronized with host development. Seedlings growing in favorable environments quickly become resistant to *Pythium* spp. (42) and presumably roots behave similarly. But juvenile roots are available for penetration during most of the growth period of herbaceous plants. As the top increases in size, roots must provide increasing amounts of water. A static root system is inadequate under anything but sustained conditions of abundant water because water cannot move to the roots by capillary action with sufficient rapidity (32). During the grand growth period the root system expands so that juvenile tissues are continuously present. I will continue to discuss seedlings, but the same phenomena occur beyond seedling stages.

Since tissues normally become resistant within relatively short periods of time, the pathogen must respond quickly to opportunity (13). *Plasmodiophora* (89), *Pythium* (77), *Aphanomyces* (31), and *Phytophthora* spp. can be within host tissues in 2–12 hr. The stationary and moving target concepts (12) are important. The seed is stationary, at least for a while, even in plants that elevate the cotyledons above ground. Root tips, in contrast, are moving targets. Griffin (41) observed that chlamydospores of *Fusarium oxysporum* began to germinate at a mean distance of 9.5 mm behind the advancing root tip, activated by exudates from the advancing root tip. Maximum germination was reached 13.0 mm behind the root tip. At this point germination of chlamydospore was essentially 100% at the root surface, 30% at 0.9 mm away, 7% at 1.5 cm, and 0 at 2 cm. Most of the germ tubes grew toward the root. Wilt fusaria enter the partially differentiated region and proceed to the protoxylem (5). Thus, the sequence—leakage, response, and subsequent penetration—enables the pathogen to attain its objective—infection of the first differentiated xylem of unwounded roots, with a minimum of host resistance.

The speed of infection by some soilborne pathogens is necessary not only to enter tissues before resistance develops, but to escape significant antagonism. *Fusarium solani* f. spp. *pisi* and *phaseoli* usually attack the epicotyl or hypocotyl, respectively, more than seeds. Food near the seed is so abundant that, though many chlamydospores germinate, most germlings are killed by severe bacterial antagonism (26, 29). Food escaping from epicotyls and hypocotyls is sufficient to overcome fungistasis but it is inadequate to sustain excessive bacterial reproduction resulting in greater germling survival and greater infection. Infection of wheat by *Fusarium culmorum* is favored by dry soil because, though fewer chlamydospores germinate in dry soil, more survive than in wetter soil (28). Seed-borne fusaria can attack seeds during germination because the fungi are established in the host when the seed is planted. Competition or antagonism from soil microorganisms is avoided.

In contrast, *Pythium* spp. react so quickly they can penetrate seed within 24 hr (74), before significant antagonism develops.

Rhizoctonia solani

Rhizoctonia solani, a widespread colonist in cultivated soils, does not fit the closeness-of-fit hypothesis. Strains most adapted to life in soil exist primarily as multinucleate, heterokaryotic cells (37). A complex nuclear balance is maintained in young hyphae. Single basidiospore cultures (homokaryons) can be virulent, yet they are seldom like the wild type (37). Some single basidiospore offspring survive better in soil (69) and some are more virulent (15) than their heterokaryon parents. If a homokaryon can have both longevity in soil and virulence, why the heterokaryotic state in nature?

The complex genetic endowment (6) of *Rhizoctonia solani* is probably essential because it is adept to some degree, both as a saprophyte (70) and as a parasite (69), because most soil strains do not produce effective specialized resting structures (35), and because it may be derived from ancestors not adapted to attacking herbaceous plants. Flentje et al (37), studying penetration, found mutants that failed to grow on the host surface, that did not adhere to the stem, that formed no infection cushion, or, even after forming infection cushions, did not penetrate. Some penetrated and were stopped by a hypersensitive response. Why these complex maneuvers when single, airborne basidiospores can effect penetration of cotton bolls, and they do so directly, with no cushions (56)?

Claviceps purpurea

Before summing up this essay, consideration of a different type of plant pathogen is useful. Ascospores of *Claviceps purpurea* fall upon stigmas and style of an open rye flower, germinate, and send their slender germ tubes to the base of the juvenile ovary (34, pp. 175–78). The fungus attaches itself securely to the rachilla and absorptive contact with the host vascular elements is made. Food normally used by a rye kernel is preempted by the fungus to form the sclerotium. The sclerotium either falls to the ground or is gathered with the grain. Sclerotia on the ground germinate the following spring to produce fertile stromata from which ascospores are forcibly ejected into air currents at the time normal host flowers are in bloom. If ascospores fall into the open flowers the cycle repeats itself.

Claviceps purpurea attaches itself so securely to the host that the sclerotium, heavier than and protruding further than normal kernels, is supported until maturity. Supportive host tissues are not weakened; if anything, they are strengthened. The vascular system delivers food to the fungus in quantities in excess of that used by a single kernel, so host vascular elements are not injured. In rye, if 10% of the kernels become sclerotia, about half of the remaining florets are sterile (Seymour & McFarland, in 44, p. 595). Does this mean that these sterile florets are killed by the fungus but that sclerotia are not formed in them, or is it evidence of a growth substance effect in which developing sclerotia dominate food reserves of would-be adjacent kernels?

The mature sclerotium, a massive food repository, falls to the ground where it could not possibly attack rye flowers, and it is far in excess of that needed to attack a delicate ovary. As the host approaches flowering, a physiologic clock initiates development of stromata in which ascospores are produced. At host anthesis, the ascospores are forcefully expelled. Thus, the fungus times its operation, it solves its position problem, and it partitions its food reserves into effective but minimal units through the production of ascospores. In nature (before tillage) the short-lived (34, p. 180) sclerotia are on the surface of the ground. Since the host blooms each year, there is little need to develop a complex dormancy mechanism within sclerotia that would enable some to germinate one year and some the next. C. purpurea is not host-specific, and my prejudice leads me to expect it to be homothallic. This well-known fungus displays a remarkable coordination of physiologic and morphologic features leading to the efficient acquisition and utilization of food.

CONCLUSIONS

A review of morphologic, physiologic, and genetic characteristics of fungal colonists of cultivated soils reinforces the premise that allocation of food reserves is an important mechanism of survival in fungi. Most successful soilborne parasites are adept at possession (19) and management of food. The size of the propagule is minimal, insuring partition of food into a maximum number of units. Release from fungistasis by nonspecific foods combined with some degree of inherent dormancy insures readiness to exploit a potential source of nutrients. Chemotropic or chemotactic responses increase the effectiveness of the propagules. Some pathogens can maintain a population by utilizing root exudates to recycle propagules, even though penetration does not occur. A broad host range is fostered by an ability to penetrate many plants without eliciting an incompatible host response. An inverse relationship between the levels of endogenous food and the amount of exogenous food required to overcome fungistasis confines germination to appropriate distances from a potential food source.

Morphologic and physiologic attributes are maintained by a minimal genetic system. Closeness-of-fit is maintained mainly by asexual or homothallic reproduction. Their seeming minimal genetic base is combined in most cases with a high degree of phenotypic plasticity. Most propagules of pioneer colonists are capable of independent survival in that they contain sufficient energy to infect a suscept and they require no mating partner.

"Genetic heterogeneity in a population produces many individuals that have less than maximal fitness. The reduction in fitness has been called genetic load; it may be considered as the price a species pays for a vital advantage, the ability to respond to a changing environment and to occupy various environmental niches" (79, p. 423). Many plant pathogens that evolved from sexual ancestors have found the genetic load of maintained recombination not worth the price: They are less variable than their ancestors, and they are adapted to rather specific environments.

The morphologic, physiologic, and genetic properties of fungi are seldom determined by chance. They provide clues to the problems faced and the means by which the problems are solved.

Acknowledgments

This paper was read by R. J. Cook, K. F. Baker, K. J. Moore, J. D. Rogers, and J. F. Schafer. Thanks are due them for useful suggestions. A minimum of documentation was used and I apologize to authors whose works could have been used to embellish the themes to greater degrees.

Literature Cited

1. Abawi, G. S., Grogan, R. G. 1975. Source of primary inoculum and effects of temperature and moisture on infection of beans by *Whetzelinia sclerotiorum. Phytopathology* 65:300–9
2. Allard, R. W. 1960. *Principles of Plant Breeding.* New York: Wiley. 485 pp.
3. Allard, R. W. 1965. Genetic systems associated with colonizing ability in predominantly self-pollinated species. In *The Genetics of Colonizing Species,* ed. H. G. Baker, G. L. Stebbins, 50–78. New York. Academic. 588 pp.
4. Allen, P. J. 1965. Metabolic aspects of spore germination in fungi. *Ann. Rev. Phytopathol.* 3:313–42
5. Anderson, M. E., Walker, J. C. 1935. Histological studies of Wisconsin Hollander and Wisconsin Ballhead cabbage in relation to resistance to yellows. *J. Agric. Res.* 50:823–36
6. Anderson, N. A., Stretton, H. M., Groth, J. V., Flentje, N. T. 1972. Genetics of heterokaryosis in *Thanatephorus cucumeris. Phytopathology* 62:1057–65
7. Armstrong, G. M., Armstrong, J. K., 1975. Reflections on the wilt fusaria. *Ann. Rev. Phytopathol.* 13:95–103
8. Ashworth, L. J. Jr., Huisman, O. C., Harper, D. M., Stromberg, L. K. 1974. Free and bound microsclerotia of *Verticillium albo-atrum* in soils. *Phytopathology* 64:563–64

9. Ayanru, D. K. G., Green, R. J. Jr. 1974. Alteration of germination patterns of sclerotia of *Macrophomina phaseolina* on soil surfaces. *Phytopathology* 64:595–601
10. Baker, H. G. 1965. Characteristics and modes of origin of weeds. See Ref. 3, pp. 147–68
11. Baker, K. F., Cook, R. J. 1974. *Biological Control of Plant Pathogens.* San Francisco: Freeman. 433 pp.
12. Baker, R., Maurer, C. L., Mauer, R. A. 1967. Ecology of plant pathogens in soil. VII. Mathematical models and inoculum density. *Phytopathology* 57:662–66
13. Benson, D. M., Baker, R. 1974. Epidemiology of *Rhizoctonia solani* preemergence damping-off of radish: inoculum potential and disease potential interaction. *Phytopathology* 64: 957–62
14. Blakeman, J. P. 1971. The chemical environment of the leaf surface in relation to growth of pathogenic fungi. In *Ecology of Leaf Surface Micro-organisms,* ed. T. F. Preece, C. H. Dickinson, 255–68. New York: Academic. 640 pp.
15. Bolkan, H. A., Butler, E. E. 1974. Studies on heterokaryosis and virulence of *Rhizoctonia solani. Phytopathology* 64:513–22
16. Boosalis, M. G., Scharen, A. L. 1959.

Methods for microscopic detection of *Aphanomyces euteiches* and *Rhiioctonia solani* and for isolation of *Rhizoctonia solani* associated with plant debris. *Phytopathology* 49:192–98

17. Boughey, A. S. 1968. *Ecology of Populations.* New York: Macmillan. 135 pp.

18. Boyle, L. W. 1961. The ecology of *Sclerotium rolfsii* with emphasis on the role of saprophytic media. *Phytopathology* 51:117–19

19. Bruehl, G. W. 1975. Systems and mechanisms of residue possession by pioneer fungal colonists. In *Biology and Control of Soil-Borne Plant Pathogens,* ed. G. W. Bruehl, 77–83. St. Paul, Minn: Am. Phytopathol. Soc. 216 pp.

20. Bruehl, G. W., Millar, R. L., Cunfer, B. 1969. Significance of antibiotic production by *Cephalosporium gramineum* to its saprophytic survival. *Can. J. Plant Sci.* 49:235–46

21. Burnett, J. H., 1965. The natural history of recombination systems. In *Incompatibility in Fungi,* ed. K. Esser, J. R. Raper. New York: Springer. 124 pp.

22. Burr, T. J., Stanghellini, M. E. 1973. Propagule nature and density of *Pythium aphanidermatum* in field soil. *Phytopathology* 63:1499–1501

23. Chinn, S. H. F. 1973. Prevalence of *Dendryphion nanum* in field soils in Saskatchewan with special reference to rape crops in the crop rotation. *Can. J. Bot.* 51:2253–58

24. Chinn, S. H. F., Tinline, R. D. 1964. Inherent germinability and survival of spores of *Cochliobolus sativus. Phytopathology* 54:349–52

25. Clark, F. E. 1965. The concept of competition in microbial ecology. In *Ecology of Soil-Borne Plant Pathogens,* ed. K. F. Baker, W. C. Snyder, 339–45. Berkeley: Univ. Calif. Press. 571 pp.

26. Cook, R. J., Flentje, N. T. 1967. Chlamydospore germination and germling survival of *Fusarium solani* f. *pisi* in soil as affected by soil water and pea seed exudation. *Phytopathology* 57:178–82

27. Cook, R. J., Ford, E. J., Snyder, W. C. 1968. Mating type, sex, dissemination, and possible sources of clones of *Hypomyces (Fusarium) solani* f. *pisi* in South Australia. *Aust. J. Agric. Res.* 19:253–59

28. Cook, R. J., Papendick, R. I. 1970. Soil water potential as a factor in the ecology of *Fusarium roseum* f. sp. *cerealis* 'Culmorum'. *Plant Soil* 32:131–48

29. Cook, R. J., Snyder, W. C. 1965. Influence of host exudates on growth and survival of germlings of *Fusarium solani* f. *phaseoli* in soil. *Phytopathology* 55:1021–25

30. Cunfer, B. M., Bruehl, G. W. 1973. Role of basidiospores as propagules and observations on sporophores of *Typhula idahoensis. Phytopathology* 63:115–20

31. Cunningham, J. L., Hagedorn, D. J. 1962. Penetration and infection of pea roots by zoospores of *Aphanomyces euteiches. Phytopathology* 52:827–34

32. Curtis, O. F., Clark, D. G. 1950. *An Introduction to Plant Physiology.* New York: McGraw-Hill. 752 pp.

33. Debach, P. 1965. Some biological and ecological phenomena associated with colonizing entomophagous insects. In *The Genetics of Colonizing Species,* ed. H. G. Baker, G. L. Stebbins, 287–303. New York: Academic. 588 pp.

34. Dickson, J. G. 1956. *Diseases of Field Crops.* New York: McGraw-Hill. 517 pp.

35. Durbin, R. D. 1959. Factors affecting the vertical distribution of *Rhizoctonia solani* with special reference to CO_2 concentration. *Am. J. Bot.* 46:22–25

36. Emmatty, D. A., Green, R. J. Jr. 1969. Fungistasis and the behavior of the microsclerotia of *Verticillium alboatrum* in soil. *Phytopathology* 59:1590–95

37. Flentje, N. T., Stretton, H. M., McKenzie, A. R. 1970. Mechanisms of variation in *Rhizoctonia solani.* In *Rhizoctonia solani: Biology and Pathology,* ed. J. R. Parmeter, Jr., 52–65. Berkeley: Univ. Calif. Press. 255 pp.

38. Garrett, S. D. 1965. Toward biological control of soil-borne plant pathogens. See Ref. 25, pp. 4–17

39. Garrett, S. D. 1970. *Pathogenic Root-Infecting Fungi.* London: Cambridge Univ. Press. 294 pp.

40. Garrett, S. D. 1973. Deployment of reproductive resources by plant pathogenic fungi: An application of E. J. Salisbury's generalization for flowering plants. *Acta Bot.* 1:1–9

41. Griffin, G. J. 1969. *Fusarium oxysporum* and *Aspergillus flavus* spore germination in the rhizosphere of peanut. *Phytopathology* 59:1214–18

42. Halpin, J. E., Hanson, E. W. 1958. Effect of age of seedlings of alfalfa,

red clover, Ladino white clover, and sweetclover on susceptibility to *Pythium*. *Phytopathology* 48:481-85

43. Hammerschlag, F., Linderman, R. G. 1975. Effects of five acids that occur in pine needles on *Fusarium* chlamydospores germination in nonsterile soil. *Phytopathology* 65:1120-24

44. Heald, F. D. 1933. *Manual of Plant Diseases*. New York: McGraw-Hill. 953 pp.

45. Hendrix, F. F. Jr., Campbell, W. A. 1971. A new species of *Pythium* with spiny oogonia and large chlamydospores. *Mycologia* 63:978-82

46. Hilty, J. W., Schmitthenner, A. F. 1962. Pathogenic and cultural variability of single zoospore isolates of *Phytophthora megasperma* var. *sojae*. *Phytopathology* 52:859-62

47. Hsu, S. C., Lockwood, J. L. 1973. Soil fungistasis: behavior of nutrient-independent spores and sclerotia in a model system. *Phytopathology* 63: 334-37

48. Kessler, K. J. Jr. 1966. Xylem sap as a growth medium for four tree wilt fungi. *Phytopathology* 56:1165-69

49. King, J. E., Coley-Smith, J. R. 1968. Effects of volatile products of *Allium* species and their extracts on germination of sclerotia of *Sclerotium cepivorum* Berk. *Ann. Appl. Biol.* 61:407-14

50. Ko, W. H., Chan, M. J. 1974. Infection and colonization potential of sporangia, zoospores, and chlamydospores of *Phytophthora palmivora* in soil. *Phytopathology* 64:1307-9

51. Kommedahl, T. 1966. Relation of exudates of pea roots to germination of spores in races of *Fusarium oxysporum* f. *pisi*. *Phytopathology* 56:721-22

52. Kraft, J. M., Burke, D. W. 1974. Behavior of *Fusarium solani* f. sp. *pisi* and *Fusarium solani* f. sp. *phaseoli* individually and in combination on peas and beans. *Plant Dis. Reptr.* 58:500-4

53. Lemke, P. A. 1973. Isolating mechanisms in fungi—prezygotic, postzygotic, and azygotic. *Persoonia* 7: 249-60

54. Lin, Y. S. 1975. *Fusarium root rot of lentils in the Pacific Northwest*. MS thesis. Washington State Univ., Pullman. 59 pp.

55. Long, M. Keen, N. T., Ribeiro, O. K., Leary, J. V., Erwin, P. E., Zentmyer, G. A. 1975. *Phytophthora megasperma* var. sojae: development of wild-type strains for genetic research. *Phytopathology* 65:592-97

56. Luke, W. J., Pinckard, J. A., Wang, S.

C. 1974. Basidiospore infection of cotton bolls by *Thanatephorus cucumeris*. *Phytopathology* 64:107-11

57. Mather, K. 1965. The genetic interest of incompatibility in fungi. See Ref. 21, pp. 113-17

58. Matuo, T., Snyder, W. C. 1972. Host virulence and the *Hypomyces* stage of *Fusarium solani* f. sp. pisi. Phytopathology 62:731-35

59. Matuo, T., Snyder, W. C. 1973. Use of morphology and mating populations in the identification of formae speciales in *Fusarium solani*. *Phytopathology* 63:562-65

60. Menzies, J. D. 1970. Factors affecting plant pathogen population in soil. In *Root Disease and Soil-Borne Pathogens*, ed. T. A. Toussoun, R. V. Bega, P. E. Nelson, 16-21. Berkeley: Univ. Calif. Press. 252 pp.

61. Meronuck, R. A., Pepper, E. H. 1968. Chlamydospore formation in conidia of *Helminthosporium sativum*. *Phytopathology* 58:866-67

62. Milholland, R. D. 1975. Pathogenicity and histopathology of *Phytophthora cinnamomi* on highbush and rabbiteye blueberry. *Phytopathology* 65:789-93

63. Miller. J. J. 1945. Studies on the *Fusarium* of muskmelon wilt. I. Pathogenic and cultural studies with particular reference to the cause and nature of variation in the organism. *Can. J. Res. C* 23:16-43

64. Mircetich, S. M., Zentmyer, G. A. 1969. Effect of carbon and nitrogen compounds on germination of chlamydospores of *Phytophthora cinnamomi* in soil. *Phytopathology* 59: 1732-35

65. Mitchell, D. J. 1975. Density of *Pythium myriotylum* oospores in soil in relation to infection of rye. *Phytopathology* 65:570-75

66. Nash, S. M., Alexander, J. V. 1965. Comparative survival of *Fusarium solani* f. *cucurbitae* and *F. solani* f. *phaseoli* in soil. *Phytopathology* 55: 963-66

67. Nash, S. M., Snyder, W. C. 1967. Comparative ability of pathogenic and saprophytic fusaria to colonize primary lesions. *Phytopathology* 57: 293-96

68. Nyvall, R. F., Kommedahl, T. 1968. Individual thickened hyphae as survival structures of *Fusarium moniliformae* in corn. *Phytopathology* 58: 1704-7

69. Olsen, C. M., Flentje, N. T., Baker, K. F. 1967. Comparative survival of

monobasidial cultures of *Thanatephorus cucumeris* in soil. *Phytopathology* 57:598–601

70. Papavizas, G. C. 1970. Colonization and growth of *Rhizoctonia solani* in soil. See Ref. 37, pp. 108–22

71. Paxton, J. D. 1975. Phytoalexins, phenolics, and other antibiotics in roots resistant to soil-borne fungi. See Ref. 19, pp. 185–92

72. Pugh, G. W., Johann, H., Dickson, J. G. 1933. Factors affecting infection of wheat heads by *Gibberella saubinetii*. *J. Agric. Res.* 46:771–97

73. Schreiber, L. R., Green, R. J. Jr. 1963. Effect of root exudates on germination of conidia and microsclerotia of *Verticillium albo-atrum* inhibited by the soil fungistatic principle. *Phytopathology* 53:260–64

74. Schroeder, H. W. 1955. *Factors affecting resistance of wheat to scab caused by Gibberella zeae (Schw.) Petch.* PhD thesis. Univ. Minnesota, St. Paul

75. Schroth, M. N., Hendrix, F. F. Jr. 1962. Influence of nonsusceptible plants on the survival of *Fusarium solani* f. *phaseoli* in soil. *Phytopathology* 52:906–9

76. Smith, S. N. Snyder, W. C. 1972. Germination of *Fusarium oxysporum* chlamydospores in soils favorable and unfavorable to wilt establishment. *Phytopathology* 62:273–77

77. Spencer, J. A., Cooper, W. E. 1967. Pathogenesis of cotton (*Gossypium hirsutum*) by *Pythium* species: zoospore and mycelium attraction and infectivity. *Phytopathology* 57:1332–38

78. Sprague, R. 1950. *Diseases of Cereals and Grasses in North America*. New York: Ronald. 538 pp.

79. Srb, A. M., Owen, R. D., Edgar, R. S. 1965. *General Genetics*. San Francisco:

Freeman. 557 pp.

80. Stanghellini, M. E. 1975. Spore germination, growth and survival of *Pythium* in soil. *Proc. Am. Phytopathol. Soc.* 1:211–14

81. Stanghellini, M. E., Burr, T. J. 1973. Effect of soil water potential on disease incidence and oospore germination of *Pythium aphanidermatum*. *Phytopathology* 63:1496–98

82. Stanghellini, M. E., Hancock, J. G. 1971. Radial extent of the bean spermosphere and its relation to the behavior of *Pythium ultimum*. *Phytopathology* 61:165–68

83. Steiner, G. W., Lockwood, J. L. 1969. Soil fungistasis: sensitivity of spores in relation to germination time and size. *Phytopathology* 59:1084–92

84. Tinline, R. D., MacNeill, B. H. 1969. Parasexuality in plant pathogenic fungi. *Ann. Rev. Phytopathol.* 7:147–70

85. Toussoun, T. A. 1975. Fusarium-suppressive soils. See Ref. 19, pp. 145–51

86. Triantaphyllou, A. C. 1973. Environmental sex differentiation of nematodes in relation to pest management. *Ann. Rev. Phytopathol.* 11: 441–62

87. Tu, C. 1930. Physiologic specialization in *Fusarium* spp. causing head-blight of small grains. *Minn. Agric. Exp. Stn. Tech. Bull. 74.* 27 pp.

88. Walker, J. C. 1950. *Plant Pathology*. New York. McGraw-Hill. 699 pp.

89. Williams, P. H., Aist, J. R., Bhattacharya, P. K. 1973. Host-parasite relations in cabbage clubroot. In *Fungal Pathogenicity and the Plant's Response*, ed. R. J. W. Byrde, C. V. Cutting, 141–58. New York: Academic. 499 pp.

Copyright © 1976 by Annual Reviews Inc.
All rights reserved

THE TUMOR-INDUCING   ✦3643
SUBSTANCE OF *AGROBACTERIUM*
TUMEFACIENS

C. I. Kado
Department of Plant Pathology, University of California, Davis, California 95616

INTRODUCTION

The shroud of the crown gall mystery was lifted when Cavara (48) isolated crown gall–inducing bacteria from diseased grapevines and when Smith et al (226) firmly established the bacterial nature of the disease. That was 80 years ago.

Ironically, in 1975 speculations on the nature of the agent in *Agrobacterium tumefaciens* that causes the induction of crown gall tumors are still being made. During this period, so many papers have appeared on the descriptive aspects of crown gall that the field is saturated. Only recently have significant achievements been made, primarily because studies were directed at the causal organism rather than at the host plant and because techniques were developed in molecular biology that were directly applicable to studies of the elusive tumorigenic substance, commonly known as the *tumor-inducing principle*. This principle, hypothesized by Braun (34, 37), was considered as (*a*) a metabolic product of *A. tumefaciens*, (*b*) a normal host constituent, converted by the bacteria into tumor-inducing substance, (*c*) a chemical fraction of the bacterial cell such as DNA, (*d*) a virus or other agent present in association with *A. tumefaciens*, or (*e*) *A. tumefaciens* cells altered in their morphology and physiology so as not to be demonstrable by current methods of isolation. A similar proposal was advanced by Gautheret (89).

Almost every major component of the crown gall bacterium has been examined in relation to tumor-inducing substance, and many of them have been implicated or at least suspected at one time or another until they failed to pass rigorous tests in control experiments. Of the *A. tumefaciens* macromolecules examined, nucleic acids continue to be the most attractive because the earlier biological observations indicate that the tumor-inducing principle is an autoreplicating "epigenetic" factor that confers abnormal regulatory growth characters on a normal cell. This is because of the peculiar nature of the crown gall disease: it is one disease that is *incited* by the causal organism, *A. tumefaciens*. Once the host plant cells are converted or "transformed" into tumor cells, the bacterial cells are no longer necessary to maintain the tumorous state. Thus, the

265

bacteria are in more or less parasitic association with the host since their status is only temporary and mainly transient; they leave the scene after triggering the necessary responses of the host cell. The tumor cells may be maintained indefinitely as callus tissues in axenic culture. The cultured tumor cells (either as single-celled clones or nonclones) may be grafted to healthy compatible hosts on which new tumors are initiated (1, 35, 39, 173, 264). Thus, the tumor-inducing principle is apparently maintained during tumor growth and transmitted to healthy parts of the plant to generate near tumor foci (173, 174).

This review focuses on recent advances in the crown gall field, particularly on the nucleic acid concept, and ties various loose ends to develop a clear picture of the field as it now stands. The nucleic acid concept postulates that a nucleic acid of *A. tumefaciens* is the tumor-inducing substance elaborated during the course of infection of higher plants by *A. tumefaciens*.

It is difficult to envision proteins, polysaccharides, lipids, etc having autoreplicative and genetic transportable potentials without some associated molecular template, as in the case of RNA and DNA. Nevertheless, macromolecules other than nucleic acids should not be completely ignored. The nucleic acid concept encompasses categories *a, b,* and *c* stated above, namely that bacterial or plasmid genes[1] specify products that directly or indirectly control regulatory functions necessary for the growth of higher cells.

Several reviews have been written in the past five years covering much of the field of crown gall, particularly that aspect dealing with the physiology of crown gall tumor development and the potential underlying mechanisms (2, 8, 155, 160, 174, 175, 243, 248, 255).

In this review I therefore confine my discussion to the action of the crown gall bacteria during infection and outline the stepwise sequence of events that the hypothetical tumorigenic substance may follow during the course of infection of the host plant culminating in the formation of crown gall tumors.

NATURE OF THE TUMOR-INDUCING PRINCIPLE

Many attempts have been made to isolate and characterize the hypothetical tumor-inducing principle. The term *tumor-inducing principle,* first coined by Braun (34, 37), is in itself an enigma. It does not strictly imply that the hypothetical tumorigenic factor elaborated by *A. tumefaciens* can be characterized chemically. As Braun (34) pointed out, the tumorigenic factor could be the crown gall bacteria themselves which enter wounded cells and subsequently develop into altered forms that are no longer demonstrable by the usual cytological procedures, a hypothesis proposed earlier by D'Herelle & Peyre (71). In spite of the relative genetic relatedness between *Agrobacterium* species and *Rhizobium* species based on DNA homologies (91, 99), and the fact that *Rhizobium* species form pleomorphic bodies (or bacteroids) in cells of root

---

[1]Bacterial gene(s) denotes genetic material in the bacterial cell and therefore includes plasmids.

nodules of legumes (195), cytological examinations at the electron microscopic level of crown gall tumor cells have detected no such altered forms of *A. tumefaciens* (90, 108, 151, 209, 257). Recently, however, thin-section analysis of young sunflower crown gall tumor cells has revealed small vesicle-like bodies in the nuclei (167). This suggests that possibly bacteroid types of *A. tumefaciens* may be present in the nucleus of crown gall cells. Other forms of vesicular bodies or inclusion bodies have been observed in the cytoplasm of crown gall cells of *Datura stramonium* (108, 257). If these nuclear and cytoplasmic vesicles are indeed pleomorphic bodies of *A. tumefaciens* that no longer revert to the normally cell-walled bacilliform types, one should expect crown gall tumor cells in axenic culture to be affected by antibiotics that are specific to bacteria. In other words, the phytohormone prototrophic status of tumor cells should diminish when the tumor cells are grown on basal medium containing bacterial-specific antibiotics. Manasse (157) compared suspension cultures of *Vinca rosea* crown gall tumor cell lines and normal *V. rosea* cells. Crown gall tumor cells decreased in growth rate in the presence of kanamycin and neomycin, antibiotics that affect protein synthesis by binding to the 30S subunit of 70S ribosomes (262) causing mistranslation of bacterial messenger RNAs (63).

One interpretation of these results is that at least ribosomes of bacterial origin were present in crown gall tumor cells. However, Manasse (157) showed that crown-gall cells, originally transformed by an *A. tumefaciens* strain resistant to neomycin, kanamycin, and vancomycin, were inhibited similarly to crown gall cells transformed by the parental wild-type strain. Thus, these aminoglycosidic antibiotics most likely are affecting the 70S mitochondrial and proplastid ribosomes of *V. rosea* instead of any bacterial ribosomes. Also, *V. rosea* crown gall cells are more permeable than normal cells to various chemicals (266) and therefore the greater in vivo concentrations of antibiotics in crown-gall cells than in normal cells may be eliciting the greater effect.

The evidence for perpetuation of crown gall tumors in the absence of bacteria seems unequivocal, based on the early work of Braun & White (39) and on the axenic growth of cloned crown gall callus tissues on media (which would easily support the rapid growth of *A. tumefaciens* if any viable cells were present). Therefore, it has been widely accepted that this bacterium elaborates, during infection, a substance that induces crown gall tumor formation. The emphasis is now on a chemically characterizable bacterial component, that is, a *tumorigenic substance* rather than the former less stringent term, *tumor-inducing principle*. Henceforth, the term *tumorigenic substance* will be used in this discussion [the term *substance* for the tumor-causing agent was commonly used in 1917 by E. F. Smith (225)].

Besides the probacteria (or altered-bacteria) hypothesis, a number of components isolated from *A. tumefaciens* cells have been implicated (Table 1). Most of the alleged tumorigenic substances do not induce crown gall tumors at will and the claims of inducing such tumors have met with opposition principally because the experiments that were performed were unsatisfactory because of: e.g. neglect of stringent sterility precautions, poor choice of host, such as *D. stramonium*

Table 1 Alleged crown gall tumor-inducing substances of *Agrobacterium tumefaciens*

Bacterial cell component	Test host(s)	Year	Reference
Ammonium ions	*Ricinus communis*	1917	225
Lipopolysaccharide	*Helianthus annuus, Pelargonium* sp.	1935	31
Endotoxin	*Datura stramonium*	1971	180, 180a, 211
Phosphatid	*Nicotiana glauca, D. stramonium*, French marigold *L. esculentum, H. annuus*	1937	149
Plastin	*D. carota*	1927	12
Virus	*N. tabacum, H. annuus, Vinca rosea*	1947	69
Ribonucleic acid	*Kalanchoë diagremontiana*	1966	41
	L. esculentum	1972	244
	D. stramonium	1974	2a, 19, 20
Deoxyribonucleic acid	*Pelargonium* sp.	1956	162
	Daucus carota	1957	127
	L. esculentum, P. tatula, D. innoxia	1958	24
	D. carota, Helianthus sp., *Bryophyllum* sp.	1959	246
	D. stramonium	1958	164, 165
	Scorzonera hispanica	1967	131
	Phaseolus vulgaris, D. carota	1971	21
	H. annuus	1972	118
	D. carota, P. vulgaris	1975	204
Of a bacteriophage	*H. annuus*	1970	147
	P. vulgaris, N. tabacum	1972	8
Of a plasmid	Not tested	1974	269

[which responds to almost any stimuli for forming pseudotumors (104)], impure components, and failure to characterize rigorously the tumors that were induced.

Nucleic Acid Concept and DNA

Although much of the search for the tumorigenic substance has focused on the nucleic acids of *A. tumefaciens* (Table 1), studies on the possible effects of other bacterial components such as glycoproteins, proteins, and lipoproteins have been neglected. It is no longer a question whether or not DNA plays a role in conferring tumorigenicity on the *A. tumefaciens* cell. The tumorigenic substance is obviously a product of a bacterial gene or sets of genes that either confer tumorigenic properties on the organism or induce tumors directly through expressed genetic products. To extend this to a point where DNA or RNA directly induces crown gall tumors is still equivocal and controversial (80). The nucleic acid concept (117) seems to be preferred since nucleic acids are autoreplicative macromolecules, whereas it is difficult to envision proteins or other complexed proteins (e.g. glycoproteins, lipoproteins) achieving the same degree of autoreplicative potential. Also, analogies can be drawn from animal oncogenic models where nucleic acids of various tumor viruses such as papilloma, polyoma, and simian virus 40 induce tumors (30, 57, 58), making the nucleic acid concept in crown gall research more attractive than alternative hypotheses.

The nucleic acid concept has been recently reviewed (80) and its various aspects recapitulated elsewhere (33, 111, 117, 160, 255). The current model of this concept is that bacterial genes specify the tumorigenic substances(s). Whether or not these gene(s) function within the host cell is unknown, but recent controversial evidence suggests that the functioning of bacterial genetic material is within the host cell (discussed in the section on *Genetic Expression*). Most of the work seems to be focused on bacterial nucleic acids as the responsible tumorigenic substance (Table 1).

The assay of biological and tumorigenic properties of purified nucleic acid preparations is the most obvious experiment to perform. A number of laboratories have attempted to induce crown gall tumors with bacterial DNA and RNA preparations of varying degrees of purity (19–21, 26, 118, 127, 162, 164, 165, 193, 204, 244, 247) and attempts have been made to "transform" or treat cells in axenic culture with DNA of *A. tumefaciens* (103, 120, 137, 259). Although the early work by Klein and co-workers (127, 246) and Manigault & Stoll (164, 165) implicated bacterial DNA by direct inoculation experiments, the design of these experiments lacked controls to show that they were actually working with a highly purified preparation of DNA free of bacterial contamination. Moreover, there was a lack of adequate characterization of the tumors that were initiated in these experiments. Control experiments, subsequently performed by Klein & Braun (126), showed that some of the "DNA" preparations contained live bacteria.

Recent investigations have employed DNA preparations of greater purity, and precautionary measures were taken to minimize contamination more strictly than in earlier work. Crown-gall "transformation" with purified bacterial DNA has been reported by various laboratories (19–21, 118, 131, 166) but other workers have shown no evidence for transforming activity in their DNA preparations (26, 103, 120, 193, 247, 259). The experiments reporting transformation have been preliminary and consequently difficult to reproduce. We have employed high molecular weight DNA obtained by lysing bacterial cells on a cesium chloride density gradient (118). The tumorigenic activity of this type of DNA and that prepared by phenol extraction could not be reproduced at will (118). We have performed numerous experiments with highly polymerized DNA to see if we can improve the efficiency of the DNA inoculation experiments. Various reagents, plant cell-wall degrading enzymes, DNA enzymes, bacterial nucleic acids, and proteins failed to improve this efficiency (Table 2). Consequently, we have viewed the induction of tumors by *A. tumefaciens* DNA with skepticism until it can be demonstrated reproducibly and at will.

It is therefore interesting that one laboratory has recently reported successful induction of tumors with DNA with an efficiency of 30% (204) confirming their previous work (21). These tumors, induced on bean stems, grew on minimal Skoog medium. On the other hand, Phillips & Butcher (193) recently reported

Table 2 Attempts to potentiate the possible tumorigenic activity of *A. tumefaciens* DNA with supplements to the DNA inoculum

Supplement	Concentration (μg/ml)	Buffer[a]	pH	DNA Concentration (μg/ml)	Tumor formation[b]
A. tumefaciens components					
Killed cells	10^8 cells/ml	KP	6.8	400	—
Total RNA	100	NP	6.8	81	—
Ribosomal RNA	85	NP	6.8	81	—
Transfer RNA	3.4	NP	6.8	81	—[c]
Soluble Protein	200	H_2O	—	187	—[c]
DNA ligase	100	KP	7.0	375	—
DNA polymerase	10	TMN	7.5	200	—
E. coli components					
Crude nucleotides	50	TM	7.5	400	—
DNA ligase	190	KP	7.0	125	—
DNA ligase	190	KP	7.0	375	—[c]
DNA	500	SSC	7.0	100	—
Cell wall degrading enzymes					
Pectinase	50	KP	7.0	537	—
Pectinase	100	KP	7.0	100	—
Pectinase	100	KP	7.0	20	—
Pectin methyl esterase	50	KP	7.0	537	—
Deoxyribonucleotides					
dAMP	50	TMN	7.5	220	—
dCMP	50	TMN	7.5	220	—
dGMP	50	TMN	7.5	220	—
dAMP	100	TMN	7.5	412	—
dCMP	100	TMN	7.5	412	—
dGMP	100	TMN	7.5	412	—
dTMP	100	TMN	7.5	412	—
Diphosphopyridine nucleotide	50	TMN	7.5	220	—
6-Amino purine	0.1	KP	6.8	81	—
α-Naphthalene acetic acid	10	TMN	7.5	460	—
Kinetin	10	TMN	7.5	460	—
Benzyladenine	10	TMN	7.5	460	—
DEAE-dextran	500	H_2O	—	21	—[c]
Inositol	500	KP	6.8	100	—
Salmon sperm DNA	100–500	SSC	7.0	100	—
Calf thymus DNA	100–500	SSC	7.0	100	—

[a]Buffers: KP = 0.01 M potassium phosphate, pH 6.8; NP = 0.01 M sodium phosphate, pH 6.8; TMN = 0.01 M Tris-Cl; 0.01 M $MgCl_2$, 0.1 M NaCl, pH 7.4; TM = 0.01 M Tris-Cl, 0.01 M $MgCl_2$, pH 7.4; SSC = 0.15 M NaCl, 0.015 M Na_3 citrate, pH 7.0.
[b]All assays performed on young sunflower plants (*H. annuus* cv. Russian Mammoth). Between 4 to 20 plants were used per assay.
[c]A few plants showed tiny tumor-like outgrowths.

negative results using *A. tumefaciens* DNA prepared either by the standard detergent-lysis chloroform extraction procedure or by phenol extraction.

Tumorigenic activity with bacteriophage DNA also has been reported. Leff & Beardsley (147) claimed that the DNA of *A. tumefaciens* phage PS8 induces tumors in sunflower and tobacco. The cut ends of the stems were placed in a solution of DNA, the plants were rooted in vermiculite, and wounded 2 days later. After 2 weeks Leff & Beardsley observed large tumors on the DNA-treated plants that were indistinguishable from those on control plants placed in a suspension of crown gall bacteria. Attempts to repeat these results were made by several laboratories. Beiderbeck et al (18) were unable to reproduce these results irrespective of the various procedures of DNA isolation and tumorigenicity tests that were used. This, coupled with the recent reports showing the absence of phage DNA in tumorigenic strains of *A. tumefaciens* (67) and in tumor cells (83, 86), strongly argues against the role of phage DNA in tumorigenesis.

Tumorigenic RNA?

Analogous to the DNA experiments, *A. tumefaciens* RNA was also examined for tumorigenic activity. Using nucleic acid depolymerizing enzymes, Braun & Wood (41) showed that ribonuclease A at relatively high concentrations (2–4 mg/ml) inhibited tumor formation in *Kalanchoë daigremontiana* whereas no inhibition was observed when deoxyribonuclease (5 mg/ml) was used. Control experiments showed that ribonuclease had no effect on the growth of *A. tumefaciens* cells in wounds and no effect on the ability of the organism to initiate tumors. The enzyme did not affect the capacity of the host cells to respond in a normal manner to the stimulus of a wound, on the basis of histological evidence. Braun & Wood (41) concluded that the tumor-inducing principle itself or an essential component of that principle is a ribonucleic acid, or that ribonuclease enters the bacteria or host cells and selectively inactivates some component essential for tumor inception. Although these experiments indirectly demonstrate the possibility of RNA as the tumorigenic substance, the possibility of altering the metabolic processes of the cell (e.g. protein synthesis by depolymerization of various RNA species) is equally likely despite the absence of differences between enzyme-treated and untreated host cells at the histological level. It is interesting to note that the rapidly growing regions and parenchyma cells are devoid of ribonuclease activity (224). However, due to the preliminary nature of the ribonuclease experiments, further studies will be necessary to elucidate whether or not bacterial RNA is involved in tumorigenesis. Nevins et al (184) showed that ribonuclease-treated *A. tumefaciens* cells are subtly modified in their macromolecular synthetic and degradative functions as manifested by increased rates of absorption and release of radioactive precursors of nucleic acids and protein. Thus, ribonuclease at the high concentrations used interferes sufficiently with the bacterial nucleic acid metabolism to interfere with tumor induction. Likewise, the enzyme is known to affect plant cells. Ruesink & Thimann (208) report deleterious effects of ribonuclease on protoplasts and cite a number of examples of distorting effects of the enzyme on plant cells.

Recently, RNA extracted from *A. tumefaciens* produced tumors in tomato plants (244). Ribonuclease-treated RNA from the tumorigenic strain B6 and RNA from a nontumorigenic strain of *A. tumefaciens* did not possess such activity. Some inconsistencies are apparent in this work when compared to a similar recent claim by Beljanski and co-workers (19, 20). Swain & Rier (244) found tumor-inducing activity only with the RNA isolated from the tumorigenic strain B6 whereas Beljanski et al (19, 20) claimed that RNA from both tumorigenic and nontumorigenic strains produced tumors on *D. stramonium* plants. Furthermore, the RNA preparations of Beljanski et al (19, 20) that were treated with ribonuclease A and ribonuclease T1 in 0.3 M NaCl-0.03 M Na citrate buffer retained their tumorigenic activity, whereas the RNA preparations of Swain & Rier (244) treated with ribonuclease A lost their tumor-inducing activity. The latter workers used RNA in 0.01 M Na pyrophosphate at pH 9, containing 0.1 M mercaptoethanol and 1% sodium lauryl sulfate. Since sodium lauryl sulfate is a strong inhibitor of ribonuclease (72), it was not clear whether or not RNA itself was actually digested by the added enzyme or whether the components of the buffer caused the effect. Beljanski et al (19, 20) showed that two fractions contained the tumorigenic RNA: RNA bound to a "RNA-directed DNA polymerase" and RNA associated with DNA. The tumor-inducing RNA is A-G rich, single-stranded with a buoyant density of 1.642 g/cm^3 in Cs_2SO_4, and 5–6S in size. It was therefore concluded that this RNA resembles "viroid RNA." It is interesting that so many new features have been claimed at once by Beljanski and co-workers, such as the presence of a RNA-directed DNA polymerase (reverse transcriptase), viroid-like RNA, and an episomal RNA.

We have recently studied some of the enzymes necessary for DNA metabolism in *A. tumefaciens* (88, 117) and find traces of reverse transcriptase-like activity in cells (7% relative to DNA-dependent DNA polymerase) of this organism. At the moment, we are uncertain about its needs by the bacterium, its role in tumorigenicity, or whether the activity is an artifact of the enzyme assay. These results tentatively support the report of Beljanski et al (19, 20) on the presence of RNA-dependent DNA polymerase in *A. tumefaciens.*

Stroun (235) has claimed that RNA polymerase of bacterial origin is necessary for transcription of bacterial DNA in plant cells, based on indirect experiments using rifamycin, an antibiotic that specifically inhibits bacterial RNA polymerase. In his experiments cut ends of eggplants (*Solanum melongena*) were put into a suspension of *A. tumefaciens* cells for 42 hr. They were then washed and transferred to buffer with or without 10 μg/ml rifamycin for 10 hr followed by a 3-hr period of labeling with [^{3}H]uridine. RNA was extracted from the cortex of the treated plants and was hybridized to DNA bound on nitrocellulose membrane filters under conditions not specified. RNA from plants treated with rifamycin hybridized with *A. tumefaciens* DNA at a level one fourth that of RNA from control plants.

These results imply that bacterial DNA enters host cells first and then bacterial RNA is synthesized. In contrast, Beljanski et al (19, 20) claimed the opposite finding, that bacterial RNA first enters host cells and then bacterial DNA is synthesized with the aid of *A. tumefaciens* reverse transcriptase.

It is therefore interesting to determine if any of these claims can be substantiated in other laboratories. Some laboratories have already shown that *A. tumefaciens* RNA preparations failed to induce tumors. Kado et al (118) tested total RNA, ribosomal RNA, and transfer RNA from two strains of *A. tumefaciens* and found no tumorigenic activity on sunflower plants. Using preparations containing both *A. tumefaciens* RNA and DNA, Phillips & Butcher (193) also were unable to induce tumors on carrot root explants.

Bacteriophage Concept

Bacteriophages were implicated in crown gall induction as early as 1927 (110). In 1947, DeRopp (69) resurrected a crown gall virus hypothesis by considering that a virus may translocate from tumor cells to normal cells based on experiments of grafting normal tissue onto tumor tissue in vitro. Since that time, a number of strains of *A. tumefaciens* were shown to be lysogenic (7, 233, 271), making the virus hypothesis very attractive, for it can be modeled after various oncogenic viral systems known in the animal field.

Reports of enhancing the infectivity of *A. tumefaciens* after exposure to ultraviolet light (82, 100, 102), and mitomycin C (101) suggest an increase in the levels of a tumorigenic substance which could likely be induced phages in the lysogenic strains of *A. tumefaciens*. Both ultraviolet light and mitomycin C induce phage production in *A. tumefaciens* lysogens (43, 256), in which some of the phages are defective and yet can be induced (25, 256). The hypothesis gained impetus when Parsons & Beardsley (188) reported bacteriophages (phage PS8) in homogenates of sunflower crown gall tumor tissues that were grown in culture. The phage hypothesis gained ground when Tourneur & Morel (249) also found infectious phages from nine different sources of crown gall tissue growing in sterile culture and none from normal or habituated tissues. Subsequently, Tourneur & Morel (250) proposed that these phages may integrate in the plant genome. Evidence in support of this hypothesis came from the report of detecting phage PS8 DNA sequences in crown gall tumors; 900 genome equivalents of this phage in crown gall DNA was estimated from cRNA-DNA hydridization experiments (219). Finally, as discussed in the section on DNA, the DNA extracted from phage PS8 was claimed to possess tumorigenic activity (147).

The phage concept becomes temptingly conclusive, particularly when it can be based on analogous plant systems where it has been claimed that callus cultures, inoculated with *E. coli* phages λ and φ 80, express *E. coli* genes (75–77, 112).

The evidence, as discussed above, seems overwhelmingly in favor of phages as being directly involved in tumor induction; however, there is evidence equally strong that rejects this hypothesis.

Bacterial lysates and purified *A. tumefaciens* phages have been tested for tumorigenic activity (18, 134, 147, 206, 219, 233; R. S. Kupor and C. I. Kado, unpublished data); to date no such activity seems to be associated with the phages tested. Despite that, phages are readily isolated from extracts of crown gall tumors (125, 188, 249); the phages so far studied apparently play no role in crown gall tumor induction. Manasse et al (158) characterized several phages from tumorigenic and nontumorigenic strains of *A. tumefaciens* and found no specific differences in their biological or biophysical properties. Also, Kurkdjian (135, 136) observed no differences in the morphology and size of phages after examination of ultrathin sections of decapitated pea seedlings inoculated with either tumorigenic or nontumorigenic strains of crown gall bacteria.

Since the same phages have been observed in both tumorigenic and nontumorigenic strains of *A. tumefaciens,* the phage hypothesis loses its attractiveness.

The most obvious experiment to test the phage hypothesis is to compare the tumorigenicity of lysogens and nonlysogens. Such an experiment has been carried out: *A. tumefaciens* strain V-1, when cured of phage LV-1, retained the same level of tumor-forming ability on pinto bean leaves (43). This cured strain (V-1c) possessed no omega-type prophage DNA (omega type = Ω, PS8, PB2A and LV-1) as judged by DNA·DNA hybridization analyses of the bacterial genome (67). Bacteria cured of phage were as pathogenic as the prophage carrying parental strain V-1.

Based on these recent findings and on the report that PS8 phage particles and its DNA failed to induce tumors (18), serious doubt is raised on the phage concept in crown gall tumorigenesis.

Plasmid Concept

The plasmid hypothesis in crown gall tumorigenesis was generated in parallel with the phage hypothesis since both may occupy the tumorigenic bacterial cell in the form of DNA. In 1955, Klein & Link (128) considered the existence of plasmids that control and regulate cell behavior in crown gall cells. Subsequently, Roussaux et al (206) noted a temporary loss of tumor-inducing ability in a fraction of surviving *A. tumefaciens* when the bacterial population was treated with acridine dyes, reagents known to eliminate plasmids (96, 107). They, therefore, hypothesized that tumor induction by the crown gall bacterium is mediated by episomes. Work in our laboratory showed that a number of *A. tumefaciens* mutants screened for temperature sensitivity at 37° C lost their virulence (B. Day and C. I. Kado, 1970, unpublished data). This suggested that an extra chromosomal element such as a plasmid may be involved in conferring the tumor-inducing ability of *A. tumefaciens*. A search for plasmids was therefore made in strain 1D135, using density analysis in CsCl containing ethidium bromide to separate plasmid from chromosomal DNA. Strain 1D135 cells were digested with lysozyme and lysed with sarkosyl directly on the CsCl gradient in an attempt to resolve any plasmid DNA. We found no plasmid DNA by this procedure and reasoned that plasmids may not be involved in tumorigenesis because the tumorigenic strain apparently lacked plasmid and those strains treated with acridine orange retained their tumorigenicity (118). However, Zaenen et al (269) recently showed that this strain among others actually contains a plasmid of very high molecular weight ($0.6-1.2 \times 10^8$d), varying in size depending on the strain. They also noted that the nontumorigenic *A. radiobacter* strains did not possess plasmid and therefore they concluded that the plasmid may itself be the tumor-inducing principle. We have subsequently resolved this plasmid along with a second plasmid in strain 1D135 met$^-$ by alkaline sucrose density gradient analysis (116). The role of these two plasmids is unclear but preliminary experiments indicate that they are attached to the 3600S nucleoid (folded chromosome) of *A. tumefaciens* by RNA linkages, since they are released from the nucleoid by ribonuclease treatment, and not by treatment with pronase (C. I. Kado and P. A. Okubara, in preparation). It is uncertain if the plasmid is integrated into the chromosome but the above studies suggest association with the chromosome.

Watson et al (258) recently showed by DNA·DNA hybridization that the DNA of nontumorigenic strains of strain C-58 did not contain any plasmid DNA sequences indicating that curing bacteria of plasmids is not due to its cryptic integration into the chromosomal DNA as an episome. No tests were made with strain C-58 chromosomal DNA that was physically separated from its plasmid.

Hamilton & Fall (98) observed that certain strains of *A. tumefaciens* (C-58 and ACH) lost their tumorigenicity when subcultured at 36° C. Eight single colony isolates of strain C-58 began to lose their tumorigenic character after 48 hr incubation at this temperature and by 120 hr all eight of the isolates showed complete irreversible loss of tumor-inducing ability. These observations were confirmed and extended by Van Larebeke et al (251) who showed that after 5 days of subculturing strain C-58 at 37° C, the 150 colonies selected at random and tested on pea seedlings were all nontumorigenic. Furthermore, a sample of 12 of these colonies showed that all had lost their plasmid. Thus it was concluded that the genetic information for the tumorigenic substance in *A. tumefaciens* is carried by one or more plasmids. We have recently shown that strain 1D135 is also curable by subculturing at 37° C and that strain C-58 is curable by ethidium bromide treatment (116; B. C. Lin and C. I. Kado, in preparation). When converted to nontumorigenic derivatives by the temperature treatment, the larger plasmid is lost. Thus nontumorigenic cultures can still possess plasmid molecules. Work at Davis also showed that glycine-attenuated nontumorigenic strains possess the smaller plasmid (205). This plasmid is believed to possess a relatively large deletion changing the molecular weight from 110.6×10^6d in the tumorigenic parental strain to 62.5×10^6d in the attenuated strain (Rogler et al, in preparation). Such deletions have been observed elsewhere recently (J. Schell, personal communication).

Several laboratories have recently shown that nontumorigenic, plasmidless strains may be converted, by still unknown means, into tumorigenic plasmid carrying strains in crown gall tumors. Following the work of Kerr (121, 122), who showed that in tumors, *A. radiobacter* strains can acquire the tumorigenic character from tumorigenic strains, Hamilton & Chopan (97) also showed that in mixed-infection crown gall tumors, the tumorigenic marker from a different strain could be transferred to a cured nontumorigenic C-58 strain. Van Larebeke et al (252) found that a cured plasmidless, nontumorigenic C-58 strain not only reacquired the tumorigenic character, but also contained a large plasmid similar in size to the wild-type tumorigenic C-58 strain. Watson et al (258) also confirmed that there is a transfer of the tumorigenic character in tomato galls and that the converted recipient cells, which were originally free of plasmid, now contained plasmid DNA.

In the results of these laboratories, there are marked variations in the frequency of transfer of the tumorigenic character. Hamilton & Chopan (97) and Van Larebeke et al (252) show between 50 to 60% conversion, whereas Watson et al (258) show about 4% conversion of nontumorigenic strains to tumorigenic strains. The relatively high frequency of transfer suggests conjugal transfer between donor and recipient bacteria in the tumors rather than through plasmid

DNA–mediated transformational exchange. However, the actual mechanism awaits further detailed studies.

Size of the Tumorigenic Substance

Although plasmid DNA has been implicated as playing a principal role in crown gall tumorigenesis, it is still uncertain whether this molecule induces tumors directly or simply confers the tumorigenic property on the crown gall bacterium. Thus, knowledge of the size of the tumorigenic substance in the plant cell would provide valuable clues for answering this question.

DeRopp (69) proposed that a tumorigenic substance such as a virus moves from tumor callus to normal callus, thus perpetuating the tumor state. The development of secondary tumors distant from the primary crown gall tumors is well known. This material moving from the primary tumor must elicit responses elsewhere in the plant host. The chemical nature of this substance is unknown and can be reasonably assumed to be the same substance that induced the primary tumor. To determine the minimum size of this substance, Bender & Brucker (23) performed grafting experiments with Sartorius® membrane filters of various pore sizes placed between the scion and stock of their test plants (*Solanum* and *Datura* species). They found that a tumorigenic substance passed through filters with a pore size of 10–20 μm. Recently Neinhaus & Gliem (185) reported a tumorigenic substance in tobacco that passed through Sartorius membrane filters of 0.2 μm pore size. They recently isolated the tumorigenic substance from tobacco and found it to be a spherical particle 20 μm in diameter containing RNA (F. Nienhaus, personal communication). Meins (174) also recently demonstrated that autonomous teratomas could be induced in Havana tobacco stem segments in axenic culture where teratoma explants were separated from the host by 0.22 μm pore size Millipore® filters.

Based on these experiments, the minimum size of the tumorigenic substance is much smaller than *A. tumefaciens* cells, capable of perpetuating itself and moving out of tumor tissues.

CELLULAR SITE OF BACTERIAL ATTACHMENT

It is well established that crown gall tumors are initiated when live *A. tumefaciens* cells are introduced into freshly made wounds on the host plant (for recent review see references 8, 155, 216). The bacteria multiply for several days at an average generation time of 10–12 hr in *D. stramonium* (161), in *V. rosea* (recalculated data of reference 34), and in *Helianthus annuus* (C. I. Kado and R. A. Langley, 1971, unpublished data). It is assumed that during bacterial growth at the inoculated site, bacterial cells become attached to a surface component of the host cell. Apparently a specific site is involved.

Stonier et al (233) noted that tumor induction could not be inhibited by using a mixed inoculum containing *A. tumefaciens* B6 and the highly virulent phage PB2$_1$; however, inhibition was achieved when the phage concentration was high (10^8–10^9 infective centers added per wound site). They, therefore, postulated that the bacterial cells become quickly located in the host tissue in a manner which makes them relatively inaccessible to phage. Cells of nontumorigenic (avirulent) strains of *A. tumefaciens* such as IIBNV6 were shown to inhibit tumor initiation when mixed at relatively equal or greater proportions with cells of the tumorigenic strain, B6 (152). Heat-inactivated cells of either IIBNV6 or B6 also inhibited tumor initiation. From such experiments, Lippincott & Lippincott (152) obtained a linear relationship between tumor inhibition and bacterial cell concentration. Their data appear to fit a one-particle dose-response curve, which indicated that a single IIBNV6 cell can prevent tumor initiation by a single B6 cell.

Similar mixed inoculation experiments showing inhibition of tumor induction were reported by Manigault (159), Beiderbeck (17), and Beaud et al (10). Beaud et al (11), on the other hand, earlier found no competition between avirulent mutant and the parental virulent strain in *D. stramonium,* whereas, Beiderbeck (17) showed that tumor induction is inhibited when *A. tumefaciens* strain B6 is mixed with high concentrations of *A. tumefaciens*-cell-wall preparations. Cell-wall preparations treated with ether or chloroform retained their inhibitory activity, suggesting that organic solvent soluble material does not partake in the attachment process.

Electron microscope studies by Schilperoort (214) and Bogers (29) showed that *A. tumefaciens* B6 and *A. radiobacter* cells arrange themselves along the inner cell wall of damaged cortical and pith cells of *Kalanchoë daigremontiana.* No such attachments were observed when *Escherichia coli* B or *Rhizobium leguminosarum* PRE were used in place of *A. tumefaciens.* Schilperoort (214) suggested that attachment of the bacteria to the plant cell wall might be one of the first stages of tumor initiation.

Surface materials on the bacterial cell are probably necessary for attachment and subsequent crown gall tumor induction. Permanent spheroplasts of *A. tumefaciens* (obtained from cultures grown in medium supplemented with D-methionine) failed to produce crown gall tumors in *Kalanchoë* and Boston daisy (9). However, small tumors were formed when these plants were inoculated with glycine-induced revertible spheroplasts. These results seem to rule out the hypothesis of possible bacterial conversion to spheroplastic forms during tumorigenesis. On the other hand, these results also suggest that bacterial cell-wall material may be essential in maintaining the tumorigenicity of *A. tumefaciens.*

The necessity for the presence of cell-wall material was further supported by studies on L-forms of *A. tumefaciens.* Cabezas de Herrera & Rubio-Huertos (45) showed that UV-induced L-forms of *A. tumefaciens* were nontumorigenic on *Phaseolus vulgaris.* The possibility of inducing UV lesions in genes critical for specifying tumorigenicity of *A. tumefaciens* was not mentioned by these workers, particularly when the bacterial cells were irradiated heavily twice and showed inability to utilize a number of carbon sources, which the wild-type strain was able to use. Manigault & Kurkdjian (163) concluded that there is no direct relation between cell-wall structure and the tumor-inducing ability of the bacteria, based on change in morphology of *A. tumefaciens* growing in a medium containing glycine or valine, without concomitant loss in tumor-inducing ability. No attempts were made to analyze the cell-wall material. The argument for the need of cell-wall surface materials for tumorigenicity seems to be negated by the report of tumorigenic L-forms (207). However, comparative amino acid analysis between wild-type *A. tumefaciens* and this L-form showed that the L-form contained diaminopimelic acid at the same level as that of the wild-type strain (22) indicating the possibility of vestigial cell-wall material in the L-form strain.

At the initial phase of pathogenesis cell-wall material appears to be required since *A. tumefaciens* cells, treated with lysozyme and ethylenediaminetetraacetate as a function of time, rapidly lost their ability to infect *K. daigremontiana* leaves (13). The addition of sucrose to the inoculum retarded this loss, indicating that cell lysis was occurring as a result of cell-wall loss due to the lysozyme treatment.

There are several indirect lines of evidence supporting surface differences between tumorigenic and nontumorigenic *A. tumefaciens* cells. Kerr and co-workers (123, 201) demonstrated a high correlation between bacteriocin sensitivity and pathogenicity of *A. tumefaciens*. Nonpathogenic strains were insensitive to the bacteriocin.

To explain this difference and because bacteriocins attach to specific receptor sites on bacterial surface structures, Roberts & Kerr (201) postulated that there must be a specific change in molecular configuration on the cell surface acting as the receptor site for the bacteriocin (bacteriocin 84). Although it is possible to have configuration changes, which imply that a protein(s) is altered in amino acid sequence, other possibilities must not be excluded; for example, the surface components of nontumorigenic *A. tumefaciens* cells could lack the bacteriocin receptors, or possess receptors of altered conformation [meaning changes at the protein level and not in amino acid sequence as opposed to "configuration" (212)], or blockage of the receptors. Nonetheless, surface components *are changed* in tumorigenic strains. These changes may be directly conferred by extrachromosomal bacterial genes, namely those genes present in the plasmid(s) recently discovered to be specifically associated with tumorigenic strains of *A. tumefaciens* (269). Recent evidence supports this contention because a direct correlation between bacteriocin sensitivity and the presence of the plasmid has been recently demonstrated (84).

There may be more than one type of change in the surface components of *A. tumefaciens*. An apparent correlation has been recently found between phage sensitivity, the presence of plasmid, and tumorigenicity. Schell and co-workers (J. Schell, personal communication; 252) showed that those *A. tumefaciens* strains with plasmid are also sensitive to a virulent *Agrobacterium* phage (phage AP1) and conversely those strains without plasmid exclude this phage. Thus, phage AP1 seems to specifically attach and infect tumorigenic strains, and hence like bacteriocin, surface components such as the presence of phage attachment sites may be specifying the differences between tumorigenic and nontumorigenic strains.

Parallel to the need of bacterial surface component(s), specific complementary host cell surface material seems to be equally necessary. Tobacco leaf mesophyll protoplasts were inoculated with *A. tumefaciens* 542 man⁻, aus 21, a mutant unable to utilize mannitol and hypersensitive to aureomycin (213). About 10^9 cells/ml of this mutant were mixed with protoplasts and incubated for 30 to 125 min at 25°C. After washing and preincubation in mannitol-containing medium, the protoplasts were suspended in 1.2% agar medium minus auxin and plated. The number of tissue colonies appearing on the plate were scored. There was no significant difference between uninfected control and infected protoplast populations. In fact, there were slightly more colonies in the control samples than in the infected samples.

Although significant, most of the above observations are mainly indirect in supporting the hypothesis that *A. tumefaciens* cells attach to specific sites on the host plant. Host-parasite interactions in other model systems seem to suggest that there are such attachment sites (3). For instance, it is well known that galactomannans of plant cells interact with bacterial polysaccharides (65). In addition, legume lectins bind to the O-antigens of *Rhizobium* and not to the exopolysaccharides of the bacteria (P. Albersheim, personal communication). Therefore, certain surface components of *A. tumefaciens* cells would have affinity for complementary surface components in the host plant.

Evidence for this argument can be derived from preliminary immunological data. Cross-reactive antigens have been observed between *A. tumefaciens* and *H. annuus* (sunflower) (221), joint vetch (52), *Nicotiana tabacum* and *N. glutinosa* (tobacco) (70).

Also, cross-reacting antigens have been observed between *A. tumefaciens* and crown gall tissue-culture antigens of tobacco (49, 70, 215, 218). Although there appear to be complementary structures between the bacteria and host cell, the pitfalls of interpreting heteroreactant immunodiffusion analyses should be pointed out. A number of bacterial polysaccharides contain acetyl and pyruvate groups including *Rhizobium* (28) and *A. tumefaciens* (270), and these are major determinants of serological specificity (81). Therefore, numerous nonspecific cross-reactions between distinct organisms have been found to occur. For example, antiserum made against the polysaccharides of the gram-negative bacteria, *Rhizobium trifolii* and *R. meliloti*, react strongly with the polysaccharides of a gram-positive *Pneumococcus* species (81). Based on these results, Dudman & Heidelberger (81) have questioned the degree of specificity of the interactions based on immunodiffusion analyses.

ELABORATION OF THE TUMOR-INDUCING SUBSTANCE

Early cytological and histological studies at the light microscope level showed that *A. tumefaciens* cells remain intercellular during infection (105, 169, 200, 203). Recent fine structure studies at the electron microscope level confirm this earlier conclusion (108, 168). It has long been accepted that the prerequisites for crown-gall tumor initiation are wounding and live *A. tumefaciens* cells. As a prelude for tumorigenesis it is a reasonable assumption that crown gall bacteria attach to the cell surface (as already discussed above) and elaborate a tumorigenic substance which in turn causes host-cell conversion or transformation.

If one assumes that the nucleic acid hypothesis is correct, fine structure analysis should reveal nucleic acid such as DNA emerging from at least some of the infecting bacterial cells that were examined. Gee et al (90) were the first to show some foreign thread-like substance being elaborated by the infecting bacteria. Their fine structure analysis of sunflower petioles (*H. annuus*), inoculated 12 hr earlier, disclosed *A. tumefaciens* cells invading wounded petiole cells and releasing fibril-like substances within the cell. Although these fibrils were not inferred as DNA, they have the dimensions characteristic of that type of molecule. Work in our laboratory showed that such substances are released from the crown gall bacteria in culture under proper conditions and are sensitive to deoxyribonuclease (Figure 1). Cellulose fibrils, remotely similar in appearance to DNA-like strands, are produced by *A. tumefaciens* (66), but are much larger and vary in diameter (200–300 Å). Also, exocellular polysaccharide material of this bacteria appears granular (46). Whether the existence of such fibrils plays a decisive role in tumor initiation is uncertain and awaits further exploration. Yajko & Hegeman (268) recently concluded that DNA is transferred from *A. tumefaciens* cells to host cells. They showed that [³H]thymidine-labeled material was specifically transferred to carrot root disks within a few hours after inoculation with *A. tumefaciens* cells labeled with [³H]thymidine and [¹⁴C] uridine, or with [³H]thymidine and [³⁵S]sulfate. On the other hand, *Escherichia coli* labeled in the same manner transferred very little [³H]thymidine. An apparent dilemma arose when they found that *A. radiobacter* also transferred [³H]thymidine but at somewhat reduced levels approximately the same as that of a poorly tumorigenic *A. tumefaciens* strain B6UC. They reasoned that a deficiency in the DNA transfer step cannot fully explain the complete lack of tumor-forming ability by *A. radiobacter*. These DNA transfer experiments have been reexamined recently by Ewing & Cooper (85) who were unable to show selective uptake of [³H]thymidine-labeled material by carrot cells.

Nevertheless, certainly biologically active DNA molecules move out of bacteria during crown gall tumorigenesis. The interesting experiments of Kerr (121, 122) showed that isolates of *A. radiobacter* biotype 1 (resistant to chloramphenicol or novobiocin) were converted to tumorigenic types when they were inoculated into sites where a tumorigenic (donor) isolate had been inoculated 3 days earlier in tomato plants. After 42 days, an average of about 38% of the recipient bacteria that were recovered from the developing galls and plated on selective media were tumorigenic.

Although no definite conclusion of a genetically mediated transfer of virulence was given by Kerr (121, 122), his experiments are suggestive of an interaction between the two bacterial isolates and the host, resulting in the movement of bacterial DNA from donor to recipient bacteria, and perhaps into plant cells. The transfer phenomenon could be in the opposite direction where antibiotic resistance is transferred to the tumorigenic donor isolate; thereby one would be selecting and scoring for antibiotic resistances in the same virulent type rather than for the tumorigenic marker. Using different multiple antibiotic markers in the recipient strains, the apparent transfer of the tumorigenic character in mixed inoculation experiments has been confirmed recently by other workers (97, 252,

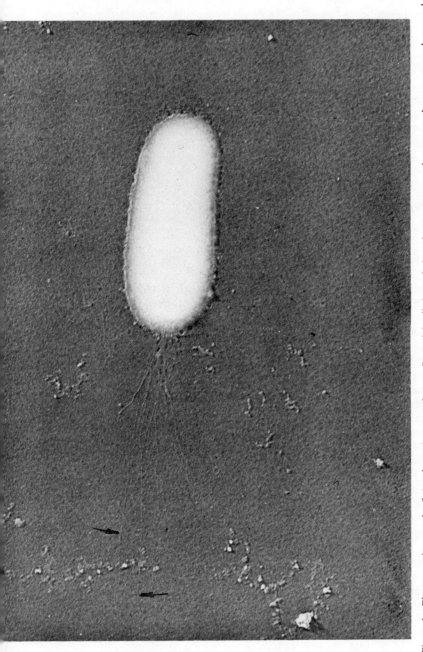

Figure 1 Electron micrograph of *Agrobacterium tumefaciens*. Bacterial cells in early log phase were spread on a cytochrome *c* monolayer and prepared for electron microscopy by the procedure of Kleinschmidt et al (130).

258). This would make the transfer of antibiotic resistance from the recipient strain to the donor strain less likely than the transfer of tumorigenicity from the specified donor to the recipient as contended. Nonetheless, experiments are necessary to rule out unequivocally the possible reverse transfer of antibiotic markers rather than the interpreted forward transfer of tumor-inducing markers(s). Furthermore, the question remains open, since it is difficult to explain the high frequency of transfer of the tumorigenic character (50–60%), even if it is plasmid mediated. The frequency of transfer of a plasmid, carrying genes for resistance to penicillin G, neomycin, and chloramphenicol, was 0.1–0.25% between *Rhizobium japonicum* and *A. tumefaciens* (55). Similarly low frequency of transfer of the RP4 plasmid from *E. coli* to *A. tumefaciens* was reported (62). Conjugational genetic transfer has not been demonstrated in *A. tumefaciens* so the frequency of transfer of any genetic marker between *Agrobacterium* species is unknown. If it can be assumed that plasmid transfer is mediated through some form of conjugation between donor and recipient in the crown gall tissue, this transfer must be highly efficient in crown gall tissue. If so, the possibility exists for the movement of bacterial plasmid DNA to the host plant cell.

UPTAKE AND FATE OF THE TUMOR-INDUCING SUBSTANCE

The uptake of foreign macromolecules by higher cells has long been of interest to genetic engineers who wish to insert new genetic information through a biochemical approach. This gained impetus since the discovery of DNA-mediated transformation in bacteria. There have been reports from time to time of genetically transforming higher cells. In the case of oncogenic systems, particularly in mammalian models, the DNA of tumor-inducing viruses such as papiloma, polyoma, and SV40 are known to induce tumor transformation (30, 57, 58). Furthermore, the DNA obtained from a rat hepatoma has been reported to induce tumors in rats (47). In another study with murine lymphoma DNA (202), lymphoma cells were able to take up as much as 5% of the exogenous host DNA. In parallel, a number of claims have been made by Ledoux & co-workers (137–141) that bacterial DNA can be taken up by plant cells (under certain conditions), migrate into the nucleus, and integrate into the chromosomal plant DNA. If bacterial DNA can indeed be taken up by plant cells, the DNA of *A. tumefaciens* could also be taken up by host cells during the infection process. Stroun, Anker & Ledoux (240, 241) have argued that *A. tumefaciens* DNA appears as hybrid DNA molecules when apical shoots of young tomato plants are dipped in a solution of bacterial DNA (200 μg/ml) for 48 hr and analyzed by density gradient centrifugation in CsCl. They believed that the hybrid DNA (DNA molecules of intermediate density with respect to host DNA and bacterial DNA in CsCl) are formed "as a consequence of the attachment of the foreign double strand DNA to the native endogenous one."

They argued that such hybrid DNA is not the result of artifacts such as bacterial contamination because: (*a*) no or very few bacteria are observed either in the final solution or in the stems; (*b*) the contaminating bacteria would have to be identical types each time the experiment was performed; (c) control experiments with homologous DNA, or heterologous DNA with GC content identical with homologous DNA, or no DNA did not show a hybrid DNA peak in the density analysis.

The above experiments by Stroun, Anker & Ledoux (240, 241) have not been reproduced in other laboratories. In fact, serious doubts have been cast on the validity of the interpretations of these experiments. Kado & Lurquin (119) have repeated their experiments using mung bean seedlings under similar conditions. Both DNA uptake and DNA replication were examined, using [^{3}H]thymidine-labeled *A. tumefaciens* DNA and [^{3}H]thymidine incorporation during presumed uptake of nonlabeled *A. tumefaciens* DNA. Instead of employing CsCl density analysis, which is insensitive for detecting very small amounts of foreign DNA, the highly sensitive DNA·DNA filter hybridization procedure was used for the detection of any foreign *A. tumefaciens* DNA. After treatment as prescribed by Stroun et al (240, 241), the cotyledons, leaves, roots, and stems were separated and the DNA of each plant part was examined for *A. tumefaciens* DNA. Except for the leaves, no *A. tumefaciens* DNA sequences were detected. The small apparent DNA·DNA homology observed with the leaf DNA samples was also shown to be merely host DNA sequences when the hybrid was examined by thermal elution chromatography on hydroxylapatite. Therefore, all of the added *A. tumefaciens* DNA was degraded and reutilized by the host-plant cells. Although the experiments of Kado & Lurquin (119) employed mung bean instead of tomato, the results should be consistent with various hosts of *A. tumefaciens* (such as mung bean). In fact, Kleinhofs et al (129) used tomato shoots and also were unable to confirm the experiments of Stroun, Anker & Ledoux (240, 241). Using the same CsCl procedure as Stroun et al, they found no DNA of intermediate density in a CsCl density gradient when tomato shoots were exposed to 200 μg of *A. tumefaciens* DNA per ml for 48 hr followed by incubating the shoots for an additional 28 hr with [^{3}H]thymidine. Lurquin et al (156a) have shown that foreign DNA in a mixture of two different DNAs cannot be resolved by density analysis in CsCl unless the foreign DNA represented more than 20% of the mixture. Thus, at least this amount of bacterial DNA was necessary in the experiments of Stroun, Anker & Ledoux (240, 241). Judging from their data where a DNA with a density of 1.707 g/ml was observed, approximately 50% of the DNA appears to represent bacterial DNA. On the basis of genome sizes (assuming 10^{12}d for tomato DNA and 10^9d for *A. tumefaciens* DNA), nearly 5000 copies of the bacterial genome should be present. As shown by Kado & Lurquin (119) no *A. tumefaciens* DNA equivalent to even one bacterial genome could be detected.

Nevertheless, all of the negative evidence of DNA uptake does not imply that no foreign DNA make their way into plant cells. Although extremely sensitive, the DNA·DNA reassociation methods of assay may still be insufficient to detect very minute amounts of foreign DNA in plant cells. Probably the efficiency of DNA uptake is extremely low and certainly not at the high levels claimed above by Stroun and co-workers. Ledoux & Huart (142) recently emphasized that highly polymerized DNA of relatively high molecular weight is required for efficient

DNA uptake into plant cells. The same argument has been put forth by Snow & McLaren (227) for DNA uptake by mouse embryonic cells. Thus, based on the recent status of DNA uptake, it is reasonable to conjecture that during infection, *A. tumefaciens* cells may be much more efficient in transferring some of its DNA (or plasmid) to host plant cells than by any of the experimental conditions used with purified DNA.

GENETIC EXPRESSION OF BACTERIAL GENES?

Subsequent to DNA uptake [assuming that there is some bacterial (or plasmid) DNA which makes its way to the nucleus and enters into the milieu of the nucleoplasm], the expression of at least a few of the foreign bacterial genes would logically be anticipated based on the role of tumor viruses in animal systems. This concept is logical and seems to be supported by plant systems.

DNA Integration and Replication at Initial Phases of Tumorigenesis

Ledoux and his collaborators have claimed integration of bacterial DNA into *Arabidopsis thaliana* (141, 144), *Hordeum sativum* cv. Proctor (barley) (138–140), and *Sinapsis alba* (138). These interpretations were based mainly on indirect evidence from density analyses in CsCl and a recent thermal renaturation experiment (143). They have recently reported supporting biological evidence for the apparent expression of such "integrated" bacterial DNA. Ledoux et al (145, 146) showed that thiamineless mutants of *Arabidopsis thaliana* could be corrected by soaking sterile seeds in DNA preparations from various bacteria, including DNA of *A. tumefaciens*. If confirmed, *A. tumefaciens* DNA can be expressed in host cells. The expression *E. coli* genes mediated through phages λ and φ 80 in haploid cell lines of *A. thaliana* (75–77), and a diploid suspension culture of *Acer pseudoplatanus* (sycamore) (112), seems to provide additional support for genetic expression of bacterial DNA in higher plant cells. Thus, in the case of *A. tumefaciens*, it would be reasonable to expect the presence of bacterial messenger RNA accompanied by a certain amount of bacterial DNA amplification if it is assumed that some bacterial DNA makes it way into the host cell. Stroun and co-workers (240, 241) concluded that *A. tumefaciens* DNA molecules reach the nucleus of tomato meristematic cells and replicate. The interpretations of Stroun et al (239–241) were based on the detection of a DNA with a buoyant density close to that of *A. tumefaciens* DNA ($\rho = 1.718$ g/cm^3) when some of the DNA extracted from DNA-treated plants that were of intermediate density in CsCl density gradients was sonicated and reanalyzed in a second CsCl density gradient. As discussed in the above section, it is questionable whether or not bacterial DNA integration had occurred since many copies of the bacterial genome would be necessary to shift DNA to a resolvable density and hence these DNAs would be easy to detect by DNA·DNA reassociation analyses. By such methods, we have recently reexamined the experiments of Stroun and co-workers (240, 241) and found that there was no replication and integration of bacterial DNA in *Phaseolus aureus* (119). The possibility of bacterial contamination may also contribute to false "hybrids" (68, 129). In fact, recent experiments conducted under stringent sterile conditions showed that such "hybrid" DNA molecules are absent in CsCl analyses (120, 129).

In a series of papers, Stroun and his co-workers (4, 236, 238, 242) reported transcription and replication of *A. tumefaciens* DNA in plants exposed to *A. tumefaciens* cells. *Solanum melogena* (eggplant), dipped in a suspension of *A. tumefaciens* cells for 41 hr and then labeled for 3 hr with [³H]uridine, contains RNA in the cortical tissues that will hybridize to *A. tumefaciens* DNA. Competitive hybridization experiments showed that 70 to 90% of the label was displaced by nonlabeled *A. tumefaciens* RNA. Conversely, unlabeled RNA from similarly treated plants displaced labeled *A. tumefaciens* RNA. Also similar results were obtained when *Lycopersicon esculentum* (tomato) shoots were dipped in a suspension of *A. tumefaciens* B6 cells (10^9 cells/ml) for 41 hr and then in a solution of [³H]uridine for 7 additional hr (242). Any contribution of [³H]RNA from the resident *A. tumefaciens* cells was claimed to be ruled out in both the eggplant and tomato experiments on the basis that most of the bacterial cells were removed mechanically by dissecting away the central cylinder and xylem vessels of eggplant (236) and that bacteria are found only in the xylem and on the epidermis of tomato plants (237, 239). This was supported by electron microscopic and radiographic studies showing that most of the label was in tissues not containing bacteria, that homogenized cortical tissues were sterile, and that a brief 1 hr exposure of plants to *A. tumefaciens* cells gave the same results as untreated control plants. To reaffirm their stand on bacterial transcription in plants, Stroun et al (239) showed that rifamycin [an antibiotic that inhibits *A. tumefaciens* but not plant-cell RNA polymerases (235)] blocks tumor development when it is administered to tomato shoots containing *A. tumefaciens* cells immediately after wounding with a sterile needle. Tumors were initiated, however, when the administration of rifamycin was delayed 10 hr or more. Stroun et al (239) analyzed labeled RNA from tomato shoots that were allowed to take up *A. tumefaciens* cells for 21 hr through the cut end of the shoot stem, followed by 15 hr of rifamycin treatment (or no rifamycin treatment) and then labeled with [³H]uridine. They found RNA only from those shoots that were not treated with rifamycin hybridized to bacterial DNA immobilized on nitrocellulose filters. Part of the DNA extracted from plants treated with bacteria had a buoyant density intermediate to the densities of bacterial DNA and plant DNA when analyzed by CsCl density gradient centrifugation. When the DNA of intermediate density was heat denatured, only a single peak in the CsCl gradient was observed whereas sheared DNA separated into two species: one at the density of bacterial DNA and the other at the density of plant DNA. As in their previous papers (4, 236–238), Stroun et al (239) have stressed that the DNA of intermediate density represents bacterial DNA linked to plant DNA and that this DNA may serve as template for the RNA species apparently detected in the above experiments. It was, therefore, concluded that crown gall tumor induction is dependent on the synthesis of bacterial RNA in plant cells and that time necessary for this induction is short. We have recently stressed that Stroun's results remain ambiguous

(80) because it is difficult to eliminate unequivocally the possibility that the RNA species being detected were synthesized by contaminating bacteria residing in the plant, and the rifamycin treatment may have killed the living bacteria known to be necessary for transformation. Also, the findings from the nucleic acid analyses were similar whether or not the plants received the wounding treatment required for tumor formation (236, 238, 239). We have also questioned whether or not the plant cells that actually undergo transformation represented a measurable fraction of the sample analyzed. Furthermore, the appearance of bacterial nucleic acids in the plants did not arise from the tumorigenic properties of *A. tumefaciens* since identical results were obtained when either *E. coli* or *Pseudomonas fluorescens* were substituted for *A. tumefaciens* in previous experiments by Stroun et al (235–238).

About the same time as the rifamycin experiments of Stroun et al (4, 239), Beiderbeck (14) also showed that rifampicin (a derivative of rifamycin) inhibits crown gall tumor formation in *Kalanchoë daigremontiana*. Treatment of inoculation sites with rifampicin inhibits tumor formation only when the antibiotic is administered within a few hr after inoculation. No inhibition of tumor development was observed when rifampicin was applied 30 hr after infection. These results were confirmed by Kurkdjian et al (136) who showed that rifampicin inhibits tumor formation in peas and that viable *A. tumefaciens* cells were not recovered after 24 hr in the presence of 3.35 μg rifampicin per wound site. Also, Anand & Heberlein (3a) reported inhibition of tumor development in *K. daigremontiana,* pinto bean leaves, and carrot discs when rifampin was applied to wound sites immediately after inoculation. Inhibition was maximal at 200 μg rifampin/ml. These results are interesting because these antibiotics inhibit specifically RNA polymerase of prokaryotic cells (261); therefore host plant cells were probably unaffected by the antibiotic but the bacteria were specifically killed when the antibiotic was administered at various intervals following inoculation. However, rifampicin (or rifamycin or rifampin) was no longer effective when administered 24–36 hr after infection, indicating that the tumor transformation process had been triggered by this time. However, the time period after inoculation during which rifampin remained inhibitory varies with the host plant used (3a). The antibiotic-mediated inhibition of crown gall induction is apparently not due to its influence on the plant cell, since rifampicin did not block tumor induction when a rifampicin-resistant mutant of *A. tumefaciens* was employed (15).

These results imply that RNA polymerase of *A. tumefaciens* is not functioning in a transcriptional capacity in inoculated plant cells (assuming that rifampicin is taken up by the plant cells). However, a brief period of transcription of host plant DNA by its own RNA polymerases seems to be a necessary function for crown gall tumor induction. The initiation of crown gall tumors was reduced by about 38% when α-amanitin [an inhibitor of nuclear RNA-polymerase II in plant cells (109, 210, 234)] was mixed with *A. tumefaciens* cells or applied 24 hr after inoculation of *Kalanchoë daigremontiana* (16). No reduction of tumor initiation was observed when α-amanitin was added 96 hr after inoculation.

It is difficult to interpret clearly all of the above experiments employing antibiotics because a number of parameters must be considered, particularly those concerning the variations in the uptake of the antibiotics by plant cells during time-course studies. The permeability of host cells should be monitored, perhaps by radio-labeled antibiotics, to determine whether or not breakdown of the antibiotic occurs or uptake of the antibiotic ceases during cell maturation after wound induction. Consequently, any firm conclusions on DNA transcriptional functions cannot be made at this time.

Replication of Bacterial DNA and Transcription in Established Crown Gall Cells

If some *A. tumefaciens* DNA makes its way into plant cells during infection, it would be reasonable to assume that the DNA replicates and is passed on to daughter cells during host-cell division, or it is degraded and its degradation products reutilized by the host cell, or it remains in limbo eventually being diluted by the proliferating cell population. Based on the first hypothesis a search was made for bacterial DNA sequences in established crown gall tumor cell lines.

Analysis of the guanosine-cytosine (percentage of GC) base composition between crown gall and normal cells might indicate a detectable difference; however, no differences in percentage of GC have been observed (247) in spite of the large difference between the percentage of GC of *A. tumefaciens* (60% GC) and that of plants (36–39% GC). Very few, if any, distinct differences would be expected if the levels of bacterial DNA are low and therefore masked by the overwhelming levels of plant DNA. Axenic cultures of crown gall cells obtained by cloning from an isolated single transformed cell would be expected to contain bacterial DNA in most of the daughter cells if it is assumed that the bacterial DNA replicates along with host DNA. If that were the case, the foreign DNA would be very easy to detect by present methods, e.g. DNA·DNA filter hybridization and DNA·DNA reassociation kinetic procedures. In 1966, Tinbergen (247) first used the hybridization procedure of McCarthy & Bolton (172) to estimate the amount of *A. tumefaciens* DNA in tobacco crown gall cells but was unable to obtain reproducible results. In a series of experiments using [³H]ribonucleotidyl polymers synthesized in vitro with *A. tumefaciens* DNA as the template and *E. coli* RNA polymerase, Schilperoort (214, 215) and co-workers (220) found 24–31% of the radioactivity bound to filters containing crown gall DNA relative to bacterial DNA whereas about 1% of the radioactivity bound to filters containing tobacco leaf DNA. We have recently pointed out that these experiments lacked two important controls (80): first, it was not shown whether the in vitro synthesized RNA (cRNA) represented a uniform transcript of the entire bacterial genome, making any quantitative conclusion questionable; and second, the base-pairing fidelity of the duplexes scored as hybrids was not analyzed.

Quétier et al (194) claimed "direct evidence for partial homology" between *A. tumefaciens* DNA and crown gall cell DNA by using DNA·DNA filter hybridization. By adding increasing amounts of [³²P] of *A. tumefaciens* to Millipore filters possessing a fixed amount of either normal or tumor cell DNA (of either tobacco or *Scorzonera*), a hybrid saturation value of 0.1% can be interpreted from their results, although a value of 0.2% was claimed. On the basis of this latter estimate, Quétier et al (194) calculated that this amount is enough DNA to account for one bacterial genome per diploid transformed cell.

Using DNA·DNA filter hybridization, Srivastava (229) also reported homologies between *A. tumefaciens* DNA and the DNA of tobacco tumor tissue grown for 3 weeks in axenic culture. He also found the bacterial DNA to reassociate with the DNA of normal tobacco tissue. The extent of homology between bacterial and tumor DNA was 0.1% and that between bacterial and normal cell DNA was 0.04%. Although Quétier et al (194) found no homologies between bacterial and normal cell DNA, they observed homologies between bacterial DNA and a satellite DNA which was only present in wounded plant tissues. Srivastava (229) found no such satellite DNA in his preparations.

Some of the pitfalls of using DNA·DNA filter hybridizations for estimating the amount of bacterial DNA in tumor cells have been discussed previously (80). The procedure involves annealing denatured pieces of radioactive *A. tumefaciens* DNA to tumor or normal cell DNA immobilized on nitrocellulose membrane filters. At saturating concentrations of radioactive DNA, the ratio of bacterial DNA annealed to the filter-bound tumor DNA provides the estimated fraction of tumor DNA that has sequences complementary to the bacterial DNA, presumably of bacterial origin. In actuality, the results of such procedures will not represent true saturation values because the rate of self-renaturation in solution is greater than the rate of annealing of the bacterial DNA with tumor DNA immobilized on the nitrocellulose filters. The experimentally derived saturation values thus are usually underestimates of the degree of base sequence homology. Saturation experiments must be calibrated accordingly with saturation values obtained in parallel experiments using model nitrocellulose filters possessing various small, known amounts of *A. tumefaciens* DNA mixed with large amounts of normal plant DNA. Besides the need for calibrating the system, it is also necessary to characterize the DNA·DNA duplexes that formed under the conditions used. In the studies by Quétier et al (194) and Srivastava (229), the DNA annealing systems were not calibrated with model filters nor were the DNA duplexes characterized (e.g. thermal stability of the DNA duplexes relative to the T_m of *A. tumefaciens* DNA). Consequently, the quantitative values derived from these preliminary experiments remain tenuous.

Using radioactive cRNA [³H-cRNA] synthesized in vitro from *A. tumefaciens* DNA and from *A. tumefaciens* bacteriophage PS8 DNA, Schilperoort et al (219) estimated that in crown gall cells there are nine genome equivalents of *A. tumefaciens* strain A6 DNA and 900 genome equivalents of *A. tumefaciens* bacteriophage PS8 DNA. They also stressed that the A6 bacterial DNA base sequences are present in the bacteriophage genome and that the bacterial specific base sequences in crown gall DNA are also present in the phage genome. Since the bacterial DNA base sequences in crown gall DNA also apparently form part of the PS8 phage genome, Schilperoort et al (219) concluded that in crown gall DNA only a small part of the A6 bacterial genome is present repeatedly as part of the nine estimated genome equivalents. It was also suggested that almost complete copies of the PS8 phage genome are present in crown gall DNA.

Close examination of their data indicated that the base pairing fidelity was low and certain contradictions in the RNA·DNA ratios were apparent. Possible explanations for these contradictions were given (219) on the basis that the efficiency of hybridization differs between presumptive repeated base sequences in crown gall DNA and the bacterial DNA, and also that the transcription of these repetitious bacterial DNA sequences in phage PS8 DNA was relatively inefficient. Judging from the thermal denaturation curves between bacterial [³H]cRNA and crown gall DNA and that of bacterial [³H]RNA and bacterial DNA (253), the base pairing fidelity seems too low to draw any decisive conclusions.

Patillon (189) recently employed DNA·DNA reassociation kinetics to estimate the amount of *A. tumefaciens* DNA in crown gall DNA of *Scorzonera hispanica*. Unlike the above experiments, he used that fraction of the crown gall DNA, which reannealed very slowly (presumptive unique sequences), to be probed for any bacterial DNA sequences. From the data resulting from reassociating the very slowly reannealing DNA fraction and [³²P]DNA of *A. tumefaciens*, Patillon concluded that each copy of the very slow fraction of crown gall DNA carries one equivalent of the *A. tumefaciens* genome.

We have recently pointed out (80) that these experiments were incomplete since normal tissue DNA was not included as a control for nonspecific sequence homologies. Also the annealing system was not properly calibrated by measuring the reassociation rate of radioactive *A. tumefaciens* DNA in the presence of small known amounts of nonradioactive *A. tumefaciens* DNA and normal callus tissue DNA. Furthermore, as in the above critiques, the base-pairing fidelity of the molecules scored as hybrids was not assessed by thermal dissociation studies.

In direct contradiction to all of the above reports claiming the presence of *A. tumefaciens* DNA and PS8 phage DNA in crown gall DNA, our studies (79, 117) and those of Loening et al (156), Chilton et al (53), Dons (74), Eden et al (83), Farrand et al (86), and Merlo & Kemp (179) showed that no DNA equivalent to one *A. tumefaciens* genome or one phage PS8 genome per diploid host cell is present in crown gall tumor cell of sunflower, tobacco, and *Vinca rosea*. Both DNA·DNA filter hybridization and DNA·DNA solution enrichment techniques were employed by Drlica & Kado (79). The filter saturation hybridization technique included model filters containing known percentages of bacterial DNA mixed with *V. rosea* leaf DNA. In the DNA solution enrichment procedure, radioactive crown gall tumor DNA that was sheared and denatured was annealed to an excess of sheared, denatured leaf nuclear DNA for 20 hr at 63°C. The single-stranded DNA remaining after this incubation were separated from double-stranded DNA by hydroxylapatite chromatography and redenatured. These denatured strands were then annealed to an excess of sheared, denatured *A. tumefaciens* DNA. Any hybrid DNA were separated from single-stranded DNA by a second chromatography step on hydroxylapatite. A control experiment was performed to show that the solution enrichment procedure could adequately resolve minute known quantities of *A. tumefaciens* [³²P]DNA in the presence of tumor [³H] DNA. A second type of enrichment experiment was performed to determine if any highly reiterated sequences of *A. tumefaciens* DNA were present in crown gall nuclear DNA and therefore possibly escaped detection by the first type of experiment. In the second experiment sheared, denatured radioactive tumor DNA was annealed with an excess of sheared, denatured *A. tumefaciens* DNA for 6 hr at 80°C, conditions that were too stringent for plant DNA reassociation but allowed 70% of the bacterial DNA to reassociate. The double-stranded fractions were collected by chromatography on hydroxylapatite. No differences were found in either enrichment experiment between tumor DNA and normal leaf DNA. The level of detection in these experiments was less than 0.1% of the tumor DNA.

In further support of these results, we (79) also found by DNA·DNA filter hybridization reconstruction experiments that no more than 0.02% of the crown gall tumor genome could be homologous to *A. tumefaciens* DNA. Any hybrids detected between *A. tumefaciens* DNA and crown gall DNA were analyzed by thermal dissociation and had a lower T_m than bona fide *A. tumefaciens* DNA, indicating that no bacterial DNA hybrids were formed. The hybridization reactions were carried out under relatively stringent conditions (18–20°C below T_m in 50% formamide). Recently, the same type of DNA·DNA filter saturation hybridization and characterization experiments were performed by Farrand et al (86). Their study also showed that very little if any *A. tumefaciens* DNA is present in tobacco crown gall tissue. Using the renaturation kinetic technique, Chilton et al (53) also concluded crown gall tumor DNA cannot contain as much as 0.01% *A. tumefaciens* DNA or one third of a bacterial genome per diploid tumor cell. Loening et al (156) employed cRNA·DNA filter saturation and also were unable to demonstrate any presence of *A. tumefaciens* genome equivalents of DNA.

Schilperoort and co-workers (217) have recently reexamined their earlier experiments and now conclude that their cRNA, synthesized in vitro, does not anneal to homologous DNA in the crown gall DNA preparations. They now consider the possibility that cRNA may have annealed to either contaminating RNA molecules present in crown gall DNA preparations or to the indiscriminant binding of poly A segments of RNA to the nitrocellulose filters. Loening et al (156) found that DNA from tumor callus tissues were always more contaminated with possibly polysaccharides than those from normal callus tissue. These contaminants were found to permit nonspecific binding of DNA. Consequently, the importance in the use of highly purified DNA cannot be overly stressed when used in DNA reassociation experiments since as discussed above the levels of foreign DNA that may exist are in the region of the noise levels of the background.

TRANSCRIPTION   If it is assumed that minute quantities of bacterial DNA such as plasmid DNA make their way into cells and retain and promote their genetic activity, one would expect a certain amount of genetic amplification of at least the transcriptional and translational products of these genes in crown gall cells. Although Dons (74) and Chilton et al (54) recently found no nucleic acid hybridization with plasmid DNA and crown gall DNA, searches have been made to detect any transcriptional products (bacterial RNA) which may be highly amplified in the tumor cells. Milo & Srivastava (182) reported that DNA of *A. tumefa-*

ciens strains 4–32, B6, and IIBNV6 showed 0.44, 0.38, and 0.24% hybridization with a 14–18S RNA fraction from tobacco crown gall callus tissue, and they found no significant hybridization with any RNA fraction from the normal callus tissue. The nature of this RNA was not established. Consequently, we have recently examined RNA species, isolated from sunflower and *Vinca rosea* crown gall callus tissue, for any bacterial RNA sequences homologous to *A. tumefaciens* DNA (117). Radioactive purified messenger RNA, isolated from sunflower crown gall callus tissue, was hybridized to bacterial DNA and immobilized on nitrocellulose filters. Although a 4–5-fold difference in hybridization was observed with the messenger RNAs of tumor cells relative to those of normal callus cells, it was concluded that no bacterial RNA sequences were present after characterizing the radioactive counts bound to the filter by thermal denaturation. We also conducted competition hybridization experiments with radioactive messenger-ribosomal RNA fraction isolated from *Vinca rosea* tumor callus tissue (117). The tumor messenger RNA-ribosomal RNA fraction was hybridized with *A. tumefaciens* DNA and with *V. rosea* leaf DNA as a function of *A. tumefaciens* B6 ribosomal RNA concentration. The bacterial ribosomal RNA competed about equally with the radioactive tumor RNA for binding sites on *V. rosea* leaf DNA or *A. tumefaciens* DNA. This suggested that DNA sequences which formed hybrids with ribosomal RNA could at best be paired nonspecifically. This was confirmed by thermal denaturation of the RNA·DNA hybrid. Therefore it was concluded that these results cannot be interpreted as showing nucleotide sequence homologies with *A. tumefaciens* DNA. Recently, Matthysse et al (171) reported that RNA synthesized in vitro from isolated nuclei of crown gall cells contained about 3% sequences complementary to *A. tumefaciens* plasmid DNA. RNA synthesized similarly from normal cell nuclei did not appear to contain any such sequences. Van Sittert et al (253, 254) examined crown gall RNA fractions by column chromatography on methylated albumin kieselguhr (MAK). A fraction that bound tightly to the MAK column was high in AMP content (which is probably associated with the presence of poly A sequences). This RNA showed a difference in labeling kinetics compared with that of the normal tissue RNA. They also noted a change in the tRNA population, with tRNA of crown gall cells showing a lower CMP content than that of normal cells.

Molecular Complementation Hypothesis

The present query of whether or not crown gall cells contain *A. tumefaciens* DNA seems to stand at a position where there either may be: (*a*) a small number of bacterial, plasmid, or phage genes present, undetectable by present techniques, or (*b*) no foreign bacterial genes, at least, not part of the permanent fixture ascribed to as genetic material.

In a somewhat analogous situation, possibly significant to crown gall, is the finding by Rake (196) who reported the lack of homology between *Glycine max* (soybean) DNA and the DNA of its *Rhizobium* symbiont, strain 61A76. It was concluded that if any genes common to the bacteria are present in soybean, their total must represent less than 0.6% of the bacterial genome. Interaction between genes of the bacterial and plant cell may have some evolutionary advantages. The haem prosthetic group of leghaemoglobin, a myoglobin-like protein found in large quantities in root nodules induced by rhizobial infection of leguminous plants, is the product of bacterial genes (60) and the plant contains the genetic determinants for the apoprotein (61, 73). Thus, in the crown gall system and *A. tumefaciens*, a relative of *Rhizobium* species [based on DNA homologies between the two bacteria (91, 99)], certain genetic exchanges or contributions may be taking place in the formation or conjugation of macromolecules necessary for tumor induction, promotion, and autonomous growth, perhaps through the induction of a provirus as in the case of tobacco (185), or through direct interactions with the plant genome. An example for the latter case is the action of substituted fluorenes on tobacco pith callus. Such compounds interact with DNA and induce a directed and heritable cellular change in tobacco cells permitting them to grow on cytokinin-free media (12a). Loening et al (156) recently observed sunflower and tobacco DNA to possess regions homologous to *A. tumefaciens* DNA. If this is confirmed, the mechanism of crown gall tumorigenesis may be one of genetic complementation whereby a tumorigenic substance is synthesized by the condensation of two molecular species, one species synthesized by the crown gall bacteria and the other by the plant to form an active component, analogous to the leghaemoglobin biosynthesis in legume root nodules.

Bacterial Gene Products in Crown Gall Cells

A number of comparative studies have been made to see if any biochemical differences exist between crown gall and normal cells, particularly from the standpoint of determining whether or not bacterial gene products occur in crown gall cells. No attempt is made here to review the literature in this area categorically, especially since the reports of observed changes in crown gall cells are likely to be simply the observation of the manifestation of the disease rather than that of foreign genetic products of the crown gall bacteria. Various aspects of the physiological differences in crown gall cells with respect to normal cells have been previously reviewed (36, 40, 128).

Most of the studies in search of biochemical differences have shown that they are merely quantitative rather than qualitative differences. Examples of such differences of various measurable cellular components are listed in Table 3. The elevated levels of various enzymes may simply reflect the higher growth rate of crown gall cells than that of normal cells, which in turn indicates a generalized increase in many of the metabolic processes, mobilizing various components essential for continuous oncogenic growth. Likewise, the decreased levels of enzyme activities are due to diversions and lowering of metabolic processes unessential for the growth of crown gall tumors.

So far no evidence is available which unequivocally demonstrates the presence of gene products of *A. tumefaciens* in crown gall tumor cells. Various claims have been made that bacterial components are detectable immunologically (49, 70, 218) but no worker has yet isolated and characterized these antigens and showed that they are identical with a given antigen of the crown gall bacteria by bio-

Table 3 Examples of observed differences in metabolic constituents of crown gall tumor cells

Component	Differences in levels of activities relative to normal cells	References
Isoenzymes		
Acid-β-glycerophosphatase	Increased	223
Phosphatases	Increased	223
Acid phosphatase	Increased	32, 44
Acid phosphatase	Decreased	199
Acetylesterase	Decreased	199
Esterase	No difference	32
Esterases	Increased	44, 87, 223
NAD + NADP-malate dehydrogenase	Increased	199
Phosphogluconate dehydrogenase	Increased	199
Isocitrate dehydrogenase	Increased	199
Alcohol dehydrogenase	Increased	44
Glucose 6-phosphate dehydrogenase	Increased	44
NADP-glutamate dehydrogenase	No difference	199
Peroxidase	Increased	32, 44
Peroxidase	Decreased	—[a]
Polyphenoloxidase	Increased	44
Polyphenoloxidase	Decreased	—[a]
Arginase	Decreased	183
Leucine aminopeptidase	Increased	199
Catalase	Increased	199
Nucleic acids, nucleic acid enzymes and components		
Total nucleic acids	Increased	228
Total DNA	Increased	132
DNA	No difference	197
Total RNA	No difference	114
Transfer RNA	No difference	56
Chromatin RNA	Increased	186
Ribonuclease	Increased	32, 198
Chromatin ribonuclease	Decreased	231
Chromatin deoxyribonuclease	Decreased	231
DNA polymerase	Increased	232
DNA polymerase	No difference[b]	88
Nucleic acid methylases	Increased	230
Chromatin protease	Decreased	232
Pseudouridylic acid synthetase	No difference	170a
Histones and polyamines		
Histones	No difference	232
Spermidine	Increased	6
Putrescine	Increased	6
Hormones and Regulators		
Adenosine 3':5'-cyclic monophosphate	No difference	78
Ribosyl *trans* zeatin ("cytokinesins")	Increased	181, 267
Cytokinin	No difference	95
Auxin	Increased	5, 36, 128

Table 3   (Continued)

Component	Differences in levels of activities relative to normal cells	References
Miscellaneous		
Syringic acid	Increased	178
Fatty acids	Increased	260
Proteinase inhibitor I	Increased	265
Endoplasmic reticulum	Increased	151
Lignin	Increased	150

[a]Decreased levels of peroxidase and polyphenoloxidase found in sunflower but not in *Vinca rosea* crown gall tumors (J. E. DeVay, personal communication).
[b]No difference observed in 6S DNA polymerase of tobacco and *Vinca rosea* crown gall cells (88).

chemical analyses (e.g. compositional analysis). Bomhoff (32) obtained a strong precipitin reaction between *A. tumefaciens* B6 antigens and crown gall antiserum, but has also stressed the pitfalls of interpreting such reactions in agar immunodiffusion analysis.

A number of new isoenzymes have been observed in crown gall cells, and these along with other components that are apparently formed solely in crown gall cells are shown in Table 4. Although the possibility exists that some of these enzymes and components may be of bacterial origin, there is equal probability that they are merely the result of the expression of normal genes which are transcribed only at certain stages of growth in normal cells but remain expressed in crown gall

Table 4   New metabolic constituents in crown gall tumor cells

Component	Reference
Isoenzymes	
Glucose-6-phosphate dehydrogenase	199
Phosphogluconate dehydrogenase	199
NAD-glutamate'dehydrogenase	199
Isocitrate dehydrogenase	199
Acid phosphatase	199
Basic peroxidase	199
Peroxidase	59
Proteins and uncharacterized antigens	
Cross-reacting antigens	70, 214, 218
Protein antigens	49
Peptides	50, 51
Arginine derivatives	
Octopine [N^2-(D-1-carboxyethyl)-L-arginine]	94, 176, 216
Nopaline [N^2-(1, 3-dicarboxypropyl)-L-arginine]	93, 216
Octopinic acid [N^2-(D-1-carboxyethyl)-L-ornithine]	177
Growth Factors	
Growth Factor I	154
Growth Factor II (γ-amino-butyric acid)	190

cells, or that they are simply normal products modified in crown gall cells. It is yet uncertain which of these mechanisms are operating in crown gall cells. Certainly the enhanced metabolic state of crown gall cells and the stage of growth would be reflected in various elevated isoenzyme patterns and synthesis. To probe blindly for any bacterial gene products in crown gall cells by current biochemical methods is nearly an insurmountable task, considering that the search will be for a few bacterial-specific products among several thousands of plant-gene products.

Some encouragement in this area of study comes from the correlation between the type of guanidino amino acids in crown gall tissues and the strain of *A. tumefaciens*. Comparative studies have shown that the arginase activity in crown gall tissues was lower than that in normal tissues and that arginine was incorporated into several guanidine derivatives in tumor tissues but not in normal tissues (176, 183). These derivatives were subsequently identified as octopine [N^2—(D-1-carboxyethyl)-L-arginine] (176); nopaline [N^2-(1,3-dicarboxypropyl)-L-arginine] (93); lysopine [N^2-(D-1-carboxyethyl)-L-lysine] (27); and octopinic acid [N^2-(D-1-carboxyethyl)-L-ornithine (177). The type of arginine derivative synthesized in the tumor seems to be specified by the strain of bacteria used (92, 191). Thus, strains A6 and B6 elicit the biosynthesis of octopine whereas nopaline is detected in tumors induced by strains T37 and C-58. The biosynthetic response is independent of the host plant (191, 216) so that the bacteria are apparently specifically modulating the biosynthesis of arginine in the host plant. This idea is supported by another correlation: *A. tumefaciens* strains inducing the biosynthesis of a particular arginine derivative in crown gall tissue also specifically catabolize that derivative in culture media (191). For example, those strains that metabolize octopine only induce tumors containing octopine (153, 192) and a similar correlation holds for nopaline utilizers (153). There are a few unusual strains that seem to utilize both octopine and nopaline (153). Nevertheless, these correlative results strongly suggest that the biosynthesis of these arginine derivatives is directed by bacterial genes transferred during tumorigenesis, a hypothesis proposed earlier by Morel (183) and by Petit et al (191). Such a hypothesis must be taken with caution since several alternatives could explain the results. The main alternative hypothesis is that these arginine derivatives are simply a shunt in the biosynthetic arginine pathway created during tumorigenesis where an overall modulation of the derepression and repression of genes is altered. This argument is apparently supported by the finding of low but detectable concentrations of lysopine (222) and octopine (113, 263) in *normal* tissues. However, unlike crown gall tissues, normal tissues were unable to synthesize lysopine in vitro (148). Furthermore, Bomhoff (32) and Schilperoort & Bomhoff (216) could not confirm these reports of octopine in normal tissues. Bomhoff (32) clearly showed that artifacts mimicking octopine and interfering compounds can be generated by the extraction procedures employed by the earlier workers (113, 263).

By far, the most significant correlative evidence on the connection between the utilization of arginine derivatives and tumorigenicity of the crown gall bacterium has come from the recent discovery that both are apparently specified by the large plasmid in *A. tumefaciens* (269). *A. tumefaciens* strains cured of the plasmids by subculturing at 36–37°C are no longer tumorigenic and also no longer are able to utilize the arginine derivative, the biosynthesis of which they normally induce in crown gall cells (32, 216, 251, 258).

These results strongly suggest that *A. tumefaciens* plasmid carries genetic information for the utilization of octopine, nopaline, and lysopine depending on the strain. The enzyme(s) necessary for the catabolism of these compounds is perhaps situated on the plasma membrane of *A. tumefaciens;* therefore, these enzymes are conferred by a plasmid gene(s). Lysopine oxidase, an inducible enzyme, has been linked to the membrane fraction of *A. tumefaciens* (115). Likewise, octopine oxidase is inducible and is found mainly in the membrane fraction (32). The oxidation of lysopine and octopine seems to be linked to the cytochrome oxidase system (32, 115). Hirata & Fukui (106) clearly showed that in *A. tumefaciens* there are at least five cytochrome types: *a, b,* $c_{552}, c_{556,}$ and *o*. The *o* type cytochrome is membrane bound and serves as the terminal oxidase. Both cytochrome c_{552} and *o* apparently function in the formation of 3-ketosugars. Whether or not cytochrome *o* participates in the oxidation of lysopine and octopine is unknown.

All of the above evidence strongly implicates the plasmid as the carrier of genetic information for tumorigenicity, octopine or nopaline biosynthesis in planta, and the means necessary for their utilization in vitro. It is still necessary to demonstrate that these rare amino acids can be synthesized in vitro by the use of plasmid DNA as the template. If this can be shown, it will be strong proof that plasmid genes are operating in crown gall cells. Less than one third of the plasmid is necessary for tumorigenesis since nontumorigenic strains carry a plasmid about two thirds the size of that in tumorigenic strains (205; B. C. Lin, P. A. Okubara, J. C. Dutra, and C. I. Kado, in preparation).

CONCLUDING REMARKS

A number of recent experiments were discussed in this review to help elucidate the molecular basis of crown gall tumorigenesis. Only in the past few years have some of the key questions been partially, if not wholly, answered.

Although the tumorigenic substance elaborated by *A. tumefaciens* remains unidentified, the genetic component conferring the tumorigenic and virulence properties on the crown gall bacteria has been characterized: it seems to rest in large plasmid DNA molecules of the bacterium. The preliminary evidence dictates that this plasmid is in itself the tumorigenic substance, but the quantitative experiments have failed to demonstate unequivocally the presence of plasmid DNA in crown gall cells. Of course, the sensitivities of such experiments may be insufficient to detect minute amounts of plasmid DNA, particularly when recent experiments have shown that less than half of the plasmid DNA sequences are needed for tumorigenicity. Alternatively, it may be an equally likely function of the plasmid simply to confer the tumorigenic properties on the bacterium, particularly when there is evidence for the alteration of surface components of the bacteria when they carry plasmid. Such surface alterations permit bacteria to attach specifically to target sites in wounded host tissues and subsequently permit the transfer of a still unidentified tumorigenic substance other than the plasmid. Because the plasmid DNA can code for roughly 50–100 or so genes, many of these genes may simply be necessary for the invasive property of the crown gall bacterium (e.g. specific attachment proteins, enzymes for growth and survival in the host). Probably only a few genes are necessary for the transformation of the host cell analogous to the oncogenic virus system where at least one gene can elicit transformation of animal cells (42, 124, 170, 187, 245). Perhaps the recent discovery of a tumorigenic RNA "virus" inducible in wounded tobacco plants may shed a different perspective on the current plasmid hypothesis.

In the main, the most pertinent questions to be answered in the near future are: (*a*) what is the limiting size of the plasmid that confers the tumorigenic property on *A. tumefaciens?* (*b*) is the plasmid itself or a specific plasmid DNA segment being introduced into host cells in order for tumorigenesis to take place? (*c*) what are the gene products of the plasmid that are necessary for tumorigenic induction? The answers to these questions await further experiments, most likely by the use of several newly devised techniques such as DNA heteroduplex analysis and the use of plasmids and phages of *E. coli* as molecular cloning vehicles for *A. tumefaciens* plasmid segments cleaved specifically by the variety of restriction endonucleases and joined by ligation enzymes now available. The results of such experiments will prove most interesting, the outcome of which may be relevant to many fields of biology.

ACKNOWLEDGMENTS

The very capable technical and library assistance of Patricia A. Okubara is appreciated. Special thanks go to Robert A. Langley, Karl A. Drlica, John M. Gardner, Bor-chian Lin, Robert S. Kupor, and Charles E. Rogler for the many stimulating discussions and experimental assistance, referred to in this review. This work was supported by grant No. CA-11526, awarded by the National Cancer Institute, Department of Health, Education, and Welfare.

Literature Cited

1. Aaron-da Cunha, M. I. 1970. Recherches sur la permanence et la transmission des propriétés tumorales des cellules de crown-gall. *Ann. Sci. Nat. Bot. Biol. Veg. Ser. 12.* 11:333–408
2. Aaron-da Cunha, M. I. 1971. Le "Crown gall", maladie cancéreuse de végétaux. *Bol. Soc. Port. Ciênc. Nat.* 13:51–72
2a. Aaron-da Cunha, M. I., Kurkdjian, A., Le Goff, L. 1975. Nature tumorale d'une hyperplasie obtenue expérimentalement. *C. R. Soc. Biol.* 169: 755–59

3. Albersheim, P., Anderson-Prouty, A. J. 1975. Carbohydrates, proteins, cell surfaces, and the biochemistry of pathogenesis. *Ann. Rev. Plant Physiol.* 26:31–52
3a. Anand, V. K., Heberlein, G. T. 1975. Effect of rifampin application time on crown-gall tumor induction by *Agrobacterium tumefaciens. Can. J. Bot.* 53:258–88
4. Anker, P., Stroun, M., Gahan, P., Rossier, A., Greppin, H. 1971. Natural release of bacterial nucleic acids into plant cells and crown gall in-

duction. In *Informative Molecules in Biological Systems,* ed. L. Ledoux, 193–200. Amsterdam: North-Holland. 466 pp.

5. Appler, H. 1951. Uber die Tumorbildung durch *Pseudomonas tumefaciens* und durch Heteroauxin an *Helianthus annuus. Biol. Zentralbl.* 70:452–69

6. Bagni, N., Serafini Fracassini, D., Corsini, E. 1972. Tumors of *Scorzonera hispanica:* their content in polyamines. *Z. Pflanzenphysiol.* 67:19–12

7. Beardsley, R. E. 1960. Lysogenicity in *Agrobacterium tumefaciens. J. Bacteriol.* 80:180–87

8. Beardsley, R. E. 1972. The inception phase in the crown gall disease. *Prog. Exp. Tumor Res.* 15:1–75

9. Beardsley, R. E., Stonier, T., Lipetz, J., Parsons, C. L. 1966. Mechanisms of tumor induction in crown gall. I. Production and pathogenicity of spheroplasts of *Agrobacterium tumefaciens. Cancer Res.* 26:1606–10

10. Beaud, R., Beardsley, R. E., Kurkdjian, A. 1972. Bacteriophages of *Agrobacterium tumefaciens,* III. Effect of attenuation of virulence on lysogeny and phage production. *Can. J. Microbiol.* 18:1561–67

11. Beaud, G., Manigault, P., Stoll, Ch. 1963. Observations sur des tumeurs végétales incomplétes. *Phytopathol. Z.* 47:25–37

12. Bechhold, H., Smith, L. 1927. Tumefaciens-plastin. *Z. Krebsforsch.* 27:97–104

12a. Bednar, T. W., Linsmaier-Bednar, E. M. 1971. Induction of cytokinin-independent tobacco tissues by substituted fluorenes. *Proc. Natl. Acad. Sci. USA* 68:1178–79

13. Beiderbeck, R. 1970. Untersuchungen an Crown-Gall. I. Quantitative Bestimmung des Infektionserfolgs verschieden vorbehandelter Bakterien mit dem Igel-Test. *Z. Naturforsch.* 25b:407–11

14. Beiderbeck, R. 1970. Untersuchungen an Crown-Gall. III. Der Einfluss von Rifampicin auf die Tumorinduktion durch *Agrobacterium tumefaciens. Z. Naturforsch.* 25b:735–38

15. Beiderbeck, R. 1970. Untersuchungen an Crown-Gall. IV. Rifampicin und ein resistenter Klon von *Agrobacterium tumefaciens* bei der Tumorinduktion. *Z. Naturforsch.* 25b:1458–60

16. Beiderbeck, R. 1972. α-amanitin hemmt die Tumorinduktion durch *Agrobacterium tumefaciens. Z. Naturforsch.* 27b:1393–94

17. Beiderbeck, R. 1973. Bakterienwand und Tumorinduktion durch *Agrobacterium tumefaciens. Z. Naturforsch.* 28c:198–201

18. Beiderbeck, R., Heberlein, G. T., Lippincott, J. A. 1973. On the question of crown-gall tumor initiation by DNA of bacteriophage PS8. *J. Virol.* 11:345–50

19. Beljanski, M., Aaron-da Cunha, M. I., Beljanski, M. S., Manigault, P., Bourgarel, P. 1974. Isolation of the tumor-inducing RNA from oncogenic and nononcogenic *Agrobacterium tumefaciens. Proc. Natl. Acad. Sci. USA* 71:1585–89

20. Beljanski, M., Manigault, P., Beljanski, M. S., Aaron-da Cunha, M. I. 1975. Genetic transformation of *Agrobacterium tumefaciens* B6 by RNA and nature of the tumor-inducing principle. *Proc. 1st Int. Congr. Int. Assoc. Microbiol. Soc.,* ed. T. Hasegawa, 1:131–44. 699 pp.

21. Beltrá, R., Rodriguez de Lecea, J. 1971. Aseptic induction of the crown-gall tumors by the nucleic acid fraction from *Agrobacterium tumefaciens. Phytopathol. Z.* 70:351–58

22. Beltrá, R., Rodriguez de Lecea, J., de la Rosa, C. 1972. A comparative study of chemical fractions isolated from *Agrobacterium tumefaciens* and from its stable L-form. *J. Gen. Microbiol.* 73:185–88

23. Bender, E., Brucker, W. 1958. Studien zur zellfreien Tumorerzeugung an Pflanzen II. *Z. Bot.* 46:121–24

24. Bender, E., Brucker, W. 1959. Weitere Versuche zur zellfreien tumorerzeugung an Pflanzen III. *Z. Bot.* 47:258–71

25. Bezdek, M., Tkadlecek, L., Kubickova, D. 1974. The presence of defective prophages in *Agrobacterium tumefaciens* strains 806 and B6S. *Neoplasma* 21:551–54

26. Bieber, J., Sarfert, E. 1968. Zur Frage der Tumorbildung durch Desoxyribonukleinsäure aus *Agrobacterium tumefaciens* (Smith and Townsend) Conn. *Phytopathol. Z.* 62:323–26

27. Biemann, K., Lioret, C., Asselineau, J., Lederer, E., Polonsky, J. 1960. On the structure of lysopine, a new amino acid isolated from crown gall tissue. *Biochim. Biophys. Acta* 40:369–70

28. Björndal, H., Erbing, C., Lindberg, B., Fohraeus, G., Ljunggren, H. 1971. Studies on an extracellular polysaccharide from *Rhizobium meliloti. Acta Chem. Scand.* 25:1281–86

29. Bogers, R. J. 1972. On the interaction of *Agrobacterium tumefaciens* with cells of *Kalanchoë diagremontiana*. *Proc. 3rd Int. Cong. Plant Pathog. Bact.*, ed. H. P. Maas Geesteranus, 239–50. Wageningen: Cent. Agric. Publ. Doc. 365 pp.

30. Boiron, M., Thomas, M., Chenaille, Ph. 1965. A biological property of deoxyribonucleic acid extracted from bovine papilloma virus? *Virology* 26: 150–53

31. Boivin, A., Marbe, M., Mesrobeanu, L., Juster, P. 1935. Sur l'existence, dans le *Bacillus tumefaciens,* d'une endotoxine capable de provoquer la formation de tumeurs chez les végétaux. *C. R. Acad. Sci.* 201:984–86

32. Bomhoff, G. H. 1974. Studies on crown gall-a plant tumor. Investigations on protein composition and on the use of guanidine compounds as a marker for transformed cells. PhD thesis. Rijksuniversiteit te Leiden, the Netherlands. 142 pp.

33. Borriss, G. H. 1972. Induktion von Crown-gall-tumoren durch Nuclein-säure-Präparationen aus *Agtobacterium tumefaceins* und seine Bakteriophagen. *Biol. Rundsch.* 10:121–22

34. Braun, A. C. 1947. Thermal studies on the factors responsible for tumor initiation in crown gall. *Am. J. Bot.* 34:234–40

35. Braun, A. C. 1953. Bacterial and host factors concerned in determining tumor morphology in crown gall. *Bot. Gaz.* 114:363–71

36. Braun, A. C. 1962. Tumor inception and development in the crown gall disease. *Ann. Rev. Plant Physiol.* 13: 533–58

37. Braun, A. C., Mandle, R. J. 1958. Studies on the inactivation of the tumor-inducing principle in crown gall. *Growth* 12:255–69

38. Braun, A. C., Stonier, T. 1958. Morphology and physiology of plant tumors. *Protoplasmatologia* 10:1–93

39. Braun, A. C., White, P. R. 1943. Bacteriological sterility of tissues derived from secondary crown-gall tumors. *Phytopathology* 33:85–100

40. Braun, A. C., Wood, H. N. 1961. The plant tumor problem. *Adv. Cancer Res.* 6:81–109

41. Braun, A. C., Wood, H. N. 1966. On the inhibition of tumor inception in the crown-gall disease with the use of ribonuclease A. *Proc. Natl. Acad. Sci. USA* 56:1417–22

42. Brugge, J. S., Butel, J. S. 1975. Role of

43. Brunner, M., Pootjes, C. F. 1969. Bacteriophage release in a lysogenic strain of *Agrobacterium tumefaciens.* *J. Virol.* 3:181–86

44. Burnett, C. 1973. Survey of isoenzymes induced by infection with *Agrobacterium tumefaciens* in Pinto bean leaves and sunflower stems. MSc thesis. Univ. Maryland, College Park. 104 pp.

45. Cabezas de Herrera, E., Rubio-Huertos, M. 1967. Algunas caracteristicas de las formas L fijas del *Agrobacterium tumefaciens* obtenidas por medio de radiacion ultravioleta. *Microbiol. Esp.* 20:131–45

46. Cagle, G. D. 1975. Fine structure and distribution of extracellular polymer surrounding selected aerobic bacteria. *Can. J. Microbiol.* 21:395–408

47. Cantarow, A., Williams, T. L., Goddard, J. W. 1966. Tumors in the rat after injection of neoplastic and preneoplastic nucleic acids. *Cancer Res.* 26:652–58

48. Cavara, F. 1897. Intorno alla cz\u00adiologia di acune malattie di piante coltivate. *Stn. Sper. Agrar. Ital. Modena* 30: 482–509

49. Chadha, K. C., Srivastava, B. I. S. 1971. Evidence for the presence of bacteria-specific proteins in sterile crown gall tumor tissue. *Plant Physiol.* 48:125–29

50. Chang, C.-C. 1972. Isolation of a peptide from plant tissues. *Bot. Bull. Acad. Sin.* 13:121–24

51. Chang, C.-C., Lin, B.-Y. 1973. Accumulation of a peptide in crown-gall tumors. *Bot. Bull. Acad. Sin.* 14: 174–79

52. Charudattan, R., Hubbell, D. H. 1973. The presence and possible significance of cross-reactive antigens in *Rhizobium*-legume associations. *Antonie van Leeuwenhoek J. Microbiol. Serol.* 39:619–27

53. Chilton, M.-D., Currier, T. C., Farrand, S. K., Bendich, A. J., Gordon, M. P., Nester, E. W. 1974. *Agrobacterium tumefaciens* DNA and PS8 bacteriophage DNA not detected in crown gall tumors. *Proc. Natl. Acad. Sci. USA* 71:3672–76

54. Chilton, M.-D., Farrand, S. K., Eden, F., Currier, T., Bendich, A. J., Gordon, M. P., Nester, E. W. 1975. Is there foreign DNA in crown gall tumour DNA? *2nd John Innes Symp.*

Mod. Inf. Content Plant Cells, ed. R. Markham, D. R. Davies, D. A. Hopwood, R. W. Horne, 297–311. Amsterdam: North Holland. 350 pp.

55. Cole, M. A., Elkan, G. H. 1973. Transmissible resistance to penicillin G, neomycin, and chloramphenicol in *Rhizobium japonicum. Antimicrob. Agents Chemother.* 4:248–53

56. Cornelis, P., Dupont-de Patoul, M. C. 1975. Characterization of the tRNA and aminoacyl-tRNA synthetases of healthy and tumorous callus tissues from *Nicotiana tabacum. Phytochemistry* 14:397–401

57. Crawford, L., Dulbecco, R., Fried, M., Montagnier, L., Stoker, M. 1964. Cell transformation by different forms of polyoma virus DNA. *Proc. Natl. Acad. Sci. USA* 52:148–52

58. Crawford, L. V., Black, P. H. 1964. The nucleic acid of simian virus 40. *Virology* 24:388–92

59. Curtis, C. R. 1971. Disc electrophoretic comparisons of proteins and peroxidases from *Phaseolus vulgaris* leaves infected with *Agrobacterium tumefaciens. Can. J. Bot.* 49:333–37

60. Cutting, J. A., Schulman, H. M. 1969. The site of heme synthesis in soybean root nodules. *Biochim. Biophys. Acta* 192:486–93

61. Cutting, J. A., Schulman, H. M. 1971. The biogenesis of leghemoglobin. The determinant in the Rhizobium-legume symbiosis for leghemoglobin specificity. *Biochim. Biophys. Acta* 229:58–62

62. Datta, N., Hedges, R. W. 1972. Host ranges of R factors. *J. Gen. Microbiol.* 70:453–60

63. Davies, J., Davis, B. D. 1968. Misreading of ribonucleic acid code words induced by aminoglycoside antibiotics. *J. Biol. Chem.* 243:3312–16

64. Dazzo, F. B., Hubbell, D. H. 1975. Cross reactive antigens and lectin as determinants of symbiotic specificity in the Rhizobium-clover association. *Appl. Microbiol.* 30:1017–33

65. Dea, I. C. M., Morrison, A. 1975. Chemistry and interactions of seed galactomannans. *Adv. Carbohydr. Chem. Biochem.* 31:241–312

66. Deinema, M. H., Zevenhuizen, L. P. T. M. 1971. Formation of cellulose fibrils by Gram-negative bacteria and their role in bacterial flocculation. *Arch. Mikrobiol.* 78:42–57

67. De Ley, J., Gillis, M., Pootjes, C. F., Kersters, K., Tytgat, R., Van Braekel, M. 1972. Relationship among

temperate Agrobacterium phage genomes and coat proteins. *J. Gen. Virol.* 16:199–214

68. Delseny, M. 1975. DNA synthesis and bacterial contamination in plants, In *Genetic Manipulations with Plant Materials,* ed. L. Ledoux, p. 582. New York: Plenum

69. De Ropp, R. S. 1947. The growth-promoting and tumefacient factors of bacteria-free crown-gall tumor tissue. *Am. J. Bot.* 34:248–61

70. DeTorok, D., Cornesky, R. A. 1971. Antigenic and immunological determinants of oncogenesis in plants infected with *Agrobacterium tumefaciens. Colloq. Int. CNRS,* No. 193: 443–51

71. D'Herelle, F., Peyre, E. 1927. Contribution á l'étude des tumeurs expérimentales. *C. R. Acad. Sci.* 185:227–30

72. Dickman, S. R., Aroskar, J. P., Kropf, R. B. 1956. Activation and inhibition of beef pancreas ribonuclease. *Biochim. Biophys. Acta* 21:539–45

73. Dilworth, M. J. 1969. The plant as the genetic determinant of leghaemoglobin production in the legume root nodule. *Biochim. Biophys. Acta* 184: 432–41

74. Dons, J. J. M. 1975. Crown gall—a plant tumor investigations on the nuclear DNA content and on the presence of *Agrobacterium tumefaciens* DNA and phage PS8 DNA in crown gall cells. PhD thesis. Rijksuniversiteit te Leiden, the Netherlands. 124 pp.

75. Doy, C. H., Gresshoff, P. M., Rolfe, B. G. 1973. Biological and molecular evidence for the transgenosis of genes from bacteria to plant cells. *Proc. Natl. Acad. Sci. USA* 70:723–26

76. Doy, C. H., Gresshoff, P. M., Rolfe, B. 1972. Transgenosis of bacterial genes from *Escherichia coli* to cultures of haploid *Lycopersicon esculentum* and haploid *Arabidopsis thaliana* plant cells. In *The Biochemistry of Gene Expression in Higher Organisms,* ed. J. Pollack and J. Lee, 21–25. Artarmon: Aust. NZ Book Co. 656 pp.

77. Doy, C. H., Gresshoff, P. M., Rolfe, B. G. 1973. Time-course of phenotypic expression of *Escherichia coli* gene Z following transgenosis in haploid *Lycopersicon esculentum* cells. *Nature New Biol.* 244:90–91

78. Drlica, K. A., Gardner, J. M., Kado, C. I., Vijay, I. K., Troy, F. A. 1974. Cyclic adenosine 3':5'-monophosphate levels in normal and trans-

formed cells of higher plants. *Biochem. Biophys. Res. Commun.* 56:753 – 59

79. Drlica, K. A., Kado, C. I. 1974. Ouantitative estimation of *Agrobacterium tumefaciens* DNA in crown gall tumor cells. *Proc. Natl. Acad. Sci. USA* 71: 3677–81

80. Drlica, K. A., Kado, C. I. 1975. Crown gall tumors: are bacterial nucleic acids involved? *Bacteriol. Rev.* 39:186–96

81. Dudman, W. F., Heidelberger, M. 1969. Immunochemistry of newly found substituents of polysaccharides of *Rhizobium* species. *Science* 164: 954–55

82. Duggar, B. M., Riker, A. J. 1940. The influence of ultraviolet irradiation on the pathogenicity of *Phytomonas tumefaciens*. *Phytopathology* 30:6 (Abstr.)

83. Eden, F. C., Farrand, S. K., Powell, J. S., Bendich, A. J., Chilton, M-D, Nester, E. W., Gordon, M. P. 1974. Attempts to detect deoxyribonucleic acid from *Agrobacterium tumefaciens* and bacteriophage PS8 in crown gall tumors by complementary ribonucleic acid/deoxyribonucleic acid-filter hybridization. *J. Bacteriol.* 119:547–53

84. Engler, G., Holsters, M., Van Montagu, M., Schell, J., Hernalsteens, J. P., Schilperoort, R. 1975. Agrocin 84 sensitivity: a plasmid determined property in *Agrobacterium tumefaciens*. *Mol. Gen. Genet.* 138:345–49

85. Ewing, J., Cooper, D. 1974. Agrobacterium DNA transfer to carrots. *65th Ann. Rept. John Innes Inst. Norwich*, p. 90

86. Farrand, S. K., Eden, F. C., Chilton, M-D. 1975. Attempts to detect *Agrobacterium tumefaciens* and bacteriophage PS8 DNA in crown gall tumors by DNA·DNA-filter hybridization. *Biochim. Biophys. Acta* 390: 264–75

87. Gahan, P. B., Sheikh, K., Maggi, V., Stroun, M., Smith, A. R. W. 1971. Electrophoretic analysis of hydrolases from crown gall tissues. *FEBS Lett.* 13:53–55

88. Gardner, J. M., Kado, C. I. 1976. Studies on *Agrobacterium tumefaciens*. VI. DNA polymerases of crown-gall tumor and normal cells, and of the bacterium. *Physiol. Plant Pathol.* In press

89. Gautheret, R. J. 1959. *La Culture des Tissus Végétaux;* Techniques et Réalisations. Paris: Masson. 863 pp.

90. Gee, M. M., Sun, C. N., Dwyer, J. D. 1967. An electron microscope study of sunflower crown gall tumor. *Protoplasma* 64:195–200

91. Gibbins, A. M., Gregory, K. F. 1972. Relatedness among *Rhizobium* and *Agrobacterium* species determined by three methods of nucleic acid hybridization. *J. Bacteriol.* 111:129–41

92. Goldmann, A., Tempe, J., Morel, G. 1968. Quelques particularités de diverses souches d'*Agrobacterium tumefaciens*. *Seances Soc. Biol. C. R.* 162:630–31

93. Goldmann, A., Thomas, D. W., Morel, G. 1969. Sur la structure de la nopaline, métabolite anormal de certaines tumeurs de crown-gall. *C. R. Acad. Sci.* 268D:852–54

94. Goldmann-Ménagé, A. 1970. Recherches sur le métabolisme azoté des tissus de crown-gall cultivés in vitro. *Ann. Sci. Nat. Bot.* 11:233–310

95. Häggman, J., Kupila-Ahvenniemi, S. 1973. Cytokinin activity in normal, wounded and crown gall-infected internodes of *Pisum sativum. Aquilo Ser. Bot.* 12:16–20

96. Hahn, F. E., Ciak, J. 1971. Elimination of bacterial episomes by DNA-complexing compounds. *Ann. NY Acad. Sci.* 182:295–304

97. Hamilton, R. H., Chopan, M. N. 1975. Transfer of the tumor inducing factor in *Agrobacterium tumefaciens*. *Biochem. Biophys. Res. Commun.* 63: 349–54

98. Hamilton, R. H., Fall, M. Z. 1971. The loss of tumor-initiating ability in *Agrobacterium tumefaciens* by incubation at high temperature. *Experientia* 27:229–30

99. Heberlein, G. T., De Ley, J., Tijtgat, R. 1967. Deoxyribonucleic acid homology and taxonomy of *Agrobacterium, Rhizobium*, and *Chromobacterium. J. Bacteriol.* 94:116–24

100. Heberlein, G. T., Lippincott, J. A. 1965. Photoreversible ultraviolet enhancement of infectivity in *Agrobacterium tumefaciens. J. Bacteriol.* 89: 1511–14

101. Heberlein, G. T., Lippincott, J. A. 1967. Enhancement of *Agrobacterium tumefaciens* infectivity by mitomycin C. *J. Bacteriol.* 94:1470–74

102. Heberlein, G. T., Lippincott, J. A. 1967. Ultraviolet-induced changes in the infectivity of *Agrobacterium tumefaciens. J. Bacteriol.* 93:1246–53

103. Heyn, R. F., Schilperoort, R. A. 1973.

The use of protoplasts to follow the fate of *Agrobacterium tumefaciens* DNA on incubation with tobacco cells. Protoplastes et fusion de Cellules Somatiques. *Coll. Int. CNRA,* No. 212:385–95

104. Hildebrand, D. C., Thompson, J. P., Schroth, M. N. 1966. Bacterial enhancement of self-limiting outgrowth formation on *Datura. Phytopathology* 56:365–66

105. Hill, J. B. 1928. The migration of *Bacterium tumefaciens* in the tissue of tomato plants. *Phytopathology* 18: 553–64

106. Hirata, H., Fukui, S. 1968. Cytochrome C_{552} in *Agrobacterium tumefaciens. J. Biochem.* 63:780–88

107. Hirota, Y. 1960. The effect of acridine dyes on mating type factors in *Escherichia coli. Proc. Natl. Acad. Sci. USA* 46:57–64

108. Hohl, H. R. 1961. Über die submikroskopische Struktur hyperplastischer Gewebe von *Datura stramonium* L. *Phytopathol Z.* 40:317–56

109. Horgen, P. A., Key, J. L. 1973. The DNA-directed RNA polymerases of soybean. *Biochim. Biophys. Acta* 294: 227–35

110. Israilsky, W. P. 1926. Bakteriophagie und Pflanzenkrebs I. Mitteilung. *Zentralbl. Bakteriol. Parasitenkd. Infektionskr. Abt. 2* 67:236–42

111. Johnson, C. B., Grierson, D. 1974. The uptake and expression of DNA by plants. *Curr. Adv. Plant Sci.* 9:1–12

112. Johnson, C. B., Grierson, D., Smith, H. 1973. Expression of λplac5 DNA in cultured cells of a higher plant. *Nature New Biol.* 244:105–7

113. Johnson, R., Guderian, R. H., Eden, F., Chilton, M.-D., Gordon, M. P., Nester, E. W. 1974. Detection and quantitation of octopine in normal plant tissue and in crown gall tumors. *Proc. Natl. Acad. Sci. USA* 71:536–39

114. Johnson, T. B., Ross, C., Baker, R. 1970. Similarity of cytokinin contents and electrophoretic banding patterns of tomato crown gall tumor and stem RNA's. *Biochim. Biophys. Acta* 199:521–24

115. Jubier, M.-F. 1972. Degradation of lysopine by an inducible membrane-bound oxidase in *Agrobacterium tumefaciens. FEBS Lett.* 28:129–32

116. Kado, C. I. 1975. Activation of plasmids in *Agrobacterium tumefaciens* and its relationship to tumorigenicity. *Ann. Proc. Am. Phytopathol. Soc.* 2:58

117. Kado, C. I. 1975. Studies on *Agrobacterium tumefaciens.* III. A concept on the role of *Agrobacterium tumefaciens* DNA in plant tumorigenesis, ed. T. Hasegawa, 1:100–30. *Proc. First Int. Congr. Int. Assoc. Microbiol. Soc. Tokyo.* 699 pp.

118. Kado, C. I., Heskett, M. G., Langley, R. A. 1972. Studies on *Agrobacterium tumefaciens:* characterization of strains 1D135 and B6, and analysis of the bacterial chromosome, transfer RNA and ribosomes for tumor-inducing ability. *Physiol. Plant Pathol.* 2:47–57

119. Kado, C. I., Lurquin, P. F. 1975. Studies on *Agrobacterium tumefaciens.* IV. Nonreplication of the bacterial DNA in mung bean (*Phaseolus aureus*). *Biochem. Biophys. Res. Commun.* 64:175–83

120. Kado, C. I., Lurquin, P. F. 1976. Studies on *Agrobacterium tumefaciens.* V. Fate of exogenously added bacterial DNA in *Nicotiana tabacum. Physiol. Plant Pathol.* 8:73–82

121. Kerr, A. 1969. Transfer of virulence between isolates of *Agrobacterium. Nature* 223:1175–76

122. Kerr, A. 1971. Acquisition of virulence by non-pathogenic isolates of *Agrobacterium radiobacter. Physiol. Plant Pathol.* 1:241–46

123. Kerr, A., Htay, K. 1974. Biological control of crown gall through bacteriocin production. *Physiol. Plant Pathol.* 4:37–44

124. Kimura, G., Itagaki, A. 1975. Initiation and maintenance of cell transformation by simian virus 40: a viral genetic property. *Proc. Natl. Acad. Sci. USA* 72:673–77

125. Klein, R. M., Beardsley, R. E. 1957. On the role of omega bacteriophage in formation of crown-gall tumor cells. *Am. Nat.* 91:330–31

126. Klein, R. M., Braun, A. C. 1960. On the presumed sterile induction of plant tumors. *Science* 131:1612–13

127. Klein, R. M., Knupp, J. L. Jr. 1957. Sterile induction of crown-gall tumors on carrot tissues in vitro. *Proc. Natl. Acad. Sci. USA* 43:199–203

128. Klein, R. M., Link, G. K. K. 1955. The etiology of crown-gall. *Q. Rev. Biol.* 30:207–77

129. Kleinhofs, A., Eden, F. C., Chilton, M.-D., Bendich, A. J. 1975. On the question of the integration of exogenous bacterial DNA into plant DNA. *Proc. Natl. Acad. Sci. USA* 72:2748–52

130. Kleinschmidt, A. K., Lang, D., Jacherts, D., Zahn, R. K. 1962. Darstellung und Längenmessungen des Gesamten desoxyribonuclein-säure-inhaltes von T_2-Bakteriophagen. *Biochim. Biophys. Acta* 61: 857–64

131. Kovoor, A. 1967. Sur la transformation de tissus normaux de Scorsonère provoquée in vitro par l'acide désoxyribonucléique d'*Agrobacterium tumefaciens. C. R. Acad. Sci.* 265D: 1623–26

132. Kupila, S., Stern, H. 1961. DNA content of broad bean (Vicia faba) internodes in connection with tumor induction by *Agrobacterium tumefaciens. Plant Physiol.* 36:216–19

133. Kurkdjian, A. 1968. Apparition de phages au cours de l'induction de tumeurs du crown gall. *J. Microsc.* 7:1039–44

134. Kurkdjian, A. 1969. Phages and tumoral induction in plants. *11th Int. Bot. Cong.*, Seattle, Washington, p. 119 (Abstr.)

135. Kurkdjian, A. 1970. Observations sur la présence des phages dans les plaies infectées par différentes souches d'*Agrobacterium tumefaciens* (Smith et Town) Conn. *Ann. Inst. Pasteur Paris* 118:690–96

136. Kurkdjian, A., Beardsley, R., Milani, V. 1974. Nature d'une action de la rifampicine sur l'expression de la virulence d'une population de Bactéries *Agrobacterium tumefaciens. C. R. Acad. Sci.* 279D:655–58

137. Ledoux, L. 1968. *Absorption des Acides Desoxyribonucléiques par les Tissus Vivants.* Liège: Vaillant-Carmanne. 88 pp.

138. Ledoux, L. 1975. Fate of exogenous DNA in plants. *Genetic Manipulations with Plant Material*, ed. L. Ledoux. New York: Plenum

139. Ledoux, L., Huart, R. 1968. Integration and replication of DNA of *M. lysodeikticus* in DNA of germinating barley. *Nature* 218:1256–59

140. Ledoux, L., Huart, R. 1969. Fate of exogenous bacterial deoxyribonucleic acids in barley seedlings. *J. Mol. Biol.* 43:243–62

141. Ledoux, L., Huart, R. 1972. Fate of exogenous DNA in plants. In *Uptake of Informative Molecules by Living Cells*, 254–76. Amsterdam: North-Holland. 416 pp.

142. Ledoux, L., Huart, R. 1975. Importance of DNA size for integration in plant materials. *Arch. Int. Physiol.*

Biochim. 83:196–97

143. Ledoux, L., Huart, R. 1975. Integration and replication of bacterial DNA in barley root cells. *Arch. Int. Physiol. Biochim.* 83:194–95

144. Ledoux, L., Huart, R., Jacobs, M. 1972. Fate and biological effects of exogenous DNA in *Arabidopsis thaliana.* In *The Way Ahead in Plant Breeding*, ed. F. G. Lupton, G. Jenkins, R. Johnson, 165–85. Darking: Adlard & Son. 269 pp.

145. Ledoux, L., Huart, R., Jacobs, M. 1974. DNA-mediated genetic corrections of thiamineless *Arabidopsis thaliana. Nature* 249:17–21

146. Ledoux, L., Huart, R., Mergeay, M., Charles, P., Jacobs, M. 1975. DNA mediated genetic correction of thiamineless *Arabidopsis thaliana.* In *Genetic Manipulations with Plant Materials*, ed. L. Ledoux, New York: Plenum

147. Leff, J., Beardsley, R. E. 1970. Action tumorigène de l'acide nucléique d'un bactériophage présent dans les cultures de tissu tumoral de Tournesol (*Helianthus annus*). *C. R. Acad. Sci.* 270D:2505–7

148. Lejeune, B. 1967. Étude de la synthèse de lysopine in vitro par des extraits de tissu de crown-gall. *C. R. Acad. Sci.* 265D:1753–55

149. Levine, M., Chargaff, E. 1937. The response of plants to chemical fractions of *Bacterium tumefaciens. Am. J. Bot.* 24:461–72

150. Lipetz, J. 1962. Calcium and the control of lignification in tissue cultures. *Am. J. Bot.* 49:460–64

151. Lipetz, J. 1970. The fine structure of plant tumors. I. Comparison of crown gall and hyperplastic cells. *Protoplasma* 70:207–16

152. Lippincott, B. B., Lippincott, J. A. 1969. Bacterial attachment to a specific wound site as an essential stage in tumor initiation by *Agrobacterium tumefaciens. J. Bacteriol.* 97:620–28

153. Lippincott, J. A., Beiderbeck, R., Lippincott, B. B. 1973. Utilization of octopine and nopaline by *Agrobacterium J. Bacteriol.* 116:378–83

154. Lippincott, J. A., Lippincott, B. B., El Khalifa, M. D. 1968. Evidence for a tumor-associated factor active in the promotion of crown-gall tumor growth on primary pinto bean leaves. *Physiol. Plant.* 21:731–41

155. Lippincott, J. A., Lippincott, B. B. 1975. The genus *Agrobacterium* and plant tumorigenesis. *Ann. Rev. Mi-*

crobiol. 29:377–405

156. Loening, U. E., Butcher, D. N., Schuch, W., Sartirana, M. L. 1975. Experiments on the homology between DNA from *Agrobacterium tumefaciens* and from crown gall tumour cells. See Ref. 54, pp. 269–80

156a. Lurquin, P., Mergeay, M., Van Der Parren, J. 1972. Banding of depolymerized DNA's in CsCl density gradients studied by computer aided simulations. See Ref. 141, pp. 47–50

157. Manasse, R. J. 1974. Physiological comparisons of transformed (crown gall) and nontransformed *Vinca rosea* L. cells. *In Vitro* 9:434–40

158. Manasse, R. J., Staples, R. C., Granados, R. R., Barnes, E. G. 1972. Morphological, biological, and physical properties of *Agrobacterium tumefaciens* bacteriophages. *Virology* 47:375–84

159. Manigault, P. 1970. Intervention dans la plaie d'inoculation de bactéries appartenant a différentes souches d'*Agrobacterium tumefaciens* (Smith et Town) Conn. *Ann. Inst. Pasteur Paris* 119:347–59

160. Manigault, P. 1975. Les acides nucleiques au cours de la transformation tumorale chez les vegetaux. *C. R. Soc. Biol.* 169:766–72

161. Manigault, P., Beaud, G. 1967. Expression de l'efficacité-de la bactérie *Agrobacterium tumefaciens* (Smith et Town) Conn. dans l'induction tumorale [*Datura stramonium* L]. *Ann. Inst. Pasteur Paris* 112:445–57

162. Manigault, P., Camandon, A., Slizewicz, P. 1956. Préparation d'un "Principe Inducteur" de la tumeur du *Pelargonium. Ann. Inst. Pasteur Paris* 91:114–17

163. Manigault, P., Kurkdjian, A. 1970. Atténuation de la virulence et changement de forme de la bactérie du crown-gall. *Agrobacterium tumefaciens* (Smith et Town.) Conn. *Zentralbl. Bakteriol., Parasitenkd. Infektionskr. Hyg. Abt. 2.* 124:733–38

164. Manigault, P., Stoll, Ch. 1958. Induction stérile d'une tumeur végétale (crown-gall). *Ann. Inst. Pasteur Paris* 95:793–95

165. Manigault, P., Stoll, Ch. 1958. The role of nucleic acids in crown-gall tumor induction. *Experientia* 14:409–10

166. Manigault, P., Stoll, Ch. 1960. Induction et croissance de Tumeurs végétales exemptes de Bactéries. *Phytopathol. Z.* 38:1–12

167. Manocha, M. S. 1970. Fine structure of sunflower crown gall tissue. *Can. J. Bot.* 48:1455–58

168. Manocha, M. S. 1970. Induction of phage-like particles in *Agrobacterium* by wounded host tissue. *Can. J. Microbiol.* 17:819–20

169. Margrou, J. 1926. Le *Bacterium tumefaciens* dans les tissus du cancer des plantes. *C. R. Acad. Sci.* 183:804–6

170. Martin, R. G., Chou, J. Y. 1975. Simian virus 40 functions required for the establishment and maintenance of malignant transformation. *J. Virol.* 15:599–612

170a. Matsushita, T., Davis, F. F. 1971. Studies on pseudouridylic acid synthetase from various sources. *Biochim. Biophys. Acta* 238:165–73

171. Matthysse, A. G., Rich, K., Kontak, C. 1975. RNA synthesis in crown gall tumor cells. *Ann. Meet. Am. Soc. Plant Physiol., Corvallis, Oregon,* p. 24 (Abstr.)

172. McCarthy, B. J., Bolton, E. T. 1963. An approach to the measurement of genetic relatedness among organisms. *Proc. Natl. Acad. Sci. USA* 50:156–64

173. Meins, F. Jr. 1973. Evidence for the presence of a readily transmissible oncogenic principle in crown gall teratoma cells of tobacco. *Differentiation* 1:21–25

174. Meins, F. Jr. 1974. Mechanism underlying tumor transformation and tumor reversal in crown-gall, a neoplastic disease of higher plants. In *Developmental Aspects of Carcinogenesis and Immunity,* ed. E. J. King, 23–39. New York: Academic. 238 pp.

175. Meins, F. Jr. 1974. Mechanisms underlying the persistence of tumour autonomy in crown-gall disease. *Tissue Culture and Plant Science,* ed. H. E. Street, 233–64. New York: Academic. 502 pp.

176. Ménage, A., Morel, G. 1964. Sur la présence d'octopine dans les tissus de crown-gall. *C. R. Acad. Sci.* 259D: 4795–96

177. Ménage, A., Morel, G. 1965. Sur la présence d'un acide aminé nonveau dans les tissus de crown-gall. *C. R. Acad. Sci.* 261D:2001–2

178. Méndez, J., Brown, S. A. 1971. Changes in the phenolic metabolism of tomato plants infected by *Agrobacterium tumefaciens. Can. J. Bot.* 49:2101–5

179. Merlo, D. J., Kemp, J. D. 1975. Attempts to detect *Agrobacterium tumefaciens* DNA sequences in tobacco

and sunflower crown gall. *Ann. Meet. Am. Soc. Plant Physiol., Corvallis, Oregon*, p. 24 (Abstr.)

180. Mesrobeanu, L., Popescu, A., Kupferberg, S. 1974. Aspects of involvement of plant growth substances in crown gall tumors. *Rev. Roum. Biol. Ser. Bot.* 19:195–204

180a. Mesrobeanu, L., Popescu, A., Movileanu, D. 1971. Tumor-inducing action of thermolabile endotoxin from *Agrobacterium tumefaciens* on *Datura stramonium* plants. *Rev. Roum. Biol. Ser. Bot.* 16:445–52

181. Miller, C. O. 1975. Cell-division factors from *Vinca rosea* L. crown gall tumor tissue. *Proc. Natl. Acad. Sci. USA* 72:1883–86

182. Milo, G. E., Srivastava, B. I. S. 1969. RNA-DNA hybridization studies with the crown gall bacteria and the tobacco tumor tissue. *Biochem. Biophys. Res. Commun.* 34:196–99

183. Morel, G. 1971. Déviations du métabolisme azoté des tissus de crowngall. *Colloq. Int. CNRS* No. 193: 463–71

184. Nevins, M. P., Grant, D. W., Baker, R. R. 1970. Ribonuclease-induced alterations in *Agrobacterium tumefaciens*. *Phytopathology* 60:381–82

185. Nienhaus, F., Gliem, G. 1973. Evidence for a new mechanically transmissible tumour-inducing principle in tobacco. *Phytopathol. Z.* 78:367–68

186. Niles, R. M., Mount, M. S. 1973. Chromatin-directed ribonucleic acid synthesis. A comparison of chromatins isolated from healthy, avirulent *Agrobacterium tumefaciens* inoculated, and crown-gall tumor tissues of *Vicia faba*. *Plant Physiol.* 52:368–72

187. Osborn, M., Weber, K. 1975. Simian virus 40 gene A function and maintenance of transformation. *J. Virol.* 15:636–44

188. Parsons, C. L., Beardsley, R. E. 1968. Bacteriophage activity in homogenates of crown gall tissue. *J. Virol.* 2:651

189. Patillon, M. 1974. Quantitative evaluation by reassociation kinetics of *Agrobacterium* DNA sequences residing in the genome of a crown-gall tissue culture. *J. Exp. Bot.* 25:860–70

190. Peters, K. E., Lippincott, J. A., Studier, M. 1974. Identification of a crown-gall tumor growth factor as GABA. *Phytochemistry* 13:2383–86

191. Petit, A., Delhaye, S., Tempé, J., Morel, G. 1970. Recherches sur les guanidines des tissus de crown gall. Mise en évidence d'une relation biochimique spécifique entre les souches d'*Agrobacterium tumefaciens* et les tumeurs qu'èlles induisent. *Physiol. Veg.* 8:205–13

192. Petit, A., Tourneur. J. 1972. Perte de virulence associée à la perte d'une activité enzymatique chez *Agrobacterium tumefaciens*. *C. R. Acad. Sci.* 275D: 137–39

193. Phillips, R., Butcher, D. N. 1975. Attempts to induce tumours with nucleic acid preparations from *Agrobacterium tumefaciens*. *J. Gen. Microbiol.* 86:311–18

194. Quétier, F., Huguet, T., Guillé, E. 1969. Induction of crown-gall partial homology between tumor-cell DNA, bacterial DNA and the G+C-rich DNA of stressed normal cells. *Biochem. Biophys. Res. Commun.* 34: 128–33

195. Quispel, A. ed. 1974. *The Biology of Nitrogen Fixation.* Amsterdam: North-Holland. New York: Am. Elsevier. 769 pp.

196. Rake, A. V. 1972. Lack of DNA homology between the legume *Glycine max* and its symbiotic Rhizobium bacteria. *Genetics* 71:19–24

197. Rasch, E. M. 1964. DNA synthesis in plant tumor cells. *Exp. Cell Res.* 36:475–86

198. Reddi, K. K. 1966. Ribonuclease induction in cells transformed by *Agrobacterium tumefaciens*. *Proc. Natl. Acad. Sci. USA* 56:1207–14

199. Reddy, M. N., Stahmann, M. A. 1973. A comparison of isozyme patterns of crown gall and bacteria free-gall tissue cultures with noninfected stems and noninfected tissue cultures of sunflower. *Phytopathol. Z.* 78:301–13

200. Riker, A. J. 1923. Some relations of the crowngall organism to its host tissue. *J. Agric. Res.* 25:119–32

201. Roberts, W. P., Kerr, A. 1974. Crown gall induction: serological reactions, isozyme patterns and sensitivity to mitomycin C and to bacteriocin of pathogenic and nonpathogenic strains of *A. radiobacter*. *Physiol. Plant Pathol.* 4:81–92

202. Robins, A. B., Taylor, D. M. 1968. Nuclear uptake of exogenous DNA by mammalian cells in culture. *Nature* 217:1228–31

203. Robinson, W., Walkden, H. 1923. A critical study of crown gall. *Ann. Bot.* 37:299–324

204. Rodríguez de Lecca, J., de la Rosa, C., Beltrá, R. 1975. Aseptic induction of

tumors by the nucleic acid fraction from fixed L forms of *Agrobacterium tumefaciens. Phytopathol. Z.* 83:57–65

205. Rogler, C. E., Thompson, W. F., DeVay, J. E. 1975. Bacterial plasmids of *Agrobacterium tumefaciens* and their role in crown gall disease. *Carnegie Inst. Yearb. Ann. Rept.,* pp. 791–93

206. Roussaux, J., Kurkdjian, A., Beardsley, R. E. 1968. Bactériophages d'*Agrobacterium tumefaciens* (Smith et Town.) Conn. 1. Isolement et caractères. *Ann. Inst. Pasteur* 114:237–47

207. Rubio-Huertos, M., Beltrá, R. 1962. Fixed pathogenic L forms of *Agrobacterium tumefaciens. Nature* 195:101

208. Ruesink, A. W., Thimann, K. V. 1965. Protoplasts from the *Avena* coleoptile. *Proc. Natl. Acad. Sci. USA* 54:56–64

209. Ryter, A., Manigault, P. 1964. Etude an microscope électronique de la formation de tumeurs provoquées par *Agrobacterium tumefaciens* chez le topinambour. *Bull. Soc. Fr. Physiol. Veg.* 10:44–56

210. Sasaki, Y., Sasaki, R., Hashizume, T., Yamada, Y. 1973. The solubilization and partial characterization of pea RNA polymerases. *Biochem. Biophys. Res. Commun.* 50:785–92

211. Savulescu, A., Mesrobeanu, L., Popescu, A., Movileanu, D. 1971. Induction de tumeurs par l'endotoxine thermostable d'*Agrobacterium tumefaciens* sur *Datura stramonium* L. *Ann. Inst. Pasteur Paris* 121:405–12

212. Schellman, J. A., Schellman, C. 1964. The conformation of polypeptide chains in proteins. *The Proteins. Composition, Structure, and Function,* ed. H. Neurath, 2:1–137. New York: Academic. 840 pp.

213. Schilde-Rentschler, L. 1973. Preparation of protoplasts for infection with *Agrobacterium tumefaciens.* Protoplastes et fusion de Cellules Somatiques. *Colloq. Int. CNRA,* No. 212: 479–83

214. Schilperoort, R. A. 1969. *Investigation on plant tumors crown gall. On the biochemistry of tumor-induction by Agrobacterium tumefaciens.* PhD thesis. Rijksuniveriteit Leiden, the Netherlands. 170 pp.

215. Schilperoort, R. A. 1972. Integration of *Agrobacterium tumefaciens* DNA in the genome of crown gall tumor cells and its expression. *Proc. 3rd Int. Conf. Plant Pathog. Bact.,* ed. H. P. Maas Geesteranus, 223–38. Wageningen: Cent. Agric. Publ. Doc. 365 pp.

216. Schilperoort, R. A., Bomhoff, G. H. 1975. Crown gall: a model for tumor research and genetic engineering. See Ref. 146, pp. 141–62

217. Schilperoort, R. A., Dons, J. J. M., Ras, H. 1975. Characterization of the complex formed between PS8 cRNA and DNA isolated from A₆-induced sterile crown gall tissue. See Ref. 54, 253–68

218. Schilperoort, R. A., Meijs, W. H., Pippel, G. M. W., Veldstra, H. 1969. *Agrobacterium tumefaciens* cross-reacting antigens in sterile crown-gall tumors. *FEBS Lett.* 3:173–76

219. Schilperoort, R. A., Van Sittert, N. J., Schell, J. 1973. The presence of both phage PS8 and *Agrobacterium tumefaciens* A₆ DNA base sequences in A₆-induced sterile crown-gall tissue cultured *in vitro. Eur. J. Biochem.* 33:1–7

220. Schilperoort, R. A., Veldstra, H., Warnaar, S. O., Mulder, G., Cohen, J. A. 1967. Formation of complexes between DNA isolated from tobacco crown gall tumours and RNA complementary to *Agrobacterium tumefaciens* DNA. *Biochim. Biophys. Acta* 145:523–25

221. Schnathorst, W. C., DeVay, J. E., Kosuge, T. 1964. Virulence in *Agrobacterium tumefaciens* in relation to glycine attenuation, host, temperature, bacteriophage, and antigenicity of pathogen and host. *Conf. Abnormal Growth Plants, Proc. Univ. Calif., Berkeley,* ed. J. E. DeVay, E. E. Wilson, 30–33. 53 pp.

222. Seitz, E. W., Hochster, R. M. 1964. Lysopine in normal and in crown-gall tumor tissue of tomato and tobacco. *Can. J. Bot.* 42:999–1004

223. Sheikh, K., Gahan, P. B., Stroun, M. 1971. A cytochemical study of acid phosphotases and esterases during the induction of crown gall in the tomato. *Histochem. J.* 3:179–83

224. Simard, A. 1973. Détection histochimique de l'activité ribonucléasique dans les tissus normaux et néoplasiques du Tabac. *Rev. Can. Biol.* 32:261–66

225. Smith, E. F. 1917. Mechanism of tumor growth in crown gall. *J. Agric. Res.* 8:165–86

226. Smith, E. F., Brown, N. A., Townsend, C. O. 1911. Crown-gall of plants: its cause and remedy. *US Dep. Agric. Bur. Plant Ind. Bull. 213,* pp. 1–200

227. Snow, M. H. L., McLaren, A. 1974.

The effect of exogenous DNA upon cleaving mouse embryos. *Expt. Cell Res.* 86:1–8

228. Srivastava, B. I. S. 1968. Patterns of nucleic acids synthesis in normal and crown gall tumor tissue cultures of tobacco. *Arch. Biochem. Biophys.* 125:817–23

229. Srivastava, B. I. S. 1970. DNA-DNA hybridization studies between bacterial DNA, crown gall tumor cell DNA and the normal cell DNA. *Life Sci.* 9(II):889–92

230. Srivastava, B. I. S. 1970. Studies on methylation of nucleic acids in normal and crown gall tumor tissue cultures of tobacco. *Life Sci.* 9(II):1315–20

231. Srivastava, B. I. S. 1971. Studies of chromatin from normal and crown gall tumor tissue cultures of tobacco. *Physiol. Plant Pathol.* 1:421–33

232. Srivastava, B. I. S. 1973. DNA polymerase and terminal deoxynucleotidyl transferase activity of normal and bacteria-free crown gall tumor tissue cultures of tobacco. *Biochim. Biophys. Acta* 299:17–23

233. Stonier, T., McSharry, J., Speitel, T. 1967. *Agrobacterium tumefaciens* Conn IV. Bacteriophage PB2₁, and its inhibitory effect on tumor induction. *J. Virol.* 1:268–73

234. Strain, G. C., Mullinix, K. P., Bogorad, L. 1971. RNA polymerases of maize: nuclear RNA polymerases. *Proc. Natl. Acad. Sci. USA* 68:2647–51

235. Stroun, M. 1971. On the nature of the polymerase responsible for the transcription of released bacterial DNA in plant cells. *Biochem. Biophys. Res. Commun.* 44:571–78

236. Stroun, M., Anker, P. 1971. Bacterial nucleic acid synthesis in plants following bacterial contact. *Mol. Gen. Genet.* 113:92–98

237. Stroun, M., Anker, P., Auderset, G. 1970. Natural release of nucleic acids from bacteria into plant cells. *Nature* 227:607–8

238. Stroun, M., Anker, P., Cattaneo, A., Rossier, A. 1971. Effect of the extent of DNA transcription of plant cells and bacteria on the transcription in plant cells of DNA released from bacteria. *FEBS Lett.* 13:161–64

239. Stroun, M., Anker, P., Gahan, P., Rossier, A., Greppin, H. 1971. *Agrobacterium tumefaciens* ribonucleic acid synthesis in tomato cells and crown gall induction. *J. Bacteriol.* 106:634–39

240. Stroun, M., Anker, P., Ledoux. L. 1967. Apparition de DNA de densités différentes chez *Solanum lycopersicum* Esc. au cours de la période d'induction d'une tumeur par la bactérie *Agrobacterium tumefaciens. C. R. Acad. Sci.* 264D:1342–45

241. Stroun, M., Anker, P., Ledoux, L. 1967. DNA replication in *Solanum lycopersicum* esc. after absorption of bacterial DNA. *Curr. Mod. Biol.* 1:231–34

242. Stroun, M., Gahan, P., Sarid, S. 1969. *Agrobacterium tumefaciens* RNA in non-tumorous tomato cells. *Biochem. Biophys. Res. Commun.* 37:652–57

243. Sun, C. N. 1969. Recent advance on crown gall investigation. *Adv. Front. Plant Sci.* 23:119–35

244. Swain, L. W., Rier, J. P. Jr. 1972. Cellular transformation in plant tissue by RNA from *Agrobacterium tumefaciens. Bot. Gaz.* 133:318–24

245. Tegtmeyer, P. 1975. Function of simian virus 40 gene A in transforming infection. *J. Virol.* 15:613–18

246. Thomas, A. J., Klein, R. M. 1959. *In vitro* synthesis of the crown-gall tumour-inducing principle. *Nature* 183:113–14

247. Tinbergen, B. J. 1966. Onderzoekingen over plantentumoren crowngall vergeligkende analyse of DNA-basis. PhD thesis. Rijksuniv. Leiden, the Netherlands. 82 pp.

248. Tourneur, J. 1972. Modèle d'Etude des transformations chez les végétaux. *Rev. Hortic.* 2309:2334–38

249. Tourneur, J., Morel, G. 1970. Sur la présence de phages dans les tissus de "Crown-gall" cultivés *in vitro. C. R. Acad. Sci.* 270D:2810–12

250. Tourneur, J., Morel, G. 1971. Bactériophages et crown-gall. *Physiol. Veg.* 9:527–39

251. Van Larebeke, N., Engler, G., Holsters, M., Van den Elsacker, S., Zaenen, I., Schilperoort, R. A., Schell, J. 1974. Large plasmid in *Agrobacterium tumefaciens* essential for crown gall-inducing ability. *Nature* 252:169–70

252. Van Larebeke, N., Genetello, Ch., Schell, J., Schilperoort, R. A., Hermans, A. K., Hernalsteens, J. P., Van Montagu, M. 1975. Acquisition of tumour-inducing ability by non-oncogenic agrobacteria as a result of plasmid transfer. *Nature* 255:742–43

253. Van Sittert, N. J. 1972. Onderzoekingen over crown gall analyse van DNA en RNA in de tumorcel. PhD

thesis. Rijksuniv. Leiden, the Netherlands. 113 pp.

254. Van Sittert, N. J., Ledeboer, A. M., Van Rijn, C. J. S., Boon, E., Schilperoort, R. A. 1975. RNA from callus cultures and leaves of *Nicotiana tabacum. Phytochemistry* 14:637–46

255. Veldstra, H. 1972. Plant tumors and crown gall, an analysis of autonomous growth. *Proc. 3rd Int. Conf. Plant Pathog. Bact.* ed. H. P. Maas Geesteranus, 213–22. Wageningen: Cent. Agric. Publ. Doc. 365 pp.

256. Vervliet, G., Holsters, M., Teuchy, H., Van Montagu, M., Schell, J. 1975. Characterization of different plaqueforming and defective temperate phages in *Agrobacterium* strains. *J. Gen. Virol.* 26:33–48

257. Von Albertini, A., Hohl, H. R., Manigault, P., Stoll, Ch., Vogel, A. 1961. Similitudes et différences entre des tumeurs expérimentales et le crown gall. *Phytopathol. Z.* 41:55–58

258. Watson, B., Currier, T. C., Gordon, M. P., Chilton, M.-D., Nester, E. W. 1975. Plasmid required for virulence of *Agrobacterium tumefaciens. J. Bacteriol.* 123:255–64

259. Watts, J. W., Cooper, D., King, J. M. 1974. Attempts to transform tobacco protoplasts with DNA of *Agrobacterium tumefaciens. Ann. Rept. John Innes Hortic. Inst.*, p. 64

260. Weete, J. D. 1971. Total fatty acids of habituated and teratoma tissue cultures of tobacco. *Lipids* 6:684–85

261. Wehrli, W., Knüsel, F., Schmid, K., Staehelin, M. 1968. Interaction of rifamycin with bacterial RNA polymerase. *Proc. Natl. Acad. Sci. USA* 61:667–73

262. Weisblum, B., Davies, J. 1968. Antibiotic inhibitors of the bacterial ribosome. *Bacteriol. Rev.* 32:493–528

263. Wendt-Gallitelli, M. F., Dobrigkeit, I.

1973. Investigations implying the invalidity of octopine as a marker for transformation by *Agrobacterium tumefaciens. Z. Naturforsch.* 28c:768–71

264. White, P. R. 1945. Metastatic (graft) tumors of bacteria-free crown-galls on *Vinca rosea. Am. J. Bot.* 32:237–41

265. Wong, P. P., Kuo, T., Ryan, C. A., Kado, C. I. 1976. Differential accumulation of proteinase inhibitor I in normal and crown gall tissues of tobacco, tomato and potato. *Plant Physiol.* 57:214–17

266. Wood, H. N., Braun, A. C. 1965. Studies on the net uptake of solutes by normal and crown-gall tumor cells. *Proc. Natl. Acad. Sci. USA* 54:1532–38

267. Wood, H. N., Lin, M. C., Braun, A. C. 1972. The inhibition of plant and animal adenosine $3':5'$-cyclic monophosphate phosphodiesterases by a cell-division-promoting substance from tissues of higher plant species. *Proc. Natl. Acad. Sci. USA* 69:403–6

268. Yajko, D. M., Hegeman, G. D. 1971. Tumor induction by *Agrobacterium tumefaciens:* specific transfer of bacterial deoxyribonucleic acid to plant tissue. *J. Bacteriol.* 108:973–79

269. Zaenen, I., Van Larebeke, N., Teuchy, H., Van Montagu, M., Schell, J. 1974. Supercoiled circular DNA in crowngall inducing *Agrobacterium* strains. *J. Mol. Biol.* 86:109–27

270. Zevenhuizen, L. P. T. M. 1971. Chemical composition of exopolysaccharides of *Rhizobium* and *Agrobacterium. J. Gen. Microbiol.* 68:239–43

271. Zimmerer, R. P., Hamilton, R. H., Pootjes, C. 1966. Isolation and morphology of temperate *Agrobacterium tumefaciens* bacteriophage. *J. Bacteriol.* 92:746–50

Copyright © 1976 by Annual Reviews Inc.
All rights reserved

ELECTROPHYSIOLOGICAL RESEARCH IN PLANT PATHOLOGY

♦3644

Terry A. Tattar

Department of Plant Pathology, University of Massachusetts, Amherst, Massachusetts 01002

Robert O. Blanchard

Department of Botany and Plant Pathology, University of New Hampshire, Durham, New Hampshire 03824

INTRODUCTION

More effective means are needed for early diagnosis of plant diseases. One possible answer is the use of nondestructive electrical diagnostic measurements such as those used extensively in human medicine. Research on the electrophysiology of plants has been conducted for almost as long as man has conducted research with electricity (4, 68, 84). The objective of this review is to encourage further research on presymptomatic and nondestructive detection of disease in living plants with the use of bioelectronic techniques. The review could encompass a wide variety of experimental techniques, but in an attempt to make it most relevant to plant pathology we limit our discussion to research on electrical measurements and applications of electrical currents to living plants, living plant tissues, and wood in service.

HISTORY OF ELECTROPHYSIOLOGICAL RESEARCH ON PLANT DISEASES

Various 18th and 19th century investigators claimed beneficial effects of electric currents on plant growth. Although experimental details often were incomplete (4, 98, 99) these early reports led several 20th-century agricultural scientists such as Clarence Warner, Asa Kinney, and George Stone to investigate possible beneficial uses of electricity in agriculture.

Warner (98) applied direct current (DC) to soil in greenhouse beds and found that lettuce plants in the electrically treated soil had a substantially lower incidence of "mildew" (probably downy mildew) than control beds. He also studied the effects of alternating current (AC) on seed germination, growth rate, and

309

yield of eight other crops (99). A positive growth stimulus was detected and significantly increased yields were observed with most of the plants tested.

Kinney (53) found that seeds of mustard, rape, and red clover would germinate more quickly when pretreated with AC. He also found that root elongation in electrically treated seed was greater than in untreated seeds. The range of effective voltages, however, was small (1-3 volts); voltages outside these limits were either nonstimulatory or inhibitory to seed germination. The stimulatory effects of the electrical treatments were not long-lasting after treatments ceased; germination and root growth of the controls soon equalled those of the treated seeds. But the original advantage in germination and growth could be maintained into the seedling stage if the electrical treatments were continued daily. Kinney did not determine whether or not his electrogermination treatments had an inhibitory effect on pathogenic soil microorganisms or a stimulatory effect on beneficial soil microorganisms or on formation of mycorrhizae in the seedlings but we present these speculations as possible explanations of his results.

Stone (83) demonstrated a stimulatory effect of low voltage electric currents on the growth of bacteria and yeasts but found that higher voltages were inhibitory. Earlier, he also found a stimulatory effect of electric currents on crop plants (82). In addition, Stone was concerned with the injury that occurred to trees adjacent to uninsulated power lines of streetcars and railroads (84). He found that periodic contact between tree branches or trunks and wires carrying DC produced more injury than similar contact with wires carrying AC. Stone formulated a theory about the range of effects of electrical current—increasing from subthreshold to stimulatory to inhibitory to lethal. Stone (84) also measured the electrical resistance properties of woody tissue in living trees. He found that: (a) electrical resistance was lowest at the cambium; (b) cambial resistance could be used as a measure of tissue viability; (c) healthy tissue was low in resistance; and (d) dead tissues were substantially higher in resistance.

It was Osterhout (68) who first published a comprehensive monograph on the study of electrical conductivity as affected by injury and death of living plant and animal tissues. His experiments, although primitive by today's standards, were used by subsequent electrophysiologists to formulate conceptual models for the electrical properties of injured and diseased tissues. He attempted to quantify changes of membrane permeability, as measured by increased electrical conductivity, with degrees of injury. He was able to relate minor injury to slight decreases in conductivity and to identify the time of cell death by large and irreversible increases in conductivity. Since altered permeability of cell membranes is an important feature of most diseases and injuries in plants, loss of selective permeability of electrolytes resulting in increased tissue conductivity became a basic principle of plant electrophysiology.

After the widespread interest in electrophysiology studies on plants from the 1890s to the early 1930s, research activity decreased until the late 1940s. The resurgence of research in this area followed rapid advances in electronic technology that still continue. Electrophysiology research on plant diseases during this period can be classified according to the principles of measurement employed.

PRINCIPLES OF ELECTROPHYSIOLOGICAL MEASUREMENTS ON PLANTS

Membranes and Ions

The passage of an electric current in a dilute solution such as found in plant tissues is by the movement of ions (6, 102). The plasma membrane controls the movement of ions or electrolytes in and out of the cell (44). A living cell is selectively permeable to certain ions such as K^+, Ca^{2+}, Cl^-, NO_3^-, $H_2PO_4^-$, and SO_4^{2-} and maintains them at high concentration within the cell (47). The differences in ion concentrations within and outside cells frequently result from differences in electrical potential that are actively produced across living membranes (47). These potential differences are caused mainly by the combined effects of Donnan potentials (charges on the surface of large molecules within the cell), concentration effects, and active ion pumps (11, 14).

The plasma membrane is a lipoprotein layer and acts as an electrical insulator to low frequency currents (6). Low frequency electric currents applied to tissues pass from cell to cell primarily through the cell walls and in the fluids within the conducting elements (28, 102). As long as a cell is metabolizing normally, its electrical properties will reflect primarily changes in metabolic rate. Upon disturbance or injury to the membrane, depolarization and/or electrolyte loss may occur. If the injury results in cell death, as is often the case during host-parasite interactions, the release of electrolytes from cells into the intercellular spaces will cause a large local increase of ion concentration (101). These changes in interstitial ion concentration will have pronounced effects on the electrical properties of the affected tissue.

Electrical Current in Plant Tissue

Flow of an electrical current in plant tissues results in the movement of charges (mobile ions) through the dilute solution such as the intercellular wall fluids and sap of plants. The mobility of an ion is determined primarily by its molecular size in the hydrated form (6). A high density of mobile ions results in a low electrical resistance (to current flow) in the solution or tissue.

The amounts of ions in the interstitial fluids of plants is also a function of the relative metabolic activity of tissues (44). Cambial area tissues of woody plants were found to be the lowest in electrical resistance and their resistance was related to seasonal changes in the activity of the tree (25). The electrical resistance of activity metabolizing meristems can be used to determine their relative growth rate (86).

In woody stems most of the tissue volume is composed of nonliving supportive tissue. When these tissues are attacked by microorganisms, increasing amounts of electrolytes accumulate in the zone of degradation (71, 73, 87). The electrical resistance of the tissues decreases as tissues are progressively discolored and decayed (87). Wood in service has also been shown to undergo the same changes in electrical resistance during microbial degradation (75).

Measurement of Electrical Resistance

Under most circumstances electrical resistance measurements of tissues involve introducing a current into the tissue (28). Ohm's law shows that resistance is a function of current flow: $E = IR$ where E is the electrical potential difference in volts, R is the resistance in ohms, and I is the current flow in amperes. Electrical conductivity of tissue is expressed in mhos and is the reciprocal of the electrical resistance: (mho) $= 1/R$. Both resistance and conductivity measurements are used extensively to obtain the electrical properties of plant tissues.

The electrical resistance of tissue is indirectly related to the cross-sectional area of tissue and directly related to the distance between electrodes (28). For this reason, resistance measurements should be compared only when using similar types of electrodes, similar sizes of tissue, and similar distances between electrodes. Comparisons of unlike measurements can be made, however, if they are converted to units of specific resistance (Sp. R). This can be accomplished by multiplying resistance (R) by cross section (A) area in centimeters and dividing by electrode distance (L) in centimeters: $Sp. R = RxA/L$ (28).

Electrical Resistance and Temperature

Electrical resistance of tissue will progressively decrease with rising temperature, if measurements are taken over a relatively short period (1-2 days), since ions become more mobile at higher temperatures. Prolonged exposure to an electrical current (several weeks), however, will result in conditioning of the tissue. Under these circumstances differences in electrical resistance will occur at the same temperature between plants that have been conditioned and those that have not (9). Since measurement of electrical resistance depends on ion movement, the change from free water to ice will dramatically affect the electrical properties of tissues (25).

Electrical Resistance and Moisture

Electrical resistance of plant tissues will be relatively independent of small changes in tissue water content as long as sufficient interstitial free water remains to allow ion movements (87). However, when free water becomes limiting due to moisture stress, electrical resistance becomes dependent on the moisture content of the tissue.

Electrical Impedance

Alternating current is most correctly described by the term *impedance* instead of *resistance,* although both terms often are used synonymously. When an alternating voltage (sinewave or otherwise) is applied to a tissue under study, the resulting current generally is not related to the voltage by Ohm's law ($E \neq IR$) (100). This is due primarily to separation of charges (ions) at tissue boundaries. This effect is called capacitance. It has been found, however (28), that Ohm's law applies approximately when sinewave voltages are used whose frequency is less than 1 kHz (cycles per second). If alternating voltages are used, the frequency should be specified in order to allow accurate comparisons.

Biopotentials

Differences in electrical potential between various parts of a plant or between a plant and the soil, are called biopotentials. These are passive electrical measurements of the differences in relative electrochemical activity between two points. These differences are called biopotentials because the differences in electrical potentials are due primarily to the metabolic activities of the living tissues. Biopotentials also can be measured across the plasma membrane by inserting glass microelectrodes, usually less than 1 μm diameter, into single plant cells (47). A reference electrode is placed into the bathing medium and the biopotentials (membrane potentials) are measured. Instantaneous changes in membrane polarity can then be detected in living cells (47). Biopotentials usually are measured in millivolts (mV).

Effects of Electrical Measurements on Tissues

All electrical measurements have some disruptive effects on the physiology of the organisms being measured. The most important effects are those caused by the insertion of electrodes into cells and tissues (32). These effects can be minimized by careful insertion of electrodes and by the use of inert metals or nonpolarizing electrodes. Polarization is the accumulation of ions around an electrode due to an induced flow of current caused by electrochemical reactions. These reactions often result in altered electrical properties or artifacts. Electrodes made from inert metals such as stainless steel, silver, platinum, and gold result in decreased polarization; in the case of most intracellular electrodes, glass micropipettes filled with an electrolyte such as 3M KCl are used. Electrodes are often "chloridized" (coated electrically with a thin layer of silver chloride) so that they will exchange silver and chloride ions with solution to maintain an electrically neutral solution around the electrodes.

Most polarizing effects in plant tissue also can be minimized by the use of low frequency alternating or pulsed currents (28). Measurements such as electrical resistance, where currents are applied to tissues, should be taken as rapidly as possible to minimize increasing polarization effects.

ELECTRICAL RESISTANCE

Cold Temperature Injury

Intolerance to cold is a major barrier to the geographic spread of many cultivated plants in temperate climates (55). Cold temperature injury is thought to result primarily from injury to cell membranes due to formation of large ice crystals in the free water of the cell and cell wall (55). A subsequent loss of ions occurs and there is a substantial lowering of electrical resistance of injured tissues. Electrical measurements may be particularly useful because the change in electrical properties of tissues can be measured immediately after injury while visual symptoms usually do not appear until 24 to 48 hr later. Frost or cold tolerance is often measured as the ability of plants to avoid such injury under various test conditions. Differences between electrical resistance of tissues before and after treatments are used to determine the extent of injury. Hardening off or induced cold tolerance is also determined in a similar manner.

Although a number of experimental approaches have been used, cold injury most commonly has been determined by measuring electrical resistance of stems and roots of woody plants before and after exposure to freezing temperatures, using low frequency (under 1 kHz) currents (34–36, 85, 91, 103). Cold injury in perennial legumes (9, 40, 42) and some other herbaceous plants (13, 96) also has been determined by obtaining resistance measurements in a similar manner as in woody plants. Some investigators (16, 35, 40, 61) have found that measuring electrical resistance at both low (1 kHz) and high (100 kHz to 1 MHz) frequencies, and using the ratio between these measurements resulted in more sensitive determinations of injury. A low: high frequency ratio in healthy tissue usually will be about 4:1 while in tissue that has undergone severe cold injury the ratio will be less than 1.5:1 (16, 61). Less severe injury will result in ratios between these extremes depending on the extent of injury. The use of ratios, however, still depends upon changes in low frequency measurements that are sensitive to the condition of the plasma membrane.

Hayden et al (45) devised a method to detect cold hardiness of potato (*Solanum tuberosum*) clones by measuring electrical resistance of tubers at 4.4°C. They found that electrical resistance was inversely proportional to hardiness of the stems at this temperature. The major advantage of this method is that after testing it allows the immediate selection of plant material for superior cold hardiness without the destructive effects of exposure to extreme cold described previously.

Zaerr (104) studied the effects of cold and herbicide injury on the electrical properties of Douglas fir needles by sending a 100 Hz square-wave current through the needles and observing the effects on that wave with an oscilloscope. By using a dual-trace oscilloscope he was able to observe the effect on square wave signals as they passed through healthy and injured tissue simultaneously. Since low frequency currents are sensitive to the condition of the plasma membrane, the square wave form was distorted progressively with increasing degree of tissue injury. Zaerr (104) used the resulting oscilloscope pattern to determine the degree of injury to the leaves.

Miscellaneous Injuries to Plant Tissues

Evert (20) found that if he used a progression of frequencies from 50 Hz to 500 Hz, he could determine the effects on the cell membranes due to heat injury by measuring changes in the phase angles. Phase angles are measures of the time relationship of current and voltage in an alternating current and they are useful in determining capacitance of materials. Healthy plasma membranes possess relatively high capacitance and act as an electrical insulator to low frequency currents. An injured membrane will have progressively decreased capacitance depending upon the extent of injury. Greenham (41) found that he could accurately evaluate the extent of bruise and pressure injury in apple (*Malus* sp.) fruits with the use of low:high frequency ratios of electric resistance of the fruits.

Extent of injury was related directly to the decrease in the ratio from noninjured fruits. Hartel & Grill (43) discovered that electrical resistance of spruce (*Picea abies*) bark extracts was a more accurate method of detecting SO_2 injury than chemical analyses of foliage. Greenham & Cole (37) and Greenham (40) correlated changes in electrical resistance and capacitance with the degree of injury from herbicides on skeleton weed (*Chondrilla juncea*), bindweed (*Convolvulus arvensis*), and potato tubers. They found electrical capacitance measurements to be the reciprocal of electrical resistance measurements.

Infectious Diseases of Plants

FUNGUS DISEASES Greenham & Muller (39) measured increases in electrical conductivity of potato tubers associated with the invading hyphae of *Phytophthora infestans* and the invading hyphae of *Pythium* sp. They used electrical conductivity measurements to study different responses of the potato tissue to compatible and noncompatible combinations of races of *P. infestans* and potato varieties. The noncompatible combinations showed increases in conductivity 4 mm in advance of the mycelium while the susceptible combinations exhibited these changes only after the mycelium had invaded the tissues. Conductivity measurements of potato tissues artificially infected with *Pythium* sp. were found to change only several millimeters behind the edge of the invading hyphae. Greenham & Muller (39) used electrical conductivity measurements to study the genetics of host-parasite interaction in potatoes infected with *P. infestans*. They also used similar measurements to demonstrate differences in the mode of attack of *P. infestans* and *Pythium* sp. on potato.

Comstock & Martinson (12) and Garroway (31) determined differences in susceptibility among Tms and N cytoplasm inbreds of corn (*Zea mays*) to *Helminthosporium maydis* races O and T by the use of conductivity measurements. Both found that the greater the increase in conductivity the greater the degree of susceptibility of the corn cytoplasms to the race T of *H. maydis*. Comstock & Martinson (12) also found that the rate of conductivity increase per hour could be related to susceptibility. In addition, they found that treatment of Tms leaves with a toxin (T toxin) specific for that cytoplasm would result in increases in conductivity. Damann et al (15) working with *H. victoriae* and oat (*Avena sativa*) tissue developed an assay for the host-selective toxin by determining electrolyte loss with conductivity measurements. Using a similar technique Van Alfen & Turner (92) found that a *Ceratocystis ulmi* toxin did not produce electrolyte loss in callus tissue of American elm (*Ulmus americana*).

Caruso et al (10) used electrical resistance measurements of the taproots to presymptomatically detect *Fusarium oxysporum* in Bonnie Best variety tomatoes (*Lysopersicon esculentum*) 19 to 22 hr before foliar symptoms of wilt became evident. They found that virulence of the *Fusarium oxysporum* f.sp. *lycopersici* isolate could be related to the rate of decrease in electrical resistance of taproots. They also found the rate of drop in electrical resistance to be directly related to the in vitro production of polygalacturonase of the isolates.

Shigo & Berry (76) were able to detect decay associated with *Fomes annosus* in red pine (*Pinus resinosa*) by measuring the electrical resistance of wood around the root collar. Decayed woody tissues were low in resistance, but resin-soaked wood that is often associated with *F. annosus* infection was of extremely high resistance. The pattern of alternating low and high resistances was indicative of *F. annosus* infection, as sound healthy wood did not show erratic and abrupt changes in resistance.

BACTERIAL DISEASES   Friedman & Jaffe (29) studied the effects of soft rot bacteria, *Erwinia* and *Pseudomonas* sp., on the electrical conductance of leaf tissue of witloof chicory (*Cichorium intybus*). Conductance of tissue increased in advance of the margins of visible lesions caused by strains of soft-rot bacteria that produced abundant pectic enzymes in culture. They concluded that the increases in electrical conductivity and subsequent development of soft rot in the tissues were associated with the production of pectic enzymes, since a mutant nonpectolytic strain of *P. marginalis* produced neither increases in conductivity nor soft rot. They were also able to distinguish between the soft rot caused by *Erwinia* sp. and that caused by *Pseudomonas* sp. by comparison of the conductivity changes. In early stages of infection, the conductances of tissues infected with both bacteria were approximately the same, but after 24 hr, tissues infected with *Erwinia* sp. had much higher conductances than those infected with *Pseudomonas* sp. Mount, Bateman & Basham (63) studied the effects of endopolygalacturonate trans-eliminase (endo-PGTE) enzyme from the soft-rot bacterium *Erwinia carotovora* on the tissues of potato. Exposure of these tissues to partially purified endo-PGTE caused rapid increases in conductivity of the bathing solution while similar experiments with autoclaved enzymes resulted in negligible changes in conductivity. Van Alfen & Turner (93) used measurements of conductivity to demonstrate the lack of membrane damage to alfalfa stem tissues when exposed to *Corynebacterium insidiosum* toxin. Measuring flow of an electrical current in the trunk, Levengood (56) determined that black cherry (*Prunus serotina*) trees infected with crown gall had lower electrical resistances than healthy trees.

VIRUS DISEASES   Ghabrial & Pirone (33) detected an increase in the conductivity of the bathing solution of roots of tabasco peppers (*Capsicum frutescens*) infected with tobacco etch virus (TEV) 24 to 48 hr before wilt symptoms occurred. No changes in conductivity occurred in noninfected control plants or plants infected with alfalfa mosaic virus (AMV) or cucumber mosaic virus (CMV). Although both viruses cause infection, neither causes wilt in tabasco pepper. Similarly no changes occurred with California Wonder variety, a susceptible variety that does not wilt when infected with TEV. Increased conductivity also preceded by 12 to 24 hr respiratory rate changes of leaves as measured by a respirometer and preceded histological changes by 24 to 48 hr.

Greenham et al (38) were able to separate nonsymptomatic potato tubers infected with virus X from healthy tubers by the use of electrical impedance measurements. Virus X–infected tubers were significantly lower in impedance than healthy tubers. The authors also measured capacitance of the tubers and found the product of resistance and capacitance to be a reliable indicator of virus X infection. Similar experiments with tubers infected with leaf roll virus resulted in a wide variation in electrical measurements. They felt these differences were due to the varied physiological response to leaf roll between the potato varieties used.

MYCOPLASMA-LIKE-ORGANISM (MLO) DISEASES   Dostalek (17) found that roots of apple trees infected with the proliferation disease MLO were much lower in electrical impedance than roots of healthy apples. The rootstocks measured were Malling 'M I' and 'M IV' with local healthy and infected scions grafted onto them. Impedance measurements of roots taken in August did not differentiate between healthy and proliferation-infected roots, but significant differences were found in November. No differences in impedance were found at either time for measurements of infected or healthy stem tissues.

NEMATODES   The presence of feeding lesion nematodes (*Pratylechus penetrans*) in the cortex of sunflower roots was detected 3–6 hr after inoculation by decreases in electrical resistance of the roots (52). Lesion formation, however, did not occur until 16–24 hr after inoculation.

DISCOLORATION AND DECAY OF WOOD   Discoloration and decay of wood in living trees often develop following injuries that expose the xylem. After a tree is injured, the host responds, a variety of microorganisms develop, and woody tissues near the wound often become discolored and eventually decayed (74). During all stages of this progression of events from early discoloration to structural breakdown or decay, the concentration of mobile ions (mainly K^+ and Ca^{2+}) in the tissues increases (71, 73, 87). This results in a progressive change in electrical resistance (79, 87, 88), current (90), and capacitance of woody tissues (89). Discolored and decayed wood has been found to be lower in electrical resistance and impedance, higher in capacitance, and to allow more flow of electrical current than nondiscolored or sound wood in the same tree. This is true only when wood is above its fiber-saturation point (wood with free water present in the vessels). Since woody tissues in living trees are most always above the fiber-saturation point, the free water necessary for ion movement and accurate determination of defect is usually present in the tissue (87). Once wood is below its fiber-saturation point, however, the electrical resistance is influenced by the moisture content of the wood (78, 81). This is the principle upon which the commercial wood moisture-meter is based. When wood is above fiber-saturation it is the concentration of mobile ions that primarily affects its electrical properties (58, 59).

Active decay of wood in service and in freshly cut wood products requires that the wood be above the fiber-saturation point; thus wood undergoing active decay has similar electrical properties to wood in living trees (75). Shigo & Shigo (75) and Shortle et al (77) demonstrated that electrical resistance measurements could be used to detect decay in creosoted and other preservative-treated utility poles.

Disease Complexes

Many plant diseases, especially those on perennial woody plants, are caused by interactions of infectious and noninfectious pathogens. These diseases are often referred to as diebacks and declines or simply as disease complexes. They involve a progressive decrease in growth rate over several growing seasons. Host vigor is often a primary consideration in evaluating the progression of a disease complex. Zhuravleva (105) devised a method for evaluating host vigor in living spruce trees by measuring the electrical resistance of cambial tissues. Cambial metabolic activity as measured by annual growth rate was related indirectly to the degree of electrical resistance. Trees with large yearly growth increments were high in vigor and had low cambial resistance while declining trees or trees with poor vigor had high cambial resistance. In a similar study on pine (*Pinus* sp.) trees Polozhentsev & Zolotov (70) also found resistance to be inversely proportional to vigor. Wargo & Skutt (97) have found similar trends in the gypsy moth-oak decline complex with the increase in electrical resistance being directly related to the severity and number of insect defoliations.

Levengood (56) measured vigor in plants by inserting polarizing electrodes of pure copper and pure iron, and measuring the amount of induced current between the electrodes on a microammeter. The two metals' being electrochemically dissimilar formed a galvanic cell in the plant. The amount of current was indirectly related to the electrical resistance, as predicted by Ohm's law. Voltage (electrical potential) is fixed by the electrical properties of the electrode metals; thus the resistance between the electrodes will determine the amount of current that flows between them. Levengood (56) found that healthy plants showed a large flow of current between electrodes and that plants stressed by lack of moisture or insect defoliation showed progressive decreases in flow of current. Moisture stress was also related to increases in electrical resistance of stems of Douglas fir (*Pseudotsuga menziezii*), European beech (*Fagus sylvatica*), and birch (*Betula pendula*) (54). Taper & Ling (86) were able to relate resistance measurements of apple scions to rootstock vigor. Levengood et al (57) were able to select bean (*Phaseolus vulgaris*) seeds for growth and vigor prior to germination with current measurements.

BIOPOTENTIALS

Biopotentials in Intact Plants

Fensom (21) found that low concentrations of the herbicide 2,4–D spray caused elevated biopotentials in the stems of red pine (*Pinus resinosa* Ait.) trees; but high concentrations resulted in depressed potentials and eventually resulted in tissue injury. These trends continued for three summers following spraying, and the experiment was then terminated. Daily rhythms in potentials caused variations in measurements on the same trees and required that measurements be taken at the same time each day.

Asher (1) related reproductive ability and host vigor to biopotentials of branches of slash pine (*Pinus elliottii* var. *elliottii*). He found that pines located in the geographic center of the natural range of slash pine produced significantly larger biopotentials than those at the edges. Asher termed this phenomena *physiological vigor*. These types of data may also be useful in investigations of plant disease incidence within the natural and introduced distribution of hosts.

The periodic and unexplained variation in potential measurements that were noted by earlier researchers was investigated in detail on trees by Lund (60) and Burr (7, 8) and on several higher plants by Fensom (22–24, 26, 27). All of these researchers found that biopotentials change in diurnal patterns. A monthly pattern that could be linked to the lunar cycle was reported by Burr (7) but was not observed by Fensom (27). Fensom (27) found a yearly pattern and also found several daily rhythms, some as short as 3–5 min apart during the growing season. Neither author, however, was able to explain in much detail the physiological significance of these rhythms. Fensom (24) formulated a theory about potentials and the circulation of hydrogen ions. If these patterns of biopotentials could be explained in terms of altered plant function, perhaps they could become useful diagnostic tools for plant pathologists as the electrocardiograph and electroencephalograph are now to medical doctors.

Biopotentials in Plant Cells

Studies of biopotentials in a single plant cell may help explain the complex patterns of biopotentials in whole plants, since the biopotential of each living cell is thought to contribute in part to the biopotentials of the whole plant (80). Research on this subject was conducted initially on the large cells of algae, but later, tissues of higher plants were used. As a result of this research, new theories about nature of membrane transport have been formulated (47, 48).

The membrane potential appears to be a relatively fixed parameter in a living cell. The interior of the cell is usually about -100 mV relative to the surrounding medium. It appears to be constant for an individual cell (47), and has passive and active components (49). Therefore, the study of membrane potentials seems to be a useful approach for the investigation of pathological alterations of cell membranes (67).

Novacky & Hanchey (65) studied the effects of victorin toxin produced by *Helminthosporium victoriae* on the membrane potentials of oat (*Avena sativa*) roots. In susceptible oat roots, they found that treatment of plants with 50 units/ml of victorin caused an initial membrane depolarization in 2–5 min and a 30% average depolarization within 10 min. Similarly treated roots of resistant plants and all roots treated with deactivated toxin produced no changes in membrane potentials. Van Sambeek et al (95) found membrane potentials depolarized when helminthosporoside, a toxin produced by *H. sacchari,* was added to leaf mesophyll cells of susceptible cultivars of sugarcane (*Saccharum* sp.) while resistant cultivars showed only slight drops in membrane potential. In contrast, the toxin produced by *Helminthosporium carbonum* caused a rapid hyperpolarization in membrane potentials in tissues from susceptible lines of corn (30). Novacky (66) found that hypersensitive reactions of cotton (*Gossypium hirsutum*) cotyledons induced by separate inoculations with *Pseudomonas pisi* and *Xanthomonas malvacearum* also resulted in membrane depolarization. Jones et al (50) used the nematode-induced giant cells as one cell type in a comparative study of membrane potentials of cells possessing widely different levels of metabolic activity. The higher metabolic activity of giant cells was not reflected in membrane potentials, an observation consistent with the notion that cells maintain membrane potentials at fixed values.

RECENT DEVELOPMENTS AND FUTURE OUTLOOK OF ELECTROPHYSIOLOGICAL RESEARCH

Introduction

Rapid advances in electronic technology have faciliated the use of electrical measurements to determine plant functions. Certain of these electronic techniques have been more useful than others, but it is doubtful that the potential value for plant disease research of any of them can be made with the limited research information available. In many cases new techniques have been applied only to one particular problem and potential applications in other areas have not been tested.

Electrostimulation and Electrotherapy

The theory that an electrotherapeutic treatment may be beneficial in the control of plant diseases is not new but it is still relatively untested under experimental conditions. Black et al (3) demonstrated through a series of experiments a positive growth stimulation and an ion uptake in tomato plants that were exposed to various low density (3–15 μA/plant) currents applied for 4.5, 12, and 24 hr/day intervals. Blanchard (5) found that the application of a one fourth ampere DC pulse at 6500 volts for 16 or 4 hr, 2 or 3 days after wounding, respectively, resulted in little or no discoloration behind the wound and almost a complete absence of invading microflora. This experiment demonstrated the possibility that various sublethal electrical treatments may affect invasion of pathogenic microorganisms and subsequent infection following wounding.

Diffusive Resistance

Duniway (18), using a diffusive resistance meter developed by Van Bavel et al (94) and Kanemasu et al (51), studied the resistance to water loss in tomato leaves of plants infected with *Fusarium oxysporum* f. sp. *lycopersici*. Using similar diffusive resistance measurements, Kaplan (52), was able to relate internal moisture stress of sunflower (*Helianthus annus*) plants to infections of low and high populations of lesion nematodes (*P. penetrans*). Diffusive resistance was measured by a moisture-sensitive coil in a closed chamber that was clamped over the leaf. The rate of moisture leaving the leaf due to stomatal and cuticular diffusion was proportional to the rate of change of current through the coil. The major advantage of this technique is that it is nondestructive, and the same leaf can be measured throughout the experiment. In any plant disease where water relations are part of the disease syndrome, a measure of the rate of loss of moisture through the leaves, like diffusive resistance, may provide useful information about the progression of the disease.

Miscellaneous Electronics Techniques

X rays (2,19) have been used to detect pockets of extensive decay inside living trees in much the same way that fractures in bones are detected with medical X rays. However, the bulky shielding necessary to protect the operator and other related equipment presently limit the field usage of these techniques. Ultrasonics, the application of ultrahigh frequency sound, have also proven useful in detection of physical defects in wood products (62) and for the control of nematodes in the field (46). Kirlian photography, the use of high positive ion discharge, was used to test seed viability of peanuts (69). It was found that healthy seeds emitted a large blue discharge halo while dead seed emitted none.

Potential Applications of Electronics to Plant Pathology

Many electronic techniques that have been used to study the properties of plant tissues or some function have not been used to study plant diseases. These techniques may or may not prove useful in plant disease research but warrant examination for their potential usefulness. We mention only two techniques to serve as examples: (a) electronic capacitance measurements and (b) magnetohydrodynamic measurements.

Electrical capacitance was related to the amount of fresh weight of forage grasses while they were growing in the field (64). Since the plasma membranes in the plant tissues act as capacitors, the greater the mass of plant tissue being measured the greater amount of capacitance that was found. Although they were nondestructively estimating standing herbage for animal feed, the measurements could be applied to a physiological stress or infectious disease that may affect the aboveground mass of plant tissue. Sheriff (72) measured sap movement of intact stems in several woody plant species by a magnetohydrodynamic technique. Magnetic lines of force were applied at right angles to the stem and the movement of sap across these lines induced a voltage across the stem. Determination of sap movement in intact plants of course would be useful in the study of a large number of plant diseases. It is hoped that these examples will indicate the diversity of electrophysiological techniques that are being developed.

CONCLUSIONS

Electrophysiology is the application of electronic technology to better understand biological function. The application of recent electronic technology to better understand the interactions between plants and their pathogens should be the objective of current and future electrophysiology research in plant pathology. New electrophysiological measurements that can be related to plant diseases should be developed and used in diagnosis as soon as they are proven reliable.

ACKNOWLEDGMENTS

The authors wish to express appreciation to Bud Etherton, D. S. Fensom, Anton Novacky, Alex L. Shigo, and H. Richard Skutt for their advice and assistance in preparation of the manuscript and also to all those who generously supplied us with research information for this review.

Literature Cited

1. Asher, W. C. 1964. Electrical potentials related to reproduction and vigor in slash pine. *For. Sci.* 10:116–21
2. Beaton, J. A., White, W. B., Berry, F. H. 1972. Radiography of trees and wood products. *Mater. Eval.* 30: 14a–17a
3. Black, J. D., Forsyth, F. R., Fensom, D. S., Ross, R. B. 1971. Electrical stimulation and its effect on growth and ion accumulation in tomato plants. *Can. J. Bot.* 49:1809–15
4. Blackman, V. H. 1924. Field experiments in electro-culture. *J. Agric. Sci.* 14:240–67
5. Blanchard, R. O. 1974. Electrotherapy: a new approach to wound healing. *Proc. Am. Phytopathol. Soc.* 1:133–34 (Abstr.)
6. Bull, H. B. 1971. *An Introduction to Physical Biochemistry.* Philadelphia: Davis. 433 pp.
7. Burr, H. S. 1945. Diurnal potentials in the maple tree. *Yale J. Biol. Med.* 17:727–35
8. Burr, H. S. 1947. Tree potentials. *Yale J. Biol. Med.* 19:311–18
9. Calder, F. W., MacLeod, L. B., Hayden, R. I. 1966. Electrical resistance in alfalfa roots as affected by temperature and light. *Can. J. Plant Sci.* 46:185–94
10. Caruso, F. L., Tattar, T. A., Mount, M. S. 1974. Changes in electrical resistance in the early stages of Fusarium wilt of tomato. *Proc. Am. Phytopathol. Soc.* 1:134 (Abstr.)
11. Clarkson, D. T. 1974. *Ion Transport and Cell Structure in Plants.* New York: Wiley. 350 pp.
12. Comstock, J. C., Martinson, C. A. 1972. Electrolyte leakage from Texas male sterile and normal cytoplasm corn leaves infected with *Helminthosporium maydis* races O and T. *Phytopathology* 62:751–52 (Abstr.)
13. Cordukes, W. E., Wilner, J., Rothwell, V. T. 1966. The evaluation of cold and drought stress of turfgrasses by electrolytic and ninhydrin methods. *Can. J. Plant Sci.* 46:337–42
14. Dainty, J. 1962. Ion transport and electrical potentials in plant cells. *Ann. Rev. Plant Physiol.* 13:379–402
15. Damann, K. E., Gardner, J. M., Scheffer, R. P. 1974. An assay for *Helminthosporium victoriae* toxin based on induced leakage of electrolytes from oat tissue. *Phytopathology* 64:652–54
16. DePlater, C. V., Greenham, C. G. 1959. A wide-range a-c bridge for determining injury and death. *Plant Physiol.* 34:661–67
17. Dostalek, J. 1973. The relation between the electric impedance of apple-tree tissues and the proliferation disease. *Biol. Plant.* 15:112–15
18. Duniway, J. M. 1971. Water relations of Fusarium wilt in tomato. *Physiol. Plant Pathol.* 1:537–46
19. Eslyn, W. E. 1959. Radiographical determination of decay in living trees by means of the thulium x-ray unit. *For. Sci.* 5:37–47
20. Evert, D. R. 1973. Factors affecting electrical impedance of internodal stem sections. *Plant Physiol.* 51: 478–80
21. Fensom, D. S. 1955. Effect of synthetic growth regulators on the electrical potentials of red pine trees. *J. For.* 53:915–16
22. Fensom, D. S. 1957. The bio-electric potentials of plants and their functional significance. I. An electrokinetic theory of transport. *Can. J. Bot.* 35: 573–82
23. Fensom, D. S. 1958. The bio-electric potentials of plants and their functional significance. II. The patterns of bio-electric potential and exudation rate in excised sunflower roots and stems. *Can. J. Bot.* 36:367–83
24. Fensom, D. S. 1959. The bio-electric potentials of plants and their functional significance. III. The production of continuous potentials across membranes in plant tissue by the circulation of the hydrogen ion. *Can. J. Bot.* 37:1003–26
25. Fensom, D. S. 1960. A note on electrical resistance measurements in *Acer saccharum. Can. J. Bot.* 38:263–65

26. Fensom, D. S. 1962. The bioelectric potentials of plants and their functional significance. IV. Changes in the rate of water absorption in excised stems of *Acer saccharum* induced by applied electromotive forces: The "flushing effect." *Can. J. Bot.* 40:405–13

27. Fensom, D. S. 1963. The bioelectric potentials of plants and their functional significance. V. Some daily and seasonal changes in the electrical potential and resistance of living trees. *Can. J. Bot.* 41:831–51

28. Fensom, D. S. 1966. On measuring electrical resistance in situ in higher plants. *Can. J. Plant Sci.* 46:169–75

29. Friedman, B. A., Jaffe, M. J. 1960. Effect of soft rot bacteria and pectolytic enzymes on electrical conductance of witloof chicory tissue. *Phytopathology* 50:272–74

30. Gardner, J. M., Scheffer, R. P., Higinbotham, N. 1974. Effect of host-specific toxins on electropotentials of plant cells. *Plant Physiol.* 54:246–49

31. Garroway, M. O. 1973. Electrolyte and peroxidase leakage as indicators of susceptibility of various maize inbreds to Helminthosporium races O and T. *Plant Dis. Reptr.* 57:518–22

32. Geddes, L. A. 1972. *Electrodes and the Measurement of Bioelectric Events.* New York: Interscience. 364 pp.

33. Ghabrial, S. A., Pirone, T. P. 1967. Physiology of tobacco etch virus-induced wilt of tabasco peppers. *Virology* 31:154–62

34. Glerum, C. 1969. The influence of temperature on the electrical impedance of woody tissue. *For. Sci.* 15:85–86

35. Glerum, C. 1970. Vitality determinations of tree tissue with kilocycle and megacycle electrical impedance. *For. Chron.* 46:63–64

36. Glerum, C., Krenciglowa, E. M. 1970. The dependance of electrical impedance of woody stems on various frequencies and tissues. *Can. J. Bot.* 48:2187–92

37. Greenham, C. G., Cole, D. J. 1950. Studies on the determination of dead or diseased tissues. I. Investigations on dead plant tissues. *Aust. J. Agric. Res.* 1:103–17

38. Greenham, C. G., Norris, D. O., Brock, R. D., Thompson, A. M. 1952. Some electrical differences between healthy and virus-infected potato tubers. *Nature* 169:973–74

39. Greenham, C. G., Muller, K. O. 1956. Conductance changes and responses in potato tubers following infection with various strains of *Phytophthora* and with *Pythium. Aust. J. Biol. Sci.* 9:199–212

40. Greenham, C. G., Daday, H. 1957. Electrical determination of cold hardiness in *Trifolium repens* L. and *Medicago sativa* L. *Nature* 180:541–43

41. Greenham, C. G. 1966. Bruise and pressure injury in apple fruits. *J. Exp. Bot.* 17:404–9

42. Greenham, C. G. 1966. The stages at which frost injury occurs in alfalfa. *Can. J. Bot.* 44:1471–83

43. Hartel, V. O., Grill, D. 1973. The conductivity of bark-extracts from spruce, a sensitive indicator for air pollution. *Eur. J. For. Pathol.* 2:205–15

44. Hayden, R. I., Moyse, C. A., Calder, F. W., Crawford, D. P., Fensom, D. S. 1969. Electrical impedance studies on potato and alfalfa tissue. *J. Exp. Bot.* 20:177–200

45. Hayden, R. E., Dionne, L., Fensom, D. S. 1972. Electrical impedance studies of stem tissue of Solanum clones during cooling. *Can. J. Bot.* 50:1547–54

46. Heald, C. M., Menges, R. M., Wayland, J. R. 1974. Efficacy of ultra-high frequency (UHF) electromagnetic energy and soil fumigation on the control of the reniform nematode and common purslane among southern peas. *Plant Dis. Reptr.* 58:985–87

47. Higinbotham, N. 1973. Electropotentials of plant cells. *Ann. Rev. Plant Physiol.* 24:25–46

48. Higinbotham, N. 1974. Conceptual developments in membrane transport, 1924–1974. *Plant Physiol.* 54:454–62

49. Higinbotham, N., Anderson, P. 1974. Electrogenic pumps in higher plant cells. *Can. J. Bot.* 52:1011–21

50. Jones, M. G. K., Novacky, A., Dropkin, V. 1975. Transmembrane potentials of parenchyma cells and nematode-induced transfer cells. *Protoplasma* 85:15–37

51. Kanemasu, E. T., Thurtell, G. W., Tanner, C. B. 1969. Design, calibration and field use of a stomatal diffusion porometer. *Plant Physiol.* 44:881–85

52. Kaplan, D. T. 1975. *The influence of lesion nematode infection on plant water relations and root tissue integrity.* MS thesis. Univ. Mass., Amherst. 40 pp.

53. Kinney, A. S. 1897. Electro-germination. *Mass. Agric. Coll. Bull. 43.* 32 pp.

54. Kitching, R. 1966. Investigating moisture stress in trees by an electrical re-

sistance method. *For. Sci.* 12:193–97

55. Levitt, J. 1973. *Responses of Plants to Environmental Stresses.* New York: Academic. 697 pp.

56. Levengood, W. C. 1973. Bioelectric currents and oxidant levels in plant systems. *J. Exp. Bot.* 24:626–39

57. Levengood, W. C., Bondie, J., Chen, C. 1975. Seed selection for potential viability. *J. Exp. Bot.* 26:911–19

58. Lin, R. T. 1965. A study on the electrical conduction in wood. *For. Prod. J.* 15:506–14

59. Lin, R. T. 1967. Review of the electrical properties of wood and cellulose. *For. Prod. J.* 17:54–61

60. Lund, E. J. 1931. Electric correlation between living cells in cortex and wood in the Douglas fir. *Plant Physiol.* 6:631–52

61. Luyet, B. 1932. Variation of the electric resistance of plant tissues for alternating currents of different frequencies during death. *J. Gen. Physiol.* 15:283–87

62. McDonald, K. A. 1973. Ultrasonic location of defects in softwood lumber. *Timber Trades J.*, Jan., pp. 17–19

63. Mount, M. S., Bateman, D. F., Basham, H. G. 1970. Induction of electrolyte loss, tissue maceration, and cellular death of potato tissue by an endopolygalacturonate trans-eliminase. *Phytopathology* 60:924–31

64. Neal, D. L., Neal, J. L. 1973. Uses and capabilities of electronic capacitance instruments for estimating standing herbage. *J. Br. Grassl. Soc.* 28:81–89

65. Novacky, A., Hanchey, P. 1974. Depolarization of membrane potentials in oat roots treated with victorin. *Physiol. Plant Pathol.* 4:161–65

66. Novacky, A. 1974. Transmembrane potentials in bacterial hypersensitivity. *Proc. Am. Phytopathol. Soc.* 1:75 (Abstr.)

67. Novacky, A., Karr, A. L., VanSambeek, J. W. 1976. Electrophysiology as a tool for study of plant disease development. *BioScience* 26:In press

68. Osterhout, W. J. V. 1922. *Injury, Recovery, and Death, in Relation to Conductivity and Permeability.* Philadelphia: Lippincott. 259 pp.

69. Pettit, R. E. 1974. Bioelectric discharge patterns of plant tissues as influenced by plant pathogens. *Proc. Am. Phytopathol. Soc.* 1:80 (Abstr.)

70. Polozhentsev, P. A., Zolotov, L. A. 1970. Dynamics of electrical resistance of bast tissues as an indicator of

changes in their physiological condition. *Sov. Plant Physiol.* 17:694–98

71. Safford, L. O., Shigo, A. L., Ashley, M. 1974. Gradients of cation concentration in discolored and decayed wood of red maple. *Can. J. For. Res.* 4:435–40

72. Sheriff, D. W. 1974. Magnetohydrodynamic sap flux meters: an instrument for laboratory use and the theory of calibration. *J. Exp. Bot.* 25:675–83

73. Shigo, A. L., Sharon, E. M. 1970. Mapping columns of discolored and decayed tissue in sugar maple, *Acer saccharum. Phytopathology* 60:232–37

74. Shigo, A. L., Hillis, W. E. 1973. Heartwood, discolored wood, and microorganisms in living trees. *Ann. Rev. Phytopathol.* 11:197–222

75. Shigo, A. L., Shigo, A. 1974. Detection of discoloration and decay in living trees and utility poles. *US For. Serv. Res. Pap. NE-294.* 11 pp.

76. Shigo, A. L., Berry, P. 1975. A new tool for detection of decay associated with *Fomes annosus* in *Pinus resinosa. Plant Dis. Reptr.* 59:739–42

77. Shortle, W. C., Shigo, A. L., Ochrymowych, J. 1976. Detection of decay in utility poles with the shigometer. *For. Prod. J.* 26:In press

78. Skaar, C. 1972. *Water in Wood.* Syracuse, NY: Syracuse Univ. Press. 218 pp.

79. Skutt, H. R., Shigo, A. L., Lessard, R. A. 1972. Detection of discolored and decayed wood in living trees using a pulsed electric current. *Can. J. For. Res.* 2:54–56

80. Spanswick, R. M. 1972. Electrical coupling between cells of higher plants: A direct demonstration of intercellular communication. *Planta* 102:215–27

81. Stamm. A. J. 1964. *Wood and Cellulose Science.* New York: Ronald. 549 pp.

82. Stone, G. E., Monahan, N. F. 1905. The influence of electrical potential on the growth of plants. *17th Ann. Rept. Hatch Exp. Stn. Mass. Agric. Coll.*, pp. 14–31

83. Stone, G. E. 1909. Influence of electricity on microorganisms. *Bot. Gaz.* 48:359–79

84. Stone, G. E. 1914. Electrical injuries to trees. *Mass. Agric. Coll. Bull. 156.* 19 pp.

85. Svejda, F. 1966. Investigations on the relationship between winterhardiness in roses and the electric impedance

measured during the growing period. *Can. J. Plant Sci.* 46:441–48

86. Taper, C. D., Ling, R. S. 1964. Relation of electrical resistance to vigour of apple scions. *Nature* 203:782–83

87. Tattar, T. A., Shigo, A. L., Chase, T. 1972. Relationship between the degree of resistance to a pulsed electric current and wood in progressive stages of discoloration and decay in living trees. *Can. J. For. Res.* 2:236–43

88. Tattar, T. A., Saufley, G. C. 1973. Comparison of electrical resistance and impedance measurements in wood in progressive stages of discoloration and decay. *Can. J. For. Res.* 3:593–95

89. Tattar, T. A., Blanchard, R. O., Saufley, G. C. 1974. Relationship between electrical resistance and capacitance of wood in progressive stages of discoloration and decay. *J. Exp. Bot.* 25:658–62

90. Tattar, T. A. 1974. Measurement of electric currents in clear, discolored, and decayed wood from living trees. *Phytopathology* 64:1375–76

91. Van den Driessche, R. 1973. Prediction of frost hardiness in Douglas fir seedlings by measuring electrical impedance in stems at different frequencies. *Can. J. For. Res.* 3:256–64

92. Van Alfen, N. K., Turner, N. C. 1975. Influence of a *Ceratocystis ulmi* toxin on water relations of elm (*Ulmus americana*). *Plant Physiol.*, 55:312–16

93. Van Alfen, N. K., Turner, N. C. 1975. Changes in alfalfa stem conductance inducted by *Corynebacterium insidiosum* toxin. *Plant Physiol.* 55:559–61

94. Van Bavel, C. H. M., Nakayama, F. S., Ehrler, W. L. 1965. Measuring transpiration resistance of leaves. *Plant Physiol.* 40:535–40

95. Van Sambeek, J. W., Novacky, A., Karr, A. L. 1975. Effect of helminthosporoside on transmembrane potentials in leaf mesophyll cells of *Saccharum. Plant Physiol.* 56 (Suppl.): 53 (Abstr.)

96. Walton, P. D. 1973. Electrical impedance of tissue of grasses receiving cold treatments. *Can. J. Plant Sci.* 53: 125–27

97. Wargo, P. M., Skutt, H. R. 1975. Resistance to pulsed electric current —an indicator of stress in forest trees. *Can. J. For. Res.* 5:557–61

98. Warner, C. D. 1892. Electricity in agriculture. *Mass. Agric. Coll. Bull. 16.* 8 pp.

99. Warner, C. D. 1893. Electricity in agriculture. *Mass. Agric. Coll. Bull. 23.* 15 pp.

100. Weber, L. J., Mclean, D. L. 1975. *Electrical Measurement Systems for Biological and Physical Scientists.* Reading, Mass.: Addison-Wesley. 399 pp.

101. Wheeler, H., Hanchey, P. 1968. Permeability phenomena in plant disease. *Ann. Rev. Phytopathol.* 6:331–50

102. Williams, E. J., Johnston, R. J., Dainty, J. 1964. The electrical resistance and capacitance of the membranes of *Nitella translucens. J. Exp. Bot.* 15: 1–14

103. Wilner, J. 1967. Changes in electric resistance of living and injured tissues of apple shoots during winter and spring. *Can. J. Plant Sci.* 47:469–75

104. Zaerr, J. B. 1972. Early detection of dead plant tissue. *Can. J. For. Res.* 2:105–10

105. Zhuravleva, M. V. 1972. Method of determining cambial activity in growing spruce trees. *Lesn. Zh.* 15:140–41 (in Russian). *For. Abstr.* 34:240–41

Copyright © 1976 by Annual Reviews Inc.
All rights reserved

RELATIONSHIPS BETWEEN   ♦3645 NEMATODE POPULATION DENSITIES AND CROP RESPONSES[1]

K. R. Barker

Department of Plant Pathology, North Carolina State University, Raleigh, North Carolina 27607

T. H. A. Olthof

Research Station, Research Branch, Agriculture Canada, Vineland Station, Ontario, Canada

INTRODUCTION

The fundamental quantitative relationships between plant-parasitic nematodes and growth and yield of annual crops are primarily a function of preplant densities. The importance of initial densities results from the nematodes' negligible motility and relatively low reproductive rate. Nematode-plant relationships may be modified by plant variety and age, nematode species or race, and environment. Responses of perennials to nematodes often are altered by reintroduction, buildup of nondetectable initial densities with time, or both. Characterization of nematode-host interactions is dependent on reliable methods of monitoring nematode populations, precise measurement of host responses, and consideration of environmental factors. Understanding these parameters has served a vital role in the development of effective control systems that reduce initial inoculum for all crops and prevent buildup on perennials. Early nematologists made very significant contributions in these areas (20, 39, 74), including the concept of tolerance of plants to disease (113). Recently, the relationships between plant yield, tolerance, and preplant nematode densities have been expressed in mathematical models (55, 91, 119, 121, 129), but few predictive models or simulators are available. Concurrently, research interests in pest management are evolving to include fundamental nematode-host relationships, ecological interactions, and economic considerations. Although many problems are encountered in elucidating nematode-plant relationships, sufficient practical information has been obtained for limited grower advisory services to function successfully (6, 93).

[1] Paper No. 4863 of the Journal Series of the North Carolina Agricultural Experiment Station, Raleigh, North Carolina.

Knowledge of the population dynamics of nematodes and host responses at the plant, cellular, and physiological levels is essential to understand the influence of nematodes on crop yield. Discussion of these topics is limited since they were reviewed recently (14, 24, 25, 27, 81, 121).

EXPERIMENTAL APPROACHES FOR DETERMINING HOST RESPONSES

The initial density (P_i) required to cause significant plant damage and yield losses varies with nematode species. Low numbers of *Meloidogyne arenaria* often cause severe damage on tobacco (28, 104), whereas extremely high numbers of *Criconemoides curvatum* may have no effect on the growth of hairy vetch (69). Damage caused by many nematode species varies with different races or pathotypes (55, 64). Expression of resistance or tolerance to nematodes is influenced immensely by host age (10, 36, 156).

The effects of plant-parasitic nematodes on host growth can be determined with greater precision in the greenhouse or growth chamber than in the field. Slight changes of systems such as size of pots in greenhouse experiments, however, alter the effects of nematodes on plant growth (96). Thus, results from the greenhouse cannot be used to relate nematode densities to the magnitude of disease under field conditions. For example, Seinhorst & Sen (130) found that the equilibrium density of *Heterodera trifolii* was low in pastures and could not be correlated with poor growth of white clover, but in the greenhouse, this nematode reproduced rapidly and caused stunting when present in excess of 50 eggs per gram of soil. Interspecific competition may limit populations of some species (15, 49, 50, 59, 141). Soil texture, structure, temperature, and moisture differentially affect motility, penetration, and buildup of nematode species (137, 146, 149).

Because of the problems in relating growth response in greenhouse experiments directly to field situations, Jones (52) used microplots with known densities of given nematodes, such as *Heterodera schachtii,* to quantify relationships between growth of plants and numbers of nematodes. Microplot methodology includes most of the advantages of greenhouse experiments such as less variation in nematode numbers and more uniform experimental conditions, yet the environmental conditions are largely comparable to those prevailing in the field.

Research with microplots generally involves only one species of nematodes. Some single species cause no damage in fumigated soil, whereas concomitant inoculations of two species may cause severe damage in the greenhouse (143). Since nematodes occur in polyspecific communities (91), experiments are still needed to determine crop responses to mixed populations of nematodes and of nematodes with associated organisms, including detrimental soil inhabitants and beneficial organisms (mycorrhizal fungi, Rhizobia). Most experiments involving such interactions have been limited to the greenhouse, although Hijink (42) determined the effects of more than one nematode species under field conditions.

Field research designed to relate crop responses to nematode population densities involves crop rotation (82) and chemical soil treatments (83, 124). Rotation systems provide the most natural experimental conditions but introduce many unknowns such as the effects of previous crops on soil structure, population levels of many organisms in the soil, and nutrient levels (43, 82, 91). Chemical soil treatments have greater limitations in these types of experiments. Fumigation with certain materials may affect the forms of nitrogen (34); some nematicides have an inherent stimulatory or inhibitory effect on plant growth (1, 80). Furthermore, certain nonfumigant nematicides alter nematode behavior without effecting death (47) thereby making meaningful nematode assays very difficult. Seinhorst (124) has provided theoretical relationships between dosage of nematicide and nematode survival, and between nematicide dosage and increases in relative yield at different initial densities.

The relationships between nematode and host in the field vary with location and with stress conditions. Thus, to quantify interactions between nematode densities and crop responses, information should be obtained from a range of experimental conditions for each nematode-crop combination.

CONSIDERATIONS IN MONITORING NEMATODE POPULATIONS

Nature of Nematode Community as Related to Sampling

The structure of the polyspecific communities of nematodes varies immensely with developmental stage of individuals and their relative density, age, activity, infectivity, and distribution (6). Plant-parasitic nematodes develop and reproduce in close association with their hosts, and are classified according to their feeding habits. Some species of nematodes are ectoparasites, others semiendoparasites, and still others are endoparasites, which increases the technical problems involved in population assays. Growing nonhost crops or fallowing often results in a dramatic decline of population densities of nematodes (52). Furthermore, the presence of a poor host may result in an even greater decline than will fallow (52, 119).

Predicting changes in populations with precision is difficult, and accuracy depends on the type of nematode as well as the environment. Changes in *Heterodera rostochiensis,* a host-specific pest with one life cycle per year, are relatively predictable in England where this nematode declines in the absence of a host at the rate of 30–40% per year (55). The observed and the predicted densities of *Ditylenchus dipsaci* in Utah were in very close agreement when season, soil temperature, and moisture were considered as parameters (140). In contrast, seasonal changes in populations of *Meloidogyne* spp. are difficult to predict (7, 28); however, further knowledge of environmental influences on species of this genus could effect a solution to this problem (75).

Nicholson (79) stated that competition, particularly intraspecific competition, tends to cause animals to reach and maintain stable densities. Some investigators, however, do not accept Nicholson's theories (2). The activity of man and the continuous changes in agricultural fields prevent nematodes from reaching this steady density (121). Oostenbrink (91) finds that this steady state is usually followed by a decline, and that nematode reproduction usually follows a "top curve," depending on time as well as increasing initial densities. The densities of a few species such as *Xiphinema diversicaudatum* remain relatively stable with time (33), whereas numbers of some species may decline below detectable levels during unfavorable conditions. This species produces eggs only for a brief period each year in England, and may require two years for development from egg to adult (33), whereas *Meloidogyne incognita* reproduces continuously in optimum (warm) conditions and completes its life cycle every 25–30 days (139). In Belgium, only one generation of *Meloidogyne naasi* develops per growing season (35), but more than 90% of the eggs overwinter and do not hatch until soil temperatures exceed 10°C in the spring. *Meloidogyne arenaria* may, on the other hand, complete several life cycles in the southeastern United States, but less than 5% of the larvae survive during the winter (28). These variations present difficulties in assessing the status of a population or community of nematodes at a given time and location. Assessing nematode densities in dry soil also requires special precautions with respect to extracting techniques (132).

Species comprising a community in a field vary in both horizontal and vertical distribution. Uneven horizontal distribution of plant-parasitic nematodes was observed by Cobb (20), who developed many of the sampling and assay techniques still in use. This type of variation seems to be greatest with species that damage their host and/or deposit high numbers of eggs in limited loci (6). The vertical distribution of plant-parasitic nematodes is highly variable, but is generally related to root distribution. *Ditylenchus dipsaci,* however, may migrate downward during certain times of the year (65), and *Trichodorus* species generally occur at greater depths than do most species (107). Wallace (145) concluded that vertical distribution patterns of nematodes are likely produced by orientation in movement to a particular soil zone where conditions are optimal for reproduction. This zone may fluctuate with habitat and season as moisture and temperature gradients change. The microclimate of the soil is also influenced by type of vegetation, penetration of the soil by plant roots, irrigation patterns, and intermittent disturbance of the cultivated layer, and in turn these affect nematode populations (91).

Proctor & Marks (102), estimated that at least seven hours of field sampling and laboratory analyses are required to obtain an estimate of the mean population density of *Pratylenchus penetrans* in a plot of 0.01 ha with the coefficient of variation below 20% of the mean at a 95% confidence level. Much of this variation with nematodes is due to aggregation, such as that of eggs in masses or cysts and small colonies in roots or root debris.

Seinhorst (125) showed that the relation between average population density of nematodes and yield depends on the frequency distribution of nematode density in the field. Overestimation of losses increases as the degree of distribution of the nematodes differs from a random one. Variation in nematode counts can be minimized by limiting sampling errors by adapting the sampling method and number of samples taken to the distribution of the population sampled and the magnitude of the error that is tolerated. J. W. Seinhorst (personal communication) reduced variation to 15–20% by collecting and bulking 80 twenty-gram soil samples/m² to a depth of 20 cm, extracting at least 500 grams, and counting at least 100 nematodes per sample.

Methods of Population Assay

The selection of suitable extraction procedures is essential in relating crop response to nematode population densities. Nematode populations may consist of motile, slightly motile, or immotile forms, or various combinations of these. Ectoparasitic species are in the soil proper, whereas high percentages of endoparasitic forms are in roots during certain periods of the year. Actively moving larvae or adults of various nematodes can be extracted by most procedures, whereas few methods are suitable for immotile species (6). Thus, the assay procedure used depends on the kinds and numbers of nematodes present and their characteristics, the nature and condition of the samples including soil type, and the time of collection. Another important consideration in nematode assays is adequate mixing of a representative 500– to 1500-cm³ soil sample before extraction of a subsample. If population densities are below a *detectable* level with laboratory techniques, an appropriate bioassay, tomato for most *Meloidogyne* spp. or soybean for *Heterodera glycines,* may be utilized.

GENERAL NEMATODE-HOST RELATIONSHIPS

The general host responses to infection by nematode species with various feeding habits have been reviewed by Krusberg (62) and Dropkin (24). Nematodes often damage plants by altering host physiology, which may result in the formation of galls or other abnormal leaf or root structures, and induce necrosis in shoot or root tissues. Nematode-induced changes in host physiology (10, 25, 97, 152, 153) apparently are responsible for tissues becoming more susceptible to associated fungi and bacteria (11, 101).

In contrast to detrimental crop responses, nematodes may stimulate plant growth under certain conditions. Low populations of *Heterodera rostochiensis, Heterodera tabacum* (18, 19, 67), *Meloidogyne hapla, Pratylenchus penetrans,* and *Heterodera schachtii* (85–87) can enhance host growth and sometimes result in increased yields. Reparative growth was implicated for enhanced growth of root systems parasitized by *Heterodera rostochiensis* (18). This stimulatory effect was short-lived and was followed by a suppression in growth of infected roots. Infection of some unsuitable "hosts" by *Meloidogyne* species may increase growth (68). Pepper grew more rapidly with low levels of *Meloidogyne javanica* than healthy plants, and greater growth rates of peanuts also resulted from moderate levels of *M. javanica* or *M. incognita* (68). Increased synthesis of growth regulators in response to nematode infection of sugar beet by *Heterodera schachtii* has been suggested as being responsible for increases in petiole length over non-infested plants (23). The wild grass, *Sporobolus poiretti,* develops a heavier top and more inflorescence when infected with *Aphelenchoides besseyi* (70). Changes in growth regulators in infected plants, as described by Viglierchio (144), could be responsible for these types of increased growth. Feeding by *Trichodorus christiei* enhances the development of secondary roots less than 1 cm in length but depresses the development of longer roots (45). Stimulatory responses attributed to nematodes may be due to the introduction of unidentified mycorrhizal fungi in some instances where methyl bromide-treated soil or steamed soil is used. The absence of endomycorrhizae in fumigated soils can cause stunting of plants (60).

Depression of plant growth by nematodes is a more common phenomenon than stimulation. *Meloidogyne* species generally induce root-gall formation, which results in heavier root systems, but concomitantly suppress shoot growth. In contrast, most pathogenic species inhibit root and shoot growth.

Concepts of Tolerance, Resistance, Susceptibility, and Intolerance

Theoretically, relationships between plant-feeding nematodes and their hosts range from neutral ecological association to different levels of parasitism and pathogenesis (81). Cobb apparently was the first to use the concept of tolerance in 1894 in reference to *rust-enduring* varieties of wheat (113). Plant pathologists use the term to refer to host plants that survive and give satisfactory yields at a level of infection that causes economic loss on other varieties of the same species (113). To understand tolerance, the reaction of the host and parasite must be examined, and not just the relationship between inoculum level and plant yield (151). Thus, it is necessary to understand the biochemical and ecological systems responsible for tolerance of a plant to a given nematode. Jones (52) and Oostenbrink (89) suggested that the term *host plant* should be replaced by *host efficiency or suitability* (measuring nematode reproduction) and *susceptibility to damage*.

Several schemes or attempts to characterize all possible nematode-host relationships have subsequently been made. Hollis (46) separated parasitic and pathogenic action and suggested that 90% of phytophagous nematodes are simply parasites, and about 10% are pathogens. Although the system presented by Hollis has merit, the practical use of his approach has limitations (4). The terminology used by Dropkin & Nelson (26), relating parasite growth and host growth on the basis of tolerance, susceptibility, resistance, and intolerance has been utilized widely. Their basic scheme has been revised extensively (21, 105). Cook (21) diagrammatically characterized relationships in terms of the concepts of *resistance* and *susceptibility* in reference to host efficiency, and in terms of *tolerance* and·*intolerance* in reference to host sensitivity. This scheme is useful; however, the term *susceptibility* is generally used in reference to host sensitivity; whereas *resistance* implies minimal host damage and reproduction of the parasite. These four terms should be used only with a qualification, indicating whether or not they are used in reference to damage or to reproduction. Although good hosts often are damaged at medium to high densities of nematodes and poor hosts are not, good hosts may also be tolerant to very high densities of the parasite, and poor hosts are sometimes damaged by them (89, 92, 120). Examples of poor hosts that may be damaged are the following: tobacco resistant to *M. incognita* (F. A. Todd, personal communication); corn infected by *Heterodera avenae* (51); resistant potato varieties infected by certain races of *Heterodera rostochiensis* (133). Corn, a good host for *M. incognita* and *Tylenchorhynchus claytoni,* can support high populations without apparent damage (K. R. Barker, unpublished).

Concepts of Tolerance Limits, Equilibrium Density, Maximum Rates of Reproduction, and Economic Threshold

The most promising scheme for quantitatively characterizing host sensitivity and efficiency involves the concepts of tolerance limit, maximum rate of reproduction, and equilibrium density (119, 120). *Tolerance limit* refers to the density below which no loss in yield occurs. A somewhat similar concept of *tolerance level* was developed by Oteifa & Elgindi (95). The status of a plant as a host for a parasitic nematode may· be summarized by *a* which is the *maximum rate of reproduction* and *E* the *equilibrium density* where birth rate equals death rate (120). For a good host, both of these constants are high; for a poor host, both are low. For an intermediate host, both are intermediate or one is fairly high and the other is relatively low. The maximum rate of reproduction and the equilibrium density of nematode populations depend on the external conditions as well as on inherent characters of the nematode and plant. The influences of a factor on one of these are not necessarily related. A change of conditions may result in an increase of one and a decrease of the other, or an increase or decrease of both.

The usefulness of tolerance limits in ecological studies of nematodes was questioned by Wallace (151) who suggested that it may be more realistic to let the farmer decide what yield loss he is willing to accept on an economic basis (action threshold). Wallace indicated that the density of nematodes associated with this loss could be called the tolerance level. Graphic tabulations of density ranges as related to expected crop damage as presented by Oostenbrink (93) should prove useful for such purposes. The term *threshold density* has been used to refer to the approximate minimal density which inhibits plant growth (6). This concept is easily expanded to *economic threshold density* (action threshold) which would reflect costs of production, and the price of a given crop with time. This concept should fulfill the practical need suggested by Wallace, and is in agreement with its general use with other pests (103). Olthof & Potter (85) introduced the concept of *economic loss threshold,* defined as the percentage of the total value of a crop equivalent to the cost of nematode control. A comprehensive summary of approximate threshold densities available is given in Table 1.

CHARACTERIZATION OF QUANTITATIVE RELATIONSHIPS

Critical investigations designed to determine relationships between nematode population densities and crop growth have been limited to a few nematode-host combinations. Relations between initial population densities (P_i) of nematodes at time of planting, and the damage they cause to plants have been described by linear regression between log P_i and growth of damaged plants (8, 40–42, 44, 52, 67, 91, 116, 150) by a regression according to a quadratic curve between these factors (98), and by linear regression between log density and probit unit of number of diseased plants (112). As Seinhorst (119) indicates, approximation of theoretical relationships to those actually observed is fair to good in all of these

Table 1   Relationship of certain nematodes to the yield or growth of plants[a]

Nematode species and plant[a]	Type of experiment[b]	Approximate threshold density (per 100 cm³[†] or gm[††] of soil)[c]	Correlation coefficient[c]	Reference
Belonolaimus longicaudatus				
Cotton	F, M	5[†]	−0.75**	6, d
Peanut	F, M	5[†]	0.82**; −0.70**	110, d
Soybean	F, M	5[†]	0.37 to −0.54*	110, d
Criconemoides ornatus				
Peanut	F, M	5–10[†]	−0.65***	110, d
Ditylenchus dipsaci				
Onion	F	0.2–1.0[††]	−0.96**	116
Potato (vs *Phoma* infection)		<2[†]		41, 91
Helicotylenchus dihystera				
Corn	M	>200	NS	d
Peanut	F, M	>250[†]	−0.50**	110, d
Soybean	F, M	>200[†]	NS	d
Heterodera avenae				
Oat (plant height)	Pot	40[††]		40
Wheat		<100[††]	−0.93	72
Heterodera rostochiensis				
Potato (length of haulms)	Pot	1500[††]		98
(Yield)	F	31 viable cysts[††]		16
Heterodera schachtii				
Cabbage	M	600 larvae[††]	−0.32**	87
Cauliflower	M	1800[††]	NS	87
Rutabagas	M	100 larvae[††]	−0.44**	87
Spinach	M	200 larvae[††]	−0.68**	87
Sugar beet	F, M	400–1000 eggs[††]; 300 larvae[††]		38, 52
Table beet	M	600[††]	−0.51**	87

Heterodera tabacum				
Tobacco	Pot	5,000 larvae††		67
Heterodera trifolii				
White Clover	Pot	5,000 eggs††		130
Meloidogyne arenaria				
Tobacco	F	<2†	−0.67**	104
Meloidogyne hapla				
Cabbage	M	1,800 larvae††	NS	85
Cauliflower	M	600 larvae††	NS	85
Lettuce	M	100 larvae††	−0.39** to NS	85, 100
Onion	M	200 larvae††	−0.46**	85
Pea	F	<10†	−0.94**	42, 91
Peanut	F, M	20–100†	−0.57*	110, d
Potato	M	100 larvae††	NS	85
Soybean	M	100–200 larvae†	−0.52*	d
Spinach	M	600–1,800 larvae††	NS	100
Strawberry	F	5 larvae†		6
Sugar beet	F	<100†	−0.92**	91
Table beet	M	600–1,800 larvae††	NS	100
Tomato				
Coastal Plain (NC)	M	<4 larvae†	−0.83**	8
Mountains (NC)	M	150 larvae†	−0.68**	8
Meloidogyne incognita				
Corn	M	>400†	NS	d
Cotton	F, M	<200†	−0.85**	d
Lettuce	M	>60 larvae†		114
Soybean	F, M	20†	−0.60*	58, d
Tobacco	F	5–10†	−0.83* to NS	6, 30
Tomato				
Coastal Plain (NC)	M	<4 larvae†	−0.99**	8
Mountains (NC)	M	40 larvae†	−0.81**	8

Table 1 (Continued)

Nematode species and plant[a]	Type of experiment[b]	Approximate threshold density (per 100 cm³[t] or gm[tt] of soil)[c]	Correlation coefficient[c]	Reference
Meloidogyne javanica				
Soybean	M	35 larvae and eggs[t]	−0.76	d
Tomato (growth)	Pot	<0.2 larvae[tt]	−0.92**	134
Pratylenchus brachyurus				
Peanut	F	5[t]	−0.36*	6, 110
Tobacco	F	>40[t]		6
Pratylenchus crenatus + Tylenchorhynchus dubius + Rotylenchus robustus				
Oat	F	33[t]	−0.94**	91
Pratylenchus penetrans				
Apple	F	20–50[tt]		44
Birdsfoot-trefoil	Pot	<200[t]	−0.99**	155
Cabbage	M	200–600[tt]	−0.43**	86
Cauliflower	M	200–600[tt]	−0.67**	86
Corn (sweet)	M	100[tt]	−0.43**	86
Daffodil		0.2–1[tt]		117
Digitalis purpurea	F	<10[tt]		90
Lettuce	M	200–600[tt]	−0.33** to 0.49**	86, 100
Onion	M	<100[t]	−0.81**	32, 86
Pea	F	100[tt]	−0.84*	91
Peach (length of shoots)	F	5[tt]		78
Potato	F	100[tt]; 200–600[tt]	−0.38**	86, 89, 116
Soybean	M	<200[t]	−0.66**	d
Spinach	M	600–1,800[tt]	−0.31**	100
Table beet	M	600–1,800[tt]	−0.35**	100
Tobacco	F, M	100–200[t]	−0.88**	63, 84
Rotylenchus robustus + P. penetrans				
Valeriana (root yield)	F	<50[t]	−0.89**	42, 91

Criconemoides xenoplax + *P. penetrans* + *R. robustus*				
Birch (seedling growth)		<100[t]	−0.93**	91
Rotylenchus uniformis				
Carrot	Pot (20°C)	700[tt]		Seinhorst (personal communication)
Carrot	Pot (8°C)	40[tt]		
Trichodorus christiei				
Corn	M	>20[t] (yield)	NS (yield); −0.72 (growth)	d
Peanut	F		−0.39*	110
Soybean	F, M	20–40[t]	−0.39* to NS	110, d
Trichodorus porus				
Pine (height)	F	<5[tt]	−0.42*	150
Tylenchorhynchus claytoni				
Azalea	M	>20[t]		3,6
Corn	M	>200[t]	NS	d
Soybean	M	>400[t]	NS	d
Tobacco	F	not damaged		3
Tylenchorhynchus claytoni + *T. christiei*				
Corn	F	<250[t]	−0.79*	104
Xiphinema americanum				
Grape	F		−0.29* to −0.39**	31

[a]Unless otherwise indicated, data on crop yield.
[b]F = field experiment; M = microplot experiment.
[c][t] = per 100 cm³ of soil; [tt] = per 100 grams of soil (some threshold values and correlation coefficients determined from publication indicated). Asterisks * and ** indicate significance at P = 0.05 and P = 0.01, respectively.
[d]Unpublished data, K. R. Barker.

studies. Additional promising means of relating naturally occurring nematode densities to crop damage include the use of isopathological curves (115) and the utilization of a nonparametric statistical test for computing rank correlation coefficients to determine trends of data which are too variable for normal statistical analyses (22). Plant growth may often be inversely correlated with midseason (P_m) and final densities (P_f) (30) as well as with changes during the growing season (31). Seinhorst (119) concluded that an exponential model, based on Nicholson's competitive curve (79), is more appropriate than linear regressions to relate crop growth to nematode density. Jones et al (53) in microplot tests found that the regression of the weight of pea tops on the log density of *Heterodera goettingiana* per gram of soil gave good fits at two sites. However, the pattern at another site gave more of a sigmoid curve. Further experiments with *H. rostochiensis* on potato (56) gave a linear relationship, but the real relationship appeared to be sigmoid.

Transformation of Nematode Population Data

Transformations ($\log_{10}$, natural log, and square root) are used to facilitate the management of nematode population data (Figures 1, 2, 3). Wallace (151) and Vanderplank (142), however, have questioned the use of such transformations in attempting to relate infection, disease development, or both, to inoculum densities. Vanderplank (142) points out that logarithms obscure relations, and data so plotted are likely to be misinterpreted. Wallace (151) concluded that the tolerance limits derived by Seinhorst & Kuniyasu (127) from the initial flat portion of the sigmoid curve from the log data of *Rotylenchus uniformis* and growth of carrots was largely the result of the $\log_{10}$ transformation. Arithmetic plotting and statistical analysis of the data by Wallace (151) indicated that the relationship between the weight of carrot leaves and nematode densities was essentially linear, and regression analyses of other data by Wallace were similar to that with *R. uniformis,* or else no statistically significant regression lines could be fitted.

Although the use of transformations will continue, we should be aware of the problems involved. Ferris (30) indicates that with root-zone sampling, there is a biologically based statistical necessity for a transformation of data, dependent upon the distribution of nematodes in the soil. Where the population density is low and variable, the sample mean will be low and the range wide, especially with a population that is aggregated. Early-season sampling may involve errors which are not normally and independently distributed, but tend to increase as means increase. This skewness decreases as the population distribution becomes more uniform. Ferris (30) found that log transformation of early to midseason population data of *Meloidogyne* spp. stabilized the variance, but tobacco yield and value sometimes were better correlated with nontransformed than transformed final densities.

Linear Regression

Regression analyses have been used so extensively to relate numbers of nematodes in greenhouse, microplot, and field experiments to crop performance that

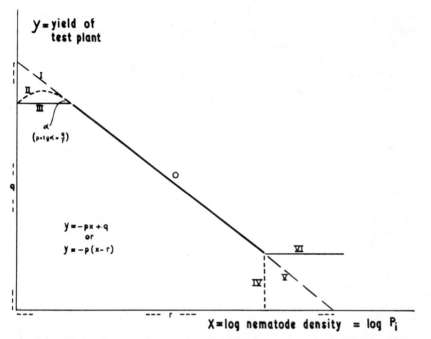

Figure 1   Model of a curve showing the relationship between nematode densities and plant yield, with the main possible variations at low densities (I, II, III) and at high densities (IV, V, VI) (after Oostenbrink, 1966).

only representative data are included in this review (Table 1). Oostenbrink (91, 92) stresses the importance of the rectilinear regression of log numbers of nematodes compared to yield, and indicates that the relationships are similar to response curves used in animal toxicology. As shown in Figure 1, he indicates that in some instances low densities of nematodes cause damage, and the main curve is extended to zero nematodes per sample (I), there may be a peak due to stimulation of plant growth at low numbers (II), or it may be flat because the plant can sustain a certain amount of damage or may compensate for it by regeneration (III). At high densities, the curve may suddenly drop to zero when plants die (IV), it may also continue in a linear fashion to zero yield (V), or the curve may become horizontal when the plants escape damage (VI). Most of the data used by Oostenbrink support his conclusions that most yield or growth responses are linear and that few exceptions to this rule occur, which explains his questioning the use of sigmoid curves in lieu of linear regressions. He further indicates that the two basic assumptions on which the sigmoid curve is built are essentially unfulfilled for plant nematodes, namely, even distribution in the soil and equal effect per nematode on a plant for a low and high density.

Highly significant negative regressions can be obtained readily in microplot experiments dealing with nematodes such as *Meloidogyne* spp. or *Heterodera* spp.

on annual crops. However, with many ectoparasites such as *Tylenchorhynchus claytoni, Helicotylenchus dihystera,* and *Trichodorus christiei,* infestations of microplots with single nematodes frequently result in no significant effect on yields (K. R. Barker, unpublished). This lack of response may often be due to poor establishment of nematodes placed in artificial milieux such as pots with soil (M. Oostenbrink, personal communication), or due to the absence of other species of nematodes (or fungi or bacteria) normally interacting with them under field conditions.

A study of perennial crops in relation to nematode densities, even with simple regression analysis, becomes complex. Reintroduction, buildup of nondetectable densities, and decline of populations after severe damage compound the problems in characterizing such interactions. Damage to cultivars of *Ilex crenata* caused by *Criconemoides xenoplax* in microplots may be correlated with nematode densities at one time, but not at subsequent samplings, especially after the plant is severely stunted (3). Further research in North Carolina indicates that older plants are more tolerant to this nematode, especially with low initial densities. Nematodes can severely damage the root systems of smaller plants, which results in declines of population densities. This type of response was encountered with *Meloidogyne incognita* on various hollies and with *Pratylenchus vulnus* on boxwood in North Carolina where these nematodes build up to relatively high densities on their respective hosts and then decline to numbers that may be near or below detectable levels as plant health declines (K. R. Barker et al, unpublished). The tolerance limit of one-year-old apple seedlings is six times that of very young plants (122).

In an attempt to relate numbers of nematodes to growth and yield of long-established grapevines, Ferris & McKenry (31) used relative changes in population densities for given growing seasons for regression analyses. Densities of *Xiphinema americanum* were negatively correlated with yield of low-vigor vines, but had no relationship to yield of high-vigor vines. Densities of *Pratylenchus hamatus* were positively correlated with yield, growth, and vigor. Many of the problems in such experiments undoubtedly involve the complex problem of population densities declining rapidly, once plants are severely damaged; the time of this event would vary greatly on long-lived plants. Thus, age of perennial crops and timing of nematode assays are very critical. The role of nematodes on forest trees is equally complex (108).

Exponential Models

The basic concepts involved in mathematical models (55, 91, 121) have raised many questions and opened up new lines of research in plant nematology. Seinhorst (117) has offered suggestions on the quantitative study of nematode-plant relations including the use of a wide range of nematode densities if amount of damage is to be predicted. He developed arbitrary equations to relate P_i or density at the time of planting to the yield [(119, 121); Figure 2]:

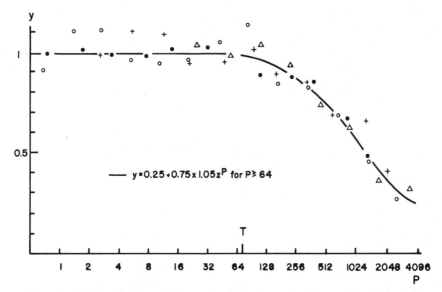

Figure 2 Relationship between log nematode density and relative weight of plants in a number of pot experiments (weight of plants in the absence of nematodes = 1). Line according to equation $y = m + (1 - m)Z^{P-T}(m = 0.25)$. The different symbols indicate different experiments (after Seinhorst, 1972).

$$Y = Z^P \qquad\qquad 1.$$
$$Y = m + (1 - m)Z^{P-T} \qquad\qquad 2.$$

which he concludes, fit such relations satisfactorily. For these models, Y is the ratio between yield and nematode density P and the ratio in the absence of nematodes; Z is a constant less than 1 (equal to the proportion of plants not attacked at density $P = 1$); P is the nematode density, and would be equal to or greater than T, and $Y = 1$ when P is equal to or less than T; m (the minimum yield) is the yield at nematode densities in which all available space is occupied by the nematodes; and T the nematode density below which no noticeable loss in yield occurs. The value of Z depends on the nematode species, the plant species, and external conditions; Z^{-T} ranges from 1.05 to 1.15.

The primary hypothesis behind the models by Justesen & Tammes (57) and Seinhorst (118), which are based on the Nicholson competition curve (79), is that the effect per nematode on rate of growth per plant decreases with increase in nematode density, since areas of root tissue damaged by single nematodes overlap as the density increases. Later models of Seinhorst (119, 121) are based on two assumptions: (*a*) the "average nematode" is the same at all densities (size and activity of the nematode is not influenced by density); and (*b*) whether a nematode will attack a plant or not is not influenced by the presence or absence of other nematodes attacking the same plant. Wallace (147) suggested six ways in

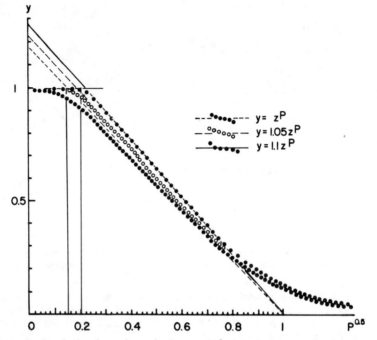

Figure 3 Graph showing values of y in $y = CZ^P$ plotted against $P^{1/2}$ (after Seinhorst, 1972).

which density of *M. javanica* could affect penetration of tomato, and Linford (66) and Baxter & Blake (9) have shown that nematodes prefer certain parts of roots. These observations appear incompatible with Seinhorst's assumptions. However, Seinhorst (121) indicated that nonrandom distribution may reflect only the relative attractiveness of portions of roots. Tolerance limits can be estimated from data on yields of plants at a limited range of nematode densities (123). When the values of equation 2 are plotted against the square root of P, the relationship between Y and the square root of P is essentially linear (Figure 3) but is dependent on the value of Z^{-T} (123).

Heterodera rostochiensis suppresses the growth of potato in pot tests in two ways: induction of root-tip necrosis by medium to high densities of second-stage larvae, and development of syncytia at lower densities (129). An apparent stimulation of growth at high densities was explained by a delay in plant development caused by the nematode that did not affect tuber growth. Seinhorst & den Ouden (129) offered the following model of the three factors that influenced the final weight of the plant (two that suppress and one that increases growth):

$$Y = [m_1 + (1 - m_1) Z^P][m_2 + (1 - m) Z^{kq}][c - (c - 1) Z^{lp}] \qquad 3.$$
for $p = P - T_1$
when $P \geq T_1$

and $p = 0$ for $P \leq T_1$
and for $q = P - T_2$
when $P \geq T_2$ and
$q = 0$ for $P \leq T_2$
and in which further k and l are constants [k = a constant
relative to growth stimulation per nematode;
$q = 1 - P$; c is a constant > 1,
indicating the maximum value of y
approached when $P \rightarrow \infty$ (ratio between yield
at $P = \infty$ and $P = 0$); l = constant]

and T_1 and T_2 are tolerance limits for the factors operating at high and low densities, respectively.

Although most nematode models have limited value for predictive purposes, Jones et al (55) utilized a population model for *H. rostochiensis* to develop rotation schemes to maximize the useful life of resistant varieties of potato.

The application of exponential models in field experiments is difficult because of the limited number of replications and the inherent variation from such test conditions. Microplot results with *Heterodera avenae* on cereals (72) and *Pratylenchus penetrans* on tobacco (86) gave sigmoid curves which fitted the model described by the equation 2. Nevertheless, it should be emphasized that results from both greenhouse and microplot experiments, regardless of how they are analyzed, need interpretation and cannot be used for extrapolation to crop responses in field situations. For example, Hanounik et al (37) found the tolerance limit for flue-cured tobacco to *Meloidogyne incognita* to be 2 and 1 eggs/1.5 cm^3 of soil for resistant and susceptible cultivars, respectively. Results in these pot experiments gave sigmoid curves that did fit equation 2. However, this species occurs primarily as larvae in the spring, and much lower numbers of larvae damage susceptible and resistant tobacco in field situations than indicated by the T values for eggs in pot tests.

Simulators

The application of systems analysis (measurement, analysis, simulation) to management of plant pests is a recent development (109, 157). Simulation models are being developed for nematodes (29, 55). Several simulators are available for insects (12, 109) and fungi (61, 131, 157).

The term *simulation* in a broad context is used to include any mimicking of a real-world situation. The simulation approach offers the possibility of analyzing the behavior of systems too complex to be solvable by conventional mathematical approaches (12). Simulators are formed by describing elements and interactions quantitatively, resulting in a series of equations linking together the system components. They can be used in research to integrate results from the laboratory, greenhouse, and field. In agriculture, models have the limitation of not being able to describe a complex situation completely (99), but they have much potential for predictive purposes in integrated pest control. Although laboratory findings are useful in making quantitative predictions about field behavior, a simulator model that is accurate in the laboratory may not be dependable in the field (61). Thus, the study of plant-pathogen interactions in the field must be given high priority.

The systems approach has much potential for integrating the numerous ecological and physiological systems which govern crop responses to nematodes into predictive simulators. Considerable information is available on major systems of some species, including infection (73), development (75), and population dynamics (121). However, there are major gaps in the information and techniques needed for modeling nematode-plant systems. Since simulators are never complete (157), they can be improved as new information becomes available.

Ultimately, crop simulators should integrate economic and land-use models (99) with those for insects, fungi, bacteria, weeds, and nematodes. The interactions of nematodes with soil-inhabiting fungi and insects warrant special consideration. The disease potential of many fungi is enhanced by nematodes, particularly *Meloidogyne* species (10, 101). Models developed for fungi (4, 142) and nematodes may be suitable for quantifying these interactions. The costs of developing multiple-pest simulators could be prohibitive, but cooperative efforts on regional and national levels have promise (61).

ROLE OF ENVIRONMENT

Plant-parasitic nematodes and host plants are very sensitive to the environment (111, 149). Most nematodes occur in much greater numbers in temperate zones than in many tropical areas where prolonged dry periods occur (9). Observations are common concerning nematode damage being more severe in light than in heavy soils. Damage is usually more severe when plants are under environmental stress, as tolerance or resistance is lowered (85, 94, 151). Furthermore, much of the estimated 5 to 12% annual yield losses to nematodes may be due to their predisposing plants to attack by various fungi, bacteria, and their serving as vectors of viruses (101, 136).

Abiotic Environment

Environmental stress, by lowering tolerance of the plant to the nematode, may be the basis of some of the more serious crop failures due to nematodes (148). Most above- and below-ground factors that affect growth of plants and other soil microorganisms (88) also influence nematode activity, and alter the relationship between numbers of nematodes and crop yield (148, 149). Damage may be limited by planting crops at times which allow satisfactory growth, but inhibit nematode activity. For example, during seasons with cool moist weather, high initial egg populations of *Heterodera goettingiana* have little effect on the growth of peas (54, 76). In microplot experiments, the damage to oats caused by *Heterodera avenae* was most severe when water supply was inadequate (48). Due to the great variation in results obtained in numerous experiments with *H. rostochiensis* on potato, Brown (17) concluded that it is impossible to relate nematode density to crop performance. *Meloidogyne incognita* caused severe damage to tomato in microplots in the coastal plain of North Carolina, whereas, it effected only slight losses in yield in the mountains where conditions (heavier soil, higher rainfall, and lower temperature) favored plant growth (8). Losses in onions due to stem nematodes (*Ditylenchus dipsaci*) vary considerably from place to place and from year to year (126). Some cultivars normally resistant to nematodes become susceptible when exposed to high temperatures (24).

Plant nutrition plays a major role in the damage caused by nematodes. High levels of potassium tend to protect crops from nematode damage although reproduction of the pathogen may increase (94). A high rate of ammonium-nitrate application supresses the development of *Heterodera glycines* on soybeans and minimizes damage (106). The interaction of nematode and host nutrition may depend on nematode density, as with *M. javanica* on tomato (13). At low densities, a deficiency in nitrogen favored nematode growth, whereas growth was depressed with low N at high densities. The effects of environment and nematode densities on nematode development were included in recent reviews (14, 138, 149).

Associated Microflora

The pathogenicity of soil nematodes per se may prove to be less important in the etiology of plant disease than interactions between nematodes and the soil microflora (77). Powell (101) suggested that the doctrine of specific etiology may have led to neglect of interaction factors that play a role in disease etiology. The numerous papers on which recent reviews (11, 101) of these interactions were based provide massive evidence that much of the damage attributed to nematodes in field situations is due to disease complexes. These interactions are not limited to pathogenic fungi or bacteria, since saprophytic organisms may also act as pathogens in the parasitized tissues, especially when the complex involves *Meloidogyne* spp. (71, 101). Infection of host plants by some fungi enhance nematode reproduction (101, 128). Nematodes also cause damage by interfering with growth of mycorrhizal fungi (5). *Helicotylenchus dihystera* predisposes shortleaf pine to *Phytophthora cinnamomi* by disrupting the fungus mantle of ectomycorrhizae, thereby providing infection courts for *P. cinnamomi* (5). Normally, the fungus symbionts provide resistance in the pine hosts against this pathogen.

In addition to the nematode genera, *Xiphinema, Longidorus,* and *Trichodorus,* which serve as vectors of viruses, various types of interactions between other plant-parasitic nematodes and viruses are involved in disease losses (136). Tobacco infected with mosaic virus and *Aphelenchoides ritzemabosi* or *Ditylenchus dipsaci* is severely damaged even where the numbers of nematodes are low (154).

Capacity of Plants to Compensate for Injury

Plants such as corn have an extremely high capacity to compensate for injury, as described by Tammes (135). The relationship between the yield of corn and root pruning gave a sigmoid curve similar to the curves encountered with yield losses due to some nematodes. Excising one fourth to one half of the corn root system to a depth of 20 cm just before and after flowering had no effect on yield (135). The capacity of corn to compensate for injury may explain why this crop frequently recovers from early stunting caused by nematodes such as *Trichodorus christiei* or *Belonolaimus longicaudatus* (K. R. Barker, unpublished data). In microplots in North Carolina, very high numbers of *T. christiei* caused an initial stunting of this crop similar to that which occurs in the field; however, after application of ammonium nitrate, plants recovered and no significant loss of yield was detected. Wallace (151) indicates that plants likely have the physiological capacity to compensate for and to repair damage. Remaining *Rhizobium* nodules of soybean infected with *Heterodera glycines* have a greater efficiency than those in healthy plants (64). Thus, plants are not passive recipients of destruction by nematodes; they can actively repair damage and regenerate tissues, apparently through hormonal regulation. Wallace (151) suggests that with nematodes such as *Meloidogyne javanica,* opposing influences (root destruction and metabolic-sink formation versus root regeneration and growth-hormone production) may be working simultaneously. The investigations showing accumulations of various elements and increased synthesis in roots (10, 25, 97), changes in auxins (144), and modified patterns of root growth (24, 45) support this hypothesis.

CONCLUSIONS

Much progress has been made during the last 20 years in characterizing the relationships between nematode densities and plant growth and yield. Damage is largely determined by preplant densities, which is the result of nematodes having limited mobility and low reproductive potentials. This relation can be altered in perennials by reintroduction, buildup of negligible initial densities, or both. Crop-nematode relationships vary with cultivars, nematode species and races, and environment. Very low population densities of certain nematodes may stimulate plant growth. Reliable methods of monitoring most species of nematodes are available, but a major barrier that limits precise estimates of nematode densities is their erratic, nonrandom distribution in the soil. Because nematodes are obligate parasites, the time span in which densities are linearly related to crop growth or yield is limited. Once damage is effected by high initial densities, the rate of reproduction declines as compared to nematodes on plants not yet damaged by lower initial populations. Experimentation under controlled environmental conditions is useful for elucidating fundamental interactions in the nematode-plant system. Since one cannot extrapolate from such experiments to the field, research in the field and in microplots is essential to practical quantitative nematology. Most field and microplot research with natural or artificial infestations has provided data which show that yields are lower with increasing densities of nematodes.

The concepts of *host efficiency* and *host sensitivity* often are characterized by the terms *resistant, susceptible, tolerant,* and *intolerant.* Since reference to two variable phenomena with single terms is often confusing, usage of either of these terms should include a qualification, indicating whether or not the reference is to damage or to reproduction. For quantitative purposes, the most promising scheme includes *tolerance limit* (*T*) (density below which no detectable damage occurs) to characterize host sensitivity and *a (maximum rate of reproduction)* and *equilibrium density* (*E*) (density at which birth rate equals death rate) to characterize hose efficiency.

Most plants, especially those with extensive root systems such as corn, can compensate for injury resulting from parasitism by nematodes. The tolerance of a crop to a given species of nematode, however, is a dynamic rather than a static phenomenon; it varies with several environmental factors such as soil type, moisture, temperature, host nutrition, and the presence or absence of microflora and microfauna, as well as with the age or size of plants and population or race of the nematode in question. Regardless of such variation, *tolerance limit* and *threshold density* as well as the practical concept of *economic threshold density* (or economic loss threshold) have important applications.

In addition to the more commonly used linear regressions, exponential models describe the nematode-host relationships that give sigmoid curves. Although there has been debate concerning the relative merits of exponential models vs regression analyses and the precision required with each, both approaches should continue to play important roles in quantitative nematology, although both have limitations. For example, some type of transformation of data is usually utilized with both, but the transformation itself can influence conclusions. The extension of mathematical modeling to simulators has much potential for integrating the many ecological and physiological interactions of nematodes with their hosts and environment into more functional systems.

Field research on quantitative relationships of crop performance to nematode densities will continue to play an indispensable role in applied agriculture since information derived therefrom is the basis for control or management of these pests. Predicting crop responses to nematode densities is extremely difficult in field situations. However, as integrated control materializes, the problems should not be insurmountable. The development of simulator models, including the concepts of pest management, will require further research in basic ecology and physiology of nematodes and associated organisms, and should integrate fundamental and practical information to the benefit of agriculture.

Literature Cited

1. Altman, J., Tsue, K. M. 1965. Changes in plant growth with chemicals used as soil fumigants. *Plant Dis. Reptr.* 49:600–2
2. Andrewartha, H. G., Birch, L. C. 1954. *The Distribution and Abundance of Animals.* Chicago: Univ. Chicago Press. 782 pp.
3. Aycock, R., Barker, K. R., Benson, D. M. 1976. Susceptibility of Japanese holly to *Criconemoides xenoplax, Tylenchorhynchus claytoni* and certain other plant-parasitic nematodes. *J. Nematol.* 8:26–31
4. Baker, R., Maurer, C. L., Maurer, R. A. 1967. Ecology of plant pathogens in soil. VII. Mathematical models and inoculum density. *Phytopathology* 57:662–66
5. Barham, R. O., Marx, D. H., Ruehle,

J. L. 1974. Infection of ectomycorrhizal and nonmycorrhizal roots of shortleaf pine by nematodes and *Phytophthora cinnamomi*. *Phytopathology* 64:1260–64

6. Barker, K. R., Nusbaum, C. J. 1971. Diagnostic and advisory programs. In *Plant Parasitic Nematodes*, ed. B. M. Zuckerman, W. F. Mai, R. A. Rohde, 1:281–301. New York: Academic. 345 pp.

7. Barker, K. R., Nusbaum, C. J., Nelson, L. A. 1969. Seasonal population dynamics of selected plant-parasitic nematodes as measured by three extraction procedures. *J. Nematol.* 1:232–39

8. Barker, K. R., Shoemaker, P. B., Nelson, L. A. 1976. Relationships of initial population densities of *Meloidogyne incognita* and *M. hapla* to yield of tomato. *J. Nematol.* 8:232–39

9. Baxter, R. I., Blake, C. D. 1967. Invasion of wheat roots by *Pratylenchus thornei*. *Nature* 215:1168–69

10. Bergeson, G. B. 1968. Evaluation of factors contributing to the pathogenicity of *Meloidogyne incognita*. *Phytopathology* 58:49–53

11. Bergeson, G. B. 1972. Concepts of nematode-fungus associations in plant disease complexes: a review. *Exp. Parasitol.* 32:301–14

12. Berryman, A. A., Pienaar, L. V. 1974. Simulation: a powerful method of investigating the dynamics and management of insect populations. *Environ. Entomol.* 3:199–207

13. Bird, A. F. 1970. The effect of nitrogen deficiency on the growth of *Meloidogyne javanica* at different population levels. *Nematologica* 16:13–21

14. Bird, A. F. 1974. Plant response to root-knot nematode. *Ann. Rev. Phytopathol.* 12:69–85

15. Bird, G. W., Brooks, O. L., Perry, C. E. 1974. Dynamics of concomitant field populations of *Hoplolaimus columbus* and *Meloidogyne incognita*. *J. Nematol.* 6:190–94

16. Brown, E. B. 1961. A rotational experiment on land infested with potato root eelworm, *Heterodera rostochiensis* Woll. *Nematologica* 6:201–6

17. Brown, E. B. 1969. Assessment of the damage caused to potatoes by potato cyst eelworm, *Heterodera rostochiensis* Woll. *Ann. Appl. Biol.* 63:493–502

18. Chitwood, B. G., Buhrer, E. M. 1946. Further studies on the life history of the golden nematode of potatoes (*Heterodera rostochiensis* Wollenweber),

season 1945. *Proc. Helminthol. Soc. Wash.* 13:54–56

19. Chitwood, B. G., Feldmesser, J. 1948. Golden nematode population studies. *Proc. Helminthol. Soc. Wash.* 15:43–55

20. Cobb, N. A. 1918. Estimating the nema-population of soil. *US Bur. Plant Ind. Agric. Tech. Circ.* 1:1–48

21. Cook, R. 1974. Nature and inheritance of nematode resistance in cereals. *J. Nematol.* 6:165–74

22. Cuany, A., Lavergne, J.-C. 1974. Répercussions biologiques et économiques d'un traitement au bromure de méthyle sur une culture de melons en présence de divers nématodes. *Phytiatr. Phytopharm.* 23:183–88

23. Doney, D. L., Whitney, E. D., Steele, A. E. 1971. Effect of *Heterodera schachtii* infection on sugarbeet leaf growth. *Phytopathology* 61:40–41

24. Dropkin, V. H. 1969. Cellular responses of plants to nematode infections. *Ann. Rev. Phytopathol.* 7:101–22

25. Dropkin, V. H. 1972. Pathology of *Meloidogyne*—galling, giant cell formation, effects on host physiology. *Eur. Plant Prot. Org. Bull. No.* 6:23–32

26. Dropkin, V. H., Nelson, P. E. 1960. The histopathology of root-knot nematode infections in soybeans. *Phytopathology* 50:442–47

27. Endo, B. Y. 1975. Pathogenesis of nematode-infected plants. *Ann. Rev. Phytopathol.* 13:213–38

28. Ferris, H. 1972. *Population dynamics of Meloidogyne spp. in relation to the epidemiology and control of root knot of tobacco*. PhD thesis. NC State Univ., Raleigh. 95 pp.

29. Ferris, H. 1976. Development of a computer-simulation model for a plant-nematode system. *J. Nematol.* 8:255–63

30. Ferris, H. 1974. Correlation of tobacco yield, value, and root-knot index with early-to-midseason, and post-harvest *Meloidogyne* population densities. *J. Nematol.* 6:75–81

31. Ferris, H., McKenry, M. V. 1975. Relationship of grapevine yield and growth to nematode densities. *J. Nematol.* 7:295–304

32. Ferris, J. M. 1962. Some observations on the number of root lesion nematodes necessary to cause injury to seedling onions. *Plant Dis. Reptr.* 46:484–85

33. Flegg, J. J. M. 1968. Life-cycle studies of some *Xiphinema* and *Longidorus* species in southeastern England. *Nematologica* 14:197–210

34. Good, J. M., Carter, R. L. 1965. Nitrification lag following soil fumigation. *Phytopathology* 55:1147–50

35. Gooris, J., d'Herde, C. J. 1972. Mode d'hivernage de *Meloidogyne naasi* Franklin dans le sol et lutte par rotation culturale. *Rev. Agric. Brussels* 25:659–64

36. Griffin, G. D., Hunt, O. J. 1972. Effect of plant age on resistance of alfalfa to *Meloidogyne hapla. J. Nematol.* 4:87–90

37. Hanounik, S. B., Osborne, W. W., Pirie, W. R. 1975. Relationships between the population density of *Meloidogyne incognita* and growth of tobacco. *J. Nematol.* 7:352–56

38. Heijbroek, W. 1973. Forecasting incidence of and issuing warnings about nematodes, especially *Heterodera schachtii* and *Ditylenchus dipsaci. Inst. Int. Rech. Better.* 6:76–86

39. Hellinga, J. J. H. 1942. De invloed van het Bietenaaltje op de opbrengst en de samenstelling van Suikerbieten. *Meded. Inst. Suikerbietenteelt* 12:163–82

40. Hesling, J. J. 1957. *Heterodera major* O. Schmidt 1930 on cereals—a population study. *Nematologica* 2:285–99

41. Hijink, M. J. 1963. A relation between stem infection by *Phoma solanicola* and *Ditylenchus dipsaci* on potato. *Neth. J. Plant Pathol.* 69:318–21

42. Hijink, M. J. 1964. Over regressies van de opbrengst van gewassen op gemengde populaties van twee of meer parasitaire nematoden. *Meded. Rijksfac. Landbouwwet. Gent* 29:818–22

43. Hijink, M. J. 1967. Fruchtwechseleffekte und Nematoden. *Mitt. Biol. Bundesanst. Land Forstwirtsch. Berlin Dahlem* 121:21–28

44. Hoestra, H., Oostenbrink, M. 1962. Nematodes in relation to plant growth. 4. *Pratylenchus penetrans* (Cobb) on orchard trees. *Neth. J. Agric. Sci.* 10:286–96

45. Högger, C. 1972. Effect of *Trichodorus christiei* inoculum density and growing temperature on growth of tomato roots. *J. Nematol.* 4:66–67

46. Hollis, J. P. 1963. Action of plant-parasitic nematodes on their hosts. *Nematologica* 9:475–94

47. Hough, A., Thomason, I. J. 1975. Effects of aldicarb on the behavior of *Heterodera schachtii* and *Meloidogyne javanica. J. Nematol.* 7:221–29

48. Jaspers, C. P., Kort, J. 1975. Herstellingsvermogen van haver bij aantasting door het graancystenaaltje. *Gewasbescherming* 6:1–4

49. Johnson, A. W. 1970. Pathogenicity and interaction of three nematode species on six bermudagrasses. *J. Nematol.* 2:36–41

50. Johnson, A. W., Nusbaum, C. J. 1970. Interactions between *Meloidogyne incognita, M. hapla,* and *Pratylenchus brachyurus* in tobacco,. *J. Nematol.* 2:334–40

51. Johnson, P. W., Fushtey, S. G. 1966. The biology of the oat cyst nematode *Heterodera avenae* in Canada. II. Nematode development and related anatomical changes in roots of oats and corn. *Nematologica* 12:630–36

52. Jones, F. G. W. 1956. Soil populations of beet eelworm (*Heterodera schachtii* Schm.) in relation to cropping. II. Microplot and field plot results. *Ann. Appl. Biol.* 44:25–56

53. Jones, F. G. W., Meaton, V. H., Parrott, D. M., Shepherd, A. M. 1965. Population studies on pea cyst-nematode *Heterodera goettingiana* Liebs. *Ann. Appl. Biol.* 55:13–23

54. Jones, F. G. W., Moriarty, F. 1956. Further observations on the effects of peas, beans and vetch upon soil population levels of pea root eelworm, *Heterodera göttingiana* Liebscher. *Nematologica* 1:268–73

55. Jones, F. G. W., Parrott, D. M., Ross, G. J. S. 1967. The population genetics of the potato cyst-nematode, *Heterodera rostochiensis:* mathematical models to simulate the effects of growing eelworm-resistant potatoes bred from *Solanum tuberosum* ssp. *andigena. Ann. Appl. Biol.* 60:151–71

56. Jones, F. G. W., Parrott, D. M., Williams, T. D. 1967. The yield of potatoes resistant to *Heterodera rostochiensis* on infested land. *Nematologica* 13:301–10

57. Justesen, S. H., Tammes, P. M. L. 1960. Studies on yield losses. I. The self-limiting effect of injurious or competitive organisms on crop-yield. *Tijdschr. Plantenziekten* 66:281–87

58. Kinloch, R. A. 1974. Response of soybean cultivars to nematicidal treatments of soil infested with *Meloidogyne incognita. J. Nematol.* 6:7–11

59. Kinloch, R. A., Allen, M. W. 1972. Interaction of *Meloidogyne hapla* and *M. javanica* infecting tomato. *J. Nematol.* 4:7–16

60. Kleinschmidt, G. D., Gerdemann, J. W. 1972. Stunting of citrus seedlings in fumigated nursery soils related to the absence of endomycorrhizae. *Phytopathology* 62:1447–53

61. Krause, R. A., Massie, L. B. 1975. Predictive systems: modern approaches to disease control. *Ann. Rev. Phytopathol.* 13:31–47
62. Krusberg, L. R. 1963. Host response to nematode infection. *Ann. Rev. Phytopathol.* 1:219–40
63. Lange, B., Meyer, J., Burmeister, P. 1964. Vorläufige Ermittlungen über Schäden an Tabak durch freilebende Nematoden und Versuche zu ihrer Bekämpfung. *Mitt. Biol. Bundesanst. Land Forstwirtsch. Berlin Dahlem* 111:85–93
64. Lehman, P. S., Huisingh, D., Barker, K. R. 1971. The influence of races of *Heterodera glycines* on nodulation and nitrogen-fixing capacity of soybean. *Phytopathology* 61:1239–44
65. Lewis, G. D., Mai, W. F. 1960. Overwintering and migration of *Ditylenchus dipsaci* in organic soils of southern New York. *Phytopathology* 50:341–43
66. Linford, M. B. 1939. Attractiveness of roots and excised shoot tissues to certain nematodes. *Proc. Helminthol. Soc. Wash.* 6:11–18
67. Lownsbery, B. F., Peters, B. G. 1955. The relation of the tobacco cyst nematode to tobacco growth. *Phytopathology* 45:163–67
68. Madamba, C. P., Sasser, J. N., Nelson, L. A. 1965. Some characteristics of the effects of *Meloidogyne* spp. on unstable host crops. *NC Agric. Exp. Stn. Tech. Bull.* 169:1–34
69. Malek, R. B., Jenkins, W. R. 1964. Aspects of the host-parasite relationships of nematodes and hairy vetch. *NJ Agric. Exp. Stn. Bull. 813:*1–31
70. Marlatt, R. B., Perry, V. G. 1971. Growth stimulation of *Sporobolus poiretii* by *Aphelenchoides besseyi.* *Phytopathology* 61:740
71. Mayol, P. S., Bergeson, G. B. 1970. The role of secondary invaders in *Meloidogyne incognita* infection. *J. Nematol.* 2:80–83
72. Meagher, J. W., Brown, R. H. 1974. Microplot experiments on the effect of plant hosts on populations of the cereal cyst nematode (*Heterodera avenae*) and on the subsequent yield of wheat. *Nematologica* 20:337–46
73. Merny, G. 1972. Les nématodes phytoparasites des rizières inondées de Côte d'Ivoire IV-Essai d'interprétation mathématique de l'intensité de l'infestation des racines par les nématodes endoparasites en fonction de l'inoculum. *Cah. ORSTOM Ser. Biol.* 16:129–40
74. Miles, H. W., Henderson, V. E., Miles, M. 1943. Field studies of potato-root eelworm, *Heterodera rostochiensis* Wollenweber, 1938–40. *Ann. Appl. Biol.* 30:151–57
75. Milne, D. L., Du Plessis, D. P. 1964. Development of *Meloidogyne javanica* (Treub.) Chit., on tobacco under fluctuating soil temperatures. *S. Afr. J. Agric. Sci.* 7:673–80
76. Moriarty, F. 1962. The effects of sowing time and of nitrogen on peas, *Pisium sativum* and on pea root eelworm, *Heterodera goettingiana* Liebscher. *Nematologica* 8:169–75
77. Mountain, W. B. 1965. Pathogenesis by soil nematodes. In *Ecology of Soilborne Plant Pathogens,* ed. K. F. Baker, W. C. Snyder, 285–301. Berkeley: Univ. Calif. Press. 571 pp.
78. Mountain, W. B., Boyce, H. R. 1958. The peach replant problem in Ontario. VI. The relation of *Pratylenchus penetrans* to the growth of young peach trees. *Can. J. Bot.* 36:135–51
79. Nicholson, A. J. 1933. The balance of animal populations. *J. Anim. Ecol.* 2:132–78
80. Nusbaum, C. J. 1960. Soil fumigation for nematode control in flue-cured tobacco. *Down Earth* 16(1):15–17
81. Nusbaum, C. J., Barker, K. R. 1971. Population dynamics. See Ref. 6, pp. 303–23
82. Nusbaum, C. J., Ferris, H. 1973. The role of cropping systems in nematode population management. *Ann. Rev. Phytopathol.* 11:423–40
83. Nusbaum, C. J., Todd, F. A. 1970. The role of chemical soil treatments in the control of nematode-disease complexes of tobacco. *Phytopathology* 60:7–12
84. Olthof, T. H. A., Marks, C. F., Elliot, J. M. 1973. Relationship between population densities of *Pratylenchus penetrans* and crop losses in flue-cured tobacco in Ontario. *J. Nematol.* 5:158–62
85. Olthof, T. H. A., Potter, J. W. 1972. Relationship between population densities of *Meloidogyne hapla* and crop losses in summer-maturing vegetables in Ontario. *Phytopathology* 62:981–86
86. Olthof, T. H. A., Potter, J. W. 1973. The relationship between population densities of *Pratylenchus penetrans* and crop losses in summer-maturing vegetables in Ontario. *Phytopathology* 63:577–82
87. Olthof, T. H. A., Potter, J. W., Peterson, E. A. 1974. Relationship between population densities of *Heterodera*

schachtii and losses in vegetable crops in Ontario. *Phytopathology* 64:549–54

88. Oort, A. J. P. 1972. The ecology of soil-borne pathogens. *Eur. Plant Prot. Org. Bull.* 1(6):121–28

89. Oostenbrink, M. 1956. Over de invloed van verschillende gewassen op de vermeerdering van en de schade door *Pratylenchus pratensis* en *Pratylenchus penetrans* (Vermes, Nematoda). *Tijdschr. Plantenziekten* 62:189–203

90. Oostenbrink, M. 1964. Harmonious control of nematode infestation. *Nematologica* 10:49–56

91. Oostenbrink, M. 1966. Major characteristics of the relation between nematodes and plants. *Meded. Landbouwhogesch. Wageningen* 66(4):1–46

92. Oostenbrink, M. 1971. Quantitative aspects of plant-nematode relationships. *Indian J. Nematol.* 1:68–74

93. Oostenbrink, M. 1972. Evaluation and integration of nematode control methods. In *Economic Nematology,* ed. J. M. Webster, 497–514. New York: Academic. 563 pp.

94. Oteifa, B. A., Diab, K. A. 1961. Significance of potassium fertilization in nematode infested cotton fields. *Plant Dis. Reptr.* 45:932

95. Oteifa, B. A., Elgindi, D. M. 1961. A critical method for evaluating tolerant levels in nematized host plants. *Plant Dis. Reptr.* 45:930–31

96. Ouden, H., den 1965. Potgrootte en schade aan aardappelen door *Heterodera rostochiensis. Jaarversl. Inst. Plziektenk. Onderzoek, Wageningen* 1965, p. 103

97. Owens, R. G., Specht, H. N. 1966. Biochemical alterations induced in host tissues by root-knot nematodes. *Contrib. Boyce Thompson Inst.* 23:181–98

98. Peters, B. G. 1961. *Heterodera rostochiensis* population density in relation to potato growth. *J. Helminthol. R. J. Leiper Suppl.,* pp. 141–50

99. Pimentel, D., Shoemaker, C. 1974. An economic and land-use model for reducing insecticides on cotton and corn. *Environ. Entomol.* 3:10–20

100. Potter, J. W., Olthof, T. H. A. 1974. Yield losses in fall-maturing vegetables relative to population densities of *Pratylenchus penetrans* and *Meloidogyne hapla. Phytopathology* 64:1072–75

101. Powell, N. T. 1971. Interactions between nematodes and fungi in disease complexes. *Ann. Rev. Phytopathol.* 9:253–74

102. Proctor, J. R., Marks, C. F. 1974. The determination of normalizing transformations for nematode count data from soil samples and of efficient sampling schemes. *Nematologica* 20:395–406

103. Rabb, R. L. 1970. Introduction to the conference. In *Concepts of Pest Management,* ed. R. L. Rabb, F. E. Guthrie, 1–5. Raleigh: NC State Univ. Press. 242 pp.

104. Rickard, D. A. 1973. *Responses of certain plant parasitic nematodes to selected chemical soil treatments.* PhD thesis. NC State Univ., Raleigh. 62 pp.

105. Rohde, R. A. 1972. Expression of resistance in plants to nematodes. *Ann. Rev. Phytopathol.* 10:233–52

106. Ross, J. P. 1959. Nitrogen fertilization on the response of soybean infected with *Heterodera glycines. Plant Dis. Reptr.* 43:1284–86

107. Rössner, J. 1972. Vertikalverteilung wandernder Wurzelnematoden im Boden in Abhängigkeit von Wassergehalt und Durchwurzelung. *Nematologica* 18:360–72

108. Ruehle, J. L. 1973. Nematodes and forest trees—types of damage to tree roots. *Ann. Rev. Phytopathol.* 11:99–118

109. Ruesink, W. G. 1976. Status of the systems approach to pest management. *Ann. Rev. Entomol.* 21:27–44

110. Sasser, J. N., Barker, K. R., Nelson, L. A. 1975. Correlations of field populations of nematodes with crop growth responses for determining relative involvement of species. *J. Nematol.* 7:193–98

111. Sayre, R. M. 1971. Biotic influences in soil environment. See Ref. 6, pp. 235–56

112. Sayre, R. M., Mountain, W. B. 1962. The bulb and stem nematode [*Ditylenchus dipsaci*] on onion in southwestern Ontario. *Phytopathology* 52:510–16

113. Schafer, J. F. 1971. Tolerance to plant disease. *Ann. Rev. Phytopathol.* 9:235–52

114. Schilt, H., Hartmann, H. D., Heimann, M. 1973. Einfluss verschiedener Populationsdichten von *Meloidogyne incognita* auf das Wachstum oberirdischer und unterirdischer Teile von Kopfsalat. *Z. Pflanzenkr. Pflanzenschutz* 80:662–70

115. Scotto la Massese, C. 1972. The use of "Isopathological curves" for estimation of parasitic pathogenicity, exemplified by *Tylenchulus semipenetrans* on citrus. *Abstr. 10th Int. Symp.*

Eur. Soc. Nematol., Reading, pp. 61–62

116. Seinhorst, J. W. 1960. Over het bepalen van door aaltjes veroorzaakte opbrengstvermindering bij cultuurgewassen. *Meded. Landbouwhogesch. Opzoekingsstn. Staat Gent* 25:1025–39

117. Seinhorst, J. W. 1961. Plant-nematode inter-relationships. *Ann. Rev. Microbiol.* 15:177–96

118. Seinhorst, J. W. 1963. Enkele aspecten van het onderzoek over plantenparasitaire aaltjes. *Meded. Dir. Tuinbouw Neth.* 26:349–58

119. Seinhorst, J. W. 1965. The relation between nematode density and damage to plants. *Nematologica* 11:137–54

120. Seinhorst, J. W. 1967. The relationships between population increase and population density in plant parasitic nematodes. III. Definition of the terms host, host status and resistance. IV. The influence of external conditions on the regulation of population density. *Nematologica* 13:429–42

121. Seinhorst, J. W. 1970. Dynamics of populations of plant parasitic nematodes. *Ann. Rev. Phytopathol.* 8:131–56

122. Seinhorst, J. W. 1971. Schade door *Pratylenchus penetrans* aan appel en vermeerdering van dit aaltje. *Jaarversl. Inst. Plziektenk. Onderz. Wageningen,* pp. 114–15

123. Seinhorst, J. W. 1972. The relationship between yield and square root of nematode density. *Nematologica* 18:585–90

124. Seinhorst, J. W. 1973. Principles and possibilities of determining degrees of nematode control leading to maximum returns. I. Protection of one crop sown or planted soon after treatment. *Nematologia Med.* 1:93–105

125. Seinhorst, J. W. 1973. The relation between nematode distribution in a field and loss in yield at different average nematode densities. *Nematologica* 19:421–27

126. Seinhorst, J. W. 1975. Dynamics of the plant nematode system. In *Nematode Vectors of Plant Viruses,* ed. F. Lamberti, C. E. Taylor, J. W. Seinhorst, 2:409–21. New York: Plenum

127. Seinhorst, J. W., Kuniyasu, K. 1969. *Rotylenchus uniformis* (Thorne) on carrots. *Neth. J. Plant Pathol.* 75:205–23

128. Seinhorst, J. W., Kuniyasu, K. 1971. Interaction of *Pratylenchus penetrans* and *Fusarium oxysporum forma pisi* race 2 and of *Rotylenchus uniformis*

and *F. oxysporum* f. *pisi* race 1 on peas. *Nematologica* 17:444–52

129. Seinhorst, J. W., den Ouden, H. 1971. The relation between density of *Heterodera rostochiensis* and growth and yield of two potato varieties. *Nematologica* 17:347–69

130. Seinhorst, J. W., Sen, A. K. 1966. The population density of *Heterodera trifolii* in pastures in the Netherlands and its importance for the growth of white clover. *Neth. J. Plant Pathol.* 72:169–83

131. Shrum, R. 1975. Simulation of wheat stripe rust (*Puccinia striiformis* West) using EPIDEMIC a flexible plant disease simulator. *Pa. State Univ. Prog. Rept. 347.* (32 pp., plus program)

132. Simons, W. R. 1973. Nematode survival in relation to soil moisture. *Meded. Landbouwhogesch. Wageningen Doct. Diss.* 85 pp.

133. Southey, J. F. 1965. Potato root eelworm. In *Plant Nematology. Tech. Bull. No. 7,* ed. J. F. Southey, 171–88. London: HMSO. 282 pp.

134. Swarup, G., Sharma, R. D. 1965. Root knot of vegetables: Part IV Relation between population density of *Meloidogyne javanica* and *Meloidogyne incognita* var. *acrita,* and root and shoot growth of tomato seedlings. *Indian J. Exp. Biol.* 3:197–98

135. Tammes, P. M. L. 1961. Studies of yield losses. 2. Injury as a limiting factor of yield. *Tijdschr. Plantenziekten* 67:257–63

136. Taylor, C. E. 1971. Nematodes as vectors of plant viruses. See Ref. 6, 2:185–211

137. Townshend, J. L. 1972. Influence of edaphic factors on penetration of corn roots by *Pratylenchus penetrans* and *P. minyus* in three Ontario soils. *Nematologica* 18:201–12

138. Triantaphyllou, A. C. 1973. Environmental sex differentiation of nematodes in relation to pest management. *Ann. Rev. Phytopathol.* 11:441–62

139. Triantaphyllou, A. C., Hirschmann, H. 1960. Post-infection development of *Meloidogyne incognita* Chitwood 1949 (Nematoda: Heteroderidae). *Ann. Inst. Phytopathol. Benaki, N. S.* 3:1–11

140. Tseng, S. T., Allred, K. R., Griffin, G. D. 1968. A soil population study of *Ditylenchus dipsaci* (Kühn) Filipjev in an alfalfa field. *Proc. Helminthol. Soc. Wash.* 35:57–62

141. Turner, D. R., Chapman, R. A. 1972. Infection of seedlings of alfalfa and

red clover by concomitant populations of *Meloidogyne incognita* and *Pratylenchus penetrans. J. Nematol.* 4: 280–86

142. Vanderplank, J. E. 1975. *Principles of Plant Infection.* New York: Academic. 216 pp.

143. Van Gundy, S. D., Kirkpatrick, J. D. 1975. Nematode-nematode interactions on tomato. *J. Nematol.* 7:330–31

144. Viglierchio, D. R. 1971. Nematodes and other pathogens in auxin-related plant-growth disorders. *Bot. Rev.* 37: 1–21

145. Wallace, H. R. 1963. *The Biology of Plant Parasitic Nematodes.* London: Edward Arnold. 280 pp.

146. Wallace, H. R. 1965. The soil environment. See Ref. 133, pp. 35–44

147. Wallace, H. R. 1966. Factors influencing the infectivity of plant parasitic nematodes. *Proc. R. Soc. Lond. Ser. B.* 164:592–614

148. Wallace, H. R. 1969. The influence of nematode numbers and of soil particle size, nutrients and temperature on the reproduction of *Meloidogyne javanica. Nematologica* 15:55–64

149. Wallace, H. R. 1971. Abiotic influences in the soil environment. See Ref. 6, pp. 257–80

150. Wallace, H. R. 1971. The influence of the density of nematode populations on plants. *Nematologica* 17:154–66

151. Wallace, H. R. 1973. *Nematode Ecology and Plant Disease.* London: Edward Arnold. 228 pp.

152. Wallace, H. R. 1974. The influence of root knot nematode, *Meloidogyne javanica,* on photosynthesis and on nutrient demand by roots of tomato plants. *Nematologica* 20:27–33

153. Wang, E. L. H., Bergeson, G. B. 1974. Biochemical changes in root exudate and xylem sap of tomato plants infected with *Meloidogyne incognita. J. Nematol.* 6:194–202

154. Weischer, B. 1969. Vermehrung und Schadwirkung von *Aphelenchoides ritzemabosi* und *Ditylenchus dipsaci* in virusfreiem und in TMV-infiziertem Tabak. *Nematologica* 15:334–36

155. Willis, C. B., Thompson, L. S. 1969. The influence of soil moisture and cutting management on *Pratylenchus penetrans* reproduction in birdsfoot trefoil and the relationship of inoculum levels to yields. *Phytopathology* 59:1872–75

156. Wong, T. K., Mai, W. F. 1973. Pathogenicity of *Meloidogyne hapla* to lettuce as affected by inoculum level, plant age at inoculation and temperature. *J. Nematol.* 5:126–29

157. Zadoks, J. C. 1971. Systems analysis and the dynamics of epidemics. *Phytopathology* 61:600–10

Copyright © 1976 by Annual Reviews Inc.
All rights reserved

FIRE AND FLAME FOR PLANT DISEASE CONTROL[1]

♦3646

John R. Hardison[2]

Agricultural Research Service, US Department of Agriculture, and Department of Botany and Plant Pathology, Oregon State University, Corvallis, Oregon 97331

INTRODUCTION

Fire was the first great force employed by man, and the discovery of a method to ignite vegetable matter looms as one of humanity's greatest achievements. The earliest evidence of deliberate use of fire dates from Peking man, or as much as 500,000 years ago, and use of fire is suspected to have been known to man thousands of years earlier (83).

Some of primitive man's reasons for the burning-over of land were to clear forest (or, rarely, grassland) for agriculture; to improve grazing land for domestic animals or to attract game; to deprive game of cover, or to drive game from cover in hunting; to kill or drive away predatory animals, ticks, mosquitoes, and other pests; to repel the attacks of enemies or to burn them out of their refuges; to expedite travel; to protect villages, settlements, or encampments from great fires by controlled burning; and to gratify sheer love of fires as spectacles (7). Indians of North America lighted bushes and trees so that the roots would burn slowly and provide a continuous source of fire (83).

Wildfire, along with soil, water, and temperature, has been one of the major forces that shaped the earth's vegetation. Most of the grasslands and many of the forests represent subclimax vegetation that from antiquity has been dependent on periodic fires for existence. The winter burning of dead grass left unconsumed from the growth of the previous season is a worldwide practice of ancient origin, particularly in humid regions where uncut grass does not cure into palatable winter forage. Grass fires set by the American Indians and by lightning main-

---

[1] Mention of a trademark, code name, or proprietary product is for identification purposes and does not constitute a guarantee or warranty of the product by the US Department of Agriculture, and does not imply its approval to the exclusion of other products that may also be suitable.

[2] Contribution of the Agricultural Research Service, US Department of Agriculture, in cooperation with the Agricultural Experiment Station, Oregon State University. Technical Paper No. 4166 of the latter.

355

tained open grazing lands for the bison and pronghorn antelope, the largest herds of grazing animals that the world has known. The burning practice was considered beneficial to the land and was continued by white men, particularly in the humid region of the southern United States as a means of preserving the open pine forests and maintaining good grazing (31). The ancient practice of woods burning passed from Indians to the early white settlers to the extent that a law of North Carolina enacted ca 1731 required the burning of pastures and rangelands each March 10 (19).

The domestication of plants and animals commencing about 8000 BC brought the need for open soil and led to the slash and burn practices. In Europe, trees were cut and burned, and the cleared land was sown with cereals until the forest regenerated and the inhabitants moved on (83). The cut and burn practice continues in the tropics as the only feasible way to obtain open land for food plant cultivation (7).

Similar to procuring ash for man's first mineral fertilizer by burning vegetation is a system of sod- or soil-burning used in India and Africa. Dry grass and weeds are heaped and covered with a mound of sod strips. Igniting the heap from below results in combustion of the vegetable matter and humus. The resulting ashes represent fast recycling of mineral nutrients that are necessary and immediately available to crops (7). The method must also reduce populations of certain insects and pathogens in the sod that is burned.

Primitive man also conceived the idea of lopping branches from nearby trees to burn on his field for the ash fertilizer, an ancient practice in Europe and one also used in Africa and some other parts of the world (7). Another outgrowth of the slash and burn method is the more modern piling and firing of brush on plant beds that heat the soil and kill weed plants and seeds, insects, nematodes, and fungi. Such methods were used extensively to prepare plant beds for seedlings, such as tobacco, and have only comparatively recently been largely replaced by various fumigants.

Contributions gratefully received from many colleagues around the world along with a review of the literature revealed that relatively little has been written on the effects of fire on plant disease control. This is readily apparent by the general lack of information on disease control in major reviews and books relating to the subject of fire (1–3, 7, 11, 15, 30, 35, 44, 49, 50, 57, 59, 83, 86, 87, 89). In addition, only one example of control by fire was mentioned in a review of plant disease control by unconventional methods (82).

This review describes some of the progress that is being made in control of plant diseases by thermosanitation through application of fire and/or flame. A complete account of every instance of inoculum reduction is not attempted. Control of some insects by fire is included because of their importance as vectors of plant viruses and other plant pathogens. Emphasis was deliberately placed on description of highly effective thermosanitation, especially that produced in Oregon grass seed fields by fire since 1948 and perhaps to be continued by limited use of fire under strict smoke management programs or by use of new mobile field sanitizers that are still being perfected.

OPEN FIRES

Forests

Wildfires shaped much of the world's forests by permitting fire-tolerant trees to dominate the vegetation. Forests were preserved by mild ground fires that periodically reduced the fuel accumulation, thereby protecting against catastrophic fires that would kill or severely damage trees. Management of fire for vegetation control and fuel reduction apparently will gradually replace the fire exclusion policy in the United States (1, 11, 57, 59, 83, 84, 89).

Despite the centuries of periodic wildfires and more recent managed fires or prescribed burning, information on benefits of burning in forest disease control is fragmentary. The prescribed annual burning of 1,215,000 ha of pine forest in the southern United States is possibly the most striking example of protective management by fire (E. V. Komarek, Sr., personal communication). Vegetation control is combined with control of several major tree diseases.

The classic example of the use of mild ground fires is the often cited control of brown spot needle blight (caused by *Scirrhia acicola*) of longleaf pine, *Pinus palustris,* in the southern United States. A prescribed ground fire generally eliminates sufficient inoculum for the two years needed to get longleaf seedlings up out of the "grass" stage, after which this fungus no longer deters growth (77). Longleaf pine is fire resistant, having a so-called asbestos terminal bud (8, 13, 15, 45, 77); even so, the time of burning should be carefully managed to avoid damage to the pine seedlings (64).

Burning reduces the incidence of fusiform rust, *Cronartium fusiforme,* the most important disease of southern pines. Burning kills the lower branches of saplings, where most of the rust cankers would arise, thus keeping them from ultimately becoming main stem cankers and killing the tree. Fire also suppresses oaks, *Quercus* spp., which harbor the uredo and telial stages of fusiform rust, while promoting growth of the more fire-resistant pine. Without fire, longleaf pine is replaced by the fire-sensitive and rust-susceptible slash pine, *Pinus elliotti* var. *elliotti,* and loblolly pine, *P. taeda.* Longleaf is highly rust resistant, whereas slash and loblolly pines are resistant to brown spot (22).

Mistletoe (*Phoradendron* spp.) and dwarf mistletoe (*Arceuthobium* spp.) may be eliminated in small patches if infested trees and additional ones on the perimeter of infection patches are felled. The cut trees and mistletoe seeds can then be destroyed by bonfires. For example, dwarf mistletoe, *Arceuthobium pusillum,* was eradicated in localized infection centers in black spruce, *Picea mariana,* in Minnesota by cutting down spruce trees to provide a layer of dry slash and by burning out the infected area when the forest was wet and burning was safe (51).

In several areas following logging, undersized infested trees that are not eliminated during logging must be killed to prevent perpetuation and spread of mistletoe to regenerated trees. Broadcast burning of logging slash is therefore recommended to destroy residual trees infected with dwarf mistletoe (6). Fire exclusion policies in the United States during the past half century apparently increased dwarf mistletoe levels, and prescribed burning as a supplemental method of dwarf mistletoe control was suggested to eliminate residues in cutover areas and to eliminate heavily infested unmerchantable stands (4). One effect of catastrophic wildfires that kill or severely damage ponderosa pine forests on some sites is the appearance of mistletoe-resistant lodgepole pine, *Pinus contorta.* By being resistant to the ponderosa mistletoe, *P. contorta* constitutes a barrier to the spread of mistletoe in the ponderosa (92).

A unique example of an adverse effect of fire is the predisposition to root rot, caused by a fungus, *Rhizina undulata (R. inflata).* In Europe, dying of pole-size conifers in groups as a result of this disease usually occurs around the sites of bonfires built by workers during thinning operations. In North America, the disease has been observed on seedlings planted on clear-cut areas where slash has been burned or in nurseries where brush has been burned. The fungal sporocarps develop on the forest floor after a fire. Because the optimum temperature for spore germination is high, 35–45°C, the spores may lie dormant in soil for two years. *Rhizina* can severely damage 2- to 4-year-old Douglas fir seedlings transplanted immediately after a slash burn or fire. The trouble can be avoided by delaying transplanting for two years after the fire or burn (5, 69). In earlier years after trees were felled in Holland, branches were burned to kill insects, but this has been discontinued as the burning stimulates development of *Rhizina* sp. on the newly planted small trees (J. de Tempe, personal communication).

An outbreak of balsam fir canker caused by *Nectria macrospora* was confined to a small area in Quebec in 1970 by felling about 300 balsam fir trees (*Abies balsamea*) and burning the affected crowns and trunks in both fall and spring. Three years later very few newly affected trees could be found in the surrounding forests, and use of this control measure was considered to be appropriate. This was especially so as it was the only rapid means to control or slow down the spread of this canker, because at that time, no effective fungicide for control of this nectria canker was available (M. Lortie, personal communication).

A wildfire or a controlled fuel reduction burn produces a temporary reduction in the population of *Phytophthora cinnamomi* in soil. The reduction lasts from two to six months and is caused by a change in soil microbial balance and not by direct heat killing of the pathogen. However, where *P. cinnamomi* causes severe root rot, recovery of the sensitive eucalyptus in Australia after a fire is seriously hindered and weaker trees are killed (G. C. Marks, personal communication).

Fire can have other undesirable effects. *Ribes* spp. may reproduce so abundantly following fire that their removal to protect white pines from blister rust, *Cronartium ribicola,* is impracticable, and the increase of *Ribes* may aid the spread of the rust (23).

Similarly, reports on various wood-rotting problems arising from fire-caused wounds far outnumber reports (29) of partial control of wood rot by burning. This is logical, because deadwood represents a nutrient supply in locations convenient for exposure to spores and possible invasion by a multitude of fungi. Therefore, the effects of prescribed underburning or ground fires should be distinguished from those catastrophic wildfires that kill or severely wound trees. Of course, with any kind of fire, damage varies widely with different tree species, according to their inherent tolerances to heat (15). Although most species of hardwood trees are heat-sensitive, fire can be beneficial by killing hardwood stumps and forcing suckers to develop from roots free of butt rot, whereas suckers arising from unburned stumps are invariably infected (8, 13, 45).

Prescribed fire reduces impact of *Fomes annosus* in plots burned twice before initial stand thinning operations and biannually after thinning. Fire seems to provide only partial protection, and it could not be recommended as a sole means of preventing loss (29).

Before modern fire control, ground fires ran through the old ponderosa pine stands on the east slope of the Cascade Mountains in the northwestern United States about every 14 years. Numerous scars are evident where mistletoe was burned out by past fires. Infected lower branches often were killed by heat, and heavily infected and highly inflammable trees were selectively destroyed. Several needle diseases, such as blight caused by *Elytroderma deformans,* apparently are becoming more prevalent with the continued absence of fire. Without ground fires, understory thickets of "weed" trees are more readily infected by these needle pathogens, and perhaps inoculum that increases on the young weed trees is spread to the large trees. Comandra rust on pine, caused by *Cronartium comandrae,* has apparently become more serious in recent years as a result of an increase in the alternate hosts, species of *Comandra,* which previously had been decreased by ground fires (L. F. Roth, personal communication).

Forest Slash Burning

Logging creates massive residues for which removal is usually not economically feasible. In the northwestern United States, an estimated 107,325 hectare (ha) of forest slash have been burned annually in Oregon and Washington (53). The logging residue ran as high as 672 metric tons per ha on old growth stands (E. H. Clarke, personal communication) with a usual range of 90 to 508 metric tons per ha (53). Thus, in the two states, from 9.07 to as much as 72.5 million metric tons of slash residue apparently were once burned annually.

Perpetuation of diseases in forest residue is primarily related to control of the root rots that survive saprophytically in larger residues such as stumps and large roots that cannot be removed with prescribed ground fires. In the Pacific Northwest area of the United States, three root pathogens, *Armillaria mellea, Fomes annosus,* and *Phellinus (Poria) weirii,* are considered responsible for most of the root rot mortality. These fungi, established in stumps and root systems, may infect surrounding trees. *Fomes annosus* can also infect stump surfaces and therefore presents an additional threat in managed stands (72).

One 14-year-old new forest of Douglas fir in Washington is seriously damaged with *Phellinus weirii* root rot, and efforts are being made to remove the old stumps and other large wood chunks that are reservoirs of the pathogen and therefore a source of infection for nearby young trees. Removal of the stumps, large roots, and various other wood during logging is thus important for disease control (L. F. Roth, personal communication). Reduction in volume of wood residue subject to slash burning is also desirable to reduce the smoke and reduce the soil heating. With these plus factors for wood removal, further reduction in volume of slash burning can be expected when the demand for wood fiber justifies its removal.

Slash burning is common in British Columbia, Canada, with vast areas burned every year. In fact, the British Columbia Forest Service requires the burning of logging slash to control insects such as Douglas fir or spruce bark beetles, to promote forest regeneration, and to reduce future fire hazard (R. Hunt, personal communication).

Fruit Crops

In addition to the burning of infected prunings from many different plants, burning is prescribed for disposing of whole trees infected with diseases such as armillaria root rot. Incineration of infected leaves in orchards might eliminate much inoculum of some tree fruit diseases. In discussing this possibility (with the writer) during 1962, W. D. Valleau recalled that a low ground fire had run through an apple orchard and burned the dry fallen leaves. The following year he observed that the apples were free of scab caused by *Venturia inaequalis,* the only time in 30 years in his experience that any apple crop in Kentucky was free of the disease (W. D. Valleau, personal communication). Perhaps a low creeping fire could be used in some other orchards for destruction of inoculum in fallen leaves with precautions such as fuel pulled away from tree trunks. Such special sanitation might be practical where an orchard is isolated from other sources of inoculum. Perhaps too, the antifungal properties of the smoke might assist in disease control, a possibility recently suggested (73).

Burning pineapple in Hawaii, where 269 metric tons of vegetative green matter occur per hectare, destroys some mealybugs and ants. However, burning is not considered economically important because rapid ingress of insects from field borders still necessitates prophylactic measures (J. W. Hurdis, personal communication).

Ornamental Plants

Renewal of heath by burning after the brush grows beyond a certain height is a fairly general practice because young shoots provide better feed for cattle, sheep, and goats (83). Burning heather moors, predominantly *Calluna vulgaris,* has long been a regular practice in Scotland to prevent growth from becoming too rank and woody. Burning improves the moors as habitat for grouse for which they are now chiefly maintained, and it improves their quality for sheep grazing. Rather drastic burning generally leads to recovery of heather affected by a dieback associated with *Marasmius androsaceus* (E. G. Gray, personal communication).

Berries

Numerous Indian tribes burned woods and bushes to improve the berry harvest in western North America (83) and in the Great Lakes region of the United States (44). As a modern extension of this age-old practice, lowbush blueberries, *Vac-*

cinium spp., are burned in the northeastern United States and Canada, primarily to prune, thin, and rejuvenate plants. As an alternative to burning, mowing the plants usually leaves an above-ground stub and other growth that interferes with raking the berries. New buds tend to develop from this stub rather than from the underground rhizomes. These above-ground buds then produce stems that are short and spindly and bear very little fruit. In contrast, burning kills the stems to the soil line, and regrowth begins from underground buds. Thus, burning produces the most vigorous new stem growth and the highest yields of blueberries. In Nova Scotia, a light burn on lowbush blueberries is recommended to avoid damage to underground rhizomes (R. A. Murray, personal communication).

Although burning is done mainly to rejuvenate the plants, it plays a large role in control of diseases of lowbush blueberries (33). Dieback (caused by *Diaporthe vaccinii*) and canker (caused by *Godronia cassandrae*) are largely controlled by burning. Red leaf or gall (caused by *Exobasidium vaccinii*) is controlled to some extent by this practice. Twig and blossom blights are reduced somewhat by burning because mummy cups are avoided and the *Monilinia-* and *Botrytis*-infected stems are destroyed by the fire. The latter two diseases also require fungicide control measures. In areas of Nova Scotia, with good air drainage where fog or mist is not a general problem, the burning process keeps most diseases in control, but all other areas require blight dusts (C. L. Lockhart, personal communication).

On the negative side, multiplying and spreading of powdery mildew (caused by *Microsphaera alni*), and blueberry rust (caused by *Pucciniastrum myrtelli*) were blamed on fire that stimulates growth of prolific stands of the host plant (24). Burning strawberry beds after the fruit was picked was formerly recommended in Scotland to control the principal leaf spots, *Mycosphaerella fragariae* and *Marssonina fragariae* (E. G. Gray, personal communication).

Cotton

After harvest of cotton fiber in several areas, all plant residue in fields is burned to kill sources of disease inoculum and the remaining stages of insects. Burning litter and debris around cotton fields after harvest had been the standard practice in the southern United States in the early 1900s. Rapid spread of the boll weevil, *Anthonomus grandis,* occurred coincident with development of forest fire prevention policies in that region. Numerous boll weevil overwinter in trash around field borders and are readily killed by burning (58).

Sugarcane

In most parts of world, sugarcane fields are fired before harvesting mainly to facilitate harvesting operations, because incinerating dead leaves and trash reduces the amount of vegetation to be hauled to the mill. Sugar yields are increased by preharvest burning as the stalk gets a better grind and sugar is extracted without interference of extraneous vegetative matter (J. W. Hurdis, personal communication). The firing also lessens the danger to man from venomous snakes in Australia (P. B. Hutchinson, personal communication), New South Wales, and Guyana (J. F. Bates, personal communication). In Barbados, burning sugarcane before harvest rids the fields of larger animals as well as snakes (D. I. T. Walker, personal communication). Where preharvest burning is not practiced, as in Fiji, the stubble is usually fired after harvest. Burning reduces inoculum of many sugarcane diseases and reduces populations of viruliferous insects, e.g. *Perkinsiella* spp., the leaf hopper vector of Fiji disease of sugarcane. Although heat destroys eggs, nymphal and adult stages, these reductions may not contribute significantly to disease control (P. B. Hutchinson, personal communication).

Sugarcane trash has been burned in Taiwan the last few years mainly to destroy perithecia of *Leptosphaeria taiwanensis,* the pathogen of leaf blight disease, but control has been incomplete (62). Leaf blight was first seen in Taiwan in 1934 and became serious in 1969 following introduction of more susceptible sugarcane varieties. Thus burning became necessary because of the intensification of the disease. Burning sugarcane after harvest is believed to remove cover for the froghopper, *Aeneolamia varia saccharina,* which is a serious pest in Trinidad and in parts of Central and South America (D. I. T. Walker, personal communication).

In Hawaii environmental aspects of sugarcane burning have received much attention. Alternates to burning proved prohibitive in cost; however, the near-constant tradewinds fortunately sweep the air clean (J. W. Hurdis, personal communication). Tests indicated that only a minor degree of air pollution was contributed by the practice, and in 1973 a decision was reached to allow continuation of sugarcane burning in Hawaii (R. S. Byther, personal communication).

Rice

Burning of rice stubble and straw is a common practice throughout the world. For example, during the past ten years, between 134,865 and 214,650 ha of rice straw and stubble have been burned annually in California (G. E. Miller, Jr., personal communication). Burning the rice fields after harvest eliminates straw and facilitates working soil, but the fire also is the most effective method for reducing inoculum in control of rice stem rot caused by *Sclerotium oryzae* (90); (R. K. Webster, personal communication). Burning of rice straw and stubble is a common practice in the Phillipines and one of the recommended control measures for control of sheath blight caused by *Corticium sasaki* and stem rot, caused by *S. oryzae* (D. B. Lapis, personal communication).

Field burning of rice stubble in Vietnam has been a common practice to eradicate certain pests and diseases such as rice stem borers, leaf blast, foot rot, and stem rot. Burning is sometimes done immediately after harvest to fumigate rats in their tunnels along ditches (Doan-Minh-Quan, personal communication).

Control of black mold of rice heads (caused by *Curvularia lunata*) apparently is difficult by sanitary practices, but rice straw stacks may be sources of infection and their elimination by burning was suggested (65). Burning all weeds and volunteer rice around fields was also recommended as part of a program to reduce numbers of *C. lunata* spores (66).

Grain Crops

Burning grain stubble is a regular practice in many countries and furnishes considerable insect control (58). In early studies of wheat culture, burning stubble controlled flag smut, caused by *Urocystis agropyri,* in Australia (34), *Pseudocercosporella (Cercosporella)* stem eyespot or foot rot in New Zealand (12), and *Gaeumannomyces (Ophiobolus)* root rot or take-all in Sweden (43). More recent reports concerning disease control by burning grain stubble generally have been less encouraging, but some of the reports describe incomplete burns with untreated islands of stubble protected by green weeds and quackgrass, *Agropyron repens,* that could explain lack of control. Also, the value of stubble burning for several successive years usually was not measured.

Burning of grain harvest trash in Western Australia is considered useful in control of scald (caused by *Rhynchosporium secalis*) and net blotch (caused by *Pyrenophora teres*) of barley; *Septoria nodorum* and *S. tritici* on wheat, *Septoria avenae* on oats, and brown spot of lupine (caused by *Pleiochaeta setosa*) (A. G. P. Brown, personal communication). In most years, however, there would be insufficient stubble material left to achieve a hot enough burn, because in Western Australia crop stubble is an important source of grazing during the virtually rainless summer, December through May. The practice of burning off what crop stubble exists is practically universal, but the uneven rather cool burn achieves very little weed or disease control (G. A. Pearce, personal communication).

Cutting wheat high and burning the stubble was recommended for control of the twist disease caused by *Dilophospora alopecuri* (10). The wheat seed nematode, *Anguina tritici,* often mentioned in association with *D. alopecuri,* should also be readily controlled by stubble burning.

Cereals are said to be burned in southern Sweden for control of *Gaeumannomyces (Ophiobolus)* and *Pseudocercosporella (Cercosporella)* foot rots. The sanitation may have some disease control value, but this has not been proven (B. Mattsson, personal communication). No substantive investigation has been made in Sweden (K. Linsten, personal communication).

Durum wheat stubble that was burned in North Dakota to reduce excessive mulch had less helminthosporium root rot (E. H. Lloyd, personal communication). However, stubble burning was insignificant in control of common rot rot caused by *Helminthosporium sativum* in Saskatchewan (60). Wheat stubble was formerly burned to combat *Gaeumannomyces (Ophiobolus) graminis* in South Africa, but the practice was discontinued (S. W. Baard, personal communication).

Hay, Pastures, and Rangeland

Although much has been written about the improvement in value of forage by burning over of rangeland or grassland, apparently disease control was rarely measured. In the southeastern United States, burning bermudagrass (*Cynodon dactylon*) pastures before green-up each spring hastens the spring growth, and forage yield and quality are greatly increased. Burning bermudagrass removes thatch, provides a measure of control of leaf spot and stem blight caused by *Helminthosporium spiciferum* and *H. rostratum,* and provides excellent control of two-lined spittle bug, *Prosapia bicinta* (85); (R. T. Gudauskas and R. B. Carver, personal communication).

Sclerotinia sclerotiorum in rape would be expected to be reduced in Germany by burning crop residues, because the sclerotia in stems would be burned (W. Kruger, personal communication).

Burning of old alfalfa stems was recommended to reduce inoculum for control of several stem and leaf diseases (25). About one third of the alfalfa seed growers in Saskatchewan burn alfalfa crop debris for control of aphids and diseases, such as black stem caused by the several *Phoma* and *Ascochyta* spp. (R. P. Knowles through J. D. Smith, personal communication). The only recommendation of this type, made in Manitoba, Canada, is to burn alfalfa seed fields to control black stem (W. C. McDonald, personal communication).

Grass Seed Production—A Case History

The value of fire as a tool to procure the seed of certain wild plants was widely known to prehistoric man. Throughout the western United States, Indians from about 50 tribes practiced intentional broadcast burning to increase the yield of seeds of wild grasses and weeds that were collected for food (83). Over the centuries, doubtless thousands of others observed the improvement in seed production of grasses after burning. Documentation of the use of fire to break the sod-bound condition and thus increase seed production in old stands of perennial grasses, however, apparently awaited publication in 1944 of the results of experiments at Tifton, Georgia. These tests were suggested by a US Forest Service range examiner who observed that certain native grasses, such as *Aristida stricta* and *Sporobolus curtissii,* which rarely flower in old sods, will flower profusely after the sod has been burned (17). Burning also increased seed yield in red fescue (*Festuca rubra*) in Pennsylvania in tests started in 1941 (71).

Burning of grass fields, now the single most important cultural practice in grass seed production in the Pacific Northwest region of the United States, was started originally for disease control 28 years ago (37–42). Postharvest burning of straw and stubble was adopted as a universal practice after the 1948 harvest to control blind seed disease caused by *Gloeotinia temulenta,* which was threatening the perennial ryegrass (*Lolium perenne*) seed industry in Oregon. About half the acreage was burned after the 1948 harvest, and nearly all fields have been burned since 1949. Excellent control was immediately obtained. The disease has been well controlled by burning through 1974 (Figure 1), except for a brief flareup in 1956 and 1957 when straw choppers added to combines spoiled straw distribution (39, 42).

Ergot (caused by *Claviceps purpurea*), a major disease of most cultivated grasses in north temperate zone countries, was controlled adequately in a few grass seed fields in western Oregon beginning in 1944, by burning straw and stubble in fields after harvest (38). Ergot has been controlled in perennial ryegrass by

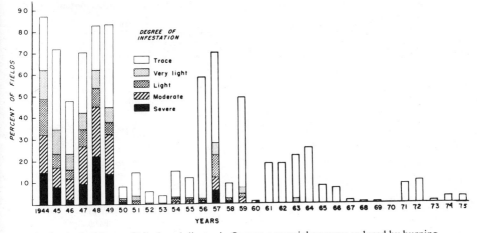

Figure 1   Incidence of blind seed disease in Oregon perennial ryegrass reduced by burning all fields after 1949.

burning since 1948 (42). When it was applied in 1949 to fields of tall fescue (*Festuca arundinacea*) burning controlled blind seed disease and ergot. Field fires destroy most of the fungus sclerotia in crop residues. Prevention of fall heading of perennial ryegrass plants is a special need, because autumn inflorescences form abundantly in this grass and are usually heavily infected with *C. purpurea.* Burning eliminates fall heading and thus prevents the secondary increase of ergot sclerotia that are important to the total supply of inoculum. Control of ergot was confirmed in Georgia where burning dormant dallisgrass (*Paspalum dilatatum*) controlled *Claviceps paspali* (91).

Grass seed nematode, *Anguina agrostis,* had caused severe losses in Chewings fescue (*Festuca rubra* supsp. *commutata*) seed crops in western Oregon. The incidence of seed nematode dropped sharply after annual field burning was adopted in 1955 and has virtually disappeared from Oregon seed production fields of Chewings fescue.

Silver top, which caused complete crop losses in chewings fescue and Highland bentgrass (*Agrostis tenuis*) during 1954 and 1955 in western Oregon, was eliminated by field burning. Silver top had been controlled earlier in Pennsylvania by burning (54), when silver top was thought to be caused by *Fusarium tricinctum* f. sp. *poae* and the mite, *Siteroptes cerealium (S. graminum).* However, later work (36) revealed that the fungus and mite were late-arriving saprophytes and suggested a return to insects as causal agents. Recognized causal insects, such as *Leptopterna (Mirus) dolabrata,* deposit eggs in the lower grass culms that are destroyed when fields are burned, thus providing excellent control of silver top.

In subsequent years, burning was adopted on other perennial grasses grown for seed in Oregon, except the heat-sensitive creeping bentgrass, *Agrostis palustris,*

Table 1 Diseases controlled, and years in which field burning started, on major perennial grass seed crops in western Oregon

Grass	Year burning started	Diseases controlled by field burning	
		Good control	Partial control
Perennial ryegrass	1948	Blind seed disease[a] Ergot[a] Silver top	Helminthosporium leaf blotch
Tall fescue	1949	Blind seed disease[a] Ergot[a] Silver top	Cercospora leaf spot
Bluegrasses	1950s	Ergot[a] Silver top[a]	Stripe smut Flag smut Rusts Powdery mildew
Chewings fescue	1950s	Silver top[a]	Red thread
Red fescue	1950s	Seed nematode[a] Ergot[a]	Septoria blotch Powdery mildew Leaf rust
Colonial bentgrass	1954	Ergot[a] Silver top[a]	Seed nematode[a] Rhizoctonia
Orchardgrass	1958	Ergot[a] Silver top[a] Multiple leaf diseases[a]	Stripe smut

[a] Diseases that were original reasons for field burning.

and velvet bentgrass, *A. canina.* In western Oregon especially, burning controlled ergot and silver top in colonial bentgrass (*A. tenuis*), red fescue (*Festuca rubra*), Kentucky bluegrass (*Poa pratensis*), and orchardgrass (*Dactylis glomerata*).

Burning also partially controlled a number of leaf and stem diseases in several grasses. Burning and flaming helped to control leaf rust, caused by *Puccinia poae-nemoralis,* in *Poa pratensis,* but later on, more effective chemicals were developed (38) to control the more serious stripe rust, caused by *P. striiformis,* that appeared suddenly in 1958. The major diseases controlled and the year field burning was started on various major grasses grown for seed in Oregon are listed in Table 1.

Pest control by burning has been effective in Canada. Burning fields of red fescue in the Peace River area of northern British Columbia was recently used to control stem eyespot caused by *Phleospora idahoensis,* and insects that cause silver top. The practice is not widespread because the amount of straw is limited, making good fire coverage difficult. In addition, adverse weather frequently results in dry and windy conditions with danger of fires getting away or in such wet conditions the fires will not burn properly (J. D. Smith, personal communication). Smith also concluded that field burning has kept the phleospora eyespot out of Oregon fescue seed fields (81).

In Saskatchewan, fall burning reduced the incidence of leaf spot caused by *Selenophoma bromigena* and increased seed yield of smooth brome, *Bromus iner-*

mis (79), and the flaming or burning of crop residues significantly reduced the incidence of leaf spot caused by *Pyrenophora bromi* (80).

Elimination of Toxins

Removal of old plant material is important in use of grasslands in many areas because of the possibility of toxin formation by various microorganisms. The saprophytic fungus, *Pithomyces chartarum,* for example, colonizes plant residues and produces the toxin, sporidesmin, that causes liver damage, loss of weight, jaundice, and photosensitivity in sheep and cattle that graze infested ryegrass pastures in New Zealand. Facial eczema describes the inflammation and scabbing of unprotected skin resulting from the photosensitivity (14). Elimination of most dead and senescent leaves should be possible with a good field fire, and this sanitation perhaps would be sufficient to reduce or to eliminate the mycotoxin problems.

Control of ergot avoids potential ergot poisoning in livestock, game animals, and birds. Burning straw and stubble in reasonably dry fields destroys most of the fungal sclerotia in crop residues and on the soil surface. In perennial ryegrass, burning eliminates autumn heads that usually are severely infected with ergot. Thus, burning prevents the possibility of abundant ergot in inflorescences that could be ingested by livestock grazing the fields.

The galls caused by the Chewings fescue seed nematode, *Anguina agrostis,* are poisonous to livestock, game animals, and birds, and many fatalities have occurred from feeding infested seed screenings in Oregon. Fortunately, field burning in consecutive years beginning in 1955 has virtually eliminated the seed nematode in Chewings fescue fields in western Oregon. Similarly, numerous fatal poisonings of cattle in Australia were associated with grazing the volunteer weed grass, Wimmera ryegrass (*Lolium rigidum*), infested with a seed nematode, *Anguina* sp. (32). Wimmera ryegrass occurs in wheat fields, and this fairly drought-resistant grass is used for grazing after the wheat is harvested. Wheat straw and stubble were formerly burned in Australia, and the nematode apparently was unknown in *L. rigidum.* Twenty or 30 years ago, burning the wheat fields was discontinued, and the nematode in Wimmera ryegrass was found as a result of investigations into the increasing and widespread livestock deaths (E. D. Higgs, personal communication). Severe losses of livestock have occurred in South Australia (67), and more than 4000 steers recently died after grazing the Wimmera ryegrass infested with the seed nematode in Western Australia (P. H. Berry, personal communication). Control of seed nematode in Wimmera ryegrass by again burning the wheat straw and stubble may now be difficult because grazing reduces the residue to the extent that effective fires may not be possible (E. D. Higgs, personal communication).

FIELD FLAMING WITH OIL OR GAS BURNERS

Application of heat by open fires may be insufficient or may not be feasible because of a lack of dry fuel in a wet climate, insufficient plant residue, or hazardous location. Even public opposition to smoke may prevent or make open burning difficult. Some of these difficulties may be overcome in part by treating the fields with flame produced by oil or gas burning devices. Flaming may be directed to green foliage or to dry fields after removing most of the crop residues to reduce the fire hazard.

Oil and gas burners apparently have been used since 1926 for weed, insect, and dodder (*Cuscuta* spp.) control in alfalfa fields in the Antelope Valley of California (Manchester Tank and Equipment Company, private publication). Control of dodder in alfalfa fields by flaming with oil burners was confirmed in Utah (61). Use of various oil and gas burners in alfalfa fields for insect and weed control has been described in many reports throughout the United States.

In 1949, G. Marcott, a seed grower near Silverton, Oregon, tried to kill perennial grasses under turkey porches using a kerosene flame thrower, but the grass plants withstood the intense flame treatment and produced a strong regrowth. Later, when an incomplete field fire left much plant material in a first-year field of *Poa pratensis* Merion, he built a propane flamer with advice of a Salem, Oregon, propane dealer and used it during 1955 to finish sanitizing his bluegrass field. Several other seed growers flamed bluegrass fields during 1957 after first removing straw and stubble with open field burning (G. Marcott, personal communication).

After the initial success of the flaming practice during 1955, flame generated by propane burners was used in western Oregon for rust control by incinerating autumn leaf growth of certain grasses (38). Leaf rust (caused by *Puccinia poaenemoralis*), stem rust (caused by *P. graminis*), and stripe rust (caused by *P. striiformis*) were temporarily eliminated in harvested *Poa pratensis* fields by open burning of straw and stubble. Spring-planted new fields, however, furnished abundant rust inoculum to reinfect the harvested fields. The rusts were killed in the new plantings by flaming and in old fields by burning, thus providing a thermal treatment for both.

Some older fields, in which straw and stubble had been removed by open burning, are now flamed for weed and disease control. The combination of open burning and one or more treatments with flame provided a more effective treatment that improved the control of ergot in bluegrass (*Poa pratensis*) fields especially. The cost of the fire and multiple flame treatments is partially offset by better control of annual bluegrass (*Poa annua*) and other "weed" grasses.

Mint growers in western Oregon were not satisfied with the performance of plant-killing chemical sprays for control of rust, *Puccinia menthae,* that overwinters on mint stubble and on wild and escaped mint (18). In the spring of 1956, K. P. Coykendall, owner-manager of a propane gas company in Albany, Oregon, conceived the idea and convinced farmers to test for possible rust control in peppermint by flaming applied with a hand-held propane burner. Because the flaming did a perfect job of rust control, and the strong new shoots that emerged after treatment in the small test plots indicated that the method was safe, he immediately brought in a field flamer and promoted large-scale flame treatment of mint in the spring of 1957. A variety of flaming devices have since been used for spring flaming of mint that quickly became standard practice with the 1957 success (K. P. Coykendall, personal communication). The value of spring flaming of mint for rust control was confirmed in scientific tests (21, 46).

Verticillium spp. have been killed in infected tissues of several plant species with heat by flaming. *Verticillium dahliae* in peppermint stems was killed by flaming the stubble (47, 68). Development of improved herbicides allowed discontinuation of plowing and cultivating that were responsible for mechanical spread of *Verticillium*. Elimination of the mechanical spread of *V. dahliae* was reinforced by autumn flaming that nearly eliminated the inoculum in the stubble. Thus, autumn flaming without plowing and cultivating has become standard practice to retard the spread of *Verticillium* in mint fields of Oregon and Washington (C. E. Horner, personal communication). Mint growers in Idaho have experimented with field flaming to control *Verticillium* in the stubble, but this practice has not been generally adopted in that state (A. Finley, personal communication).

Flaming of potato vines and harvest debris in an effort to kill *Verticillium* was initiated in south central Oregon in the late 1950s by A. E. Gross, while superintendent of the Klamath Branch Experiment Station. Laboratory tests revealed that *V. albo-atrum* was killed in the potato vines by flaming with a propane burner (74). In addition to direct killing of the fungus in vines, light flaming left a certain amount of vine stubs, which facilitated removal of more diseased roots by raking after digging the tubers than could be achieved following more intense flaming. This reduction in verticillium inoculum was considered important in slowing build-up of the disease in fields (74). Annual preharvest burning of potato vines to destroy the microsclerotia in stem tissue also increased yields after two successive years of treatment. However, there was no delay in reduction of soil populations of *V. albo-atrum* and reduction of disease appearance until vine burning was performed for three consecutive years (27).

The superficial black internal spot of potato tubers, caused by *Alternaria solani*, is reduced by preharvest vine burning to remove all infected foliage and thereby reduce inoculum. In Idaho the highest percentage of clean potato tubers resulted from a good fungicide spray program coupled with preharvest vine burning (J. Garner, personal communication). Similarly, if late blight (caused by *Phytophthora infestans*) was present in a potato field, killing the vines as quickly as possible with chemical vine killers or vine burners was recommended in Wisconsin (75).

Flaming the banks of irrigation canals and ditches and drainage ditches in a 57 km² area of central Washington killed weed virus hosts and reduced green peach aphids, thereby effectively controlling beet yellows and beet western yellows viruses in sugar beets (88). Flaming is therefore recommended to kill weeds that serve as both reservoirs of the virus and food plants for the insects (63).

Flaming of winter wheat stubble to kill inoculum of *Pseudocercosporella (Cercosporella) herpotrichoides* was considered too laborious and too expensive to be recommended for practical use in Norway (H. A. Magnus, personal communication). Treatment of stubble with high-output flame cultivators did not significantly reduce the incidence of either take-all or eyespot in Scotland (76).

Flame cultivation of wheat stubble, before or after plowing, did not decrease the incidence of eyespot in July in the following winter wheat crop (78). In contrast, in eastern and central Idaho, flaming winter wheat seedlings eliminated some of the foliage and surface residue, and this reduction in organic residues before winter significantly reduced damage from snow mold, *Typhula idahoensis* and *Fusarium nivale* (48).

Flaming of individual hills of hop plants (*Humulus lupulus*) with a hand-held propane nozzle has been used to destroy dwarfed sucker shoots (spikes) systemically infected with *Pseudoperonospora humuli*. These shoots are primary sources of inoculum. Contact herbicides are just as effective as flame, however, and are more generally used (A. M. Finley and C. E. Horner, personal communication).

Flaming of rhizomatous iris beds in late fall has reduced the iris borer (*Macronoctua onusta*) and leaf spot (caused by *Didymellina macrospora*) in Wisconsin and has been adopted as standard practice by one seed company since 1970. The practice is to wait until the ground is frozen before fall flaming and, if possible or appropriate, to flame again in the following spring (G. W. Worf, personal communication).

Burning over diseased patches after spraying with some inflammable liquid was recommended for control of the leaf-gall nematode, *Anguillulina tumefaciens*, in Bradley grass (*Cynodon transvalensis*) turf in Transvaal, South Africa (20). A flame treatment should be ideal for eradication of *A. tumefaciens* and other leaf-gall nematodes, such as *A. graminophila* in *Agrostis* spp.

CHARRING WITH SUPERHEATED AIR

The application of a blast of superheated air, heated with propane burners, charred and killed sagebrush plants (*Artemisia* spp.) without ignition in Montana (9), and this method should be explored for possible disease control. Conversion of flame to hot air, however, may be an inefficient complication for thermal sanitation in grass seed fields or other crops. Field burning is basic to weed control because it kills many weed seeds on the soil surface and provides a clean surface that aids distribution of chemicals and eliminates organic matter that prevents action of fall-applied, soil-active herbicides. A clean soil surface would be similarly helpful in control of some grass diseases, such as blind-seed disease and ergot, by chemicals when effective fungicides become available. The application of superheated air would not remove the debris and residues, and living and dead plant material may prevent the heat from reaching weed seeds and disease propagules.

MOBILE FIELD SANITIZERS

Public opposition to smoke from open field burning in western Oregon has hastened the search for alternative ways to sanitize fields. Flaming, of course, is too hazardous and therefore impossible in a dry field of straw and stubble, and the expense of removing straw and stubble and then flaming becomes too costly for low-income crops.

What is needed is some type of grass burner that will set fire to the stubble and then continue the fire by means of a draft of air such as a fan at the rear discharging forward upon the grass as the device moves over the ground thus burning a strip equal in width to the width of the machine. The device should consist of a bottomless enclosure secured to a suitable frame mounted upon wheels, so that it may be drawn over the fields . . . and is provided with means for starting and continuing a fire upon the grass or stubble that is at the time enclosed within the confines of the machine, the machine being provided with aprons of fireproof material, which will prevent spreading of the fire to one side of the machine.

The above description outlines what is needed in 1976, but surprisingly, this composite description was taken from a 76-year-old patent application filed with the US Patent Office in 1899 (70). Many other applications for patents on grass burners, land clearing devices, and prairie fire fighters apparently were filed during the nineteenth and first half of the twentieth centuries. The wild prairie fires that struck fear in the early settlers of the Great Plains of North America and part of Australia stimulated the invention of a variety of grass burners designed to cut a lane or fire break in thick grass to stop the advance of frightening wildfires or to clear land safely.

In 1970, agricultural engineers at Oregon State University designed and built a mobile incinerator that would remove the straw and stubble with a minimum of smoke while sanitizing the fields (55, 56). Although the OSU model, as diagrammed in Figure 2, still contained several operational defects, trials with this and later models demonstrated that an effective high-temperature treatment could be applied to the soil surface by an enclosed fire, utilizing only the grass residue as fuel. During treatment trials the temperature at the soil surface varied from less than 93 to over 538°C depending especially on the amount of fuel, moisture content of soil and fuel, and velocity of air supplied at ground level and above the fire. Residue was almost completely removed and most weed seeds were killed, thus assuring chemical control of winter annual weeds by herbicides.

Removal of straw before treatment with the mobile incinerator was highly desirable, to increase the ground speed of the machine and to reduce damage to

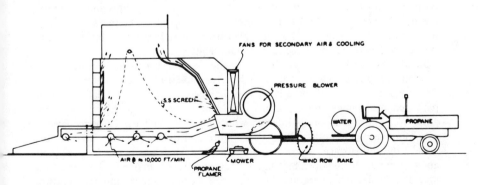

Figure 2 Schematic section view of an early model OSU mobile field sanitizer that burned straw and stubble residue directly on the ground. (Photo courtesy of D. E. Kirk.)

metals. The stubble alone was found to be sufficient fuel for the sanitizer, and adequate temperatures were generated at the soil surface to kill three major pathogens of grass seed crops, *Gloeotinia temulenta* in infected seeds, *Claviceps purpurea* in sclerotia, and *Anguina agrostis* in seed galls at the soil surface. Effective treatments were 10- to 15-seconds exposure at a variety of temperature regimes that peaked in excess of 149°C.

Tests comparing the mobile sanitizer with open burning showed still other advantages: more uniform treatment, less burnout of grass plants, and more effective treatment with probably better control of diseases and weeds. Treatment with a mobile sanitizer apparently could provide better control of ergot in grasses, such as *Poa pratensis,* than could field burning.

Severe problems have hindered the development of the mobile field sanitizer, including lateral escape of fire, metal fatigue, excessive temperatures, exhaust discharge of firebrands that started fires outside the machine, and smoldering crowns after the machine has passed. Solutions to some of these difficulties were apparent during 1975, and the development work in Oregon may result in usable sanitizer designs (T. R. Miles and B. L. Rose, personal communications).

The need for methods to apply thermal sanitation to substitute for open burning continues to be urgent. Continuation of field burning on a reduced scale with good smoke management may be a reasonable alternate. An operational field sanitizer may be developed that will provide thermal sanitation at the soil surface in grasses and other field crops; however, economic feasibility for use in production of various grass seed crops remains to be demonstrated.

TEMPERATURE AND SMOKE

Disease control by incineration of propagules is obvious and can be dismissed. The effects of exposure to a wide range of temperatures, the possible effect of smoke, and effects of the ash, however, deserve additional consideration.

Temperature

Extreme temperatures, 538 to 1260°C, associated with forest fires and the burning of massive quantities of fuel, were reported (3, 15). At the soil surface, temperatures may exceed 204°C or higher from the radiated heat, but below the surface, temperatures decrease rapidly with depth (15).

In contrast, with open grass fires, only slight soil heating occurs during combustion of the relatively light fuel load such as 4.48 to 8.96 metric tons per ha. Because most grass leaf material is consumed by the fire, most leaf and stem pathogen propagules above ground should be incinerated by a clean-burning grass fire. Many grass seeds infected with *Gloeotinia temulenta,* Claviceps sclerotia, and seed-nematode galls shatter before and during harvest and are situated at the soil surface, where the temperature during fires is much lower than above ground. During open burning of straw and stubble in one field of perennial ryegrass, the flame temperature was above 538°C, but the maximum temperature at the soil surface was measured at only 88 to 132°C.

Smoke

Where the vegetation is entirely burned and inoculum is inactivated by severe heating or incineration, the possible effects of smoke on disease control may be relatively insignificant. In other cases where inoculum is situated on limbs above a low fire or at distances beyond a fire, the effects of smoke need to be determined. Recently, various periods of exposure to smoke inhibited spore germination, mycelial growth, and infection with several diverse fungi (73). Although food can be preserved by long exposures to heat and smoke (28), the practical significance of wood smoke toxicity to plant pathogens remains to be determined.

RECYCLING NUTRIENTS

Burning plant residue returns potassium, phosphorus, calcium, and various other minerals to the soil, and burning provides an additional benefit. Urea added to the soil surface is apparently reduced by urease, and much of the nitrogen may be lost as gaseous NH_3 to the atmosphere. Burning grass stubble may have destroyed or altered the urease because urea applied just after burning was 97% as efficient a nitrogen fertilizer as NH_4NO_3 (52).

ADVANTAGES OF THERMOSANITATION

Thermosanitation by Fire

Some of the possible advantages of thermosanitation from fire applied at the soil surface for plant disease control include the following.

1. When the propagules are situated above ground, heat is broadly effective against widely different pests and pathogens, including fungi, bacteria, and nematodes; insects; and weed plants and seeds.
2. Treatment with fire is inexpensive.
3. Plant pathogens have not developed resistance to heat such as has happened with chemicals (K. F. Baker, personal communication).
4. Thermosanitation furnishes a core treatment for integrated pest control by providing a basis for effective use of fungicides, herbicides, and some insecticides.
5. The fuel is supplied by plant material, a natural resource renewable annually, thereby conserving fossil fuels.
6. Treatment with fire is nonchemical; thus, it avoids pesticide residue problems.
7. Burning is specially important to eliminate inoculum in infected plant tissue, particularly of pathogens, such as *Cephalosporium gramineum,* that survive after burial by production of an antibiotic in infected grass or grain tissue (16).
8. Burning conserves energy by removing refuse, destroying pests and weeds, and allowing reseeding with rangeland seeders in a nontillage culture.

9. Heat may kill 95–99% of weed seed at the soil surface for direct control and provides a basis for excellent herbicide distribution and action by removal of plant residue.

10. Burning has increased seed yield 20 to 1000% in various perennial grasses.

11. Burning kills adult insects and eggs deposited in plant material or in the soil surface.

12. Burning kills some incidental grasses and other weeds that harbor various weed pathogens, especially smuts, whose spores contaminate seed crops and reduce seed quality by the presence of seedborne spores even though they are not pathogenic to the crop in question.

13. Open burning lends itself to wide adoption thereby making likely universal use that will gain the benefits of area-wide treatment that prevents cross-contamination from untreated fields.

14. Burning in place recycles various nutrients needed for plant growth especially in forests and relatively infertile grasslands.

15. Prescribed burning of forest and fields or grasslands reduces extreme danger of wildfires in excessive fuel accumulation and thus, is preventative fire control.

16. Burning eliminates dead and senescent leaves thereby eliminating the substrate for growth of saprophytic fungi that produce mycotoxins.

17. Burning can prevent production of toxins by control of diseases, such as ergot and various seed-nematode galls, in grasses that are poisonous to livestock, game animals, and birds.

Additional Advantages from Flamers and Mobile Sanitizers

Some crop or plant situations are not amenable to open fire. In situations where a flame treatment or enclosed fire in some type of mobile sanitizer is feasible, the following additional advantages are possible:

1. Improved flamers and mobile sanitizers can be used where fire is impossible and thus represent new opportunities for tests on problems of many field crops.

2. Mobile sanitizers may provide thermosanitation with very little smoke or particulate matter discharge.

3. Heat can be adjusted according to plant tolerance and control requirements.

4. Combinations of liquified petroleum gas or other fossil fuel with plant residue fuel offer nearly unlimited potential for heat application in numerous crops wherein fire is not possible and flaming with fossil fuel alone would be too expensive.

5. The soil surface is very little affected by light burning or light flame treatment. Therefore, flaming can be an effective tool in studies of the relative importance of inoculum above and below ground.

CONCLUSIONS

Thermosanitation apparently is largely under-utilized in plant disease control and therefore represents one of the remaining elements that might be utilized to increase productivity of crops. By inoculum reduction, thermosanitation has both a direct and solitary control effect on some diseases and/or an indirect control effect by providing a basis for feasible use of fungicides. For still other diseases, thermal sanitation may make chemicals more efficient, thus permitting a reduction in the amount of chemicals needed, thereby enhancing protection of the environment and reducing production costs.

The ancient uses of fire to purge forests, restore grasslands, or enhance berry production are reminders that many plants not only can withstand mild ground fires, but may respond with greater yields or quality or purity of stand. Because of their underground roots and stems, many plants are less vulnerable to surface heat treatment than certain pathogens. The value of nearly complete reduction of inoculum above ground inside the field compared to other inoculum sources in control of plant diseases will determine whether thermal sanitation by open fire, flame, mobile incinerators (sanitizers), or other means will be desirable and economically feasible.

Whether thermosanitation is appropriate must be decided for individual diseases, and no blanket recommendation for use of fire, flame, or mobile sanitizers is intended in this chapter. Burning forests for control of diseases in seedling trees, as is done for pine forests in the southeastern United States, for example, by itself cannot support a universal recommendation to burn all young forests, because longleaf pine is one of the few trees known to be fire resistant in the seedling stage (15). However, possible use of thermal sanitation needs to be restudied in production of certain crops where it is not now employed, because results of some previous tests of the method apparently suffered from faulty technique, especially incomplete burning of residues above ground. In addition, improved flaming devices and new mobile field sanitizer designs that may become operational should be tested for possible application to many field crops.

An apparatus that combines the best designs of mobile incinerators with the best type of flamers perhaps could provide a thermal treatment apparatus using mainly plant refuse as fuel, with a minimum discharge of smoke and particulate matter. If an improved incinerator-flamer could project the hot gases forward so that the plant material is dried to become part of the fuel, such machines might be feasible to apply effective thermal sanitation on crops other than grasses. Alfalfa, strawberries, and certain other field crops have plant refuse that is insufficient or too green to carry an open fire. However, this limited plant material may sustain an enclosed fire and aid in providing a thermal treatment. When functioning only with plant material, the mobile incinerators would thus use a renewable resource instead of depending on fossil fuel. Thermal sanitation at the soil surface, applied with these field incinerator-flamers or with improved field burning, combined with the application of newer fungicides, should improve disease control in the production of grass seed and some other field crops.

Diseases apparently are increasing in some crops after recent adoption of minimum tillage systems. In a no-tillage or minimum-tillage program in which rye was seeded following discing after the corn harvest and corn was planted in the spring after the rye cover crop was killed with herbicide, an increase was noted in gray leaf spot, caused by *Cercospora zeae-maydis,* in corn in Virginia (C. Roane, personal communication). Similarly, following adoption and several years' use of minimum tillage in Indiana, anthracnose (caused by *Colletotrichum graminicola*) has recently become a serious problem in sweet corn production and represents a threat to dent corn production (H. L. Warren, personal communication through D. M. Huber).

Minimum tillage or no tillage rotation schemes are not all condemned, however, because they may reduce soil erosion and some have reduced the incidence of stalk rot (mainly caused by *Fusarium moniliforme*) in grain sorghum in a winter wheat-grain sorghum-fallow rotation (26). However, when clean plowing is no longer practiced as formerly recommended, and infected crop residues are left above ground in a short rotation or monoculture with a particular crop, and if leaf and stem diseases become serious, investigation of above-ground inoculum reduction by thermal sanitation might be appropriate.

More information is needed to evaluate the potential of burning and flaming. Therefore, foresters and farmers should be encouraged to watch for beneficial effects from burning, including improved plant growth, that may be the result of disease control. Foresters will be specially well situated for observations in the next several years as the fire exclusion policy of the past 50 years in the United States is replaced with prescribed burning for vegetation management and fuel reduction. Farmers are always in a good position to observe benefits from burning or flaming because of their frequent examinations of crops. In this regard, the writer will appreciate receiving information on additional cases of plant disease control by fire and flame and additional documentation for the ones mentioned.

The demand for innovative methods for disease control is strong, but better use of old, neglected methods is equally justified. Control programs that integrate disease-free seed, good cultural practices, resistant varieties, and fungicides should all be coordinated where possible with application of thermosanitation that will destroy primary inoculum for maximum disease control and maximum yields. Judicious use of the age-old and time-honored practices of prescribed burning should continue to be a legitimate part of tree, food, and fiber culture, especially as high-intensity management for maximum production becomes mandatory by the needs of an ever-increasing world population.

Literature Cited

1. Ahlgren, C. E. 1974. Effects of fires on temperate forests: North Central United States. In *Fire and Ecosystems,* ed. T. T. Kozlowski, C. E. Ahlgren, 195–223. New York: Academic. 542 pp.
2. Ahlgren, I. F. 1974. The effects of fire on soil organisms. See Ref. 1, pp. 47–72
3. Ahlgren, I. F., Ahlgren, C. E. 1960.

Ecological effects of forest fires. *Bot. Rev.* 26:483–533

4. Alexander, M. E., Hawksworth, F. G. 1975. Wildland fires and dwarf mistletoes: a literature review of ecology and prescribed burning. *US Dep. Agric. For. Serv. Rev. Tech. Rept. RM-14.* 12 pp.

5. Baranyay, J. A. 1972. Rhizina root rot of conifers. *Can. For. Serv. For. Insect Dis. Surv. Pest Leafl. 56.* 5 pp.

6. Baranyay, J. A., Smith, R. B. 1972. Dwarf mistletoes in British Columbia and recommendations for their control. *Can. For. Serv., Victoria BC.* 18 pp.

7. Bartlett, H. H. 1956. Fire, primitive agriculture and grazing in the tropics. In *Man's Role in Changing the Face of the Earth,* ed. W. L. Thomas, Jr. et al, 692–714. Chicago: Univ. Chicago Press. 1193 pp.

8. Baxter, D. V. 1952. *Pathology in Forest Practice.* New York: Wiley. 601 pp. 2nd ed.

9. Bellusci, A. V., Hallman, R. G. 1973. Using heat for sagebrush control. *US Dep. Agric. For. Serv. Bull. PNW 73-103.* 9 pp.

10. Bevilacqua, I. 1935. La micosi del Grano. *Istria Agric. NS.,* 15:317–19. *Rev. Appl. Mycol.* 14:749–50 (Abstr.)

11. Biswell, H. H., Kallander, H. R., Komarek, R., Vogl, R. J., Weaver, H. 1973. Ponderosa fire management. A task force evaluation of controlled burning in ponderosa pine forests of central Arizona. *Tall Timbers Res. Stn. Misc. Pub. No. 2.* 49 pp.

12. Blair, I. D. 1954. Nutrition and survival of *Cercosporella herpotrichoides* Fron. and aspects of the wheat eyespot disease. *NZ J. Sci. Technol. Sect. A* 36:207–20

13. Boyce, J. S. 1961. *Forest Pathology.* New York: McGraw-Hill. 572 pp. 3rd ed.

14. Brooks, P. J., White, E. P. 1966. Fungus toxins affecting mammals. *Ann. Rev. Phytopathol.* 4:171–94

15. Brown, A. A., Davis, K. P. 1973. *Forest Fire: Control and Use..* New York: McGraw-Hill. 686 pp.

16. Bruehl, G. W., Millar, R. L., Cunfer, B. 1969. Significance of antibiotic production by *Cephalosporium gramineum* to its saprophytic survival. *Can. J. Plant Sci.* 49:235–46

17. Burton, G. W. 1944. Seed production of several southern grasses as influenced by burning and fertilization. *J. Am. Soc. Agron.* 36:523–29

18. Campbell, L. 1955. Control of mint rust by premerge treatment with dinitro amine. *Down Earth* 10:6–7

19. Carrier, L. 1923. The beginnings of agriculture in America. New York: McGraw-Hill. 323 pp.

20. Cobb, N. A. 1932. Nematosis of a grass of the genus *Cynodon* (*sic*) caused by a new nema of the genus *Tylenchus* Bast. *J. Wash. Acad. Sci.* 22:243–45

21. Coykendall, I. J. 1968. Application of LP-gas flaming for disease control in mint and grass seed production. *Proc. 5th Ann. Symp. Thermal Agric.* 75 pp.

22. Czabator, F. 1971. Fusiform rust of southern pines—a critical review. *US Dep. Agric. For. Serv. Res. Pap. SO-65.* 39 pp.

23. Davis, K. P., Klehm, K. A. 1939. Controlled burning in the western white pine type. *J. For.* 37:399–407

24. Demaree, J. B., Wilcox, M. S. 1947. Fungi pathogenic to blueberries in the eastern United States. *Phytopathology* 37:487–506

25. Dickson, J. G. 1956. *Diseases of Field Crops.* New York: McGraw-Hill, 517 pp. 2nd ed.

26. Doupnik, B. Jr., Boosalis, M. G., Wicks, G., Smika, D. 1975. Ecofallow reduces stalk rot in grain sorghum. *Phytopathology* 65:1021–22

27. Easton, G. D., Nagle, M. E., Bailey, D. L. 1975. Residual effect of soil fumigation with vine-burning on control of Verticillium wilt of potato. *Phytopathology* 65:1419–22

28. Frazier, W. C. 1967. *Food Microbiology.* New York: McGraw-Hill. 537 pp. 2nd ed.

29. Froelich, R. C., Dell, T. R. 1967. Prescribed fire as a possible control for *Fomes annosus. Phytopathology* 57:811 (Abstr.)

30. Garren, K. H. 1943. Effects of fire on vegetation of the southeastern United States. *Bot. Rev.* 9:617–54

31. Greene, S. W. 1935. Effect of annual grass fires on organic matter and other constituents of virgin longleaf pine soils. *J. Agric. Res.* 50:809–22

32. Gynn, R., Hadlow, A. J. 1971. Toxicity syndrome in sheep grazing Wimmera ryegrass in Western Australia. *Aust. Vet. J.* 47:408

33. Hall, I. V., Aalders, L. E., Jackson, L. P., Wood, G. W., Lockhart, C. L. 1972. Lowbush blueberry production. *Can. Dep. Agric. Publ. 1477.* 42 pp.

34. Hamblin, C. O. 1921. Flag smut and

its control. *Agric. Gaz., NSW* 32:23.
[*Rev. Appl. Mycol.* 58:1923 (Abstr.)]
35. Hanson, H. C. 1939. Fire in land use and management. *Am. Midl. Nat.* 21: 415–34
36. Hardison, J. R. 1959. Evidence against *Fusarium poae* and *Siteroptes graminum* as causal agents of silver top of grasses. *Mycologia* 51:712–28
37. Hardison, J. R. 1960. Disease control in forage seed production. *Adv. Agron.* 12:96–106
38. Hardison, J. R. 1963. Commercial control of *Puccinia striiformis* and other rusts in seed crops of *Poa pratensis* by nickel fungicides. *Phytopathology* 53:209–16
39. Hardison, J. R. 1963. Control of *Gloeotinia temulenta* in seed fields of *Lolium perenne* by cultural methods. *Phytopathology* 53:460–64
40. Hardison, J. R. 1964. Justification for burning grass fields. *Proc. 24th Ann. Meet., Oregon Seed Growers League,* pp. 93–96
41. Hardison, J. R. 1969. Status of field burning and alternate methods for disease control in perennial grasses. In *Agricultural Field Burning in the Willamette Valley,* ed. Air Resources Center, OSU, 18–24. Corvallis: Oregon State Univ. 38 pp.
42. Hardison, J. R. 1974. Disease problems without burning. *Proc. 34th Ann. Meet., Oregon Seed Growers League,* pp. 9–11
43. Hedén, Å. 1954. Undersökning rörande förekomsten av rotdödarsvampen (*Ophiobolus graminis* Sacc.) i Skaraborgs län 1953. *Växtskyddsnotiser Stockholm* 4:50–52. [Investigation concerning the occurrence of the root-killer fungus (*Ophiobolus graminis* Sacc.) in the Skaraborg district 1953.] *Rev. Appl. Mycol.* 34: 444–45 (Abstr.)
44. Heinselman, M. L. 1973. Fire in the virgin forests of the Boundary Waters Canoe Area, Minnesota. *Quat. Res.* 3:329–82
45. Hepting, G. H. 1971. Diseases of forest and shade trees of the United States. *US Dep. Agric. Handb. No. 386.* 658 pp.
46. Horner, C. E. 1965. Control of mint rust by propane gas flaming and contact herbicide. *Plant Dis. Reptr.* 49: 393–95
47. Horner, C. E., Dooley, H. L. 1965. Propane flaming kills *Verticillium dahliae* in peppermint stubble. *Plant Dis. Reptr.* 49:581–82

48. Huber, D. M., Anderson, G. R. 1976. Effect of organic residues on snow mold of winter wheat. *Phytopathology* 66: In press
49. Humphrey, R. R. 1958. The desert grassland—a history of vegetational change and an analysis of causes. *Bot. Rev.* 24:193–252
50. Humphrey, R. R. 1974. Fire in the deserts and desert grassland of North America. See Ref. 1, pp. 366–400
51. Irving, F. D., French, D. W. 1971. Control by fire of dwarf mistletoe in black spruce. *J. For.* 69:28–30
52. Jackson, J. E., Burton, G. W. 1962. Influence of sod treatment and nitrogen placement on the utilization of urea nitrogen by coastal bermudagrass. *Agron. J.* 54:47–49
53. Jemison, G. M., Lowden, M. S. 1974. Management and research implications. In *Environmental Effects of Forest Residues Management in the Pacific Northwest, a State-of-Knowledge Compendium, US Dep. Agric. For. Serv. Gen. Tech. Rept. PNW–24,* ed. O. P. Cramer, A–1–33. Portland: Pacific NW For. Range Exp. Stn. 517 pp.
54. Keil, H. L. 1942. Control of silvertop of fescue by burning. *Plant Dis. Reptr.* 26:259
55. Kirk, D. E. 1973. Field combustion machines for soil treatment and residue removal. *Proc. 9th Ann. Symp. Thermal Agric.,* pp. 24–36
56. Kirk, D. E., Bonlie, R. W. 1973. Report on development and testing of a mobile field sanitizer. *Oregon State Univ. Agric. Exp. Stn., Corvallis.* 15 pp.
57. Komarek, E. V. 1965. Fire ecology—grasslands and man. *Proc. 4th Ann. Tall Timbers Fire Ecol. Conf.,* pp. 169–220
58. Komarek, E. V. 1970. Insect control—fire for habitat management. *Proc. 2nd Tall Timbers Conf. Ecol. Anim. Control Habitat Manage.,* pp. 157–71
59. Komarek, E. V. 1974. Effects of fire on temperate forests and related ecosystems: southeastern United States. See Ref. 1, pp. 251–77
60. Ledingham, R. J., Sallans, B. J., Wenhardt, A. 1960. Influence of cultural practices on incidence of common rootrot in wheat in Saskatchewan. *Can. J. Plant Sci.* 40:310–16
61. Lee, W. O., Timmons, F. L. 1958. Dodder and its control. *US Dep. Agric. Agric. Res. Serv. Farmers Bull.*

1161. 20 pp.

62. Leu, L. S., Hsieh, W. S. 1969. Outbreak of leaf blight in Taiwan. *Sugarcane Pathol. Newsl. No. 2,* pp. 15–16

63. Maloy, O. C. 1970. Flaming for disease control in Washington. *Proc. 7th Ann. Symp. Thermal Agric.,* pp. 43–44. 77 pp.

64. Maple, W. R. 1975. Mortality of longleaf pine seedlings following a winter burn against brown-spot needle blight. *US Dep. Agric. For. Serv. Res. Note SO-195.* 3 pp.

65. Martin, A. L. 1939. Rice straw stacks as a source of infection with the black kernel disease. *Plant Dis. Reptr.* 23: 247–49

66. Martin, A. L., Altstatt, G. E. 1940. Black kernel and white tip of rice. *Texas Agric. Exp. Stn. Bull. 584.* 14 pp.

67. McIntosh, G. H. et al 1967. Toxicity of parasitized Wimmera ryegrass, *Lolium rigidum,* for cattle and sheep. *Aust. Vet. J.* 43:349–53

68. McIntyre, J. L., Horner, C. E. 1973. Inactivation of *Verticillium dahliae* in peppermint stems by propane gas flaming. *Phytopathology* 63:172–75

69. Morgan, P. D., Driver, C. H. 1972. Rhizina root rot of Douglas-fir seedlings planted on burned sites in Washington. *Plant Dis. Reptr.* 56:407–9

70. Morrison, D. 1900. Grass or stubble burner. *US Patent No. 641,529*

71. Musser, H. B. 1947. The effect of burning and various fertilizer treatments on seed production of red fescue, *Festuca rubra* L. *J. Am. Soc. Agron.* 39:335–40

72. Nelson, E. E., Harvey, G. M. 1974. Diseases. See Ref. 53, S-1–11

73. Parmeter, J. R. Jr., Uhrenholt, B. 1975. Some effects of pine-needle or grass smoke on fungi. *Phytopathology* 65:28–31

74. Powelson, R. L., Gross, A. E. 1962. Thermal inactivation of *Verticillium albo-atrum* in diseased potato vines. *Phytopathology* 52:364 (Abstr.)

75. Schoenemann, J. A. 1968. Growing Wisconsin potatoes. *Wis. Ext. Circ. 440.* 18 pp.

76. Shipton, P. J. 1972. Influence of stubble treatment and autumn application of nitrogen to stubbles on the subsequent incidence of take-all and eyespot. *Plant Pathol.* 21:147–55

77. Siggers, P. V. 1944. The brown spot needle blight of pine seedlings. *US Dep. Agric. Tech. Bull. 870.* 36 pp.

78. Slope, D. B., Etheridge, J., Callwood, J. E. 1970. The effect of flame cultivation on eyespot disease of winter wheat. *Plant Pathol.* 19:167–68

79. Smith, J. D. 1966. Diseases of bromegrass in Saskatchewan in 1966. *Can. Plant Dis. Surv.* 46:123–25

80. Smith, J. D. 1968. Control of *Pyrenophora bromi* in *Bromus inermis* by burning crop residues. *Can. J. Plant Sci.* 48:329–31

81. Smith, J. D. 1971. Phleospora stem eyespot of fescues in Oregon and the *Didymella* perfect stage of the pathogen. *Plant Dis. Reptr.* 55:63–67

82. Stevens, N. E. 1938. Departures from ordinary methods in controlling plant diseases. *Bot. Rev.* 4:429–45

83. Stewart, O. C. 1956. Fire as the first great force employed by man. See Ref. 7, pp. 115–33

84. Stewart, O. C. 1963. Barriers to understanding the influence of use of fire by aborigines on vegetation. *Proc. 2nd Ann. Tall Timbers Fire Ecol. Conf.,* pp. 117–26

85. Suber, E. F., French, J. C. 1973. Two-lined spittle bug damage to coastal bermudagrass. *Univ. Ga. Coop. Ext. Serv. Leafl. 28*

86. Viro, P. J. 1974. Effects of forest fire on soil. See Ref. 1, pp. 7–45

87. Vogl, R. J. 1974. Effects of fire on grasslands. See Ref. 1, pp. 139–94

88. Wallis, R. L, Turner, J. E. 1968. Burning weeds in drainage ditches to suppress populations of green peach aphids and incidence of beet western yellows disease in sugarbeets. *J. Econ. Entomol.* 62:307–9

89. Weaver, H. 1974. Effects of fire on temperate forests: western United States. See Ref. 1, pp. 279–319

90. Webster, R. K., Bolstad, J., Wick, C. M., Hall, D. H. 1976. Vertical distribution and survival of *Sclerotium oryzae* under various tillage methods. *Phytopathology* 66:97–101

91. Wells, H. D., Burton, G. W., Jackson, J. E. 1958. Burning of dormant dallisgrass shows promise of controlling ergot caused by *Claviceps paspali* Stev. & Hall. *Plant Dis. Reptr.* 42:30–31

92. Wicker, E. F. 1974. Fire and dwarf mistletoes (Arceuthobium) in the Northern Rocky Mountains. *Proc. Intermt. Fire Res. Counc. Tall Timbers Res. Stn. Fire Land Manage. Symp., Missoula, Montana.* In press

Copyright © 1976 by Annual Reviews Inc.
All rights reserved

ECONOMICS OF DISEASE-LOSS MANAGEMENT[1,2]

♦3647

G. A. Carlson

Department of Economics, North Carolina State University, Raleigh, North Carolina 27607

C. E. Main

Department of Plant Pathology, North Carolina State University, Raleigh, North Carolina 27607

INTRODUCTION

Both economic forces and biological processes are generating increased interest in managing diseases, insects, weeds, and crop losses. Changes in farm cultural practices, farm structure, and relative prices of food and farm inputs are involved (4, 15, 54). This calls for closer coordination of the work of plant scientists and economists.

Cultural practices designed to maximize short-term profits often influence the vulnerability of a crop to diseases and insects. Increasingly, homogeneous crops and uniform varieties of a given crop are grown over wide areas (84). Lack of spatial heterogeneity and crop diversity is encouraged by reduced crop rotation and increasing monoculture made possible by lower prices of agricultural chemicals (fertilizer and pesticides) relative to land and labor. The development and adoption of high-yielding varieties of grain and vegetables have contributed to use of uniform genetic stock over contiguous areas. Specialized commodity processing and marketing facilities combined with higher transportation costs leave farmers with a limited choice of crops in many areas. Uniformity increases the potential rate of spread of crop diseases, especially new races and biotypes. In-

---

[1] Paper No. 4912 of the Journal Series of the North Carolina Agricultural Experiment Station, Raleigh, North Carolina.

[2] No attempt was made to include all references that might be considered relevant to the review. For supplementary bibliographic material (28 references, 3 pages) order NAPS document 02864 from ASIS, c/o Microfiche Publications, 440 Park Ave. S., New York, NY 10017. Remit in advance for each NAPS accession number $1.50 for microfiche or $5.00 for each photocopy (15¢ for each additional page over 30). Make checks payable to Microfiche Publications.

creased use of fertilizer and irrigation generally favor some diseases. Low-till crop production and double cropping may also increase disease-loss potential.

Selection and development of resistant pathogen races by use of varieties with specific resistance or use of selective fungicides also increase the need for a disease-management philosophy. The use of highly toxic pesticides, as the less toxic persistent ones are cancelled from legal use, may increase the percentage of pathogens killed and increase the rate of selection of resistant organisms. Farmers often realize that resistance is developing, but feel they cannot individually sacrifice present yields (by using crop rotation for example) for future yields if their neighbors do not do likewise.

Human populations are increasing rapidly in tropical areas of the world. To meet food demands, use of pesticides in developing countries has grown dramatically in the past 10 years (86). Higher incomes, fewer regulations, and absence of foreign exchange to import food will encourage further growth in local food demand, and hence the desire for disease control.

Several evaluations of economic aspects of disease control are available (11, 58, 72). Ordish & Dufour (58) clarified how crop price elasticities affect the share of loss reductions farmers must bear. However, complexities of the biological systems involved have not been sufficiently integrated into most economic models of crop production. It is common to attribute to unfavorable weather many of the stresses on crop growth caused by crop diseases. Many of the present economic models are only directed toward insect control (26, 29, 32, 79). This paper emphasizes the farm-level and area-management possibilities of crop diseases. Although economists have concentrated efforts in efficient insect management, there are a number of areas where economists and plant pathologists have reached common ground.

The first section deals with short-term disease treatment criteria and forecast systems. The second section examines economic criteria for long-term resource investments such as resistant varieties and crop rotations. The prevention of crop-disease spread is treated separately because of the importance of farmer groups or government units. The last section suggests ways in which data collection for economic analysis of crop-disease management can be facilitated by coordinated efforts between economists and plant pathologists.

SHORT-TERM CROP-LOSS MANAGEMENT

There have been several attempts to unify thought on methods of selecting optimal crop-protection actions. Stern and associates (74) emphasized the importance of pest densities as an indicator for the timing of pest control within a growing season. They recognized that most crops can tolerate significant levels of pest damage without large reductions in yield. Stern used the term *economic threshold* to refer to the critical or threshold level of pest damage calling for crop-protection measures. Headley (29) refined the economic threshold concept to explicitly account for the rising costs of obtaining larger and larger reductions in crop losses. He defined the economic threshold as that pest intensity at which incremental costs of damage reduction are equal to the incremental crop returns gained from the damage reduction. This leads to a profit-maximizing level of use of crop protection resources for individual producers. It will commonly lead to levels of crop protection which permit some crop damage.

When multiple crop-protection treatments and various degrees of treatment are possible, both the timing and the level of treatment need to be considered. Hall & Norgaard (26) presented a model which gives conditions for both amount (dosage) and timing of a single application of a pesticide. Both application costs (machinery, labor, management time) and pesticide material costs must be considered when multiple pesticide applications are being considered. A grower would save money in maintaining crop growth throughout the season by making fewer, more intense, pesticide applications when costs of application (say for applicator wages) rose relative to pesticide prices.

Several economists and biomathematicians have experimented with pest-control simulation models which have implications for disease-loss management (69, 79). The usual components of these models are the following: (a) a pest-growth function such as $P(t) = P_o e^{rt}$ which gives the pest-population level [$P(t)$] at various points of physiological time (t), as a function of the initial population P_o and a growth parameter (r). Some models permit the growth rate of the pest population to change following use of early-season pesticides (to simulate the effect of reduction of natural enemies); (b) kill-efficiency function or a dose-response relationship like those from Finney (21). In the case of changing pest susceptibility to a control (such as insecticide-resistance development) the kill-function parameters are related to factors affecting susceptibility such as number and intensity of pesticide use; (c) a crop-growth function which indicates the biomass of the crop at various points in time (35); (d) a pest damage function, which shows the effect of each pest for each unit of time on the final crop value: (e) a pest-control cost function. Such models have mainly been applied to insect control with insecticides. They have shown economic interrelationships such as the effects of changes in the pest population growth rate on number and intensity of treatments. Talpaz & Borosh (79) show that, for cotton insects, an increase in pest-population growth rate which implies fewer treatments with higher dosages per application, will maximize profits. Higher cotton prices should call for more treatments of less material per treatment. These models generally are not thought to include enough of the complex problems associated with multiple pests, random weather factors, and pest mobility to be used for recommendations on individual farms. However, they do highlight shortcomings in knowledge about biological relationships which indicate directions for research.

Establishment of Treatment Criteria

For individual farmers with little pathogen dissemination from surrounding farms, the major components of treatment decisions are: (a) the cost of disease control; (b) the value of the crop losses that are preventable; (c) the forecasts of disease intensity for determining timing and location of control efforts. Interdependencies of control decisions between time periods are taken up in the next section.

Costs of most disease-control efforts for individual farmers are easy to compute. Chemical purchases, hired labor, and application equipment are obvious costs, and are usually small compared with crop values. Steiner (73) points out that pest-management training costs are also an important cost component. Lewis & Hickey (47) summarize fungicide costs as a component of fruit orchard operation costs. They note how rapidly orchard labor and machinery costs have risen in recent years. It is important to recall that the important economic aspect of disease control is cost relative to income, and what alternatives are available for reducing costs per dollar of crop harvested.

Management time (hours) for deciding on type and timing of disease control is more difficult to compute, since it is often done along with other grower activities. Management involves observing crop phenology, assessing disease progress, interpreting weather reports, and reading about and discussing alternate controls (27, 89). One way to determine cost of management time is to determine its best alternative activity, e.g. no pest control, work on another crop, or leisure activities. Methods for reducing disease-monitoring cost are important to farm managers. Several plant pathologists, foresters, and entomologists have examined the time savings from sequential sampling of disease and insect intensities (67, 78).

The second major component of disease-control criteria is the level of crop damage avoided by use of fungicides, crop rotation, resistant varieties, and other controls (90). Crop-loss reduction should be continued until the value of disease loss avoided equals the incremental cost of attaining the loss reduction. The reviews of James (34), Smith (72), and Butt & Royle (8) indicate ways of associating changes in pathogen and insect density with yield and quality loss.

Treatment criteria change with prices of crops protected, costs of disease control, effectiveness of control, and the ability and willingness of growers to withstand loss (42). Criteria for control usually change through the growing season as susceptibility of the crop to the pathogen changes. For example, critical levels of soybean pests are much lower during the pod filling state (83). Bacterial fruit rots of stone fruit do not commonly damage green fruit.

Forecasting and the Value of Forecast Information

The number of crop-disease detection and forecasting services is rapidly increasing (44). This reflects the importance of saving management time by monitoring pathogens over a wide area. Forecasts for potato late blight (45), apple scab (43, 82), Stewart's disease of corn (12), *Cercospora* leafspot of peanuts (59), and fire blight of pears (5) are available in the United States. Often these systems are essentially automation of well-founded pathogen relationships with weather. By use of computers, forecasts are formulated and rapidly made available to many farmers and on a more individualized basis. They usually include a prediction of disease level and recommendations for treatment. The nematode advisory services of Holland, Great Britain, and the United States are important examples of very individualized advisory services (3). Increasing amounts of disease-control resources are being devoted to advisory and forecast systems. Economic methods are needed to determine the benefits from investments in these systems.

There are several ways to measure benefits of forecast or advisory services. Hall, Norgaard & True (27) compared the yields and pesticide-use patterns of *users* and *nonusers* of pest-management consultants in California. They found that *user* yields were higher for both citrus and cotton growers, while insecticide use was less for farmers using consultants. However, they could not attribute all of the yield difference to the advisory service per se. The users of advisory services may allocate other resources differently and have higher yields for other reasons. It is necessary to account for other factors that may affect yields before assigning benefits to the pest-advisory service. This may be done by determining changes in yields of users and nonusers from periods when all were nonusers (89). Another difficulty is determining the value of information to nonsubscribers who "borrow" the information from subscribers.

A second way to measure benefits of advisory services is to estimate the expected yields from following forecasts and compare them with what growers are now doing (9). For example, in forecasts designed for the peach-growing area of California, it was determined that 38% of the growers routinely applied fungicides to control brown rot (*Monilinia fructigena*) on all varieties during the last three weeks prior to harvest. Forty-four percent of the growers said they did not apply fungicides at all, since the brown rot disease occurred infrequently. The expected value $E(V)$ of the forecast to treat would be

$$E(V) = \Sigma_{i=1}^{n} (\tilde{N}R_1 - NR_i) \, P(Z_i) \, K + \Sigma_{i=1}^{n} (\tilde{N}R - \hat{N}R)_i \, P(Z_i) \, (1 - K), \qquad 1.$$

where

$\tilde{N}R_i$ = the net revenue per acre from following fungicide treatments based on forecast of disease level Z_i,

NR_i = the net revenue from not treating at all,

$\hat{N}R_i$ = the net revenue from treating all acres routinely,

$P(Z_i)$ = the probability of receiving forecasts Z_i,

K = proportion of acreage not treated at all.

The first term gives expected savings from preventing disease losses, while the second term reflects possible fungicide savings from not following a routine schedule. For the California peach orchards, given a forecast that was about 80% accurate, the gains were about \$25–40 per acre (about 20% of net returns).

A third way to evaluate a forecast system was demonstrated by Hull (33). He examined the record of the sugar beet yellows spray-warning scheme in England over an 8-year period. The spray-warning criterion was the presence of 0.25 aphids per plant, which was taken to indicate that 20% of the plants would be infected with yellows in August. Growers used these counts, plus current yellows levels, and weather forecasts to make pesticide-treatment decisions. Since the relationship between aphid counts and intensity of yellows is not exact, there is chance for error. For the 17 areas and 8 years there were 121 correct and 15 incorrect forecasts. Of the incorrect forecasts there were 9 cases of making a forecast for treatments when they were not justified, and 6 cases of not forecasting treatment when they were justified. Of course the latter type of error can be costly in terms of excessive disease loss. Such ex-post evaluation can indicate whether a change in the forecast criteria is justified.

Each of the forecast-benefit measurement methods has data requirements that plant pathologists may not routinely collect. Historical information on yields, and factors other than advisory service that could affect yields are needed for the first method. Probability distributions of disease incidence is critical to method two. The last method, which is similar to method two, requires records of forecasts given, and actual disease incidence over several years and locations. Evaluation of programs to provide disease information is essential for allocation of resources between various control resources.

Risk Analysis

The high degree of spatial and temporal variability of crop diseases causes uncertain decisions on control of pathogens. The degree of variability of damages, as well as mean damage (and profits), enters into farmers' disease-control decisions. The best-known implication of the aversion of farmers to crop-loss risks is that they tend to use large amounts of precautionary or "insurance" pesticides. Third-party insurance contracts for disease losses are usually not available because of difficulties in establishing disease-loss probabilities in monitoring of disease-loss claims (39).

Risk analysis, as originally developed for conventional business decisions under uncertainty (28), can be applied to pest-control decisions (9, 23, 31, 36, 45, 50). An example taken from the control of peach brown rot can illustrate the principles (9). About 90% of the peaches for canning in the United States are grown in the central valleys of California (39). The dry summer climate does not favor brown rot development except when rain occurs during the period just prior to harvest, at which time fruit is particularly susceptible. In the past this has occurred in about one year in five. About 55% of the fruit is routinely protected with fungicides, while the remaining growers do not apply fungicides. Growers can either (a) apply $25 per acre of fungicides to reduce yield losses by about 70% when the disease occurs (denote as action a_1), or (b) not apply fungicides and hope that this is not the one year in five when an epidemic will reduce marketable fruit yield by 40% (denote no application by a_2). Assume a known sales price of $75 per ton with an average yield of 12 tons per acre, $640 per acre for other production costs, and $13 per ton to harvest fruit for a 100 acre orchard.[3] The fungicide-treatment choices and net returns per acre are summarized in Table 1. $P(\Theta)$ represents the probabilities of the disease and nondisease occurrences or states (Θ_1 and Θ_2). Calculating expected net returns results in a higher expected return from not applying fungicides than from applying them ($80 > 77.60).

---

[3] Product price can be predicted by futures prices or contract prices. Effects of disease on product quality and prices can be established by market experiments.

Table 1 Possible payoffs and expected payoffs per acre with two disease-control actions and two levels of incidence of brown rot of peach

Disease status Actions	θ_1 Disease	θ_2 No disease	Expected value
a_1—apply fungicides	$72	$100	$77.60
a_2—do not apply fungicides	−$50	$125	$80.00
P(θ)	0.2	0.8	

However, Figure 1 shows why many growers may prefer to apply fungicides. The utility (utility is a measure of satisfaction) from various levels of income in dollars is given by the $U(\$)$ function. The choice is between CD (spraying) or AB (not spraying). The income axis of Figure 1 is 100 times that of the entries of Table 1 to reflect the decision of a 100 acre farm. Even though the mean value of AB (shown at F) has a higher level of income than that for spraying (shown at E), many growers will prefer spraying because it has higher expected utility: $E(U_2) > E(U_1)$. The utility function $U(\$)$ is particular to each decision maker. The degree of concavity of U is a measure of the degree of risk aversion relative to the decision maker. The point G (at about $5000) shows how low expected income from not spraying would have to be to equal utility of the spray alternative.

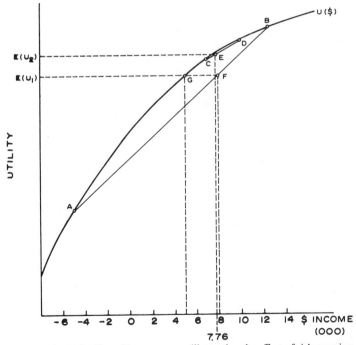

Figure 1 Hypothetical utility of income curve, illustrating the effect of risk aversion on expected utility by maximizing the choice between risky disease-control alternatives.

A disease-incidence forecast for a particular time and place can modify the farmer's personal disease severity probabilities $P(\Theta)$. Above, the objective was to choose the disease prevention action which would maximize expected utility, $E(U)$:

$$E(U) = {}^{max}_{a} [\Sigma^2_{i=1} U(a,\Theta) P(\Theta)], \qquad 2.$$

where $U(a,\Theta)$ = payoff (utility) derived from each action (a) disease state (Θ) pair,

$P(\Theta)$ = probability distribution of disease level (Θ).

With a forecast the probability to be used becomes $P(\Theta/Z)$, or the probability that disease level Θ will occur, given the disease forecast Z_k, where k is one of several possible forecasts. This conditional probability combines information from the long-run relative frequency of disease incidence with information from the forecast. Different actions might be taken with different disease-loss forecasts. In the peach brown rot case forecasts of cloudy weather, or losses greater than 5% in two different forecast models, would shift the optimal action from not applying fungicides to making one or two fungicide applications. If the forecast was for fair weather or less than 5% loss, then the optimal action would be not to treat. Since the fair-weather forecast occurred very frequently (about 90% of the time) it is possible to reduce use of fungicides by following the forecasts.

Gilmour & Fawcett (23) applied risk analysis to barley mildew (*Erysiphe graminis*) control. They calculated the expected cost of making an incorrect fungicide-application decision (recommending control when it is not needed, or not recommending control when the pathogen is present). They used the futures price of barley at the treatment decision time to determine the value of the crop to be protected. Yield responses to fungicides varied with disease incidence as affected by crop variety, weather (in different years), and planting date. Using a normal distribution of response to treatment, they calculated opportunity costs for various true responses and the associated expected opportunity costs (risks). Minimizing costs of incorrect decisions is equivalent to maximizing expected payoffs when the losses involved are small enough that the utility function for returns (as in Figure 1) can be assumed to be linear. This study also raises the important research question of how long we should continue to experiment in determining the disease-treatment threshold (or break-even point, in their terminology). Each fungicide trial gives information about the variation in response. When does the additional cost of more accurate estimates of disease incidence and disease abatement response equal the marginal returns to this information? Anderson & Dillon (1) have examined the economics of this question for fertilizer response. It thus seems appropriate to ask how many resources should be used to refine disease-treatment criteria. This question applies to all types of disease-control resources, and is not limited to chemical control.

LONG-TERM ADJUSTMENTS

It is obvious that one of the major ways to reduce crop damage in a given area is to change the commodity grown to one that is less susceptible to the pathogen.

Very little economic evaluation of the effect of disease losses on aggregate acreages of crops grown has been attempted. Spectacular epidemics which have led to changes in commodities have been noted by agriculturalists (15, 24, 75). Some economists have tried to relate farmer estimates of probable disease losses in one year to the acreages of that crop grown in an area in the following years. Proxy variables for disease losses might be acreages not harvested, or yield deviations below the trend (13).

For many crops, adjustments in seed quality, rotation of land, and selection of seed varieties have played a major role in disease control. These activities often take several years to complete. Since these expenditures are for investments in long-run improvement in disease control, the methods of capital budgeting are central to their evaluation.

Resistant Varieties

Economic models applicable for resistant-variety decisions are related to: (a) estimating likely adoption rates of varieties and associated technologies; (b) determining optimum replacement principles for varieties; (c) choosing models that can be used to examine optimum rates of decrease of resistance; (d) measuring costs of genetic gains in efforts to develop new resistant varieties.

Griliches (25) provided an economic structure for examining the returns from development and adoption of biological innovations. He examined the time stream of costs of discovery and development of hybrid corn in the United States. Benefits of the innovation were taken to be yield increases from hybrid seed times acres of hybrid seeded corn planted each year. A logistic adoption transformation of hybrid-corn acreages planted, by state, was fitted by multiple regression to determine the factors affecting adoption and to predict future hybrid acreages. Given the time streams of benefits and costs, rates of return on the investment were calculated, using discounting equations. A similar investment return analysis has been made for poultry research in the United States (60). W. Willey (unpublished) used a variation of the logistic model to examine adoption of private crop pest-control advisory-services by a sample of California farmers. He found that those who were more averse to risk and those who had larger farms were more likely to hire these services.

Adoption of disease-resistant varieties does not occur in isolation from acceptance of other modern agricultural practices. A simultaneous model of adoption of resistant varieties, pesticide use, and new land-clearing practices for rice culture in the Philippines were developed by Nerlove & Press (55). They used a three-equation multiple regression model in which the dependent variable was a zero-one variable (adopt or not adopt for each of the three practices). They found that each practice, plus educational and economic variables, affected the degree of acceptance of the resistant varieties. This study provides a method for understanding the joint acceptance of several innovations. Policies such as credit incentives, educational programs, and investment levels in crop-disease-control innovations depend upon an understanding of factors affecting adoption of modern farming practices.

The longer a variety will last before it is replaced, for whatever reason, the larger the net returns to the owner of that variety. Novel varieties are protected from unauthorized use by the US Plant Variety Protection Act of 1970 (85). This law is much like patent laws in that it grants exclusive property rights to the discoverer of the variety for 17 years. This law provides a government-sanctioned monopoly to the developer of a novel variety to encourage research and development. If there were no variety protection, there would be less incentive to undertake variety research on a private basis. For the US law, severe food-supply problems can result in a variety losing its protected status, provided compensation is paid to the patent owner. Variety life can also be limited by development of pathogen resistance. Such resistance development is affected by the nature of the pathogen and the selection pressure. Selection pressure on pathogen populations increases as higher proportions of host crops are planted to resistant varieties (87), or as higher dosages of fungicides are used (6). The case of rust, bunt, and root-rot diseases on wheat in the Pacific Northwest over the past 25 years represents one of the best documented examples of this phenomenon (2). A recent worldwide survey of plant breeders and pathologists revealed an estimated frequency of race changes (e.g. variety longevity) for stem, leaf, and stripe rust organisms of wheat of 5.3, 5.6, and 5.5 years, respectively (40).

Managing disease-resistant varieties to preserve effectiveness of resistance to pathogen race populations is a complex undertaking involving the optimum utilization of the resistance resource over time. It is similar to preserving the susceptibility of target insect species to a particular insecticide. The economics of the insecticide resistance problem was depicted by Carlson & Castle (11) as a capital investment and maintenance problem for insecticide materials. The rate of development of resistant strains of pests, and the rate of supply of new pesticide products are involved in the optimum-use rates of pesticides by growers. Headley (30) showed how capital-budgeting techniques could be used to determine the value of a pesticide over its useful life. His formula for the present value (PV) of income streams from a pesticide that induces pest resistance is that of an asset whose useful life is shortened from N years to T years.

$$PV = \Sigma_{t=1}^{T} B_t/(1 + i)^t - \Sigma_{t=1}^{T} C_t/(1 + i)^t, \hspace{2cm} 3.$$

where

B_t = the returns from sales of chemical in year t,

C_t = $E_t + KA_{it} + SA_{it}$ = costs in year t,

E_t = annual costs of manufacture, distribution, and application of the chemical,

A_{it} = annual capital charge per dollar in year t, at interest rate i,

K = present value of pesticide research and development costs,

S = amount of money invested at year 1 sufficient to grow in T years at interest rate i to be equal to the stream of costs of introducing the replacement chemical at time T rather than at time N,

N = years of useful life of the pesticide with no resistance development,

T = years of useful life of the pesticide with resistance development.

It is not difficult to see that S represents a cost of using the current pesticide in a way that hastens pesticide depreciation. In the same way high rates of use of resistant varieties of crops can hasten the onset of pathogen races that can damage host crops.

Hueth & Regev (32) have developed one of the most complete economic models of insecticide resistance. They showed how optimal current applications of pesticides (dosages, numbers of treatments) implies management of both current pest-population numbers and the associated stock of susceptibility to the pesticide held by the pests. The management criteria in their model is to continue using an insecticide until marginal benefits are equal to marginal costs:

$$\beta_{11}X_{12} + \beta_{12}X_{22} + \beta_{22}X_{22} = C + \beta_{13}X_{32} + \beta_{23}X_{32}, \qquad\qquad 4.$$

The left-hand terms are the marginal benefits from increasing plant growth $(\beta_{11}X_{12})$ and marginal benefits from reducing pest growth $(\beta_{12}X_{22})$ and terminal pest populations $(\beta_{22}X_{22})$. A reduction in pest growth at any time within the season means less damage over the remainder of the season, and an increase in current-year profits. The three terms on the right side of equation 4 comprise the marginal cost of insecticide use. The first term is the marginal unit cost of materials (C). The other two terms are losses [current season $(\beta_{13}X_{32})$ and future seasons $(\beta_{23}X_{32})$, respectively] resulting from the depletion of the factors for susceptibility by an incremental increase in insecticide use in the current period. Determining an optimal time path for depletion of a resource like susceptibility of an insect population to an available insecticide involves interdependent decisions. Analytical techniques such as dynamic programming or control theory are useful for such problems (68, 80).

Use of substitute disease controls affects the rate of depletion of a particular control. Growers can switch control materials and extend the useful life of materials to which pathogen populations are becoming less susceptible. Rotation of antibiotics has been a common recommendation in the control of bacterial diseases such as fire blight of pears, caused by *Erwinia amylovora* (66). The underlying principle is that pathogen controls are renewable resources, since susceptible races will increase relative to nonsusceptible ones in the absence of the selective controls. Geoghiou (22) evaluated the rate of such "regression" of insecticide resistance and found it to be slower than the increasing phase of development, especially in isolated populations not subject to dilution by migrants from untreated areas. Jones (38) used simulation to show how eelworm (*Heterodera rostochiensis*)-resistant and nonresistant varieties of potatoes could be alternated to maximize long-term returns. Part of the plan involves not growing only resistant potato varieties so as to prolong the lives of resistant ones. Martens (49) presented data on wheat-rust virulence and wheat-variety selection by farmers in

Kenya. There seem to be several wheat varieties (African Mayo, S.A. 43, Goblet) that regain their resistance to a large proportion of rust races after several years of little or no planting. African Mayo production decreased from 17% of total wheat acreage down to zero, and then back up to 15%.

Planned use of disease resistance to prolong useful life of varieties may not have been practiced for several economic reasons. First, genetic gains in yield and other traits may render older varieties obsolete. Also "cross-resistance" may exist, in which virulent races of the pathogen are "resistant" to old as well as new varieties. Finally, for "susceptible" races to increase there must be coordinated deployment of varieties over wide areas. This deployment may be limited by variety selection decisions of private farmers and seed companies (refer to the section on managing crop-disease spread).

The development and use of crop varieties resistant to diseases are especially important for crops that are relatively land-extensive such as the cereals and grasses. Many decisions are involved by plant breeders, deployment agencies, private seed companies, and farmers in the use of this type of disease control. Plant breeders have multiple objectives or traits to select for when developing a cultivar. Sometimes the traits are not easily converted into yield or dollar equivalents. However, criteria must be established for breeding decisions that reflect the market value of various traits.

Some economic analyses of breeding programs for tree improvement are available (17, 61). Genetic improvement (G_t) for each trait can be obtained by either selection of superior parent plants from wild stands (G_1) or by subsequent roguing of the seed orchard on the basis of progeny tests (G_2). This gain for trait t, such as resistance to fusiform rust, (*Cronartium fusiforme*) can be expressed as

$$G_t = G_1 + G_2$$
$$= i_1 h_1^2 \sigma_p + \frac{1}{2} i_2 h_2^2 \sigma_p, \qquad\qquad 5.$$

where

i_1 = wild stand selection intensity equal to the change in the mean value of the trait between wild and selected populations, divided by the standard deviations of the unselected population (σ_p),

i_2 = roughing selection intensity equal to the proportion of seed clones removed from the seed orchard,

h_1^2 = the parent-progeny heritability for the trait (20),

h_2^2 = heritability estimate based on progeny tests (53),

σ_p = standard deviation of the trait in the test-crossed progeny (53).

Given estimates of heritabilities (h^2) and trait variabilities, estimates of the effect of various selection intensities on genetic gains can be made. However, individual traits such as yield, disease resistance, and ease of harvest are frequently inherited in a certain pattern. There are genetic correlations that limit independent selection for traits. For example, forest geneticists (77) found that the genetic correlation between fusiform rust–free trees and those trees with high volume was

−0.29. That is, in selecting for high volume or yield one must select trees that are more susceptible to fusiform rust.

The benefits from a resistant variety will vary, depending upon the frequency of infestation by the pathogen(s), and the yield and price loss when an infestation does result. Selection intensities should differ for development of varieties for different areas. Porterfield (61) attempted to determine the optimum combination of selection intensities for five traits under four levels of fusiform-rust threat. His optimization model was goal programming, a type of linear programming (Figure 2). As expected, selection intensity for fusiform-rust resistance becomes more important as the potential for disease intensity increases. The 0.8 level of selective intensity for the medium infestation level for the fusiform rust-resistance trait is equivalent to selecting candidate trees only from wild stands that are rust free. Selection for volume and specific gravity becomes more important as disease losses are higher, whereas straightness and crown shape (harvest-cost characteristics) should be of less significance. Because of the −0.29 genetic correlation

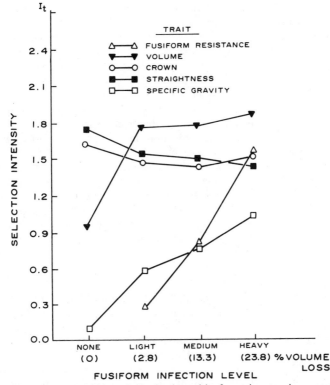

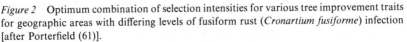

Figure 2   Optimum combination of selection intensities for various tree improvement traits for geographic areas with differing levels of fusiform rust (*Cronartium fusiforme*) infection [after Porterfield (61)].

Table 2   Yield and quality reduction comparisons of breeding lines of flue-cured tobacco resistant and susceptible to five diseases[a]

Resistance	Yield reduction/ha[b]	Price reduction
	Percentage below susceptible lines[c]	
Tobacco mosaic	5.9	1.5
Fusarium wilt	6.9	1.9
Mosaic plus fusarium wilt	9.9	1.9
Mosaic plus bacterial wilt	7.1	2.7
Mosaic plus root knot nematode	5.0	1.8
Fusarium wilt plus bacterial wilt	10.3	4.0
Fusarium wilt plus black shank	6.6	3.2

[a]Based on data given in Chaplin (14).
[b]No significant yield difference between resistant and susceptible lines for root-knot, black shank, or bacterial wilt alone.
[c]All percentages are significant at the 5% level, except for the 1.9% price reduction for mosaic and fusarium wilt.

between yield (volume) and rust resistance, finding a tree that is rust free with the required volume is difficult, especially in heavily infected areas. Porterfield (61) calculated that the probability of finding trees qualified on the basis of both volume and fusiform resistance in the heavily infested area was one in 1000. If one accounts for the negative correlation of these traits, the probability declines to 1 in 2500.

Development of disease-resistant varieties in annual-crop species frequently results in yield reduction relative to susceptible varieties. Chaplin (14) compared yields of various resistant and susceptible tobacco cultivars in a disease-free situation. Table 2 shows the percentage reductions in yield and price of various resistant lines. The yield reduction (6.9%) and price reduction (1.9%) from fusarium wilt (caused by *Fusarium oxysporum* f. sp. *nicotiana*) resistance was greater than for any other type of resistance tested. When multiple resistance (to two separate diseases) was involved, these lines have more yield and price reduction than for single-disease resistance. The combination of resistance to two types of wilt produced the largest yield reduction (10.3%) and largest price reduction (4.0%). Economic evaluation of the costs of disease loss and genetic gains of other plant-breeding programs is needed (48).

Certified Seed Investments

Certified-seed production and use is another long-term method of reducing losses to crop diseases. Minimum levels of germination, seedling vigor, and disease control are certified by rules for seed-production sanitation, seed grading, and handling. Purchasers of certified seed are buying both higher average quality and less quality variability. Third parties such as government officials inspect and verify seed quality. Often government or quasi-government corporations are involved in the production and dissemination of certified seed. Decisions to invest in certification programs are frequently based on investment analysis and availability of loans from international organizations. Investment analysis of seed certification programs includes determining the time stream of increased yields and increased costs from using certified seed compared with agricultural production without certified seed. The present value (PV) of a certification investment can be written as:

$$PV = \Sigma_{t=1}^{T} Y_t H_t (P_t - \mathrm{L})/(1+i)^t - \Sigma_{t=1}^{T} (C_t T_t + D_t)/(1 + i)^t, \qquad 6.$$

where

Y_t = the change in yield per hectare from changing to certified seed in year t,

L = the additional harvest cost of the larger crop,

H_t = change in hectares using certified seed in year t,

P_t = sales price of crop in year t,

C_t = the increased price per ton of certified over regular seed,

T_t = tonnage of certified seed used = $H_t \cdot S_t$, where S_t is the seeding rate per hectare,

D_t = direct costs of seed production and distribution.

i = interest rate of loan, or the opportunity cost of the money used for the certification program.

Agronomists and plant pathologists are frequently asked to provide estimates of the expected yield change per hectare from use of certified seed (Y_t). If additional fertilizer, water, or other resources are used when the certified seed is used, the additional yields from these resources must be deleted. Frequently the critical factor in the efficacy of the certification program will be how rapidly adoption occurs. As with adoption of disease-resistant varieties, the adoption rate of certified seed (H_t) will depend upon the expected increase in returns ($Y_t P_t$) and the increased costs of seeds (C_t), as perceived by individual growers.

Rotation

Control of diseases by crop rotation has a cost. It is the stream of income given up by not being able to produce the single, highest-value crop continuously. This cost might be small when alternative crops for an area also have high net returns. This will partly depend upon the ability to substitute land, machinery, management skills, and other resources to other commodities. When choosing among crop rotation, chemical control, resistant varieties, and cultural disease controls it is essential that the costs of rotation are known. One set of studies that provides some indication of the costs of crop rotations are linear programming evaluations of cropping patterns (70). The objective of these models is usually to select the combination of crop and livestock enterprises that will maximize net returns, given restraints on the amount of land, labor, machinery, water, and other resources available in the short term. One set of restraints on enterprise selection are the crop-rotation restraints, or the maximum proportion of the total land area which can be planted with any crop at one time without reducing crop yield or quality. For example, a 33% restraint on sugar beet acreage would mean that growing sugar beets more frequently than one in three years on the same plot would reduce tonnage or sugar yield due to the buildup of nematodes, weeds, or diseases. Estimates of such restraints are usually obtained from plant pathologists and weed specialists. When the rotation restraints are combined with the other restraints, optimum crop shares (but usually not crop sequences) can be determined. The marginal cost of the rotation restraints (available from the dual of the linear-programming solution) indicates the dollar amount which could be spent on other forms of disease, nematode, or weed control in order for one more acre of that crop to be planted. Similarly, it can show how much additional gross income is possible if other disease control allows a rotation restraint to be relaxed by a small percentage (say from sugar beets once in 5 years to once in 4 years). In column 1 of Table 3, Shumway et al (70) show what maximum acreage shares were used in a major crop-allocation study in California. Column 2 gives the associated annual rotation costs from the linear programming model. These were available for each of the ninety-five homogenous production areas of California (only the range is indicated here). The largest rotation costs were $88 for an additional acre of cotton in the Imperial Valley and $61 for cotton in the San Joaquin Valley. A zero cost (such as for all areas growing tomatoes, corn, and grain sorghum) indicates that, even though there is a rotation limit on how frequently a crop can be grown, this has no cost because other crops just as profitable can be grown. Even relatively minor constraints on rotation on low-value crops (alfalfa permitted four years in five) have some cost in areas with few substitute crops (0–$22).

The cost of lengthening rotations (periods without alternative host crops) has been examined by Empson & James (19) for control of potato eelworm (*Heterodera rostochiensis*). They used estimates of nematode growth-rates which differed

Table 3 Effect of rotation restraints for pest control on crop income per acre in California agriculture[a]

Crop	Rotation restraint[b]	Range of cost of rotation[c]
Potatoes	50	0–$12
Tomatoes, processing	67	0
Cotton	33	0–$88
Sugar beets	33	0-$37
Dry beans	33	0–$35
Alfalfa hay	80	0–$22
Safflower	50	0–$17
Barley-grain sorghum[d]	50	0–$14
Corn	80	0
Grain sorghum	80	0

[a]Data from Shumway et al (70).
[b]Maximum percentage of the acreage in local homogeneous production areas that can be planted to the specified crop without reducing crop yield or quality.
[c]Per acre reduction in net returns from reducing acres in host crop by rotation restraint. Range for various production acres.
[d]Double crop.

by starting population, rate of pathogen decline under nonhost cultivation, and a relationship of starting nematode population to potato-yield loss to derive the long-term relationship between host-crop interval and host-yield loss. To determine the profit-maximizing host-crop interval one must consider the cost of growing alternative crops and the gain in yield in the host crop from rotation. A mathematical expression for optimal host-crop interval (n*) is

$$n^* = Y \cdot L \cdot S \cdot P + (C + V) Z + F, \qquad\qquad 7.$$

where

  Y = yield potential of host crop without nematode damage,

  L = change in yield loss per unit area as interval (n) changes by one year,

  S = change in area in host crop as interval changes by one year for a given size farm,

  P = price per unit for sale of host crop,

  C = reduction in variable cost per unit area from producing the nonhost crop,

  V = returns above fixed costs per unit area for the nonhost crop,

  Z = change in area in nonhost crops as n changes by one year,

  F = annualized transition costs of moving from interval n to $n + 1$.

Annualized transition costs are an investment to get to a higher permanent sequence of host-crop yields. The magnitude of these costs depends upon yield gain per year of host-crop interval (L), costs of growing nonhost crops ($V + C$), and interest charges over the transition period.

Optimum host-crop interval (n^*) will be lower for farms with high quality soil and for farmers with high managerial ability. The interval should also shorten as the price of the host crop rises relative to the price of nonhost crops. The increase in yield as interval lengthens (L) may not be constant over all interval changes. Jones and associates (38) have used simulation of potato eelworm population levels to examine optimal sequences of nonhost crops, resistant varieties, and nematicide treatments. Multiperiod linear programming might also be attempted to find host-crop sequences that maximize returns, subject to not exceeding land, labor, machinery, and other restraints.

MANAGING CROP-DISEASE SPREAD

Many crop diseases are commonly spread from one farm to another by wind, equipment, insects, and other vectors. This leads to a special class of problems called *externalities* by economists (10, 31, 36). If the crop-disease levels of one farmer are highly dependent upon what disease controls surrounding farms use, then efficient disease-control decisions by each independent farmer will usually result in too little disease control for the whole area. Somehow, those protected from getting the disease must also assist in paying for control. Quarantine activities, area-wide control districts, and special laws are used to overcome interdependencies in disease control.

Quarantine activities and disease-control statutes appropriate individual rights for the benefit of all producers and consumers of a given crop. For example, a 1914 Virginia statute empowers the state pathologists to condemn and destroy without compensation red cedar trees which are hosts of apple rust (*Gymnosporangium juniperi-virginianae*) (64). Of course, cedar trees have value to their owners and others. What the statute, and a court case that upheld it, illustrates is the necessity for the government to make choices between two types of property —cedar trees and apple trees.

Growers must bear some of the costs of spread-control themselves. Control of spread of sugar beet cyst nematode (*Heterodera schachtii*) in California depends upon inspection of soil samples from beet harvest trucks as they unload. Those fields that are determined to be infested by the soil survey are only permitted to grow sugar beets (or other host crops) one year in four. Noninfested fields are permitted to be planted to sugar beets a maximum of 4 of 10 years and only two years in succession. As Thomason (81) has shown, this program has reduced the percentage of infested acres for the program area of the Imperial Valley, California from about 20% in 1960 to about 8% in 1968. Similarly, the Beet Eelworm Order of Great Britain ensured that sugar beets were grown no oftener than once in four years in infested fields. Although the measure helped avoid all but small losses, it did not prevent the nematode from spreading (37).

Measuring the benefits from crop-disease-spread control are essential if resources are to be efficiently allocated to area-wide control programs. Rough estimates based on potential damage of the organism per hectare and estimates of the ecological range of the organism are common in evaluating quarantine programs (65). Other considerations are costs of control if infestation occurs, probability that the quarantine effort will detect and reduce infestation, and rate of spread once a quarantine is by-passed (71). Several interesting benefit-cost analyses for animal pest-spread control are available (36, 62).

Attempts to evaluate the effectiveness of presently used control procedures might need to be expanded to determine what factors affect establishment of new foci rather than what factors limit spread within an individual focus (88). Merrill (51) used survey data on incidence of oak wilt (caused by *Ceratocystis fagacearum*) in Pennsylvania and West Virginia to estimate rates of spread of that organism. However, because there were no foci for which sanitation programs were not being applied, it was not possible for Merrill to draw inferences as to what the disease incidence (number of foci) would be in the absence of sanitary programs. Carlson (10) used multiple regression of state-level data to determine whether insecticide applications had slowed the spread of a mobile pest—the red imported fire ant (*Solenopsis invicta*).[4] Only insecticide treatments in areas with colder

---

[4]The model was $Z_t = b_o + b_1 A_{t-1} + b_2 A_{t-2} + b_3 t$, where Z_t was the percentage of change in infested acreage in year t, and A_{t-1} and A_{t-2} were acreages treated once with insecticides in two previous years.

climates seemed to slow the spread of the pest. Regressions of the form $\log_e$ $(x/1(1 - x)) = a + bt$ have been estimated and used in unpublished benefit-cost analysis of various USDA Department of Agriculture quarantine programs.

For some crop pathogens carried by vectors, groups of farmers have formed districts for coordinating and carrying out disease-control activities. Certified-seed-growing districts are examples where attempts have been made to measure their benefits. Analysis of the reduction in crop-loss levels for an area as district control activities are altered, is possible if information for several districts for several years is available (18). Miller and associates (52) evaluated the spread and intensification of Dutch elm disease (caused by *Ceratocystis ulmi*) for a city sanitation program which went through several changes in intensity. Care must be taken when control expenditures are quickly increased in response to an increase in disease threat, so that analysis of time-series data does not lead one to infer that increased abatement necessarily implies increased pathogen levels. Developing models of the simultaneous effects of controls on crop diseases, and levels of diseases on the controls used, is a reasonable way to analyze such systems (18, 46).

Regional deployment of disease-resistant varieties is frequently mentioned as a way to manage regional disease incidence, and to conserve resistance (7, 41, 87). The annual movements of wheat stem rust and oat crown rust pathogens in North America has permitted information on its movement to be collected each year. Area-wide control proposals usually suggest dividing the "Puccinia Path" into three south-to-north zones, and releasing different genes for specific resistance in the three areas. It may be necessary for some compensation to be paid to seed companies for restricting the range in which they may sell their seed, in order to get cooperation.

CONCLUDING REMARKS

There are market indications for many of the effects of various disease controls. Land prices fall when a disease causes reductions in yield and farmers shift to other nonhost crops. The difference in market price of certified and other seed indicates farmers' expression of disease-control benefits from cleaner seed. Value of genetic traits in a variety will be indicated by market-price differentials, such as for resistant tobacco varieties or for soybean varieties with higher protein content. These indicators can help plant breeders, farmers, and farm disease-control supply firms make decisions. Economists may need to assist in the interpretation of these market forces for the various levels of decision makers.

Chiarappa (16) has argued that there are many host-disease situations where closely supervised, short-term plant-disease management will be excessively expensive. Crops with low yields and low gross returns per unit area do not lend themselves to intense fungicide treatment. Disease management and economic analysis of these crop diseases as affected by changes in crop rotations, use of resistant varieties, and quarantines are critical. Economists have models and expertise which can assist in these areas. Resistance and land resources are scarce resources and should be managed just as fungicide use is.

Data on disease incidence are critical to many of the models suggested in this review. Stevens (76) showed that data on corn grades, carefully chosen as to time, can indicate incidence of corn ear rots by location and year. Powers and associates (63) showed how sampling on several frames can provide estimates of incidence and financial effect of fusiform rust in pine forests. Data for rotation changes can be gathered from long-term small-plot data or by surveys of current cropping practices (37, 56, 57). Evaluation of resistant varieties is carried out by breeders, release committees, seed companies, extension workers, and farmers. Information comes primarily from small-plot trials. The discussion above indicates how economists can assist in this evaluation.

Many plant pathologists believe that their role is to develop the best technical controls of plant diseases that their skills permit. Technological advances will make the controls low enough in cost to eventually be widely adopted. Perhaps, economic analysis can speed this process.

ACKNOWLEDGMENT

The authors are grateful to the many persons who provided information and offered suggestions.

Literature Cited

1. Anderson, J. R., Dillon, J. L. 1968. Economic considerations in response research. *Am. J. Agric. Econ.* 50:130–42
2. Baker, K. F., Cook, R. J. 1974. *Biological Control of Plant Pathogens.* San Francisco: Freeman. 433 pp.
3. Barker, K. R., Nusbaum, C. J. 1971. Diagnostic and advisory programs. *Plant Parasitic Nematodes. Morphology, Anatomy, Taxonomy, and Ecology,* ed. B. M. Zuckerman et al, I:281–301. New York: Academic. 345 pp.
4. Barnes, E. H. 1964. Changing plant disease losses in a changing agriculture. *Phytopathology* 54:1314–19
5. Barnett, W. W. 1974. Status and prospects for use of modeling in integrated pest management—a private agricultural advisor's view. *US/USSR Integrated Pest Management Symp. Long-Term and Short-Term Prediction Models of Insects, Phytopathogens and Weed Populations as They Relate to Crop Loss, Mich. State Univ., Oct. 15–17, 1974*
6. Berger, R. D. 1973. Disease progress of *Cercospora apii* resistant to benomyl. *Plant Dis. Reptr.* 57:837–40
7. Browning, J. A., Simons, M. D., Frey, K. J., Murphy, H. C., 1969. Regional deployment for conservation of oat crown-rust resistance genes. *Iowa Agric. Exp. Stn. Spec. Rept.* 64:49–56
8. Butt, D. J., Royle, D. J. 1974. Multiple regression analysis in the epidemiology of plant diseases. In *Epidemics of Plant Diseases: Mathematical Analysis and Modeling.* ed. J. Kranz, 78–114. New York: Springer. 170 pp.
9. Carlson, G. A. 1970. A decision theoretic approach to crop disease prediction and control. *Am. J. Agric. Econ.* 52:216–23
10. Carlson, G. A. 1975. Control of a mobile pest: the imported fire ant. *South. J. Agric. Econ.* 7:35–41
11. Carlson, G. A., Castle, E. N. 1972. Economics of pest control. In *Pest Control Strategies for the Future,* ed. R. L. Metcalf, 79–99. Washington DC: NAS. 376 pp.
12. Castor, L. L., Ayers, J. E., MacNab, A. A., Krause, R. A. 1975. Computerized forecasting system for Stewart's bacterial disease on corn. *Plant Dis. Reptr.* 59:533–36
13. Castro, R., Seagraves, J. A. 1974. The supply of winter green peppers in Florida. *NC State Univ. Dep. Econ.*

Bus. Econ. Res. Rept. No. 31:1–73
14. Chaplin, J. F. 1970. Associations among disease resistance, agronomic characteristics, and chemical constituents in flue-cured tobacco. *Agron. J.* 62:87–91
15. Chester, K. S. 1950. Plant disease losses: their appraisal and interpretation. *Plant Dis. Reptr. Suppl.* 193: 190–362
16. Chiarappa, L. A. 1974. Possibility of supervised plant disease control in pest management systems. *FAO Plant Prot. Bull.* 22:65–68
17. Davis, L. S. 1967. Investment in loblolly pine clonal seed orchards. Production costs and economic potential. *J. For.* 65:882–87
18. DeBord, D. V., Carlson, G. A., Axtell, R. C. 1975. Demand for and cost of coastal salt marsh mosquito abatement. *N.C. Agric. Exp. Stn. Tech. Bull.* 232:1–85
19. Empson, D. W., James, P. J. 1966. An economic approach to the potato root eelworm problem. *Gr. Brit. Natl. Agric. Advis. Serv. NAAS Q. Rev.* No. 73:22–29
20. Falconer, D. S. 1960. *Introduction to Quantitative Genetics.* New York: Ronald. 365 pp.
21. Finney, D. J. 1971. *Probit Analysis.* London: Cambridge Univ. Press. 333 pp. 3rd ed.
22. Georghiou, G. P. 1971. Resistance of insects and mites to insecticides and acaricides and the future of pesticide chemicals. *Agricultural Chemicals-Harmony or Discord for Food, People and the Environment,* ed. J. E. Swift, 112–27. Univ. Calif., Div. Agric. Sci. 151 pp.
23. Gilmour, J., Fawcett, R. H. 1973. A risk analysis of the control of barley mildew with fungicides. *Proc. 7th Br. Insectic. Fungic. Conf.* 7:1–10
24. Grainger, J. 1968. Disease assessment and the prevention of crop loss. *World Crops* 20(6):21–28
25. Griliches, Z. 1958. Research costs and social returns: hybrid corn and related innovations. *J. Polit. Econ.* 66:419–31
26. Hall, D. C., Norgaard, R. B. 1973. On the timing and application of pesticides. *Am. J. Agric. Econ.* 55: 198–201
27. Hall, D. C., Norgaard, R. B., True, P. K. 1975. The performance of independent pest management consultants in San Joaquin cotton and citrus. *Calif. Agric.* 29(10):12–14

28. Halter, A. N., Dean, G. W. 1971. *Decisions Under Uncertainty, With Research Applications* Cincinnati: South-Western. 266 pp.
29. Headley, J. C. 1972. Defining the economic threshold. See Ref. 11, pp. 100–8
30. Headley, J. C. 1972. Economics of agricultural pest control. *Ann. Rev. Entomol.* 17:273–86
31. Headley, J. C. 1973. Environmental quality and the economics of agricultural pest control. *Org. Eur. Med. Prot. Plants Bull. Stuttgart* 3(3):51–61
32. Hueth, D., Regev, U. 1974. Optimal agricultural pest management with increasing pest resistance. *Am. J. Agric. Econ.* 56:543–52
33. Hull, R. 1968. The spray warning scheme for control of sugar beet yellows in England—summary of results between 1959–66. *Plant Pathol.* 17:1–10
34. James, W. C. 1974. Assessment of plant diseases and losses. *Ann. Rev. Phytopathol.* 12:27–48
35. James, W. C., Shih, C. S., Hodgson, W. A., Callbeck, L. C. 1972. The quantitative relationship between late blight of potato and loss in tuber yield. *Phytopathology* 62:92–96
36. Johnston, J. H. 1975. Public policy on cattle tick control in New South Wales. *Rev. Mark. Agric. Econ.* 43: 3–39
37. Jones, F. G. W. 1972. Management of nematode populations in Great Britain. *Proc. Tall Timbers Conf. Ecol. Anim. Control Habitat Manag.* 4: 81–107
38. Jones, F. G. W., Parrott, D. M., Ross, G. J. S. 1967. The population genetics of the potato cyst-nematode, *Heterodera rostochiensis:* mathematical models to simulate the effects of growing eelworm-resistant potatoes bred from *Solanum tuberosum* spp. *andigena. Ann. Appl. Biol.* 60:151–71
39. Jones, L. A. 1973. Insuring crop and livestock losses caused by restricted pesticide use: an appraisal. *US Dept. Agric. Econ. Res. Serv. ERS Publ.* 512: 1–7
40. Kilpatrick, R. A. 1975. New wheat cultivars and longevity of rust resistance, 1971–75. *US Dept. Agric. Agric. Res. Serv.* NE-64:1–20
41. Knott, D. R. 1974. Using race-specific resistance to manage the evolution of plant pathogens. *J. Environ. Qual.* 1:227–31

42. Kranz, J. 1974. The role and scope of mathematical analysis and modeling in epidemiology. See Ref. 8, pp. 7–54

43. Kranz, J., Mogk, M., Stumpf, A. 1973. EPIVEN: ein Simulator für Apfelschorf. Z. Pflanzenkr. Pflanzensch. 80:181–87

44. Krause, R. A., Massie, L. B. 1975. Predictive systems: modern approaches to disease control. Ann. Rev. Phytopathol. 13:31–47

45. Krause, R. A., Massie, L. B., Hyre, R. A. 1975. Blitecast: A computerized forecast of potato late blight. Plant Dis. Reptr. 59:95–98

46. Lee, J. Y., Langham, M. R. 1973. A simultaneous equation model of the economic-ecologic system in citrus groves. Sou. J. Agric. Econ. 5:175–80

47. Lewis, F. H., Hickey, K. D. 1972. Fungicide usage on deciduous fruit trees. Ann. Rev. Phytopathol. 10:399–428

48. Main, C. E., Nusbaum, C. J., Chaplin, J. F. 1972. The effect of the brown spot disease (Alternaria alternata) on yield, quality and market value of flue-cured tobacco—a comprehensive assessment approach. CORESTA Inf. Bull. 1972 Spec., pp. 60–61

49. Martens, J. W. 1975. Virulence dynamics in the wheat stem rust population of Kenya. Plant Dis. Reptr. 59:763–67

50. Marty, R. 1966. Economic guides for blister-rust control in the east. US Forest Serv. Res. Paper NE-45:1–14

51. Merrill, W. 1967. The oak wilt epidemics in Pennsylvania and West Virginia: an analysis. Phytopathology 57:1206–10

52. Miller, H. C., Silverborg, S. B., Campana, R. J. 1969. Dutch elm disease: relation of spread and intensification to control by sanitation in Syracuse, N. Y. Plant Dis. Reptr. 53:551–55

53. Namkoong, G., Snyder, E. B., Stonecypher, R. W. 1966. Heritability and gain concepts for evaluating breeding systems such as seedling orchards. Silvae Genet. 15:76–84

54. Natl. Res. Counc. 1975. Agricultural Production Efficiency. Committee on agricultural production efficiency. Washington DC: NAS. 199 pp.

55. Nerlove, M., Press, S. J. 1973. Univariate and Multivariate Log-linear and Logistic Models. Rand Corp. R-1306-EDA/NIH. 134 pp.

56. Nusbaum, C. J., Ferris, H. 1973. The role of cropping systems in nematode population management. Ann. Rev. Phytopathol. 11:423–40

57. Oostenbrink M. 1972. Evaluation and integration of nematode control methods. In Economic Nematology, ed. J. W. Webster, 497–514. New York: Academic. 563 pp.

58. Ordish, G., Dufour, D. 1969. Economic bases for protection against plant diseases. Ann. Rev. Phytopathol. 7:31–50

59. Parvin, D. W. Jr., Smith, D. H., Crosby, F. L. 1974. Development and evaluation of a computerized forecasting method for Cercospora leafspot of peanuts. Phytopathology 64:385–88

60. Peterson, W. L. 1967. Return to poultry research in the United States. J. Farm Econ. 49:656–69

61. Porterfield, R. L. 1974. Predicted and potential gains from tree improvement program—a goal programming analysis of program efficiency. N. C. State Univ. Sch. For. Resour. Tech. Rept. 52:1–112

62. Power, A. P., Harris, S. A. 1973. A cost-benefit evaluation of alternative control policies for foot-and-mouth disease in Great Britain. J. Agric. Econ. 24:573–600

63. Powers, H. R. Jr., McClure, J. P., Knight, H. A., Dutrow, G. F. 1974. Incidence and financial impact of fusiform rust in the south. J. For. 72:398–401

64. Samuels, W. J. 1971. Interrelations between legal and economic processes. J. Law Econ. 14:435–50

65. Schieber, E. 1972. Economic impact of coffee rust in Latin America. Ann. Rev. Phytopathol. 10:491–510

66. Schroth, M. N., Thomson, S. V., Hildebrand, D. C., Moller, W. J. 1974. Epidemiology and control of fire blight. Ann. Rev. Phytopathol. 12:389–412

67. Shepard, M., ed. 1974. Insect Pest Management: Readings. New York: MSS Inf. Corp. 269 pp.

68. Shoemaker, C. 1973. Optimization of agricultural pest management. II. Formulation of a control model. Math. Biosci. 17:357–65

69. Shoemaker, C. 1973. Optimization of agricultural pest management. III. Results and extensions of a model. Math. Biosci. 18:1–22

70. Shumway, C. R., King, G. A., Carter, H. O., Dean, G. W. 1970. Regional resource use for agricultural produc-

tion in California, 1961–65 and 1980. *Giannini Found. Monogr. Calif. Exp. Stn.* 25:1–118

71. Smith, H. S., Essig, E. O., Fawcett, H. S., Peterson, G. M., Quayle, H. J., Smith, R. E., Tolley, H. R. 1933. The efficacy and economic effects of plant quarantines in California. *Calif. Agric. Exp. Stn. Bull. 553.* 276 pp.

72. Smith, R. F. 1971. Economic aspects of pest control. *Proc. Tall Timbers Conf. Ecol. Anim. Control Habitat Manag.* 3:53–83

73. Steiner, H. 1973. Cost-benefit analysis in orchards where integrated control is practiced. *Org. Eur. Med. Prot. Plant Bull. Stuttgart* 3(1):27–36

74. Stern, V. M., Smith, R. F., van den Bosch, R., Hagen, K. S. 1959. The integrated control concept. *Hilgardia* 29: 81–101

75. Stevens, N. E. 1934. Records of agricultural projects known to have failed through plant diseases. *Plant Dis. Reptr.* 18:7–16

76. [Stevens, N. E.] 1935. Incidence of ear rots in the 1916–1933 corn crops. *Plant Dis. Reptr.* 19:71–93

77. Stonecypher, R. W., Zobel, B. J., Blair, R. 1973. Inheritance patterns of loblolly pines from a nonselected natural population. *NC Agric. Exp. Stn. Tech. Bull.* 220:1–60

78. Strandberg, J. 1973. Spatial distribution of cabbage black rot and the estimation of diseased plant populations. *Phytopathology* 63:998–1003

79. Talpaz, H., Borosh, I. 1974. Strategy for pesticide use: frequency and applications. *Am. J. Agric. Econ.* 56:769–75

80. Taylor, C. R., Headley, J. C. 1975. Insecticide resistance and the evalua-

tion of control strategies for insect population. *Can. Entomol.* 107:237–42

81. Thomason, I. J. 1972. Integrated control of the sugar beet cyst nematode in the Imperial Valley. *Calif. Beet Growers Assoc., Ltd.* 4 pp.

82. Thompson, W. W. ed. 1973. *Michigan Apple Pest Management—Second Annual Report.* Mich. State Univ., East Lansing. 57 pp.

83. Turnipseed, S. G. 1972. Response of soybeans to foliage losses in South Carolina. *J. Econ. Entomol.* 65:224–29

84. U.S.D.A. 1973. *Monoculture in Agriculture: Extent, Causes and Problems—Report on the Task Force on Spatial Heterogeneity in Agricultural Landscapes and Enterprises.* Washington DC: USDA 64 pp.

85. US Dep. Agric. 1973. *United States Plant Variety Protection Act—Regulations and Rules of Practice. December 24, 1970 (84 Stat. 1542) (7 U.S.C. 2321 Et Seq.).* 44 pp.

86. US Dep. Agric. 1974. *Pesticide Review—1974.* 58 pp.

87. Vanderplank, J. E. 1968. *Disease Resistance in Plants.* New York: Academic. 206 pp.

88. Vanderplank, J. E. 1975. *Principles of Plant Infection.* New York: Academic. 216 pp.

89. von Rümker, R., Carlson, G. A., Lacewell, R. D., Norgaard, R. B. Parvin, D. W. Jr. 1975. Evaluation of pest management programs for cotton, peanuts, and tobacco in the United States *Contract Rept. Environ. Prot. Agency Counc. Environ. Qual.* 589 pp.

90. Zadoks, J. C. 1973. Schade. *Gewasbescherming* 4(5):71–75

Copyright © 1976 by Annual Reviews Inc.
All rights reserved

ACQUIRED RESISTANCE TO FUNGICIDES

◆3648

J. Dekker

Laboratory of Phytopathology, Agricultural University, Binnenhaven 9, Wageningen, the Netherlands

The term *acquired resistance* is used when in a population, which is normally sensitive to a biocide, forms arise that are less sensitive to this biocide. This resistance may have a genetic basis, and then as a rule be stable, or it may be caused by a physiological adaptation and disappear again rapidly when the organism is no longer exposed to the toxicant. Some authors call the latter type of transient-reduced sensitivity *tolerance,* and the former type *resistance.* In this chapter both terms are used interchangeably, as there is no sound biological basis to distinguish between the two when dealing with the relation between fungus and fungicide (30). This discussion deals primarily with stable, genetically determined resistance, because transient resistance will seldom reach a high level or cause problems in practice.

In an earlier review in this periodical, almost a decennium ago, Georgopoulos & Zaracovitis (49) concluded that tolerance to organic fungicides used in the control of fungal plant diseases or storage rots had created virtually no difficulties in practice. One of the exceptions was the group of aromatic hydrocarbon fungicides, whose practical use is limited. As has been outlined in recent reviews, the situation has changed dramatically since that review by Georgopoulos & Zaracovitis (30, 41). Many new organic fungicides with systemic action have come into practical use, which, in contrast to most conventional fungicides, such as copper compounds and dithiocarbamates, exert a more selective and specific action. Development of resistance to several of these compounds has been encountered in practice, in some cases to such an extent that the use of the compound concerned had to be restricted severely, or even abandoned in the treatment of certain diseases. Resistance to many of these new systemic fungicides could be induced readily in the laboratory, by treatment with UV irradiation or other mutagenic agents.

Both from a practical as well as a scientific standpoint it is important to gain insight into this phenomenon. This can best be achieved by biochemical, genetic, and epidemiological investigations of resistant strains. This review deals with the principles involved, illustrated by selected examples, and considers the practical

405

implications. No attempt has been made to cover all reports of acquired fungicide resistance. The accent is on fungicides, which are used in practice or show promise in this respect; experimental compounds are included in the discussion only when this is desirable for an illustration of the principles concerned.

MECHANISM OF RESISTANCE

For an understanding of the biochemical and physiological background of the phenomenon of acquired resistance to fungicides, knowledge of the mode of action of the fungicides concerned is indispensable. On the other hand the study of strains with acquired fungicide resistance may help to elucidate the mechanism of fungicidal action.

Development of resistance to a fungicide may be due to changes in the fungal cell that inhibit the fungicide, to a greater or lesser extent, from reaching the site of action. Such changes include a decreased permeability of the protoplast membrane to the fungicide, or an increased detoxication before the site of action is reached. If the fungicide is able to reach this site, resistance may be due to a decreased affinity between the chemical and the reactive site, to circumvention of the blocked site by operation of an alternate pathway, or to compensation for the effects of the inhibition. Examples of the various types of resistance are discussed in this chapter.

Decreased Permeability

Blasticidin-S was among the first antibiotics in Japan used against rice blast, caused by *Piricularia oryzae*. In vitro it inhibits spore germination and mycelial growth of this fungus at a concentration of 1–10 μg/ml. Although no resistance problems were encountered in the field, it was possible, by successive transfers of this fungus on potato dextrose agar containing increasing concentrations of blasticidin-S, to obtain two isolates which tolerated 1000 and 4000 μg/ml in the medium, respectively (91). The antibiotic is known to be an inhibitor of protein synthesis in sensitive microorganisms (68). Incorporation of ^{14}C-amino acids into protein was strongly reduced by 1 μg/ml of the antibiotic in intact cells of the wild strain, but not in those of the two resistant isolates, not even at concentrations of 100 and 1000 μg/ml, respectively. In cell-free systems from sensitive and tolerant strains, however, protein synthesis was inhibited equally. It was concluded that the antibiotic is unable to reach the sites of protein synthesis in the intact cell, presumably because of decreased permeability of the protoplast membrane (67).

An analogous case is reported for antibiotics of the polyoxin group, which are used in Japan against rice sheath blight, caused by *Pellicularia sasakii* and several fungal diseases of fruit, such as pear black spot, caused by *Alternaria kikuchiana*. Polyoxins are known to interfere with cell wall chitin synthesis, by competitive inhibition of the enzyme chitin synthetase (65). When, after a short period of practical use, the effect against pear blight spot remarkably decreased, strains of *A. kikuchiana* were isolated from polyoxin-B–sprayed pear orchards, which appeared tolerant to the antibiotic (94). Comparative studies in the laboratory revealed that 10 $\mu g/ml$ of this antibiotic caused about 70% inhibition of N-acetylglucosamine incorporation in the chitin of the wild strain, versus about 10% of that of the most resistant isolates. In cell-free systems of *A. kikuchiana,* however, the enzyme chitin synthetase of both the sensitive and tolerant isolates was inhibited to the same degree by polyoxin-B. This indicates that resistance is not due to the presence of a chitin synthetase with a decreased affinity to the antibiotic. It is suggested that a decreased permeability of the cell membrane prevents the antibiotic from reaching the enzyme site (64).

Conversion into an Inactive Compound

Detoxication of a fungicide by biological conversion has been reported in several instances. Insensitivity of *Fusarium oxysporum* f. sp. *lycopersici* to ascochitin was attributed to a conversion of this antibiotic into a less fungitoxic dihydroderivative (97). Conversion of pentachloronitrobenzene (PCNB) into pentachloroaniline and pentachlorothioanisole by the same fungus is also considered as a detoxication mechanism; the PCNB-sensitive *Rhizoctonia solani* shows only little conversion of the fungicide into these compounds (92). Differences in ability to convert PCNB into less fungitoxic compounds seems the basis for variation in sensitivity among strains of *Botrytis cinerea* (93). Insensitivity of *Fusarium solani* f. sp. *phaseoli* to dodine may be due to conversion of this fungicide into less active compounds (5). All these examples concern an already existing difference in sensitivity among genera, species, or strains of fungi; it seems plausible, however, that acquired resistance also could be based on such a phenomenon. Evidence was presented that development of dodine resistance in *Nectria haematococca* f. sp. *cucurbitae,* after UV irradiation, might be related to an increased capacity of tolerant strains to convert dodine into less fungitoxic compounds (71).

Lack of Conversion into a Fungitoxic Compound

In rare cases an inactive compound in itself may be converted by a fungus into a fungicide (lethal synthesis). The decreased ability of the fungus to perform this conversion may then result in resistance, which was illustrated by work with the experimental systemic fungicide 6-azauracil (AzU). For fungitoxic activity AzU has to be converted, via 6-azauridine (AzUR), into 6-azauridine-5'-phosphate (AzUMP). Only the latter compound, and not AzU, is a competitive inhibitor for orotidine-5'-phosphate decarboxylase (OMP decarboxylase), an enzyme that functions in pyrimidine biosynthesis de novo (33). By UV irradiation of conidia of the sensitive fungus *Cladosporium cucumerinum,* strains were obtained that were resistant to AzU, but not to AzUR and AzUMP. The AzU-resistant strains appeared unable to convert AzU into AzUR, because of lack of the enzyme that catalyzes this reaction (25).

Sensitivity of fungi to pyrazophos, which is predominantly used against powdery mildew diseases, may depend on ability of the fungus to convert this compound into 2-hydroxy-5-methyl-6-ethoxycarbonylpyrazolo(1,5-a)pyrimidine (PP), which is considered to be the fungitoxic principle. In contrast to *Piricularia oryzae,* which is sensitive to the fungicide, the insensitive fungi *Pythium debary-anum* and *Saccharomyces cerevisiae* were unable to convert pyrazophos into PP (129). Theoretically a decrease in the ability to carry out this conversion might result in tolerance. UV irradiation of *P. oryzae* did, however, not result in pyrazophos-resistant mutants (129).

Detoxication by Binding

Failure to control *Pyrenophora avenae* (stat. con. *Drechslera avenae*) on oats with organic mercury compounds, which has been reported from various countries, has been attributed to development of strains with decreased sensitivity to these fungicides. It was claimed (51) that a correlation exists between the rate of production of red anthraquinone pigments by the fungus and resistance, and it was suggested that the antifungal activity of organic mercury was reduced by binding to these pigments. These results could not be reproduced by others (102). Ross & Old (103) postulated that increase of a pool of thiol compounds within the cell was responsible for acquired resistance, but the existence of consistent differences between sensitive and resistant strains in this respect was not corroborated (101).

To my knowledge no clear-cut examples of this type of acquired resistance have yet been reported.

Decreased Affinity at the Reactive Site

Various examples show the validity of this principle as a basis for acquired resistance. The systemic fungicides benomyl and thiophanate-methyl are converted into methylbenzimidazole-2-yl carbamate (MBC, carbendazim), which is considered to be the main toxic principle of these fungicides (17, 110). It was shown that MBC in fungi inhibits the separation of chromosomes at mitosis (22, 58). This antimitotic activity of MBC in *Aspergillus nidulans* showed resemblance with that of colchicine in human cell cultures, which prevents assembly of the spindle by binding to protein subunits of the spindle microtubules.

Development of resistance to the above-mentioned fungicides has been reported in many instances, either spontaneously in the field or after mutagenic treatment in the laboratory. For studies on the nature of acquired resistance to MBC, use was made of a number of UV-induced mutants with a changed sensitivity to MBC, namely a resistant strain and a mutant even more MBC-sensitive than the wild-type fungus. Experiments with MBC, added to cell-free extracts of these strains showed a close correlation between MBC sensitivity and the degree of binding of MBC to some macromolecules; this binding was almost nil in the resistant strain, moderate in the wild-type strain, and high in the very sensitive strain. The properties of this macromolecule appear identical with microtubular protein. It was postulated that binding of MBC to these proteins might prevent the assembly of the microtubular subunits into functional spindle fibers, resulting in inhibition of mitosis. The development of resistance might then be ascribed to a change in the protein subunits of the spindle microtubules, resulting in a decreased affinity to MBC (23, 24). Resistance to thiabendazole seems to be governed by the same mechanism (122).

Evidence has been presented from studies with *Saccharomyces cerevisiae* that at least one type of acquired resistance to the antibiotic cycloheximide, an inhibitor of protein synthesis in fungi, is due to a slight change in the ribosomes, which results in a decreased affinity of these particles to the antibiotic (19).

Recently various resistant strains of *Piricularia oryzae* have been isolated from rice fields in Japan, which were frequently treated with kasugamycin. Development of resistance to this antibiotic, an inhibitor of protein synthesis, might also be related to changes in affinity of the ribosomes to the antibiotic. A study of acquired tolerance to this antibiotic, which shows antibacterial and antifungal activity, in *Escherichia coli* revealed that resistance was located in the 16S RNA unit of the ribosome. In the wild strain one of the nucleotides of this unit contains dimethyladenine, whereas in the resistant strain unaltered adenine was found. It seems that mutation confers resistance to kasugamycin by preventing cells from methylating this adenine residue to yield dimethyladenine. Apparently the affinity of the ribosome for the antibiotic has been reduced or lost by this change, without impairment of the capacity to synthesize proteins (63). The significance of differences between ribosomes for the phenomenon of acquired resistance to inhibitors of protein synthesis was indicated earlier by research with streptomycin, an antibiotic with systemic activity against various bacterial and some fungal diseases. In *E. coli* streptomycin attaches to the 30S unit of the ribosome, and inhibits protein synthesis by causing a misreading of the genetic code. It was possible to break this unit down into a 16S RNA unit and 19 proteins, and to reassemble these components again into a functioning ribosome. By reconstitution of the ribosome from components of resistant and sensitive strains, the factor responsible for resistance appeared to be a change in only one of the 19 proteins (96).

In a comparable way a decreased affinity of a particular enzyme to an inhibitor, due to a slight change in this enzyme that does not impair its functioning, may lead to increased tolerance. An example is the resistance acquired to ethionine, an antimetabolite of methionine in *Coprinus lagopus* (79). In the wild-type fungus, ethionine interferes with the so-called methionine-activating enzyme; in the resistant strain this enzyme had been changed in such a way that it lacked affinity to ethionine, but not to methionine.

These examples indicate that decreased affinity to a fungicide at the reactive site may be a common cause for the development of tolerance to this fungicide.

Circumvention

Ustilago maydis is normally resistant to antimycin-A, an inhibitor of electron transport in the respiratory chain, which acts at a site between cytochrome *b* and *c*. This resistance has been attributed to the operation of an electron transport from a site preceding cytochrome *b* to an alternate terminal oxidase, thus bypassing the antimycin-A blocked site (112). A mutant that lacked this alternate pathway became sensitive to the antibiotic (47). In this case, however, it is the wild type that is resistant, whereas the mutant has become sensitive by loss of the bypass.

The reverse of such a phenomenon might be the cause of acquired resistance. Evidence was obtained that development of carboxin resistance in *Ustilago hordei* depends, at least partially, on the development of alternate metabolic pathways, such as the glyoxylate cycle or carboxin-tolerant phosphate uptake (8).

Compensation

Increase of tolerance may occur when a sensitive organism changes its metabolism in such a way that less of the product of the inhibited reaction is required or that the level of the inhibited enzyme is increased. The occurrence of this type of acquired resistance is indicated by a 6-azauracil-resistant mutant of *Cladosporium cucumerinum,* which produced at least three times as much of the target enzyme as the sensitive wild-type fungus (29).

Other Effects

Counteraction of fungicidal activity may possibly also be obtained by mechanisms other than the ones mentioned above. For example, acquired tolerance of strains of *Verticillium malthousei,* growing on agar medium to benomyl and a few other, unrelated chemicals was attributed to the capacity of these strains to acidify the medium, which may influence the degree of uptake and binding of the chemicals (77).

ORIGIN OF RESISTANCE

Reports on development of resistance to conventional fungicides have been scarce. Except for the outstanding work of Georgopoulos and his co-workers (39, 48, 49) on tolerance to fungicides belonging to the aromatic hydrocarbon group only a few reports have appeared on genetic aspects of acquired resistance to some antibiotics and other experimental fungicides.

When this situation changed after the introduction of systemic fungicides, an increasing number of cases of acquired resistance in practice stimulated the study of this phenomenon in the laboratory. For these studies fungi are selected which are easily accessible for genetic analysis, such as *Aspergillus nidulans (Emericella nidulans), Neurospora crassa,* and *Saccharomyces cerevisiae.* In addition to these non-plant–pathogens, *Venturia inaequalis,* the causal agent of apple scab; *Ustilago maydis* and *U. hordei,* pathogens of corn and barley respectively; and *Hypomyces solani* (perf. *Nectria haematococca*), pathogenic to *Cucurbita* spp., are being used for this purpose.

In almost all cases where resistance to agricultural fungicides has been analyzed, it has appeared to result from mutation of chromosomal genes. Development of resistance is brought about mainly by mutation in a single gene. In one fungal species, more than one locus for fungicide resistance often could be demonstrated, resistance being mostly additive in recombinants. Varying degrees of resistance usually exist for different loci. Further influences may result from the existence of more than one allelomorph at each locus determining the degree of resistance or sensitivity, and also from the presence of modifying genes. Resistance may be dominant, semidominant, or recessive. In haploid cells mutations for resistance will be expressed immediately, but in heterokaryotic and in heterozygous diploid cells the situation is more complicated.

In bacteria, genes for resistance may be located in plasmids, extrachromosomal DNA molecules that can be transmitted between different bacteria. Transfer of resistance by plasmids might hamper contrɔl of bacterial diseases. Recently it appeared possible to trạnsfer antibiotic tolerance in *Pseudomonas aeruginosa,* via *E. coli,* to *Pseudomonas glycinea* and *P. phaseolicola* both in vitro and in the plant (76). Plasmids have not been detected in fungi.

Until the present, genetic work on fungicide resistance is rather limited. Most attention has been paid to aromatic hydrocarbons, benzimidazoles, and oxathiins.

Aromatic Hydrocarbons

The aromatic hydrocarbon group includes some aromatic or condensed nuclear hydrocarbons and some substituted derivatives, such as tetrachloronitrobenzene (TCNB) and pentachloronitrobenzene (PCNB), diphenyl, sodium orthophenylphenate, 2,6-dichloro-4-nitroalinine(dicloran), and 1,4-dichloro-2,5-dimethoxybenzene(chloroneb). This group comprises the first fungicides which encountered development of a high degree of fungicide resistance during practical use. Georgopoulos (38) established the mutational origin of tolerance to PCNB and one of the isomers of TCNB in *Hypomyces solani* f. sp. *cucurbitae.* This was shown by crossing tolerant and sensitive strains and demonstrating a Mendelian segregation in the progeny. Resistance appeared to be controlled by a single gene. A study of 12 mutants of this pathogen resistant to PCNB and TCNB revealed the presence of three loci for resistance (39). Acquired resistance to both compounds in *Aspergillus nidulans* was controlled by two closely linked genetic loci (118). Tolerance to chloroneb in *Ustilago maydis* followed the pattern for single gene resistance (119). Cross-resistance between the members of the aromatic hydrocarbon group appears to be the rule (48); tolerance to diphenyl developed by mutation at any of the loci, governing tolerance to chlorinated nitrobenzenes (128), and a chloroneb-resistant mutant of *U. maydis* was also resistant to other aromatic hydrocarbons (119).

In multinucleate fungi heterokaryosis may serve as the mechanism in maintaining low levels of fungicide tolerance, and tolerant strains may arise when selection is exerted by the chemical. Tolerance of *Botrytis cinerea* to 2,6-dichloro-4-nitroalinine and chlorinated nitrobenzenes was assumed to develop in this way (36, 132). Also emergence of resistance in *Thanatephorus cucumeris* to chloronitrobenzenes was attributed to heterokaryotic phenomena (85). As the mechanism of action of aromatic hydrocarbons has not yet been elucidated in detail, it is not yet clear in what way genes for resistance come to expression in metabolic changes in the fungal cell responsible for resistance to these compounds. As all aromatic hydrocarbons induce sectoring in *A. nidulans* diploids, action on chromosomes and mitosis is suggested (46).

Dodine

A decreased control of apple scab by dodine appeared to be due to the emergence of dodine-resistant strains of *Venturia inaequalis* (50, 116). The genetics of dodine resistance was studied by Kappas & Georgopoulos (71) using *Hypomyces solani* f. sp. *cucurbitae* as a test organism. They found four unlinked loci for resistance, which displayed varying degrees of tolerance. In recombinants the genes interacted to give higher resistance levels. Indications were obtained for the presence of modifiers. As dodine acts by disrupting the protoplast membrane, it is assumed that genes for resistance govern relevant properties of the membrane.

Cycloheximide

By UV irradiation, cycloheximide-tolerant strains of the nonpathogens *Saccharomyces cerevisiae* and *Neurospora crassa* were obtained. In *S. cerevisiae* the occurrence of six genes for resistance was demonstrated, one of which appeared to be dominant, three semidominant, and two recessive. Various levels of resistance for genes at different loci and additive effects after recombination were observed (66, 127, 134). These genes may govern accessibility of the antibiotic into the fungal cell or changes at the site of action, the ribosomes (19).

Oxathiins and Related Compounds

Although oxathiins were among the first systemic fungicides introduced in practice, hardly any problems with respect to resistance have been encountered in the field. One recent report concerns the acquirement of tolerance to oxycarboxin by *Puccinia horiana* (69), but no cases of development of carboxin tolerance in practice have been reported. In laboratory studies, however, by UV irradiation of sporidia, resistance to carboxin was readily obtained in *Ustilago maydis* (42, 43) and in *Ustilago hordei* (7). The induced resistance in *U. hordei* appeared to be controlled monogenically or polygenically (8), and strains were obtained that displayed a varying degree of resistance. In mutants of *U. maydis* resistant to carboxin three genes for resistance were shown to be present (43). Although *Aspergillus nidulans* is less sensitive than the above-mentioned fungi, increase of tolerance also was obtained by UV irradiation. Three independent genes for resistance were found, located on linkage groups VII (*carA*), VIII (*carB*), and III (*carC*). CarA was located 21 units from *benC*, the gene carrying benomyl resistance (120). Tolerance to carboxin was reported to be dominant in *U. hordei* (7), and semidominant in *U. maydis* (43).

Benzimidazoles and Thiophanates

Strains of various fungi resistant to these fungicides have been frequently obtained in laboratory studies, usually after UV irradiation, or from the field. Hastie & Georgopoulos (62) demonstrated the presence of two genes for resistance in strains of *A. nidulans,* obtained after UV irradiation of conidia. One of these mutations, conferring relatively high resistance,. was located on linkage group VIII (*benA* locus) and the other, conferring much lower resistance, on linkage group II (*benB* locus). Van Tuyl (121) confirmed this, and in addition discovered a third gene for resistance, located on linkage group VII (*benC* locus).

In *A. nidulans,* resistance to benomyl from mutation in the *benA* gene and the *benB* gene appeared to be additive (121). All three genes for resistance were semidominant (J. M. van Tuyl, unpublished). In *Ustilago hordei,* benomyl resistance was reported to be dominant (7). Cross-resistance between benomyl and thiabendazole appeared to be the rule, but not without exceptions. In rare cases benomyl resistance without concomitant thiabendazole resistance was found, and more frequently the reverse, thiabendazole resistance without benomyl resistance. Thiabendazole-resistant strains appeared, which were more sensitive to benomyl than the wild strain. It was indicated that these types of mutants are due to mutation at one locus, and that multiple alleles occur (122). An exception of cross-resistance between benomyl and thiabendazole was also found in *Ustilago maydis, Penicillium expansum,* and in the imperfect fungus *A. niger,* which may be analyzed genetically by means of the parasexual cycle. In eight benomyl-resistant strains of *A. niger,* including one not resistant to thiabendazole, one gene is probably responsible for resistance. Indications were obtained that the mutation site might be very small (121). It seems plausible to assume that this gene in *A. niger* and the *benA* locus in *A. nidulans* code for tubulin and that mutations lead to slight changes in affinity of these proteins to MBC, resulting in a change in sensitivity to this compound.

Selective or Mutagenic Action

Gene mutation may occur in all living organisms spontaneously, or deliberately induced by subjecting them to mutagenic agents. It is known that the majority of these mutants disappears rapidly or gradually from the population, unless environmental factors are present which discriminate between the mutant and the wild strain in favor of the mutant. This may be the case when a fungal population, in which all kinds of mutations continuously occur, is subjected to a fungicide, which selectively kills the sensitive individuals, so that those rare individuals which by chance harbor a mutation for resistance to the particular fungicide may multiply without competition from the normal wild population. It seems plausible that this is the most common way of emergence of fungicide resistance in the laboratory or in the field.

Virulence and other competitive qualities of resistant mutants, emerging in a population of pathogens, may show a wide variation (106). In the absence of the fungicide, such mutants that arise at random usually will lose the battle for survival in the competition with the wild strains, which had the opportunity to adapt to the host and the environment during a long period of time. This, however, should not necessarily be so in all cases. If the effect of a mutation would be limited exclusively to the degree of susceptibility to a fungicide, without interfering with other properties of the organism, such a gene for resistance might stay present in the population. The possibility cannot be excluded that in rare cases a resistant mutant appears with increased competitive ability, so that it will eventually replace the original sensitive population.

These considerations imply that mutants resistant to a fungicide may have emerged before the introduction of this compound into practice, and, probably in rare cases, have survived in the absence of the fungicide. Evidence for this phenomenon has been obtained. Thus a benomyl-resistant strain of *Verticillium malthousei* has been described, which was isolated in 1958, ten years before the introduction of this fungicide (130). Further, multinucleate fungi may, within a single species, show a wide variation in fungicide tolerance, because of the distribution of genes for resistance in the nuclei. In the above-mentioned cases tolerant strains may arise or the number of resistant isolates may increase, when selection is exerted by the chemical concerned (36, 132).

The question arises whether fungicides, beside a selective action, may exert also mutagenic activity. Benzimidazole and thiophanate fungicides induce instability of diploid cells of *Aspergillus nidulans* (61, 73). When recessive or semidominant genes for resistance are present in these cells, exposure to benomyl results in resistant mitotic segregants (62). The reported genetic effect of benomyl on *Fusarium oxysporum* (21) might possibly be explained in a similar way. No evidence, however, has been obtained that benomyl induces point mutations in *A. nidulans*. One report indicates a mutagenic effect of benomyl on *Fusarium oxysporum* (21). MBC caused mutations in *Salmonella typhimurium,* but here rather high doses were applied (109). Until the present no unequivocal examples have been obtained of systemic fungicides that induce mutations in fungi at doses used in practice. The possibility, however, that certain fungicides may also be mutagenic cannot be excluded and should be considered carefully for each new fungicide.

EMERGENCE OF RESISTANCE IN VITRO

The potential of a fungus to develop resistance to a fungicide depends on the type of fungicide and on the nature of the pathogen. Usually it can be studied easily in laboratory experiments. In order to increase the part of this potential which is realized in emergence of resistance, use can be made of chemical or physical mutagenic agents.

Type of Fungicide

Attempts to obtain fungal isolates with stable, acquired resistance to conventional fungicides have met little success, even if mutagenic agents were used. Most of these fungicides are known to be nonspecific in their mode of action, reacting at many sites in the fungal cell. Mutation leading to changes at all sites involved is unlikely or impossible, and a significant increase in tolerance due to changes in the permeability of the protoplast membrane or detoxication of the fungicide does not seem to occur frequently with these fungicides. Resistance may develop to several antibiotics and experimental fungicides which, in contrast to the conventional fungicides, act at specific sites in the fungal metabolism (27). This also holds for several systemic fungicides, introduced during the last decennium, which may raise the question whether all new systemic fungicides eventually may encounter development of resistance. In order to study this question, a comparison may be made of various types of selective fungicides which act at different sites in the fungal cell. We may distinguish between compounds that interfere with various biosynthetic processes (proteins, nucleic acids, sterols, chitin), energy production, permeability, and mitosis.

PROTEIN SYNTHESIS   Among the selective inhibitors of protein synthesis are the antibiotics cycloheximide, kasugamycin, and streptomycin. As has been pointed out above, resistance to the first two compounds may develop easily in various fungi. The mainly antibacterial antibiotic streptomycin affects few fungi, and development of resistance to fungi has not been reported. Resistance in bacteria, however, has been very frequently encountered. For all these antibiotics, resistance of at least some of the mutants was associated with the functioning of the ribosome (19, 63, 96). In these cases, apparently, the protein-synthesizing apparatus may, due to mutation, change in such a way that it shows less affinity to the inhibitor without appreciable loss of biosynthetic capacities.

NUCLEIC ACID SYNTHESIS   Most of the antifungal compounds known to interfere primarily with nucleic acid synthesis are antibiotics and analogs of purines and pyrimidines, which interfere with the synthesis of nucleotides (6-azauracil, 8-azaguanine, aristeromycin, azaserine) or with the replication and transcription of DNA (bleomycin, phleomycin, lomofungin, edeine, thiolutine, phytoactin, actinomycin, and related compounds) (32). Although a certain degree of tolerance to these types of compounds may develop due to factors that prevent the chemical from reaching the site of action in an active form, mutational modifications at the reactive site leading to fungicide resistance seem to be rare. This may be somewhat unexpected as it has been shown that most of these compounds interfere in a very specific way with nucleic acid metabolism. It might be, as the basic processes involved in nucleic acid synthesis are similar for all living organisms, that little or no flexibility for change exists, so that mutations in loci responsible for these processes tend to be lethal.

STEROL BIOSYNTHESIS   The fungicides triarimol and triforine have been reported to inhibit sterol biosynthesis. Although these compounds are chemically not related, cross-resistance occurs; two mutants of *Cladosporium cucumerinum* selected for resistance to triarimol appeared tolerant also to triforine (113). It has been suggested that triforine-resistant mutants are less virulent (37). Spores of triforine-resistant mutants showed a decreased ability to germinate. Addition of ergosterol restored the germinating capacity. Presumably the fungus has acquired resistance due to loss of the triforine-sensitive site in sterol synthesis, which apparently decreases viability of the spores that are formed (A. Fuchs, to be published).

CHITIN SYNTHESIS   Resistance to polyoxin-D, an antibiotic that was shown to inhibit chitin synthesis in *Alternaria kikuchiana,* appeared to be due to a decreased permeability of the protoplast membrane for this drug (65). No conclusions, therefore, can be drawn with respect to the possibilities of development of resistance due to site modification.

ENERGY PRODUCTION   The oxathiins carboxin and oxycarboxin interfere with energy production in sensitive organisms, predominantly Basidiomycetes (84, 133). It has been suggested that the basis for this selective action is the rate of uptake and binding by the fungal cell (83). From studies with *Ustilago maydis,* Georgopoulos et al (43), conclude that these compounds prevent respiration by preventing the transport of electrons in the mitochondrial respiratory chain, and that their main site of action lies between succinate and coenzyme Q in the complex II. Development of fungicide resistance may be due to a change at this site, as a result of single gene mutation. A second site of inhibition is the alternate pathway for electron transport in *Ustilago maydis,* which is responsible for naturally occurring tolerance to antimycin-A, an antibiotic that inhibits electron transport between cytochromes *b* and *c*. Increase of resistance may also be due to mutation, resulting in elimination of this pathway (47). From studies with *U. hordei* it is suggested that development of carboxin tolerance in this fungus depends on the development of alternate metabolic pathways such as the glyoxylate cycle, and only partly on carboxin-tolerant dichlorophenolindophenol (DCIP) reductase (8). By UV irradiation of *U. maydis* (42) and *U. hordei* (7) carboxin-resistant mutants have been readily obtained, but no resistant mutants of the latter fungus were observed in cultures not treated with a mutagenic agent. In *Aspergillus nidulans,* resistant mutants emerged with a frequency of one in 5.10^6 conidia surviving UV irradiation, whereas 95% of the conidia were killed by this treatment (120).

Evidence has been obtained that also PP, the active principle of pyrazophos, might be an inhibitor of energy production. Here, however, attempts have failed to obtain resistant mutants of *Piricularia oryzae* or *U. maydis,* after UV irradiation (45, 129).

MITOSIS   As has been discussed above, MBC and thiabendazole are inhibitors of mitosis in various fungi. Resistance to these fungicides occurs readily in many fungi, either spontaneously or induced by mutagenic treatment. From 14.10^8 conidia of *A. nidulans,* of which 6% survived after UV irradiation, we obtained 211 benomyl-resistant colonies (122); UV irradiation of spores of *Fusarium oxysporum,* of which 95% were killed, yielded one resistant mutant per 4.6×10^5 survivors (4). Resistance was attributed to single-gene mutations, leading to a changed affinity of tubular protein constituents of the spindle for these fungicides. Mutation for resistance, apparently, may occur readily.

Also another fungicide, the antibiotic griseofulvin, seems to interfere with mitosis. A study of the effect of griseofulvin on a diploid strain of *Aspergillus nidulans* showed that spindles were damaged and distorted within 30 sec to 1 min after treatment (20). Inhibition of mitosis by griseofulvin was also reported for the fungus *Basidiobolus ranarum* (55). From studies with brain extracts it appeared that griseofulvin did not inhibit the assembly of microtubules; it was suggested that the antibiotic affects some aspect of microtubule function, possibly a process vital to the sliding of microtubules, necessary for separation of chromosomes (54). Attempts to obtain griseofulvin-resistant mutants of fungi by

UV irradiation have failed. The details of the mechanism of action should be further investigated, in order to obtain insight in the question why two fungicides, which probably act at closely related sites, show such a striking difference with respect to emergence of fungicide resistance.

PERMEABILITY   Polyene macrolide antibiotics, such as filipin, pimaricin, and amphotericin, act primarily and specifically on the protoplast membrane, facilitating the loss of essential cytoplasmic constituents (28). The fungicidal activity appeared related to the affinity for certain sterols in the membrane (78, 107). The sterol content of the protoplast membrane of resistant strains, obtained in laboratory experiments, was reduced or changed (2, 53, 90, 98). Although some polyene macrolides have been in use for medical purposes during a number of years, no resistance has emerged in practice. Evidence has been obtained that increased tolerance is associated with diminished virulence (57).

COMPARISON OF DIFFERENT FUNGICIDES   From the above-mentioned data it appears that the rate of emergence of resistance may vary widely for different types of fungicides. Often, however, it is difficult to compare the data exactly, because not only the fungi but also the application of mutagenic agents and the experimental conditions are different in most studies. For a comparison of different fungicides with respect to development of resistance, the same fungi should be used, when possible, and the experimental conditions and mutagenic treatment should preferably be the same. This was done by J. M. van Tuyl (unpublished) for *Aspergillus nidulans* (Table 1). Comparable differences were found in experiments with other fungi. UV irradiation of sporidia of *Ustilago maydis* readily yielded mutants resistant to carboxin (47), but not to PP, a metabolite responsible for the fungitoxic action of the systemic fungicide pyrazophos (45). In studies with *Cladosporium cucumerinum,* strains resistant to 6-azauracil were readily obtained by UV irradiation (25), but no pimaricin-tolerant strains emerged (J. Dekker, unpublished work). Both compounds are systemic fungicides, which interfere at specific sites in the fungal metabolism; 6-azauracil, after conversion to its nucleotide, acts at one site in pyrimidine biosynthesis (26) and the polyene macrolide antibiotics to which group pimaricin belongs combine with certain sterols in the protoplast membrane (28). It appears that the potential for develop-

Table 1   Number of mutants resistant to various fungicides after UV treatment of conidia of *Aspergillus nidulans* (per 10^7 survivors)

	Number	Wild strain (ED_{50} in $\mu g/ml$)	Mutant
Benomyl	5	0.7	2–30
Carboxin	2	9	100–200
Chloroneb	25	10	> 500
Cycloheximide	40	150	>2000
Imazalil	100	0.5	2–5
Pimaricin	1	1.5	3–6

ment of fungicide resistance in powdery mildew fungi differs very much for two compounds which probably interfere at specific sites in mitosis; strains with high resistance to benomyl have been frequently isolated, but no reports have appeared on acquired resistance to griseofulvin, neither in powdery mildew fungi nor in dermatophytic fungi.

It seems therefore that also those fungicides that act primarily and specifically at one site in the fungal metabolism may differ widely with respect to the frequency with which resistant mutants arise.

Type of Pathogen

In addition to the type of fungicide, the type of pathogen might be important with respect to development of resistance. One report even mentions that no benomyl-resistant mutants of *Cercosporella herpotrichoides* could be obtained with or without mutagenic treatment (15). From studies at our laboratory, however, it appeared that by UV irradiation benomyl resistance could be induced in all fungi tested (121) (Table 2).

Table 2   UV-induced resistance to benomyl in various fungi; number of mutants per 10^7 survivors. ED_{50} in $\mu g/ml$ of the wild and the most resistant strains

	Number	ED_{50} wild	ED_{50} resistant
Aspergillus nidulans	5	0.7	30
A. niger	25	0.6	>500
Aureobasidium bolleyi	2	0.04	40
Cercosporella herpotrichoides	100	0.1	>500
Cladosporium cucumerinum	1.2	0.15	40
Fusarium nivale	5	0.1	100
Penicillium expansum	10	0.15	>500
Phialophora cinerescens	30	0.1	40
Rhodotorula rubra	50	2.0	>500
Ustilago maydis	5	1.5	100

The impression is obtained that when mutation for resistance to a particular fungicide occurs readily in one fungus we may expect it also in other fungi. The readiness with which such mutations come to expression may, however, depend on the type of fungus. In haploid pathogens, mutations for tolerance will be immediately expressed, and in diploid pathogens when tolerance is dominant to sensitivity. Expression of resistance may further be delayed, when dealing with pathogens in a dikaryotic phase, such as many of the Basidiomycetes, or with fungi possessing multinucleate cells.

ACQUIRED RESISTANCE IN THE FIELD

Even when mutants resistant to a fungicide can be readily obtained in the laboratory, with or without mutagenic treatment, this does not necessarily mean that resistance problems will arise in practice. This may among others depend on the type of disease and the characteristics of the resistant mutants involved.

Type of Disease

A resistant mutant of a pathogen which sporulates abundantly on aerial parts of a crop may spread very rapidly after elimination of the sensitive fungal population by the fungicide. An example is the rapid development of resistance to dimethirimol in cucumber powdery mildew in Dutch greenhouses (6). Resistant mutants of slowly spreading pathogens that do not produce aerial spores may not lead rapidly to problems in practice as has been suggested for benomyl resistance in *Ceratocystis ulmi* (13). Control of strawberry wilt with benomyl and thiophanate methyl for two years has not shown evidence of increased resistance, although fungicide-tolerant strains of the causal organism, *Verticillium dahliae,* were obtained in laboratory experiments (3). It seems therefore that the type of disease is of importance with respect to development of resistance problems in practice. Moreover, if it concerns a type of disease where the pathogen is difficult to eliminate by the fungicide, so that complete eradication of the sensitive wild-type strains does not occur, competition with the latter strains may counteract the spread of the resistant mutants.

Fitness of Resistant Mutants

Of prime importance for survival and spread of strains which have acquired fungicide resistance is their fitness, which comes to expression in virulence and other factors determining competitive ability. Mutation in a gene which governs sensitivity to a fungicide may influence fitness, and moreover, mutations in other genes may do so, independent of the mutation for resistance. Evidence was obtained that triforine-resistant mutants of *Cladosporium cucumerinum* form spores with a decreased ability to germinate, which might explain why development of resistance to triforine in plant pathogens has not yet been observed in practice (37). Also resistance to blasticidin-S in *Piricularia oryzae* has been found in the laboratory (91, 124), but not in the field. The same may hold for carboxin-tolerant mutants of *Ustilago* spp. which were found to be less competitive than the wild strain (7). In several other cases no direct relation between resistance and fitness of mutants was found, which means that a great variation in fitness may be observed for different resistant mutants. In studies with 6-azauracil for control of cucumber scab it was found that 2 out of 15 resistant strains of *Cladosporium cucumerinum* had lost their pathogenicity, that a few others were less virulent, but that the majority was as virulent as the wild type (31). Also in polyoxin-tolerant strains of *Alternaria kikuchiana,* isolated from the field (106) or in dodine-tolerant strains of *Nectria haematococca,* a relation between resistance and pathogenicity was not found (72).

Although many of the factors that determine fitness may be investigated in laboratory experiments, the latter do not provide all the information necessary to predict the behavior of resistant strains under field conditions. The theory of genetic homoeostasis indicates that, although particular individuals with a characteristic such as fungicide tolerance, may be selected, the whole population does not easily shift in this direction since there are many characteristics in the whole population which are held in a complex balance in the existing environmental conditions (136). As a consequence the population of a fungus with acquired resistance may shift back from resistant to sensitive, when application of the fungicide is stopped. Recent examples of this phenomenon are kasugamycin resistance in *Piricularia oryzae* (87, 89), benomyl resistance in *Cercospora beticola* (105), and polyoxin resistance in *Alternaria kikuchiana* (75). It might be of interest to investigate how a fungal population which has shifted back from resistant to sensitive will react to a renewed use of the fungicide concerned. With respect to insecticide resistance Keiding (74) concluded that even after a complete disappearance of resistant individuals, renewed use of the insecticide led to a faster development of resistance than before, which indicates that resistance, at least against insecticides, is essentially a "one-way street." If this would be true also for fungicides, it should, however, not preclude the use of the fungicide concerned in combination with other fungicides.

If no direct relation exists between fungicide resistance and fitness the possibility remains, however, that mutants arise, which possess the combination of tolerance and pathogenicity necessary for successful competition with sensitive forms of the fungus even in absence of the fungicide. Georgopoulos (41) reports that the population of benomyl-resistant strains of *Cercospora beticola* on sugar beet three years after application of benomyl was stopped did not show signs of declining. Even the occurrence of a benomyl-resistant strain of *Verticillium fungicola* (syn. *V. malthousei*) in Dutch mushroom houses has been reported, which showed a higher virulence than the wild strain (12). The possibility therefore exists that in such a case the whole population of a fungus may acquire resistance, even in the absence of the fungicide.

Examples of development of fungicide resistance in various crops are given in Table 3.

USE OF FUNGICIDES

Development of New Fungicides

It is still impossible, on the basis of the chemical structure of a newly developed fungicide, to predict whether fungi may acquire resistance to this compound. Even when the biochemical mechanism of action of a fungicide has been elucidated, which usually requires a time-consuming investigation, it remains difficult to estimate the chance that fungi have of acquiring resistance to the compound. It is, however, of great value to obtain an impression about the potential of pathogens to acquire resistance to a promising chemical before the chemical is released for practical use. This can easily be done by simple laboratory experiments with the help of UV irradiation or other mutagenic treatment. If no resistant mutants are obtained in this way, it seems unlikely that they will emerge under practical conditions. If resistance emerges in these experiments, greenhouse experiments may provide further information about the virulence and fitness of these resistant strains. As these do not provide all the information necessary to predict the behavior of resistant strains in the long run, monitoring for development of fungicide resistance in the field will then become desirable. The

Table 3   Acquired resistance to fungicides in practice

Pathogen	Crop	Fungicide	Year	Country	Reference
Alternaria kikuchiana	pear	polyoxin	1973	Japan	114
A. mali	apple	polyoxin	1974	Japan	35
Botrytis cinerea	strawberry	polyoxin	1975	Japan	69
Botrytis cinerea	cyclamen	benomyl	1971	Netherlands	11
Botrytis cinerea	chrysanthemum	benomyl	1973	USA	131
Botrytis cinerea	lettuce	benomyl	1974	England	86
Botrytis cinerea	tomato	benomyl	1974	England	86
Botrytis cinerea	raspberry	benomyl	1973	Scotland	70
Botrytis cinerea	strawberry	benomyl	1973	Scotland	70
Botrytis cinerea	cucumber	benomyl	1975	Japan	69
Botrytis spp.	eggplant	thiophanate-methyl	1975	Japan	117
Cercospora apii	celery	benomyl	1973	USA	9
C. arachidicola	peanut	benomyl	1974	USA	80
C. beticola	sugar beet	benomyl	1973	Greece	44
C. beticola	sugar beet	benomyl	1974	USA	104, 105
Cercosporidium personatum	peanut	benomyl	1974	USA	16
Colletotrichum musae	banana	benomyl	1973	West Indies	52
Erysiphe graminis	barley	ethirimol	1971	England	136
E. graminis	*Poa pratensis*	benomyl	1973	USA	126
E. cichoracearum	eggplant	thiophanate-methyl	1975	Japan	69
E. cichoracearum	cantaloupe	benomyl	1972	USA	99
Fusarium oxysporum f. sp. *gladioli*	gladiolus	benomyl	1974	USA	82
Penicillium corymbiferum	lily	benomyl	1971	Netherlands	10
P. digitatum	citrus	biphenyl	1964	USA	59
P. digitatum	citrus	thiabendazole	1972	USA	60
P. italicum	citrus	thiabendazole	1972	USA	60
P. italicum	citrus	thiabendazole	1974	Israel	56
P. italicum	citrus	thiophanate-methyl	1975	Japan	69
P. fructigenum	citrus	thiophanate-methyl	1975	Japan	69
Puccinia horiana	chrysanthemum	oxycarboxin	1974	Japan	1
Pyrenophora avenae	oats	organic mercury	1966	Scotland	95
Piricularia oryzae	rice	kasugamycin	1973	Japan	88
Rhizoctonia solani	cotton	quintozene	1963	USA	111
Sclerotium cepivorum	onion	dicloran	1969	USA	81
S. rolfsii	soybean	chlorinated nitrobenzene	1964	USA	40
Sclerotinia homoeocarpa	turf	benomyl	1974	USA	130
Sclerotinia homoeocarpa	turf	cadmium succinate	1974	USA	18
Septoria leucanthemi	daisy	benomyl	1974	USA	100
Sphaerotheca fuliginea	cucumber	benomyl	1969	USA	108
Sphaerotheca fuliginea	cucumber	benomyl	1972	Netherlands	30
Sphaerotheca fuliginea	cucumber	benomyl	1975	Japan	69
Sphaerotheca fuliginea	cucumber	benomyl	1973	DDR	14
Sphaerotheca fuliginea	cucumber	dimethirimol	1971	Netherlands	6
S. humuli	strawberry	benomyl	1975	Japan	69
S. humuli	strawberry	thiophanate-methyl	1975	Japan	69
Venturia inaequalis	apple	dodine	1969	USA	115
Venturia inaequalis	apple	thiophanate-methyl	1975	Japan	69
Venturia inaequalis	apple	benomyl	1974	Australia	135
Verticillium malthousei	champignon	benomyl	1974	USA	138
Verticillium malthousei	champignon	benomyl	1975	Netherlands	12
V. dahliae	strawberry	benomyl	1973	England	3

widespread use of a new fungicide, which has not been tested with respect to the possibility of development of fungicide resistance, should be discouraged.

In the search for new compounds to control fungal plant diseases, it seems worthwhile to pay attention to the possible use of nonfungicidal chemicals, which may increase the resistance of the host plant. Of course, the possibility cannot be excluded that the pathogen will overcome this type of host resistance, especially if it is caused by a change at one site in the host metabolism. If, however, chemically induced host resistance would depend on a number of changes in the plant metabolism, there might be an analogy with horizontal or polygenic resistance of a crop to a pathogen, which is not readily broken by a new physiologic race of the pathogen. And even when a pathogen could overcome this host resistance, the selection pressure may be less than that exerted by a highly fungicidal chemical, so that selection and survival of resistant mutants is much less favored.

Means of Application

Survival and spread of fungicide-resistant strains can be influenced by the way in which fungicides are applied. By application of a high dose of fungicide the whole fungal population will be killed except for those cells that have mutated for resistance. When mutation results in different levels of tolerance, more mutants will survive at lower doses of the fungicide. As in these cases all competition from sensitive strains is eliminated by high selection pressure, the buildup of a resistant population will be favored. At sublethal doses of the chemical, not only the resistant mutants but all the less sensitive cells of the wild-type population will survive. Although in the last case the population of the resistant mutants may build up less rapidly, there will also be a shift towards increased tolerance by the elimination of the most sensitive cells. In order to reduce the chance of emergence of resistant mutants, the dose of the fungicide should be as high as possible, of course with regard to phytotoxicity, economic, and environmental factors. If, however, resistant mutants have appeared, frequent application of high doses of fungicide will maintain a high selection pressure, which favors the increase of the resistant population. In that case, repeated application of high doses of the same fungicide should be avoided. Combination with a second fungicide should be considered with, in view of possible cross-resistance, a different mechanism of action. In experiments with a combination of benomyl and dimethirimol against cucumber powdery mildew no development of strains resistant to both compounds has been observed (34). Also a conventional fungicide may be of value in such a combination, since it may inhibit or reduce the sporulation and spread of the resistant mutants.

It would be of value if a combination of compounds with negatively correlated cross-resistance could be used so that mutants resistant to one of the fungicides would be killed by the other compound, and the reverse. A negatively correlated cross-resistance is described in *Piricularia oryzae* between phosphoramidate and phosphorothiolate fungicides (123, 125) and in one type of mutant of *Ustilago maydis* between carboxin and antimycin-A (47). Also the work with benomyl and thiabendazole indicates that possibilities in this direction may exist (122). The use of one fungicide, to which resistance can develop, over such a wide area that the total sensitive population is eliminated should be avoided. In that case the resistant population may persist even after the discontinuation of the application of the particular fungicide (44). A comparable phenomenon may occur by frequent application of the same fungicide in an area, which is geographically or artificially (greenhouse) isolated from other fields of the same crop. The sensitive population may then, under high selection pressure, be rapidly replaced by a resistant population (6). Selection and survival of resistant mutants may also be favored by persistence of the chemical in the crop. This holds true especially for fungicides, which penetrate into the plant tissue and are not easily washed off by rain. This type of fungicide can also be applied as an in-furrow or seed treatment, and then protect the crop for a considerable length of time. A gradual decrease in concentration due to breakdown and dilution may give a chance for survival to those mutants which possess a low level of resistance. In view of this, the seed treatment of winter barley against powdery mildew with ethirimol in England has been discontinued, in order to increase the proportion of the nonselective versus the selective periods in the asexual cycle, so that stabilizing selection may better operate (137).

CONCLUSIONS

Development of resistance to conventional fungicides, most of which are multisite inhibitors has been rare. The rate of mutation towards resistance to systemic fungicides, which are site-specific, varies widely, depending on the type of compound and the fungus. The mutability of the genes concerned and the flexibility of the fungal metabolism to allow mutational changes in relevant genes without appreciable loss of fitness, are of importance.

Normally resistant mutants arise spontaneously, whereupon selection occurs by the fungicide.

Whether resistance problems will arise in practice, depends further on the fitness and virulence of the resistant mutants, the type of disease, environmental conditions, and the persistence and method of use of the fungicide. A continuous selection pressure by one type of fungicide may favor the buildup of a resistant population and should therefore be avoided. New types of fungicides should not be released for practical use before the potential of fungi to develop resistance to these compounds has been investigated in vitro and on the plant.

ACKNOWLEDGMENTS

The author is indebted to Ir. L. C. Davidse and Ir. J. M. van Tuyl for a critical reading of the manuscript, to Ir. T. Hijwegen for help with the literature and to Dr. S. G. Georgopoulos for a preview of his chapter, "Development of fungal resistance to fungicides" in *Antifungal Compounds* (41).

Literature Cited

1. Abiko, K., Kishi, K., Yoshioka, A. 1975. Emergence of plantvax resistant strains of chrysanthemum rust fungus *Puccinia horiana. Ann. Phytopathol. Soc. Jpn.* 41:100 (Abstr.)
2. Ahmed, K. A., Woods, R. A. 1967. A genetic analysis of resistance to nystatin in *Saccharomyces cerevisiae. Genet. Res.* 9:179–93
3. Anonymous. 1973. *Ann. Rept. East Malling Res. Stn., 147*
4. Bartels-Schooley, J., MacNeill, B. H. 1971. A comparison of the modes of action of three benzimidazoles. *Phytopathology* 61:816–19
5. Bartz, J. A., Mitchell, J. E. 1970. Evidence for the metabolic detoxification of n-dodecylguanidine acetate by ungerminated macroconidia of *Fusarium solani* f. sp. *phaseoli. Phytopa-*

thology 60:350–54

6. Bent, K. J., Cole, A. M., Turner, J. A. W., Woolner, M. 1971. Resistance of cucumber powdery mildew to dimethirimol. *Proc. 6th Br. Insectic. Fungic. Conf.* 1971:274–82

7. Ben-Yephet, Y., Henis, Y., Dinoor, A. 1974. Genetic studies on tolerance of carboxin and benomyl at the asexual phase of *Ustilago hordei. Phytopathology* 64:51–56

8. Ben-Yephet, Y., Dinoor, A., Henis, Y. 1975. The physiological basis of carboxin sensitivity and tolerance in *Ustilago hordei. Phytopathology* 65: 936–42

9. Berger, R. D. 1973. Disease progress of *Cercospora apii* resistant to benomyl. *Plant Dis. Reptr* 57:837–40

10. Bollen, G. J. 1971. Resistance to benomyl and some chemically related compounds in strains of *Penicillium* species. *Neth. J. Plant Pathol.* 77: 187–93

11. Bollen, G. J., Scholten, G. 1971. Acquired resistance to benomyl and some other systemic fungicides in a strain of *Botrytis cinerea* in cyclamen. *Neth. J. Plant Pathol.* 77:83–90

12. Bollen, G. J., van Zaayen, A. 1975. Resistance to benzimidazole fungicides in pathogenic strains of *Verticillium fungicola. Neth. J. Plant Pathol.* 81:157–67

13. Brasier, C. M., Gibbs, J. N. 1975. MBC tolerance in aggressive and non-aggressive isolates of *Ceratocystis ulmi. Ann. Appl. Biol.* 80:231–35

14. Burth, U. 1973. Resistenz von Gurkenmehltau (*Sphaerotheca fuliginea* [Schlecht.] Salmon) gegenüber Systemfungiziden aus der Benzimidazol-Gruppe. *Arch. Phytopathol. Pflanzenschutz* 9:411–13

15. Chidambaram, P., Bruehl, G. W. 1973. Lack of benomyl tolerance in *Cercosporella herpotrichoides. Plant Dis. Reptr.* 57:935–36

16. Clark, E. M., Backman, P. A., Rodriguez-Kabana, R. 1974. Cercospora and Cercosporidium tolerance to benomyl and related fungicides in Alabama peanut fields. *Phytopathology* 64:1476–77

17. Clemons, G. P., Sisler, H. D. 1971. Localization of the site of action of a fungitoxic benomyl derivative. *Pestic. Biochem. Physiol.* 1:32–43

18. Cole, H., Taylor, B., Duich, J. 1968. Evidence of differing tolerances to fungicides among isolates of *Sclerotinia homoeocarpa. Phytopathology* 58:683–86

19. Cooper, D., Banthorpe, D. V., Wilkie, D. 1967. Modified ribosomes conferring resistance to cycloheximide in mutants of *Saccharomyces cerevisiae. J. Mol. Biol.* 26:347–50

20. Crackower, S. H. B. 1972. The effects of griseofulvin on mitosis in *Aspergillus nidulans. Can. J. Microbiol.* 18: 683–87

21. Dassenoy, B., Meyer, J. A. 1973. Mutagenic effects of benomyl on *Fusarium oxysporum. Mutat. Res.* 21: 119–20

22. Davidse, L. C. 1973. Antimitotic activity of methyl benzimidazol-2-yl carbamate (MBC) in *Aspergillus nidulans. Pestic. Biochem. Physiol.* 3:317–25

23. Davidse, L. C. 1975. Mode of action of methyl benzimidazol-2-yl carbamate (MBC) and the mechanism of resistance against this fungicide in *Aspergillus nidulans. Proc. Symp. Systemic Fungic.,* May 1974, Reinhardsbrunn, DDR, pp. 137–43

24. Davidse, L. C. 1975. Antimitotic activity of methyl benzimidazol-2-yl carbamate in fungi and its binding to cellular protein. *Proc. Symp. Microtubules Microtubular Inhibitors,* 2–5 Sept. 1975, Beerse, Belgium, pp. 317–25

25. Dekker, J. 1967. Conversion of 6-azauracil in sensitive and resistant strains of *Cladosporium cucumerinum.* In *Wirkungs mechanismen von Fungiziden und Antibiotika,* ed. M. Girbardt, 333–39. Berlin: Akad.-Verlag. 443 pp.

26. Dekker, J. 1968. The development of resistance in *Cladosporium cucumerinum* against 6-azauracil, a chemotherapeutant of cucumber scab, and its relation to biosynthesis of RNA-precursors. *Neth. J. Plant Pathol.* 74 (1968 Suppl. 1):127–36

27. Dekker, J. 1969. Acquired resistance to fungicides. *World Rev. Pest Control* 8:79–85

28. Dekker, J. 1969. Antibiotics. In *Fungicides,* ed. D. C. Torgeson, 2:579–635. New York: Academic. 742 pp.

29. Dekker, J. 1971. Selective action of fungicides and development of resistance in fungi to fungicides. *Proc. 6th Br. Insectic. Fungic. Conf.* 1971:715–23

30. Dekker, J. 1972. Resistance. In *Systemic Fungicides.* ed. R. W. Marsh, 155–74. Longman: London. 321 pp.

31. Dekker, J. 1975. Acquired resistance to systemic fungicides. *Proc. Symp. Systemic Fungic.,* May 1974, Reinhardsbrunn, DDR, pp. 49–54

32. Dekker, J. 1976. Effect of fungicides

on nucleic acid synthesis and nuclear function. In *Antifungal Compounds*, ed. H. D. Sisler, M. R. Siegel, Vol. 2. New York: Dekker. In press

33. Dekker, J., Oort, A. J. P. 1964. Mode of action of 6-azauracil against powdery mildew. *Phytopathology* 54:815–18

34. Ebben, M. H., Spencer, D. M. 1973. The integrated use of two systemic fungicides for the control of cucumber powdery mildew. *Proc. 7th Br. Insectic. Fungic. Conf.* 1973:211–16

35. Eguchi, J., Mogami, M., Okuda, S. 1974. Properties of insensitive strains of leaf spot fungus of apple, *Alternaria mali*, to polyoxins. *Ann. Phytopathol. Soc. Jpn.* 40:220 (Abstr.)

36. Esuruoso, O. F., Wood, R. K. S. 1971. The resistance of spores of resistant strains of *Botrytis cinerea* to quintozene, tecnazene and dicloran. *Ann. Appl. Biol.* 68:271–79

37. Fuchs, A., Viets-Verweij, M. 1975. Permanent and transient resistance against triarimol and triforine in some pathogenic fungi. *Meded. Fac. Landbouwwet. Rijksuniv. Gent* 40:699–706

38. Georgopoulos, S. G. 1962. Genetic nature of tolerance of *Hypomyces solani* f. *cucurbitae* to penta- and tetrachloronitrobenzene. *Nature* 194:148–49

39. Georgopoulos, S. G. 1963. Tolerance to chlorinated nitrobenzenes in *Hypomyces solani* f. *cucurbitae* and its mode of inheritance. *Phytopathology* 53:1086–93

40. Georgopoulos, S. G. 1964. Chlorinated nitrobenzene tolerance in *Sclerotium rolfsii*. *Ann. Inst. Phytopathol. Benaki, N. S.* 6:156–59

41. Georgopoulos, S. G. 1976. Development of fungal resistance to fungicides. See Ref. 32

42. Georgopoulos, S. G., Alexandri, E., Chrysayi, M. 1972. Genetic evidence for the action of oxathiin and thiazole derivatives on the succinic dehydrogenase system of *Ustilago maydis* mitochondria. *J. Bacteriol.* 110:809–17

43. Georgopoulos, S. G., Chrysayi, M., White, G. A. 1975. Carboxin resistance in the haploid, the heterozygous diploid, and the plant parasitic dicaryotic phase of *Ustilago maydis*. *Pestic. Biochem. Physiol.* 5:543–81

44. Georgopoulos, S. G., Dovas, C. 1973. A serious outbreak of strains of *Cercospora beticola* resistant to benzimidazole fungicides in Northern Greece.

Plant Dis. Reptr. 57:321–24

45. Georgopoulos, S. G., Geerligs, J. W. G., Dekker, J. 1975. Sensitivity of *Ustilago maydis* to pyrazophos and one of its conversion products, and failure to induce resistance with UV-treatment. *Neth. J. Plant Pathol.* 81:38–41

46. Georgopoulos, S. G., Kappas, A., Hastie, A. C. 1976. Induced sectoring in diploid *Aspergillus nidulans* as a criterion of fungitoxicity by interference with hereditary processes. *Phytopathology* 66:217–20

47. Georgopoulos, S. G., Sisler, H. D. 1970. Gene mutation eliminating antimycin A-tolerant electron transport in *Ustilago maydis*. *J. Bacteriol.* 103:745–50

48. Georgopoulos, S. G., Vomvoyanni, V. E. 1965. Differential sensitivity of diphenyl-sensitive and diphenyl-tolerant strains of fungi to chlorinated nitrobenzenes and to some diphenyl derivatives. *Can. J. Bot.* 43:765–75

49. Georgopoulos, S. G., Zaracovitis, C. 1967. Tolerance of fungi to organic fungicides. *Ann. Rev. Phytopathol.* 5:109–30

50. Gilpatrick, J. D., Blowers, D. R. 1974. Ascospore tolerance to dodine in relation to orchard control of apple scab. *Phytopathology* 64:649–52

51. Greenaway, W. 1971. Relationship between mercury resistance and pigment production in *Pyrenophora avenae*. *Trans. Br. Mycol. Soc.* 56:37–44

52. Griffee, P. J. 1973. Resistance to benomyl and related fungicides in *Colletotrichum musae*. *Trans. Br. Mycol. Soc.* 60:433–39

53. Grindle, M. 1973. Sterol mutants of *Neurospora crassa:* their isolation, growth characteristics and resistance to polyene antibiotics. *Mol. Gen Genet.* 120:283–90

54. Grisham, L. M., Wilson, L., Bensch, K. G. 1973. Antimitotic action of griseofulvin does not involve disruption of microtubules. *Nature* 244:294–96

55. Gull, K., Trinci, A. P. J. 1974. Effects of griseofulvin on the mitotic cycle of the fungus *Basidiobolus ranarum*. *Arch. Microbiol.* 95:57–65

56. Gutter, Y., Yanko, U., Davidson, M., Rahat, M. 1974. Relationship between mode of application of thiabendazole and its effectiveness for control of green mold and inhibiting fungus sporulation on oranges. *Phytopathology* 64:1477–78

57. Hamilton-Miller, J. M. T. 1974. Fungal sterols and the mode of action of the polyene antibiotics. *Adv. Appl. Microbiol.* 17:109–34

58. Hammerschlag, R. S., Sisler, H. D. 1973. Benomyl and methyl-2-benzimidazole carbamate (MBC):biochemical, cytological and chemical aspects of toxicity to *Ustilago maydis* and *Saccharomyces cerevisiae. Pestic. Biochem. Physiol.* 3:42–54

59. Harding, P. R. 1964. Assaying for biphenyl resistance in *Penicillium digitatum* in California lemon packing houses. *Plant Dis. Reptr.* 48:43–46

60. Harding, P. R. 1972. Differential sensitivity to thiabendazole by strains of *Penicillium italicum* and *P. digitatum. Plant Dis. Reptr.* 56:256–260

61. Hastie, A. C. 1970. Benlate-induced instability of *Aspergillus* diploids. *Nature* 226:771

62. Hastie, A. C., Georgopoulos, S. G. 1971. Mutational resistance to fungitoxic benzimidazole derivatives in *Aspergillus nidulans. J. Gen. Microbiol.* 67:371–73

63. Helser, T. L., Davies, J. E., Dahlberg, J. E. 1971. Change in methylation of 16S ribosomal RNA associated with mutation to kasugamycin resistance in *Escherichia coli. Nature New Biol.* 233:12–14

64. Hori, M., Eguchi, J., Kakiki, K., Misato, T. 1974. Studies on the mode of action of polyoxins. VI. Effect of polyoxin B on chitin synthesis in polyoxin-sensitive and resistant strains of *Alternaria kikuchiana. J. Antibiot.* 27:260–66

65. Hori, M., 'Kakiki, K., Suzuki, S., Misato, T. 1971. Studies on the mode of action of polyoxins. III. Relation of polyoxin structure to chitin synthetase inhibition. *Agric. Biol. Chem.* 35:1280–91

66. Hsu, K. S., 1963. The genetic basis of actidione resistance in *Neurospora. J. Gen. Microbiol.* 32:341–47

67. Huang, K. T., Misato, T., Asuyama, H. 1964. Selective toxicity of blasticidin S to *Pyricularia oryzae* and *Pellicularia sasakii. J. Antibiot.,* Ser. A 17:71–74

68. Huang, K. T., Misato, T., Asuyama, H. 1964. Effect of blasticidin S on protein synthesis of *Pyricularia oryzae. J. Antibiot.,* Ser. A 17:65–70

69. Iida, W. 1975. On the tolerance of plant pathogenic fungi and bacteria to fungicides in Japan. *Jpn. Pestic. Inf.* 23:13–16

70. Jarvis, J. R., Hargreaves, A. J. 1973. Tolerance to benomyl in *Botrytis cinerea* and *Penicillium corymbiferum. Plant Pathol.* 22:139–41

71. Kappas, A., Georgopoulos, S. G. 1970. Genetic analysis of dodine resistance in *Nectria haematococca* (syn. *Hypomyces solani). Genetics* 66:617–22

72. Kappas, A., Georgopoulos, S. G. 1971. Independent inheritance of avirulence and dodine resistance in *Nectria haematococca* var. *cucurbitae. Phytopathology* 61:1093–94

73. Kappas, A., Georgopoulos, S. G., Hastie, A. C. 1974. On the genetic activity of benzimidazole and thiophanate fungicides on diploid *Aspergillus nidulans. Mutat. Res.* 26:17–27

74. Keiding, J. 1967. Persistence of resistant populations after the relaxation of the selection pressure. *World Rev. Pest Control* 6:115–30

75. Kohmoto, K., Miyake, H., Nishimura, S., Udagawa, H. 1974. Distribution and chronological population shift of polyoxin resistant strains of black spot fungi of Japanese pear, *Alternaria kikuchiana,* in field. *Ann. Phytopathol. Soc. Jpn.* 40:220 (Abstr.)

76. Lacy, G. H., Leary, J. V. 1975. Transfer of antibiotic resistance plasmid RP1, into *Pseudomonas glycinea* and *Pseudomonas phaseolicola in vitro* and *in planta. J. Gen. Microbiol.* 88: 49–57

77. Lambert, D. H., Wuest, P. J. 1975. Increased sensitivity to zineb for *Verticillium malthousei* strains tolerant to benomyl. *Phytopathology* 65:637–38

78. Lampen, J. O., Arnow, P. M., Borowska, Z., Laskin, A. I. 1962. Location and role of sterol at nystatin-binding sites. *J. Bacteriol.* 84:1152–60

79. Lewis, D. 1963. Structural gene for the methionine-activating enzyme and its mutation as a cause of resistance to ethionine. *Nature* 200:151–54

80. Littrell, R. H. 1974. Tolerance in *Cercospora arachidicola* to benomyl and related fungicides. *Phytopathology* 64: 1377–78

81. Locke, S. B. 1969. Botran tolerance of *Sclerotium cepivorum* isolants from fields with different Botran-treatment histories. *Phytopathology* 59:13 (Abstr.)

82. Magie, R. O., Wilfret, G. J. 1974. Tolerance of *Fusarium oxysporum* f. sp. *gladioli* to benzimidazole fungicides. *Plant Dis. Reptr.* 58:256–59

83. Mathre, D. E. 1968. Uptake and binding of oxathiin systemic fungicides by

resistant and sensitive fungi. *Phytopathology* 58:1464–69

84. Mathre, D. E. 1970. Mode of action of oxathiin systemic fungicides. I. Effect of carboxin and oxycarboxin on the general metabolism of several basidiomycetes. *Phytopathology* 60:671–76

85. Meyer, R. W., Parmeter, J. R. Jr. 1968. Changes in chemical tolerance associated with heterokaryosis in *Thanatephorus cucumeris*. *Phytopathology* 58:472–75

86. Miller, M. W., Fletcher, J. T. 1974. Benomyl tolerance in *Botrytis cinerea* isolates from glasshouse crops. *Trans. Br. Mycol. Soc.* 62:99–103

87. Misato, T. 1975. The development of agricultural antibiotics in Japan. *Proc. First Intersect. Congr. Int. Assoc. Microbiol. Soc.*, 1–7 September 1974, Tokyo 3:589–97

88. Miura, H., Ito, H., Kimura, K., Takahashi, S. 1973. Resistance of *Pyricularia oryzae* to kasugamycin. 1. Decline in effect of kasumin in Shonai district. *Ann. Phytopathol. Soc. Jpn.* 39:239–40 (Abstr.)

89. Miura, H., Ito, H., Takahashi, S. 1975. Occurrence of resistant strains of *Pyricularia oryzae* to kasugamycin as a cause of the diminished fungicidal activity to rice blast. *Ann. Phytopathol. Soc. Jpn.* 41:415–17

90. Molzahn, S. W., Woods, R. A. 1972. Polyene resistance and the isolation of sterol mutants in *Saccharomyces cerevisiae*. *J. Gen. Microbiol.* 72:339–48

91. Nakamura, H., Sakurai, H. 1968. Tolerance of *Piricularia oryzae* Cavara to Blasticidin-S. *Bull. Agric. Chem. Insp. Stn.*, pp. 21–25

92. Nakanishi, T., Oku, H. 1969. Metabolism and accumulation of pentachloronitro benzene by phytopathogenic fungi in relation to selective toxicity. *Phytopathology* 59:1761–62

93. Nakanishi, T., Oku, H. 1970. Mechanism of selective toxicity of fungicide: absorption, metabolism and accumulation of pentachloronitrobenzene by phytopathogenic fungi. *Ann. Phytopathol. Soc. Jpn.* 36:67–73

94. Nishimura, M., Kohmoto, K., Udagawa, H. 1973. Field emergence of fungicide-tolerance strains in *Alternaria kikuchiana* Tanaka. *Rept. Tottori Mycol. Inst.* 10:677–86

95. Noble, M., Macgarvie, Q. D., Hams, A. F., Leafe, E. L. 1966. Resistance to mercury of *Pyrenophora avenae* in Scottish seed oats. *Plant Pathol.* 15:23–28

96. Nomura, M. 1970. Bacterial ribosome *Bacteriol. Rev.* 34:228–77

97. Oku, H., Nakanishi, T. 1964. Reductive detoxification of an antibiotic, ascochitine, by an insensitive fungus, *Fusarium lycopersici* Sacc. *Naturwissenschaften* 51:538

98. Patel, P. V., Johnston, J. R. 1968. Dominant mutation for nystatin resistance in yeast. *Appl. Microbiol.* 16:164–65

99. Paulus, A. O. et al 1972. Fungicides and methods of application for the control of cantaloup powdery mildew. *Plant Dis. Reptr.* 56:935–38

100. Paulus, A. O., Netzer, D. 1974. Resistance in *Septoria leucanthemi* to benzimidazole type fungicides. *Proc. Am. Phytopathol. Soc.* 1:45 (Abstr.)

101. Ross, I. S. 1974. Non-protein thiols and mercury resistance of *Pyrenophora avenae*. *Trans. Br. Mycol. Soc.* 63:77–83

102. Ross, I. S., Old, K. M. 1973. Mercuric chloride resistance of *Pyrenophora avenae*. *Trans. Br. Mycol. Soc.* 60:293–300

103. Ross, I. S., Old, K. M. 1973. Thiol compounds and resistance of *Pyrenophora avenae* to mercury. *Trans. Br. Mycol. Soc.* 60:301–10

104. Ruppel, E. G., Scott, P. R. 1974. Strains of *Cercospora beticola* resistant to benomyl in the USA. *Plant Dis. Reptr.* 58:434–36

105. Ruppel, E. G. 1975. Biology of benomyl-tolerant strains of *Cercospora beticola* from sugar beet. *Phytopathology* 65:785–89

106. Sakurai, H., Shimada, T. 1974. Studies on polyoxin resistant strains of *Alternaria kikuchiana* Tanaka. Part 3. Characters on media and pathogenicity. *Bull. Agric. Chem. Insp. Stn.* 14:54–65

107. Schloesser, E. 1965. Sterols and the resistance of *Pythium* species to filipin. *Phytopathology* 55:1075

108. Schroeder, W. T., Provvidenti, R. 1969. Resistance to benomyl in powdery mildew of cucurbits. *Plant Dis. Reptr.* 53:271–75

109. Seiler, J. P. 1972. The mutagenicity of benzimidazole and benzimidazole derivatives. I. Forward and reverse mutations in *Salmonella typhimurium* caused by benzimidazole and some of its derivatives. *Mutat. Res.* 15:273–76

110. Selling, H. A., Vonk, J. W., Kaars Sijpesteijn, A. 1970. Transformation of the systemic fungicide methyl thio-

phanate into 2-benzimidazole carbamic acid methyl ester. *Chem. Ind. London* 1970:1625–26

111. Shatla, M. N., Sinclair, J. B. 1963. Tolerance to pentachloronitrobenzene among cotton isolates of *Rhizoctonia solani. Phytopathology* 53:1407–11

112. Sherald, J. L., Sisler, H. D. 1970. Antimycin A-resistant respiratory pathway in *Ustilago maydis* and *Neurospora sitophila. Plant Physiol.* 46:180–82

113. Sherald, J. L., Ragsdale, N. N., Sisler, H. D. 1973. Similarities between the systemic fungicides triforine and triarimol. *Pestic. Sci.* 4:719–27

114. Shimada, T., Sakurai, H., Yoshida, K. 1973. Studies on polyoxin resistance of the strain Y-33 of *Alternaria kikuchiana* Tanaka. *Bull. Agric. Chem. Insp. Stn.* 13:39–42

115. Szkolnik, M., Gilpatrick, J. D. 1969. Apparent resistance of *Venturia inaequalis* to dodine in New York apple orchards. *Plant Dis. Reptr.* 53:861–64

116. Szkolnik, M., Gilpatrick, J. D. 1973. Tolerance of *Venturia inaequalis* to dodine in relation to the history of dodine usage in apple orchards. *Plant Dis. Reptr.* 57:817–21

117. Tezuka, N., Kiso, A. 1975. Emergence of resistant strains of *Botrytis* sp. to thiophanate-methyl on egg plants. *Ann. Phytopathol. Soc. Jpn.* 41:303–4 (Abstr.)

118. Threlfall, R. J. 1968. The genetics and biochemistry of mutants of *Aspergillus nidulans* resistant to chlorinated nitrobenzenes. *J. Gen. Microbiol.* 52:35–44

119. Tillman, R. W., Sisler, H. D. 1973. Effect of chloroneb on the growth and metabolism of *Ustilago maydis. Phytopathology* 63:219–25

120. Tuyl, J. M., van 1975. Acquired resistance to carboxin in *Aspergillus nidulans. Neth. J. Plant Pathol.* 81:122–23

121. Tuyl, J. M., van 1975. Genetic aspects of acquired resistance to benomyl and thiabendazole in a number of fungi. *Meded. Fak. Landbouwwet. Rijksuniv. Gent* 40:691–98

122. Tuyl, J. M., van, Davidse, L. C., Dekker, J. 1974. Lack of cross resistance to benomyl and thiabendazole in some strains of *Aspergillus nidulans. Neth. J. Plant Pathol.* 80:165–68

123. Uesugi, Y. 1973. Development of fungicide resistance in rice blast fungus, *Pyricularia oryzae* Cavara. *Jpn. Agric. Res. Q.* 7:185–88

124. Uesugi, Y., Katagiri, M., Fukunaga, K. 1969. Resistance in *Pyricularia*

oryzae to antibiotics and organophosphorus fungicides. *Bull. Natl. Inst. Agric. Sci. C* 23:93–112

125. Uesugi, Y., Katagiri, M., Noda, O. 1973. Negatively correlated cross-resistance and synergism between phosphoramidates and phosphorothiolates in their fungicidal actions on rice blast fungi. *Agric. Biol. Chem.* 38:907–12

126. Vargas, J. M. Jr. 1973. A benzimidazole resistant strain of *Erysiphe graminis. Phytopathology* 63:1366–68

127. Vomvoyanni, V. 1974. Multigenic control of ribosomal properties associated with cycloheximide sensitivity in *Neurospora crassa. Nature* 248:508–10

128. Vomvoyanni, V. E., Georgopoulos, S. G. 1966. Dosage-response relationships in diphenyl tolerance of *Hypomyces. Phytopathology* 56:1330–31

129. Waard, M. A., de 1974. Mechanisms of action of the organophosphorus fungicide pyrazophos. *Meded. Landbouwhogesch. Wageningen* 74-14:1–97

130. Warren, C. G., Sanders, P., Cole, H. 1974. *Sclerotinia homoeocarpa* tolerance to benzimidazole configuration fungicides. *Phytopathology* 64:1139–42

131. Watson, A. G., Koons, C. E. 1973. Increased tolerance to benomyl in greenhouse populations of *Botrytis cinerea. Phytopathology* 63:1218 (Abstr.)

132. Webster, R. K., Ogawa, J. M., Bose, E. 1970. Tolerance of *Botrytis cinerea* to 2,6-dichloro-4-nitroaniline. *Phytopathology* 60:1489–92

133. White, G. A. 1971. A potent effect of 1,4-oxathiin systemic fungicides on succinate oxidation by a particulate preparation from *Ustilago maydis. Biochem. Biophys. Res. Commun.* 44:1212–19

134. Wilkie, D., Lee, B. K. 1965. Genetic analysis of actidione resistance in *Saccharomyces cerevisiae. Genet. Res.* 6:130–38

135. Wicks, T. 1974. Tolerance of the apple scab fungus to benzimidazole fungicides. *Plant Dis. Reptr.* 58:886–89

136. Wolfe, M. S. 1971. Fungicides and the fungus population problem. *Proc. 6th Br. Insectic. Fungic. Conf.* 1971:724–34

137. Wolfe, M. S., Dinoor, A. 1973. The problem of fungicide tolerance in the field. *Proc. 7th Br. Insectic. Fungic. Conf.* 1973:11–19

138. Wuest, P. J., Cole, H. Jr., Sanders, P. L. 1974. Tolerance of *Verticillium malthousei* to benomyl. *Phytopathology* 64:331–34

Copyright © 1976 by Annual Reviews Inc.
All rights reserved

WHITEFLY-TRANSMITTED PLANT DISEASES

♦3649

A. S. Costa[1]

Seção de Virologia, Instituto Agronômico, Campinas, São Paulo, Brazil

INTRODUCTION

Whiteflies are known to transmit bacterial and fungus diseases, but the present review is restricted to their role as vectors of a group of disease-inducing agents that have been included among the viruses. Various aspects of these diseases have been dealt with in previous reviews (26, 28, 125).

Whitefly-transmitted diseases and their causal agents represent an important area of plant pathology, both for their economic importance and for posing as a fertile field of scientific problems that await clarification. They are present mainly in tropical countries (8, 92, 109, 113, 121), but are also known to occur in subtropical and temperate agricultural areas (32, 53, 61, 69, 80). Their impact on such crops as cassava (*Manihot esculenta*), cotton (*Gossypium* spp.), sweet potatoes (*Ipomea batatas*), tobacco (*Nicotiana tabacum*), and various legume crops has been recognized (10, 12, 61, 80, 92, 115, 120, 121). In recent years, whitefly-transmitted diseases have become important on beans (*Phaseolus vulgaris*) and other legume crops, and on tomatoes (*Lycopersicon esculentum*) in various parts of the world and particularly in the Americas (18, 27, 33, 50, 93, 101).

Disease agents transmitted by whiteflies have been included among the plant viruses and are thus considered in this review, though conclusive and comprehensive evidence of their nature is still lacking. Discoveries that certain plant diseases that were formerly attributed to viruses are associated with mycoplasmas, spiroplasmas, viroids, and bacterium-like organisms brought some doubt about the viral nature of the disease agents transmitted by whiteflies. The elucidation of their nature and of some particular types of relationships found between the disease agents, their whitefly vector, and host plants represents a challenge to plant pathologists and virologists alike. Another challenge is to find adequate control measures for some of the whitefly-transmitted plant diseases that have become important factors in reducing the yield of many crops in some areas of the world.

[1] Fellow, Conselho Nacional de Desenvolvimento Cientifico e Tecnologico of Brazil.

SYMPTOMS OF DISEASES TRANSMITTED BY WHITEFLIES

Whitefly-transmitted diseases have been reported mostly on herbaceous plants, more rarely on shrubs (86, 126) and trees (119). The symptoms associated with these diseases fall into three main categories: (*a*) the mosaic diseases, (*b*) the leaf curl diseases, and (*c*) the yellowing diseases.

Diseases of the Mosaic Type

Whitefly-transmitted diseases characterized by light green mosaic usually are less abundant than those of the bright yellow or golden mosaic type. Species of Compositae, Convolvulaceae, Labiatae, Leguminosae, Malvaceae, and Solanaceae, which are natural vegetation in many tropical areas, may show the intense bright yellow type of symptom that makes diseased plants very conspicuous among normal ones (8, 26, 125). The yellow color sometimes appears along the veins, inducing yellow net, or the yellow areas may be limited by the veins.

The bright yellow mosaic leaf areas are associated with striking changes in the lamellar system of chloroplasts; they appear vesiculated and occur also in abnormal shapes. These changes are accompanied by destruction or elimination of chlorophyll (E. W. Kitajima, 1975, personal communication).

The yellow or golden mosaic symptoms tend to become less apparent as the foliage ages. In some diseases, yellow areas may turn reddish late in the season, as in cotton common mosaic (22). *Malva parviflora* infected with the disease agent from *Abutilon thompsonii* shows strong mosaic initially and witches broom as a late symptom (48). Another case of initial symptoms of mosaic followed by witches broom of the growing tips was described for a whitefly-transmitted disease of *Acer* (119). Nene (87) reported seed coat mottling of some mung bean cultivars infected with yellow mosaic.

Necrotic symptoms have rarely been reported in association with whitefly-transmitted diseases (25, 32, 87, 110). In the leaf-crumpling disease of bean (25) necrotic spotting is present and responsible for the abnormal shape of leaves on infected plants. Necrosis of the growing point follows systemic invasion of *Nicotiana glutinosa* infected with the *Euphorbia* mosaic disease agent (32). Necrosis might also occur at the vector feeding sites. In India a type of chlorotic ringspot disease transmitted to jasmine by *Bemisia tabaci* was described (126).

Diseases of the Leaf Curl Type

These are not so frequent as the mosaic types. They are characterized by strong curling of leaves without typical mosaic symptoms, thickening of veins, and sometimes presence of leaf enations. Tobacco leaf curl in India and South Africa (80, 92), and cotton leaf curl in Africa (61, 120), exemplify the group. Other types of leaf curl transmitted by whiteflies are usually associated with yellow areas or mosaic symptoms. Bean dwarf mosaic (25) might resemble a leaf curl disease, being similar to bean infection with the beet curly top virus in the western United States (131). However, initial symptoms are of the yellow spot type, and if stunting is not too strong, new growth made by the infected plant has short internodes, small leaves, and a mosaic pattern.

Cotton leaf curl of Africa, described for Egyptian cotton (*Gossypium barbadense*), was reported as inducing yellow mosaic symptoms when transmitted to hirsutum varieties (61). Later work (56) indicated that possibly mixed infections with two whitefly-transmitted diseases were present; one of them may have been related to *Abutilon* mosaic (32).

Diseases of the Yellowing Type

Beet pseudo-yellows (45) is one of the diseases transmitted by whiteflies that is characterized by yellowing symptoms. These symptoms were of the same yellowing type on a number of different species to which the beet disease was transmitted by *Trialeurodes vaporariorum*. Since the disease may be confused with yellowing diseases induced by aphid-transmitted viruses, or with nutritional disorders (45), the cases associated with whitefly transmission may be more frequent and have escaped observation. Tomato plants colonized by *Bemisia tabaci* bred on *Sida* plants infected with some sources of *Abutilon* mosaic showed no symptoms soon after inoculation, but a yellowing of the mature leaves developed later (A. S. Costa, unpublished results). It has not yet been found whether this is a reaction to certain sources of the disease agent of *Abutilon* mosaic or to another whitefly-transmitted agent that was present in the possibly double-infected source plants. A disease of tomato transmitted by *B. tabaci* (26) was included in the leaf-curl diseases discussed above, because it is characterized by stunting, leaf curl, and some interveinal yellowing of the lower leaves. This disease, though rather intermediary, might fit the yellowing group better than the leaf curl group.

DISEASE AGENT-VECTOR RELATIONSHIPS

Bemisia tabaci is the most important species of whiteflies reported as a vector of plant diseases considered to be due to viruses. It is the vector of more than 25 different diseases (26, 125). Since the disease agent in most cases has been reported as circulative in the vector, a certain degree of specificity between the insect and the disease agent is to be expected. However, different species of whiteflies or even species belonging to different genera have been reported as vectors of the same disease (41, 70, 127). In one case this has not been confirmed (67) and this apparent lack of specificity may not be true for the other cases.

Adult whiteflies generally require a minimum period of about 10–15 min or longer (3, 4, 14, 21, 30, 45, 50, 61, 87, 97, 123) to acquire the disease agents from infected plants. This is probably because they are mostly acquired in the phloem tissues and the insects need a minimum of 10–30 min to reach this region of the mesophyll with their stylets that follow an intercellular course (9, 96). Twenty minutes were required for the acquisition of the cucumber vein-yellowing disease agent (52), which has a semipersistent type of relationship with the vector.

There is a definite incubation (or latent) period of the agent in the vector in most cases of whitefly-transmitted diseases. It varies from a few (30, 123, 125) up to 20–21 hr for cotton leaf crumple (67) and the agent of tomato yellow leaf curl disease (21). Because of this relationship between the whitefly vectors and the plant disease agents that they transmit, the agents are considered circulative in the insects, on the assumption that they are all viruses (3, 4, 30, 45, 47, 123, 124). In the case of the cucumber vein-yellowing disease agent, its acquisition and transmission could be completed in 60 min. This indicates that the vector has a rapid circulative mechanism or that it is stylet borne (52).

The minimum feeding period required for whitefly vectors to transmit disease agents was often shorter than that required for acquisition (123, 125). Varma (125) interpreted this to mean that acquisition depended on the vector reaching the phloem tissues, whereas infection was achieved in less time because the stylet of the vector reached susceptible tissues more rapidly than it reached the phloem. Another interpretation could be that to become infective the vector has to feed for a longer period to acquire enough of the disease agent so as to increase its chance of transmitting it when feeding on a healthy plant. During infection feeding a single or few infective particles of the disease agents placed in the same kind of leaf tissue would be enough to initiate infection.

Individual adult whiteflies have been shown to transmit cotton and tobacco leaf curl, cassava mosaic, *Euphorbia* mosaic, tomato yellow leaf curl, and practically all diseases that they are able to vector (3, 14, 21, 30, 45, 61, 91, 97, 106, 123, 129). Transmission efficiency of individual vectors varied with the disease concerned, source plant, and test plant. Golden mosaic of bean was transmitted by 9 out of 10 individual whiteflies that were tested on the bean variety Top Crop (50). One female out of 9, and 1 male out of 12, transmitted the cucumber vein-yellowing disease agent (52).

Persistence of the disease agent in the adult whitefly that acquired maximum inoculum has varied from a few days to about three weeks. In two cases it was retained for life (124, 125). Transmission efficiency of the same insect is intermittent and erratic, and there is loss of infectivity (21, 30, 50, 124). An exception to persistence in the vector was reported in the transmission of cucumber vein-yellowing disease in Israel. The relationship of its causal agent to the vector was first described as nonpersistent (89) and later as semipersistent (52). After a 24 hr acquisition period, the infective insects lost their ability to transmit the disease after feeding for 6 hr on a healthy plant; if they fasted for the same period, they retained some of their ability.

A particular type of relationship between *B. tabaci* and the disease agent was described in the transmission of tomato yellow leaf curl of Israel (18). Insects that became fully charged lost infectivity within 10–12 days. Infectivity could not be maintained or restored during this period if they fed on diseased plants again. However, vector infectivity was restored when the insects fed on diseased plants after loss of infectivity. This type of relationship has been called periodic acquisition (18). It is attributed to the periodic formation in the insect body of inhibitory factors that directly or indirectly affect the disease agent in vivo (16, 17, 77). Another factor found in the whitefly, mixed with tobacco mosaic virus in vitro, reduced lesion counts on *Datura stramonium* (19).

Except for the first instar larvae prior to feeding and the adult, all developmental stages of *B. tabaci* are sedentary; this makes it difficult to determine whether the disease agent persists through the early moultings or whether it is transmitted by different instars of the vector (52). Pupae can be transferred without too much injury, and tests with adult whiteflies from pupae formed on diseased plants, but which did not feed on diseased plants after emergence, have given positive results (21, 30, 53, 61, 106, 125), indicating retention through the last ecdysis. An exception was reported for cotton leaf crumple (67).

So far, evidence of multiplication of the disease agents transmitted by whiteflies in their vectors has been lacking. Transovarial passage of the causal agents to offspring has been negative (3, 12, 14, 21, 24, 30, 61). The fact that infectivity of the insects is generally reduced or lost after acquisition is considered as evidence against multiplication. However, it has been suggested for leafhoppers, aphids, and whiteflies (26, 34, 73–75) that loss of infectivity of insect vectors might occur even if the disease agents increase in them. Initial increase might be followed by recovery or suppression of the disease agent in the vector as the result of a defense mechanism. One of these could be the formation of anti-agent factors, as was reported for *Bemisia tabaci* in the transmission of tomato yellow leaf curl (16, 17). This seemed to be Cohen's view when he stated that synthesis of the inhibitor in the vector possibly occurred simultaneously with replication of the disease agent (17). Increase of the agent in *B. tabaci* is not mentioned in a later paper (19).

A vein-clearing disease of chili transmitted by *Bemisia inconspicua* (= *B. tabaci*) was reported in Ceylon (46). Infection was latent in several species of plants. The disease agent was present in all colonies of the vector and none could be freed of the agent. These results suggest multiplication of the disease agent in the vector and its transovarial passage. They could also be a toxicogenic effect of the insect on certain plants and not on others, as reported by other investigators (3, 30, 96).

Female *B. tabaci* have shown greater ability than males to transmit several diseases (21, 30, 91). This may indicate that most transmission tests ought to be carried out only with female whiteflies (21). This difference between comparable female and male insects was greater with some test plants than with others (29).

MECHANICAL TRANSMISSION OF WHITEFLY-VECTORED PLANT DISEASES

As a group, these diseases have been difficult to transmit mechanically and in most cases results were negative (3, 11, 12, 54, 80). *Euphorbia* mosaic was transmitted by mechanical inoculation to *E. prunifolia* and *Datura stramonium* (30); the latter may show chlorotic or necrotic local lesions after being rubbed on young leaves (32). Also, local lesions are produced on *Nicotiana glutinosa* that resemble small lesions induced by TMV (A. M. Costa, unpublished results). Other cases of mechanical transmission of disease agents vectored by whiteflies have been reported (31, 110), though in most cases it was initially rather difficult to achieve it. An exception was the transmission of cucumber vein yellowing in Israel (20).

It is easy to understand the difficulty in obtaining mechanical transmission of the leaf curl or yellowing diseases associated with whiteflies, since the causal agent may be restricted to internal phloem tissues. In mosaic diseases, where superficial parenchyma leaf tissues are affected, mechanical transmission should be feasible. Though the properties of the disease agents indicate that they are not very stable in vitro, they seem to be resistant enough to stand the stress involved in routine mechanical inoculation. Of the several factors involved, the test plant and the plant source of the infectious agents seem to be very important. After many years of negative results on the mechanical transmission of *Abutilon* mosaic, mechanical transmission was finally achieved when *Malva parviflora* was used in the tests (31). The *Euphorbia* mosaic from Brazil is easier to transmit from *Euphorbia* to *Datura stramonium* than to the same species (32). The same is true for a type of *Euphorbia* mosaic found in Puerto Rico (7).

Yellow leaf curl of tomato could not be transmitted mechanically in Israel (21) or in Brazil. A type of golden mosaic of tomato in Brazil, also transmitted by *B. tabaci,* and apparently similar to tomato yellow leaf curl of Israel, does not pass mechanically from tomato to tomato or does so only in very rare instances. The same agent, however, is easily transmitted mechanically when the inocula are obtained from and infected into *Nicotiana glutinosa* (tobacco) *Physalis* sp., and *Datura stramonium,* but not when tomato is the test or source plant (37). We have found recently that *Solanum pennellii* and derivatives of crosses between *L. esculentum* × *S. pennellii* can be infected mechanically with relative ease (A. S. Costa, unpublished results). Tomato varieties may be found that will be infected when mechanically inoculated.

Bean golden mosaic was not transmitted mechanically in Brazil (25). A similar disease present in El Salvador could be passed by routine methods of mechanical inoculation (81). Transmission was improved by increasing greenhouse temperatures. Bean golden mosaic from Puerto Rico was first reported as nontransmissible mechanically (6). Later, mechanical transmission was achieved with the help of sodium diethyldithiocarbamate (DIECA). Recently, J. Bird and collaborators (1975, personal communication) obtained 100% transmission with inocula from young infected leaves expressed in 0.1 M phosphate buffer applied to Top Crop and Diablo bean plants. Inoculum from lima beans was poor under comparable conditions. Galvez & Castaño (49) reported good mechanical transmission of the bean golden mosaic agent from Colombia and El Salvador by preparing the inocula in phosphate or hepes buffers at 0.1 M, pH 7.5, with 1% 2-mercaptoethanol.

Recent work carried out in Brazil by the author (unpublished) has still failed to give success in the mechanical transmission of bean golden mosaic. Various techniques were tried on many different species of source and test plants, pre- and postinoculation treatments, ways of extracting the inocula and obtaining them from different plant parts. The results were all negative but are not considered conclusive.

PHYSICAL PROPERTIES IN VITRO

The physical properties determined with crude sap from infected plants indicate that the whitefly-transmitted agents occur generally in low concentrations in the diseased tissues and are not resistant to heat or aging in vitro. The causal agent of *Abutilon* mosaic was still infectious after 10 minutes at 50°C and in some tests after 10 minutes at 55°C; its dilution endpoint was close to 1:625 and the longevity in vitro was 24 hr in water extracts prepared with phosphate buffer plus sodium sulfite (31). Rather similar properties were reported for the *Euphorbia* mosaic agent found in Brazil (32) and in Puerto Rico (7). In this case, activity was still present after 5 weeks in inocula obtained from infected *Datura* leaves desiccated over calcium chloride at room temperature. The beet pseudo-yellowing and cucumber vein-yellowing agents (20, 45), as well as the agent of tomato golden mosaic in Brazil (A. S. Costa, unpublished work), have properties similar to the other agents studied. The bean golden mosaic agent has a thermal death point of 55°C and a dilution endpoint of 1:10, and was still active after 48 hr at room temperature (49).

By high speed centrifugation Sharp & Wolf (108) concentrated virus-like particles that were considered the cause of tobacco leaf curl in Venezuela. The disease agent of bean golden mosaic was apparently not easily sedimented by high centrifugation (81), but this has been achieved recently (49). The tomato golden mosaic agent from Brazil could be sedimented and was still infectious after 3 cycles of high-low centrifugations; purified preparations were active after eight days at 4°C (79). The *Euphorbia* mosaic agents that occur in Brazil and Puerto Rico were concentrated by ultracentrifugation (5, 79).

NATURE OF WHITEFLY-TRANSMITTED DISEASE AGENTS

The evidence on the nature or morphology of the disease agents transmitted by whiteflies has been rather fragmentary and not very comprehensive. Sharp & Wolf (107, 108) worked on the purification of the whitefly-transmitted tobacco leaf curl agent in Venezuela and reported it as a virus with an isometric particle 39 nm in diameter. Attempts to reproduce the disease with the purified preparations were not made. Sunn (117) observed the presence of spheroidal particles 80 nm in diameter in the chloroplasts of *Abutilon striatum* infected with *Abutilon* mosaic, not present in normal leaves.

Kitajima & Costa (62) reported negative results when examining dip preparations for electron microscopy made from dried cassava leaves infected with the African cassava mosaic, transmitted by whiteflies. Negative results in finding virus-like particles in ultrathin sections of cassava leaves infected with mosaic were reported by Galvez & Kitajima, cited in Thurston (46), and by Kitajima & Costa (64). Plavsic-Banjac & Maramorosch (95) carried out light and electron microscopic examination of leaf sections of field-grown cassava plants infected with the African mosaic. They reported the presence of inclusions in parenchyma cells that consisted of aggregates of rigid rods that were hexagonal in cross section. These particles were later found to be cell ferritin (K. Maramorosch, 1973, personal communication). African cassava mosaic was attributed to mycoplasma infection on the basis of dodder transmission from a diseased clone to cassava, pepper, and other plants on which yellowing and witches broom symptoms were produced (44). Mycoplasma-like organisms (MLO) were seen in ultrathin sections of the phloem elements of dodder-infected pepper plants. As only one variety of cassava was used as a source of the mosaic agent and as mycoplasma diseases of cassava are known (35), the results could be due to double infection. A mycoplasma disease of cassava present in Minas Gerais, Brazil is claimed to have been transmitted experimentally by *Aleurothrixus manihotis* (O. Drumond, 1951, personal communication).

Another case of mycoplasma-like organisms associated with a whitefly-transmitted disease was observed by G. T. N. de Leeuw (personal communication, 1974). This investigator examined ultrathin leaf sections of variegated *Abutilon striatum* sold as ornamentals in Europe and found MLOs in some. They were not found in normal plants. This variegated ornamental plant was introduced in Europe in the last century (65) and is maintained by vegetative propagation. Thus, some of the existing clones may have become infected with MLOs that would be perpetuated as doubly infected material. A mycoplasma which causes witches broom in wild Malvaceae in the absence of mosaic symptoms, was described by Kitajima & Costa (63).

E. W. Kitajima and I. J. B. Camargo, formerly electron microscopists in our virus research group, examined leaf dips and ultrathin sections of several species of plants infected with whitefly-transmitted diseases over a period of years, but were unable to observe any type of particle consistently associated with these diseases (unpublished). Later Kitajima & Costa (64) described particles 20–25 nm in diameter in the lumen of sieve tube elements of leaves from several species of plants infected with whitefly-transmitted diseases, but found none in parenchyma cells. Failure to find any type of particle associated with bean golden mosaic from El Salvador in negatively stained leaf dip preparations and also in ultrathin sections was reported by Meiners et al (81). Their results were also negative in the examination of ultrathin sections of bean leaves infected with the whitefly-transmitted disease agent of *Euphorbia prunifolia* yellow mosaic from Costa Rica.

Evidence that some of the agents involved in whitefly-transmitted diseases are viruses has been accruing more recently. Bird et al (5, 7) studied a disease of *Euphorbia prunifolia,* transmitted by *Bemisia tabaci* in Puerto Rico, and obtained concentrated preparations that contained isometric particles 30 nm in diameter, similar in morphology to viruses of the same particle size group. The particles could also be seen in dip preparations and in ultrathin leaf sections. These results seem to confirm earlier findings of Sharp & Wolf (108) with tobacco leaf curl in Venezuela. The larger diameter, 39 nm, found by these investigators could be due to particle flattening on the collodion film (108).

Galvez & Castaño (49) reported that bean golden mosaic infectivity is associated with a nucleoprotein, present in infected plants as a dimer type of particle 32 × 19 nm, the single component being 15–20 nm in diameter. Nucleoprotein particles were found to be associated with tomato golden mosaic in Brazil; they are also dimers or siamese twin particles 25–30 × 12–15 nm, the single component being 12–15 nm in diameter. The two components of the dimers appear to be one pentagonal and the other hexagonal in profile (79). *Euphorbia* mosaic in Brazil is also associated with a twin type of particle of about the same size (79). A sweet potato disease in east Africa, transmitted by whiteflies, is associated with a rod-shaped virus [cited by Galvez & Castaño in (49)].

The experimental evidence just described indicates that some whitefly-transmitted disease agents consist of particles that are morphologically identical with those of other known plant viruses. In addition, indirect evidence indicates that in their physical properties, relationship to vectors, and in the symptoms induced on host plants, they share many of the characteristics of other plant viruses.

The green or yellow mosaic patterns associated with many whitefly-transmitted diseases are similar to the green or aucuba mosaic patterns induced on various host plants by potato Y, alfalfa mosaic, tobacco mosaic, or cucumber mosaic viruses. The yellow net type of symptom is not different from that found in some of the aphid-transmitted virus diseases. The curl type of symptom found in whitefly-transmitted diseases of tobacco (80, 92), cotton (61), and tomato (101) is similar to those induced by viruses of the beet-curly-top group. The symptoms of the beet pseudo-yellowing disease on beets and several other host plants duplicate those induced by aphid-transmissible viruses, such as beet yellows, beet western yellows, and malva yellows virus (43). On the other hand, the symptoms described for the whitefly-transmitted diseases are generally different from those recorded for diseases induced by mycoplasmas, spiroplasmas, or viroids.

The relationships between the whitefly vectors and the plant disease agents that they transmit is circulative or persistent but also nonpersistent or semipersistent (52, 89). The agent is not lost during pupation (21, 30, 53, 61, 67, 106). There is an incubation period in the vector in most cases. Thus, in their relationships to the vectors, the whitefly-transmitted disease agents behave more like viruses than viroids, for which no insect vector with these relationships are known.

With one exception (20), disease agents transmitted by whiteflies have been difficult to transmit mechanically (30–32, 110). Mechanical transmission is easier with some test plants or if certain techniques are employed (7, 30, 31; 49, 110). The physical properties of the viruses in crude sap indicate that their concentration is low in infected tissues and that they are not very stable in such preparations. In these respects they do not differ from the unstable tobacco streak and tomato spotted-wilt viruses.

The possibility that the infectious agent of whitefly-transmitted diseases could be a viroid has been considered by several investigators. The fact that several whitefly-transmitted disease agents are easily sedimented apparently would indicate that they are heavier molecular units than viroids, though sedimentation by high speed centrifugation has also been reported for potato spindle tuber and *Chrysanthemum* stunt viroids (42, 68). As a group, however, viroids are more easily transmitted mechanically and are more stable than the whitefly-transmitted agents. Other differences between the two groups of disease agents have been mentioned before.

Recent work and the bulk of evidence obtained by most investigators favor the assumption that the majority of whitefly-transmitted diseases are probably caused by viruses. It is probable that these insects might also vector mycoplasmas, spiroplasmas, or other disease agents as the leafhoppers do. A similar view has been expressed by Matthews (78) and Maramorosch (76).

RELATIONSHIP BETWEEN DISEASE AGENTS TRANSMITTED BY WHITEFLIES IN VARIOUS PARTS OF THE WORLD

Disease agents transmitted by whiteflies apparently constitute a series of related complexes distributed throughout various parts of the world. They appear to be very plastic and, to often evolve new strain complexes, which are adapted to different groups of host plants (28, 36). In Brazil there are various complexes host adapted to Compositae, Euphorbiaceae, Labiatae, Leguminosae, and Solanaceae. There is, of course, no sharp delimitation of host range by families, but a tendency for the complexes to become adapted to groups of plants and mostly to species that represent the natural vegetation of cultivated areas.

The existence of similar plastic complexes in other parts of the world is suggested by the similarity found between many diseases of crop plants and weeds transmitted by *Bemisia tabaci* or other whitefly vectors from many different areas (4, 10, 11, 25, 30, 66, 80, 92, 111, 121, 123, 127); this indicates that their causal agent probably evolved in the same general direction from natural complexes (28). This is the case in Asia, Africa (125), and Brazil (26). In Puerto Rico, disease agents transmitted by whiteflies are host adapted to Convolvulaceae, Euphorbiaceae, Leguminosae, and Malvaceae (7, 8). A possible explanation for the worldwide distribution of apparently related whitefly-transmitted disease agents could be the transport of infected plants used as ornamentals as happened with the variegated *Abutilon*.

It is not unusual in certain areas of Brazil to find representatives from different whitefly-transmitted disease agents also occurring in populations of natural weeds. *Euphorbia prunifolia, Leonurus sibiricus, Phaseolus longepedunculatus,* and species of *Sida* or other Malvaceae exhibiting rather similar bright yellow mosaic symptoms may be present in the same weed patches. The observation of these different species showing similar symptoms is misleading because it suggests that they are infected with the same disease agent. Cross inoculation tests have established that they differ, though they all have a common vector, *Bemisia tabaci* (32). Certain plant species, such as *Datura stramonium, Malva parviflora, Nicotiana glutinosa,* and others, have been shown experimentally to be susceptible to infection by components from many different complexes and represent adequate test plants for comparative studies.

EPIDEMIOLOGY OF WHITEFLY-TRANSMITTED DISEASES

Seed transmission of whitefly-transmitted plant disease has been reported in a few cases (59, 60), but these results are questionable. When enough testing was carried out with diseases of this group, it was found that the causal agent was not seed borne (11, 12, 22, 24, 25, 93, 100, 111, 112, 125).

Whitefly-transmitted diseases spread from infected sources to new plantings or to weeds solely by means of adult vectors. Older plantings (13, 28), ratoon or volunteer plants (1, 61), and mostly susceptible weeds (8, 23, 70, 97, 125) are usually the main sources from which the whitefly vectors acquire the disease agents and introduce them in the new plantings. In the case of cassava mosaic (115) and a sweet potato disease attributed to virus B (109), perpetuation occurs through cuttings and roots. Tubers from potato plants infected with *Abutilon* mosaic perpetuated the disease only in rare cases (A. S. Costa, unpublished results).

It has been shown that in many instances (23, 97) the whitefly vector is efficient in the transmission of the causal agent when acquiring it from certain natural plant reservoirs, but that it is a poor transmitter within the crop. These diseases are usually spread less efficiently in the planting and may show a border effect. Diseases caused by agents easily acquired by the vector from the infected crop plants are generally widespread in the plantings and may induce great losses. Crop plants vary greatly in their efficiency as disease agent reservoirs or in susceptibility to infection (87).

Diseases of tobacco of the leaf curl type were described from various parts of the world and are caused by different, but possibly related, whitefly-transmitted disease agents. In most cases, weeds, and in some, cultivated plants, are considered as reservoirs for the disease agent and serve as source for the vector that infect tobacco (8, 80, 92, 98, 122, 127).

Malvaceous species of the genus *Sida* are important reservoirs of the disease agent transmitted by *B. tabaci* to bean, cotton, okra, soybean, and other plantings in Brazil (23, 24). Bean golden mosaic spreads from wild species of *Phaseolus* and other legume plant sources to beans in Brazil. In Central America, *Calopogonium mucunoides* seems to be an important source of the bean golden mosaic agent (81). In Jamaica (93), lima bean and *Macroptilium lathyroides* are the alternate hosts of bean golden mosaic in the island. Species of *Rhyncosia* are host reservoirs of the agent that induces a disease of soybeans in Colombia (G. E. Galvez, 1974, personal communication). In Puerto Rico, *R. minima, Ipomoea quinquefolia,* and other weeds were found to be sources of different disease agents transmitted by *B. tabaci* to several cultivated plants (8).

Nene (87) indicates that certain weeds, *Brachiaria ramosa, Eclipta alba,* and *Xanthium strumarium* have been found to be infected naturally with mung bean yellow mosaic and might serve as reservoirs of the disease agent in northern India. Pigeon pea *(Cajanus cajan)* is also a source and may be responsible for carrying it over the winter months. Wild bhendi *(Abelmoschus manihotus), Althea rosea,* and old infected bhendi plants are the sources of yellow vein mosaic in India (11, 13).

Disease agents are more effectively transmitted by *Bemisia tabaci* when the insects feed on young growing leaves (3, 30, 115); transmission is less effective when feeding on the cotyledons (24, 30, 67). Young leaves are more susceptible to infection in mechanical inoculation tests with *Euphorbia* mosaic and tomato golden mosaic in Brazil (A. S. Costa, unpublished results). The former disease agent induces local chlorotic lesions on *Datura stramonium* and necrotic lesions on *Nicotiana glutinosa* when applied to young leaves, but not on medium or mature leaves. The same results were obtained with the tomato golden mosaic agent that induces local yellow lesions on young leaves of *N. glutinosa.* Also, systemic invasion occurred after local symptoms, but not when only older leaves were inoculated. Sastry & Nariani (103) reported that heavy applications of nitrogen increased susceptibility of tobacco plants to leaf curl infection; phosphorus reduced it.

Soybeans, poinsettias, sweet potatoes, and other cultivated plants or weeds (33, 87, 93, 96) have been reported as excellent breeding hosts of *Bemisia tabaci.* High density populations breed on these plants under field conditions, and migration of the whitefly increases the insect population on hosts of the natural vegetation. Thus, these cultivated annuals or perennials augment the change of (*a*) a greater carry-over of the insect among weeds in the period between two successive crops and (*b*) a higher incidence of disease in new plantings even for crops that are not good hosts for the insect vector. In certain areas of Brazil, high density populations of *B. tabaci* bred on soybeans increased the incidence of bean golden mosaic in the next bean crop (28). There was also an increase of cotton mosaic in late plantings.

Crop losses induced by whitefly-transmitted diseases are usually greater when spread occurs early in the plantings and negligible if it occurs late in the season. Nene (87) reported that various pulse crops in India suffer no measurable losses in yield if infection with mung bean yellow mosaic takes place 60 days after sowing. Pierre (93) compared the yield of field bean plants infected with golden mosaic at different growth stages and determined that if infection occurred within 17 days after planting, yield losses were about 57%; if after 32 days, less than 25%. Yield comparisons between healthy and diseased bean and lima bean plants infected with golden mosaic were carried out by Gamez (51) in the greenhouse. He reported an increase of the vegetative cycle for beans and yield increase for both species when infected, concluding that losses from golden mosaic in bean plantings result from plants not being ripe at harvest time. It is doubtful that these results would be reproduced in the field where plants are subject to other stress factors. Yield reduction due to bean golden mosaic was about 37.7% under field conditions in El Salvador (102). Early attacks of bhendi yellow vein mosaic may lead to total yield losses, whereas late spread will reduce it only about 25% (11).

High temperatures favor spread of tomato leaf curl in Sudan (128). Spread of bean golden mosaic is higher on the Pacific coastal plains of Central America than at higher altitudes where the temperature is lower (50, 130). Sheffield (109) also found that spread of sweet potato virus B is slight at high altitudes and attributed it to lower population density of the vector. In Jamaica Pierre (93) found that bean golden mosaic is rarely found at higher elevations and prevalence of the disease was greater during the warmer months (April to October). But, the temperature differential between the warmer and colder months in the island seems too small to be significant. Better results in the mechanical transmission of bean golden mosaic at higher temperatures were reported by Meiners et al (81).

High temperatures hasten the developmental stages of *Bemisia tabaci,* increasing its population density more quickly. Disease spread, however, might be more dependent on whitefly migration and existence of adequate disease reservoirs than on temperature. In certain areas of Brazil (São Paulo and Paraná states) spread of bean golden mosaic is greater in the dry season planting than in plantings made during the rainy season when the temperatures are higher. That has been attributed to mass migration of the vector from soybean and other summer crops to beans that are planted at the end of the season (33); likewise, spread of tomato golden mosaic is great in the fall months on unstaked tomatoes for the canning industry.

CONTROL

Control of whitefly-transmitted plant diseases has been approached from many angles, such as exclusion, avoidance of the disease by selection of planting time or site, elimination of disease agent reservoirs, breeding of resistant varieties, vector control by chemicals or repellents, and others.

Highest priority is given to preventing introduction of the disease agent in noninfested areas. Quarantine regulations should impede movement of cassava vegetative material from Africa to American countries to avoid the introduction of cassava mosaic in the western hemisphere. An additional risk would be the introduction of *Bemisia tabaci* races better adapted to breed on cassava than those existing in the Americas (38). Introduction of sweet potato parts or vegetative material from other plants that might be infected with whitefly-transmitted diseases should be avoided. Since no case of a seed-borne disease transmitted by whiteflies has been definitely proven, introduction of desirable germplasma can be made with true seeds. In the case of cassava mosaic from Africa, noninfected clones were obtained by meristem tissue culture (58) or heat treatment (15). The disease agent associated with whitefly-transmitted yellow dwarf of sweet potatoes was also eliminated from infected material by heat (55). Heat and chemotherapy eliminated the leaf curl disease agent from infected Malvaceae (85).

Choice of planting time and location, and other agronomical measures to control bean golden mosaic have been discussed elsewhere (28) and are applicable to most whitefly-transmitted diseases.

A satisfactory level of resistance to cassava mosaic in Africa was not found in *Manihot esculenta,* but is present in *M. glaziovii;* backcross derivatives and hybrids between these species produced clonal material resistant to the disease; their use on land previously cleared of diseased cassava plants, complemented by roguing, controlled the disease in East Africa (57).

Resistance to cotton leaf curl was found in some strains of Egyptian cotton (*Gossypium barbadense*). Moustafa (84) analyzed the leaves of the variety X 1730 A which was resistant to cotton leaf curl and found a higher content of total phenols, leucoanthocyanins, and flavonols than in those of the susceptible variety BAR 14/25. He suggested a possible relationship between resistance and leaf flavonol content.

Over 200 varieties of field and garden types of *Dolichos lablab* were screened
for yellow mosaic resistance by grafting in Coimbatore, India (99, 116); all varieties were susceptible, but some showed only mild symptoms. Large collections
of bean varieties have been screened for resistance to golden mosaic, but resistant types have not been found (28, 50, 93). Resistance to golden mosaic has been
found only in *Phaseolus* species of Asiatic origin that do not cross with *vulgaris*
(50).

Screening tests with 50 tomato varieties for resistance to leaf curl in Sudan
showed that they all were susceptible, but Pearl Harbor had a tendency to escape
infection; a few other varieties showed tolerance (128). A line of *Lycopersicon
pimpinellifolium* homozygous for resistance to tomato yellow leaf curl in Israel
was found and showed incomplete dominance in crosses with the susceptible
cultivar Pearson (94). Resistance of this *pimpinellifolium* line was apparently correlated with lower concentration of the disease agent in the plant.

Some whitefly-transmitted diseases can be controlled by avoiding overlapping
crops or having a crop-free period prior to the main crop. Okra yellow vein mosaic was controlled at Poona, India by enforcing a crop-free period and roguing
alternate source plants (125). Cotton crumpling disease practically disappeared
from California cotton fields when a crop-free period was enforced to control
pink boll worm (R. C. Dickson, 1973, personal communication).

Control of whitefly-transmitted diseases by the application of insecticides has
been tried with varying degrees of success. Adult whiteflies are easier to kill than
the larval forms (118). Since young, actively growing leaves are more susceptible
to infection and greater losses are generally induced when infection occurs early,
vector control should be carried out more frequently in the early phases of the
crop.

It is to be expected that vector control in the planting will be more effective
when whitefly-transmitted diseases are spread mostly within the crop and less so
when the disease agent is borne by insects that move in from outside. There is
practically no insecticide that will kill whiteflies rapidly enough to prevent inoculation of viruses or other disease agents in the plants they feed on, though
Nene (87) reported results indicating that mineral oil might kill whiteflies within
15 min and other insecticides in 30 min. Insecticides might also be effective in
controlling whitefly-transmitted diseases whose spread mostly depends on vectors that move in from the outside and have to acquire the disease agent from infected crop plants.

Nene (87, 88) tested different insecticides to control *Bemisia tabaci*. He recom-
mends two spray mixtures for the control of mung bean yellow mosaic in India.
Nene also emphasized that mineral oil sprays are potentially of great value for
controlling *B. tabaci* and reducing diseases spread by this insect. Favorable
results in the control of whitefly-transmitted leaf curl and yellow leaf curl of tomatoes with insecticides were reported from Israel (82), Sudan (128), Egypt (90),
and India (125). Control of okra yellow vein mosaic was obtained in India with
insecticides applied as sprays in the early phases of the crop or with a systemic
compound at sowing time (104, 105). The evaluation of insecticides for the con-

trol of *Bemisia tabaci* and bean golden mosaic has been carried out by several investigators (40, 71, 72, 102). In some experiments, control of the vector and disease was not translated into increased yield because the treatment had a depressing effect on the plants. Application of DDT on cotton to control jassids increased the importance of *B. tabaci* as a pest of the crop (96).

Nitzany et al (89) pointed out that endrin sprays were effective to control tomato yellow leaf curl in Israel. In this case, the agent has a persistent relationship with the vector; similar treatments failed to control a cucumber disease vectored by the same whitefly in a semipersistent relationship.

Bemisia tabaci is attracted by the blue ultraviolet and yellow wave lengths of the light spectrum (83). The use of attractive colors for traps, or repellent ones to avoid landing of the insects on plants, might be advantageous for controlling diseases transmitted by this whitefly, especially in seed or plant beds, or small plantings. A reduction of the *B. tabaci* population was recorded by Avidov (2) on tomato seedlings in mulched seed beds. He attributed this effect to temperature increase in the air layer above the mulch compared to that in nonmulched beds. Smith & Webb (114) and Costa (39) pointed out that the sawdust effect could also be attributed to repellency. A repellent effect on *B. tabaci* was noticed in straw-mulched cucumber plantings (89) and delayed the appearance of the cucumber vein-yellowing disease.

CLOSING REMARKS

The causal agent of plant diseases transmitted by whiteflies, and particularly by *B. tabaci,* is only partly determined. Most evidence indicates that they are viruses. Isometric, rod-shaped, and dimers, or siamese twin particles, have been associated with infectivity. This last type of particle is probably predominant for the whitefly-transmitted disease agents. Obviously, whiteflies are potential vectors of other plant disease agents, especially of those that occur in the phloem tissues.

In many tropical and subtropical areas of the globe, whitefly-transmitted plant diseases of the natural vegetation or crop plants are very similar in their symptomatology, manner of spread, and other characteristics; but their disease agents may differ slightly as to transmission, host range, and other properties when studied experimentally. These facts suggest that they possibly have a common origin or have evolved in a parallel manner in different parts of the world. Some of the differences encountered may be due to varieties of test plants used or variations in techniques. The effort made during the Grain Legume Disease Workshop held in Puerto Rico in June 1974 to promote standardization of techniques in the study of the whitefly-transmitted diseases is very commendable and should result in a better understanding of the problem.

Since diseases transmitted by whiteflies are apparently not seed-transmitted, the occurrence of similar agent complexes in different parts of the globe is puzzling. One possible explanation would be that their distribution took place through the movement of infected plants of ornamental value as in the case of variegated *Abutilon.* Introduction of vegetative material, whether of ornamentals, or for germplasma collections in the case of crop plants, should be curtailed, but seeds can be used when necessary.

The control of diseases transmitted by whiteflies through the application of insecticides is being investigated in many places. Though knowledge concerning chemicals effective against these vectors is desirable as an additional weapon for the grower, investigations in this area should not be carried out to the detriment of other approaches. Studies on the epidemiology of whitefly-transmitted diseases and on the biology of the vectors deserve a great deal of attention. Possibly, many effective control measures for diseases of this group may be obtained through agronomic practices that can be developed based on knowledge acquired in those areas. Biological control of the insect vector is still an untapped area of investigation.

The periodic occurrence of antifactors in *Bemisia tabaci* discovered by the Israeli workers brings out the question as to whether this is a phenomenon related to this whitefly species or is of more general occurrence among insect vectors.

The fact that the nature of a large group of plant diseases is still open to investigation should be an attractive challenge to research workers and graduate students, especially for those who work in regions where whitefly-transmitted diseases are found in everyone's garden or backyard.

ACKNOWLEDGMENTS

The author thanks Drs. J. Bird, G. E. Galvez, R. Gamez, and G. T. N. de Leeuw for supplying unpublished information used in this review, and Miss Ilka M. C. Guimarães for clerical help.

Literature Cited

1. Allen, R. M., Tucker, H., Nelson, R. A. 1960. Leaf crumple disease of cotton in Arizona. *Plant Dis. Reptr.* 44:246–50
2. Avidov, Z. 1957. Bionomics of the tobacco white fly (*Bemisia tabaci* Gennad.) in Israel. *Isr. J. Agric. Res.* 7:25–41
3. Bird, J. 1957. A whitefly-transmitted mosaic of *Jatropha gossypifolia*. *Agric. Exp. Stn. Univ. P. R. Tech. Pap.* 22:1–35
4. Bird, J. 1958. Infectious chlorosis of *Sida carpinifolia* in Puerto Rico. *Agric. Exp. Stn. Univ. P. R. Tech. Pap.* 26:1–23
5. Bird, J., Kimura, M., Monllor, A. C., Rodriguez, R. L., Sanchez, J., Maramorosch, K. 1975. Etiologia del mosaico de *Euphorbia prunifolia* Jacq. en Puerto Rico. *XXI Reunión An. Programa Coop. Centroam. Mejoramiento Cultiv. Alimenticios,* San Salvador (Abstr.)
6. Bird, J., Pérez, J. E., Alconero, R.,

Vakili, N. G., Meléndez, P. L. 1972. A whitefly-transmitted golden-yellow mosaic virus of *Phaseolus lunatus* in Puerto Rico. *J. Agric. Univ. P. R.* 56:64–74
7. Bird, J., Rodriguez, R. L., Sanchez, J., Monllor, A. C. 1975. Transmisión y hospederas del agente que causa mosaico de *Euphorbia prunifolia* Jacq. en Puerto Rico. *XXI Reunión An. Programa Coop. Centroam. Mejoramiento Cultiv. Alimenticios,* San Salvador (Abstr.)
8. Bird, J., Sanchez, J. 1971. Whitefly-transmitted viruses in Puerto Rico. *J. Agric. Univ. P. R.* 55:461–66
9. Capoor, S. P. 1949. Feeding methods of the cotton white-fly. *Curr. Sci.* 18:82–83
10. Capoor, S. P., Varma, P. M. 1948. Yellow mosaic of *Phaseolus lunatus* L. *Curr. Sci.* 17:152–53
11. Capoor, S. P., Varma, P. M. 1950. Yellow vein-mosaic of *Hibiscus esculentus* L. *Indian J. Agric. Sci.* 20:

217–30
12. Capoor, S. P., Varma, P. M. 1950. A new virus disease of *Dolichos lablab*. *Curr. Sci.* 19:248–49
13. Capoor, S. P., Varma, P. M. 1953. Bhendi mosaic and its control in Poona. Reprinted from *Indian Farming*, March issue. 3 pp.
14. Chant, S. R. 1958. Studies on the transmission of cassava mosaic virus by *Bemisia* spp. (Aleyrodidae). *Ann. Appl. Biol.* 46:210–15
15. Chant, S. R. 1959. A note on the inactivation of mosaic virus in cassava (*Manihot utilissima* Pohl) by heat treatment. *Emp. J. Exp. Agric.* 27: 55–58
16. Cohen, S. 1967. The occurrence in the body of *Bemisia tabaci* of a factor apparently related to the phenomenon of "periodic acquisition" of tomato yellow leaf curl virus. *Virology* 31: 180–83
17. Cohen, S. 1969. In vivo effects in whiteflies of a possible antiviral factor. *Virology* 37:448–54
18. Cohen, S., Harpaz, I. 1964. Periodic, rather than continual acquisition of a new tomato virus by its vector, the tobacco whitefly (*Bemisia tabaci* Gennadius). *Entomol. Exp. Appl.* 7:155–66
19. Cohen, S., Marco, S. 1970. Periodic occurrence of an anti-TMV factor in the body of whiteflies carrying tomato yellow leaf curl virus. *Virology* 40: 363–68
20. Cohen, S., Nitzany, F. E. 1960. A white-fly transmitted virus of Cucurbits in Israel. *Phytopathol. Mediterr.* 1:44–46
21. Cohen, S., Nitzany, F. E. 1966. Transmission and host range of the tomato yellow leaf curl virus. *Phytopathology* 56:1127–31
22. Costa, A. S. 1937. Nota sobre o mosaico do algodoeiro. *Rev. Agric. Piracicaba, Braz.* 12:453–70
23. Costa, A. S. 1954. Identidade entre o mosaico comum do algodoeiro e a clorose infecciosa das malváceas. *Bragantia* 13:XXIII–XXVII
24. Costa, A. S. 1955. Studies on Abutilon mosaic in Brazil. *Phytopathol. Z.* 24: 97–112
25. Costa, A. S. 1965. Three whitefly-transmitted virus diseases of beans in São Paulo, Brazil. *FAO Plant Prot. Bull.* 13:121–30
26. Costa, A. S. 1969. White-flies as virus vectors. In *Viruses, Vectors, and Vegetation*, ed. K. Maramorosch, 95–119. New York: Interscience. 666 pp.

27. Costa, A. S. 1974. Moléstias do tomateiro no Brasil transmitidas pela mosca-branca *Bemisia tabaci*. *VII Congr. An. Soc. Bras. Fitopatol.*, Brasília, DF (Abstr.)
28. Costa, A. S. 1975. Increase in the populational density of *Bemisia tabaci* a threat of widespread virus infection of legume crops in Brazil. In *Tropical Diseases of Legumes*, ed. J. Bird, K. Maramorosch, 27–50. New York, San Francisco & London: Academic. 171 pp.
29. Costa, A. S. 1975. Comparação de machos e fêmes de *Bemisia tabaci* na transmissão do mosaico dourado do feijoeiro. *VIII Congr. An. Soc. Bras. Fitopatol.*, Mossoró, Brazil (Abstr.)
30. Costa, A. S., Bennett, C. W. 1950. White-fly transmitted mosaic of *Euphorbia prunifolia*. *Phytopathology* 40: 266–83
31. Costa, A. S., Carvalho, A. M. B. 1960. Mechanical transmission and properties of the *Abutilon* mosaic virus. *Phytopathol. Z.* 37:259–72
32. Costa, A. S., Carvalho, A. M. B. 1960. Comparative studies between *Abutilon* and *Euphorbia* mosaic virus. *Phytopathol. Z.* 38:129–52
33. Costa, A. S., Costa, C. L., Sauer, H. F. G. 1973. Surto de moscabranca em culturas do Paraná e São Paulo. *An. Soc. Entomol. Bras.* 2:20–30
34. Costa, A. S., Duffus, J. E., Bardin, R. 1959. Malva yellows, an aphid-transmitted virus disease. *J. Am. Soc. Sugar Beet Technol.* 10:371–93
35. Costa, A. S., Kitajima, E. W. 1972. Studies on virus and mycoplasma diseases of the cassava plant in Brazil. *Proc. Cassava Mosaic Workshop*, 18–36. Nigeria: Int. Inst. Trop. Agric.
36. Costa, A. S., Kitajima, E. W. 1974. Evolução de vírus de plantas para adaptação a diferentes grupos de hospedeiras. *VII Congr. An. Soc. Bras. Fitopatol.*, Brasília, DF (Abstr.)
37. Costa, A. S., Oliveira, A. R., Silva, D. M. 1975. Transmissão mecânica do agente causal do mosaico dourado do tomateiro. *VIII Congr. An. Soc. Bras. Fitopatol.*, Mossoró, Brazil (Abstr.)
38. Costa, A. S., Russell, L. M. 1975. Failure of *Bemisia tabaci* to breed on cassava plants in Brazil (Homoptera-Aleyrodidae). *Ciênc. Cult. São Paulo* 27:388–90
39. Costa, C. L. 1972. *Emprego de superfícies reflectivas repelentes aos afídios vectores, no controle das moléstias de vírus de plantas*. PhD thesis, Univ. São

Paulo Piracicaba, Brazil. 94 pp.
40. Diaz, L. R. E. 1969. Evaluación de insecticidas en el control de la mosca blanca *Bemisia tabaci* (Genn.) en frijol. *XV Reunión An. Programa Coop. Centroam. Mejoramiento Cultiv. Alimenticios,* San Salvador (Abstr.)
41. Dickson, R. C., Johnson, M. McD., Laird, E. F. 1954. Leaf crumple, a virus disease of cotton. *Phytopathology* 44:479–80
42. Diener, T. O. 1974. Viroids: the smallest known agents of infectious disease. *Ann. Rev. Microbiol.* 28:23–39
43. Dinghra, K. L., Nariani, T. K. 1962. Tobacco yellow net virus. *Indian J. Microbiol.* 2:21–24
44. Dubern, J. 1972. A contribution to the study of African cassava mosaic disease. *Proc. Cassava Mosaic Workshop,* 13–17, IITA, Nigeria
45. Duffus, J. E. 1965. Beet pseudo-yellows virus, transmitted by the greenhouse whitefly (*Trialeurodes vaporariorum*). *Phytopathology* 55:450–53
46. Fernando, H. E., Peiris, J. W. L. 1957. Investigations on the Chilli leaf curl complex and its control. *Trop. Agric.* 113:305–23
47. Flores, E., Silberschmidt, K. 1958. Relations between insect and host plant in transmission experiments with infectious chlorosis of Malvaceae. *An. Acad. Bras. Cienc.* 50:535–60
48. Flores, E., Silberschmidt, K. 1967. Contribution to the problem of insect and mechanical transmission of infectious chlorosis of Malvaceae and the disease displayed by *Abutilon thompsonii*. *Phytopathol. Z.* 60:181–95
49. Galvez, G. E., Castaño, M., J. 1975. Purification of the whitefly-transmitted bean golden mosaic virus. *Turrialba.* In press
50. Gamez, R. 1971. Los virus del frijol en Centroamérica. I. Transmisión por moscas blancas (*Bemisia tabaci* Gen.) y plantas hospedantes del virus del mosaico dorado. *Turrialba* 21:21–27
51. Gamez, R. 1973. Changes in development of bean plants (*Phaseolus* spp.) associated with golden mosaic virus infection. *XIII Ann. Meet. Phytopathol. Soc.* (Caribb. Div.) (Abstr.)
52. Harpaz, I., Cohen, S. 1965. Semipersistent relationship between cucumber vein yellowing virus (CVYV) and its vector, the tobacco whitefly (*Bemisia tabaci* Gennadius). *Phytopathol. Z.* 54:240–48
53. Hildebrand, E. M. 1959. A white-fly, *Trialeurodes abutilonea*, an insect vector of sweetpotato feathery mottle in

Maryland. *Plant Dis. Reptr.* 43:712–14
54. Hildebrand, E. M. 1960. The feathery mottle virus complex of sweet potato. *Phytopathology* 50:751–56
55. Hildebrand, E. M., Brierley, P. 1960. Heat ·treatment eliminates yellow dwarf virus from sweetpotatoes. *Plant Dis. Reptr.* 44:707–9
56. Hutchinson, J. B., Knight, R. L., Pearson, E. O. 1950. Response of cotton to leaf-curl disease. *J. Gen.* 50:100–11
57. Jennings, D. L. 1972. Breeding for resistance to cassava viruses in East Africa. *Proc. Cassava Mosaic Workshop,* 40–42, IITA Nigeria
58. Kartha, K. K., Gamborg, O. L. 1975. Elimination of cassava mosaic disease by meristem culture. *Phytopathology* 65:826–28
59. Keur, J. Y. 1933. Seed transmission of the virus causing variegation of *Abutilon*. *Phytopathology* 23:20
60. Keur, J. Y. 1934. Studies of the occurrence and transmission of virus diseases in the genus *Abutilon*. *Bull. Torrey Bot. Club* 61:53–76
61. Kirkpatrick, T. W. 1931. Further studies on leaf curl of cotton in the Sudan. *Bull. Entomol. Res.* 22:323–63
62. Kitajima, E. W., Costa, A. S. 1964. Elongated particles found associated with cassava brown streak. *East Afr. Agric. For. J.* 30:28–30
63. Kitajima, E. W., Costa, A. S. 1971. Corpúsculos do tipo micoplasma asociados a diversas moléstias de plantas do grupo amarelo, no Estado de São Paulo. *Ciênc. Cult.* 23:285–91
64. Kitajima, E. W., Costa, A. S. 1974. Microscopia electrônica de tecidos foliares de plantas afetadas por vírus transmitidos por mosca-branca. *VIII Congr. An. Soc. Bras. Fitopatol.,* Brasília, DF (Abstr.)
65. Köhler, E. 1934. Viruskrankheiten. In *Handbuch der Pflanzenkrankeiten,* ed. O. Appel, 329–511. Berlin: Parey
66. Kunkel, L. O. 1930. Transmission of *Sida* mosaic by grafting. *Phytopathology* 20:129–30
67. Laird, E. F., Dickson, R. C. 1959. Insect transmission of the leaf-crumple virus of cotton. *Phytopathology* 49:324–27
68. Lawson, R. H. 1968. Some properties of chrysanthemum stunt virus. *Phytopathology* 58:885 (Abstr.)
69. Loebenstein, G., Harpaz, I. 1960. Virus diseases of sweet potatoes in Israel. *Phytopathology* 50:100–4
70. Lourens, J. H., Laan, P. A. van der, Brader, L. 1972. Contribution à l'é-

tude d'une "mosaique" du cottonier au Tchad: distribution dans un champ; *Aleurodidae* communs; essais de transmission de cottonier à cottonier par les *Aleurodidae*. *Coton. Fibres. Trop.* 27:225–30

71. Mancia, J. E. et al 1972. Insecticidas sistemicos para el control de mosca blanca (*Bemisia tabaci* Genn.), en infeccion virosa en frijol. *XVIII Reunión An. Programa Coop. Mejoramiento Cultiv. Alimenticios,* Managua (Abstr.)

72. Mancia, J. E. et al 1974. Utilizacion de insecticidas sistemicos granulados en el control de mosca blanca *Bemisia tabaci* Genn. en infeccion virosa en frijol comum. *Soc. Ing. Agron. El Salvador* 3:77–81

73. Maramorosch, K. 1955. Multiplication of plant viruses in insect vectors. *Adv. Virus Res.* 3:221–49

74. Maramorosch, K. 1964. Virus-vector relationships: vectors of circulative and propagative viruses. In *Plant Virology,* ed. M. K. Corbett, H. D. Sisler, 175–93. Gainesville: Univ. Florida Press

75. Maramorosch, K. 1966. Mechanical transmission of curly top virus to its insect vector by needle inoculation. *Virology* 1:286–300

76. Maramorosch, K. 1975. Etiology of whitefly-borne diseases. See Ref. 28, pp. 71–77

77. Marco, S., Cohen, S., Harpaz, I., Birk, Y. 1972. *In vivo* suppression of plant virus transmissibility by an anti-TMV factor occurring in an inoculative vector's body. *Virology* 47:761–66

78. Matthews, R. E. F. 1970. *Plant Virology*. New York: Academic. 778 pp.

79. Matyis, J. C., Silva, D. M., Oliveira, A. R., Costa, A. S. 1975. Purificação e morfologia do vírus do mosaico dourado do tomateiro. *Summa Phytopathol.* 1:267–74

80. McClean, A. P. D. 1940. Some leaf curl diseases in South Africa. *Sci. Bull. S. Afr. Dep. Agric.* 225:1–70

81. Meiners, J. P., Lawson, R. H., Smith, F. F., Diaz, A. J. 1975. Mechanical transmission of whitefly (*Bemisia tabaci*)-borne disease agents of beans in El Salvador. See Ref. 28, pp. 61–69

82. Melamed-Madjar, V., Cohen, S., Yunis, H. 1970. Control of tobacco whitefly. *Hassadeh* 50:1033–35

83. Mound, L. A. 1962. Studies on the olfaction and colour sensitivity of *Bemisia tabaci* (Genn.). *Entomol. Exp. Appl.* 5:99–104

84. Moustafa, E. M. 1961. Leaf curl virus

and the flavonoid content of resistant and susceptible strains of cotton. *Nature* 191:415

85. Mukherjee, A. K., Raychaudhuri, S. P. 1966. Therapeutic treatments against leaf curl of some malvaceous plants. *Plant. Dis. Reptr.* 50:88–90

86. Nair, R. R., Wilson, K. I. 1970. Leaf curl of *Jatropha curcas* L. in Kerala. *Sci. Cult.* 36:569

87. Nene, Y. L. 1972. A survey of viral diseases of pulse crops in Uttar Pradesh. *G. B. Plant Univ. Agric. Technol., Pantnagar Res. Bull.* 4:191 pp.

88. Nene, Y. L. 1973. Control of *Bemisia tabaci* Genn., a vector of several plant viruses. *Indian J. Agric. Sci.* 43:433–36

89. Nitzany, F. E., Geisenberg, H., Koch, B. 1964. Tests for the protection of cucumbers from a white fly-borne virus. *Phytopathology* 54:1059–61

90. Nour-Eldin, F., Mazyad, H. M., Hassan, M. S. 1969. Tomato leaf curl virus disease. *Agric. Res. Rev.* 47: 49–54

91. Orlando, A., Silberschmidt, K. 1946. Estudos sobre a disseminação natural do vírus da clorose infecciosa das malváceas (*Abutilon* vírus 1 Baur) e a sua relação com o inseto-vector *Bemisia tabaci* (Genn.) (Homoptera-Aleyrodidae). *Arq. Inst. Biol. São Paulo* 17:1–36

92. Pal, B. P., Tandon, R. K. 1937. Types of tobacco leaf-curl in northern India. *Indian J. Agric. Sci.* 7:363–93

93. Pierre, R. E. 1975. Observations on the golden mosaic of bean (*Phaseolus vulgaris* L.) in Jamaica. See Ref. 28, pp. 55–59

94. Pilowsky, M., Cohen, S. 1974. Inheritance of resistance to tomato yellow leaf curl virus in tomatoes. *Phytopathology* 64:632–35

95. Plavsic-Banjac, B., Maramorosch, K. 1973. Cassava mosaic virus. *Phytopathology* 63:206 (Abstr.)

96. Pollard, D. G. 1955. Feeding habits of the cotton whitefly. *Ann. Appl Biol.* 43: 664–71

97. Pruthi, H. S., Samuel, C. K. 1939. Entomological investigations on the leaf-curl disease of tobacco in northern India. III. The transmission of leaf-curl by white-fly, *Bemisia gossypiperda* to tobacco, sann-hemp and a new alternate host of the leaf-curl virus. *Indian J. Agric. Sci.* 9:223–75

98. Pruthi, H. S., Samuel, C. K. 1942. Entomological investigations on the leaf-curl disease of tobacco in northern India. V. Biology and population of the white-fly vector (*Bemisia tabaci*

Gen.) in relation to the incidence of the disease. *Indian J. Agric. Sci.* 12:35–57

99. Ramakrishnan, K. et al 1973. Investigations on virus diseases of pulse crops in Tamil Nadu. *Final Tech. Rept.* Dept. Plant Pathol., Tamil Nadu Agric. Univ., Coimbatore, India. 53 pp.

100. Rao, D. G., Varma, P. M. 1964. Studies on yellow vein mosaic of *Malvastrum coromandelianum* Garcke in India. *An. Acad. Bras. Cienc.* 36: 207–15

101. Retuerma, M. L., Pableo, G. O., Price, W. C. 1971. Preliminary study of the transmission of Phillipine tomato leaf curl virus by *Bemisia tabaci. Phillip. Phytopathol.* 7:29–34

102. Rodas, R. S. C. 1975. Evaluacion de insecticidas sistemicos para el control de mosca blanca (*Bemisia tabaci* Genn.) vector del virus del mosaico dorado. *XXI Reunión An. Programa Coop. Centroam. Mejoramiento Cultiv. Alimenticios,* San Salvador (Abstr.)

103. Sastry, K. S. M., Nariani, T. K. 1962. Effect of host plant nutrition on growth and susceptibility of tobacco plants to infection with tobacco leaf curl virus. *Indian J. Agric. Sci.* 32: 288–93

104. Sastry, K. S. M., Singh, S. J. 1973. Restriction of yellow vein mosaic virus spread in okra through control of vector, whitefly *Bemisia tabaci. Indian J. Mycol. Plant Path.* 3:76–80

105. Sastry, K. S. M., Singh, S. J. 1973. Field evaluation of insecticides for the control of whitefly (*Bemisia tabaci*) in relation to the incidence of yellow vein mosaic of okra (*Abelmoschus esculentus*). *Indian Phytopathol.* 26:129–38

106. Schuster, M. F. 1964. A whitefly-transmitted mosaic virus of *Wissadula amplissima. Plant Dis. Reptr.* 48:902–5

107. Sharp, D. G., Wolf, F. A. 1949. The virus of tobacco leaf curl. *Phytopathology* 39:225–30

108. Sharp, D. G., Wolf, F. A. 1951. The virus of tobacco leaf curl. II. *Phytopathology* 41:94–98

109. Sheffield, F. M. L. 1957. Virus disease of sweet potato in East Africa. II. Transmission to alternative hosts. *Phytopathology* 48:1–6

110. Sheffield, F. M. L. 1958. Virus diseases of sweet potato in East Africa. II. Transmission to alternative hosts. *Phytopathology* 48:1–6

111. Silberschmidt, K. 1943. Estudos sobre a transmissão experimental da "clorose infecciosa" das malváceas. *Arq. Inst. Biol. São Paulo* 14:105–56

112. Silberschmidt, K. 1948. Infectious chlorosis of *Phenax sonneratii. Phytopathology* 38:395–98

113. Silberschmidt, K., Tommasi, C. R. 1955. Observações e estudos sobre espécies de plantas suscetíveis à clorose infecciosa das malváceas. *An. Acad. Bras. Cienc.* 27:195–214

114. Smith, F. F., Webb, R. E. 1969. Repelling aphids by reflective surfaces, a new approach to the control of insect-transmitted viruses. See Ref. 26, pp. 631–39

115. Storey, H. H., Nichols, R. F. W. 1938. Studies of the mosaic diseases of cassava. *Ann. Appl. Biol.* 25:790–806

116. Subramanian, K. S., Samuel, G. S., Janarthanan, R., Kandaswamy, T. K. 1971. A note on the varietal resistance studies of *Dolichos lab-lab* L. to yellow mosaic disease. *S. Indian Hortic.* 19: 96–97

117. Sun, C. N. 1964. Das Auftreten von Viruspartikeln in *Abutilon*-Chloroplasten. *Experientia* 20:497–498

118. Suplicy, F⁰. N., Takematsu, A. P., Costa, A. S. 1974. Ensaio contra a mosca-branca *Bemisia tabaci* (Homoptera-Aleyrodidae) na cultura da soja. *XXVI Congr. An. Soc. Bras. Progr. Ciênc.,* Recife, Brazil (Abstr.)

119. Szirmai, J. 1972. An Acer virus disease in maple trees planted in avenues. *Acta Phytopathol. Acad. Sci. Hung.* 7:197–207

120. Tarr, S. A. J. 1964. Virus diseases of cotton. *Misc. Publ.* No. 18. Kew, England: Commonw. Mycol. Inst. 23 pp.

121. Thung, T. H. 1932. De krul en kroepoek-ziekten van tabak en de oorzaken van hare verbreiding. *Proefstn. Vorstenland. Tabak. Meded.* Vol. 72. 54 pp.

122. Thung, T. H. 1934. Vestrijding der krul-en kroepoek-siekten van Tabak. *Proefstn. Vorstenland. Tabak. Meded.* Vol. 78. 18 pp.

123. Varma, P. M. 1952. Studies on the relationship of the Bhendi yellowvein-mosaic virus and its vector, the whitefly (*Bemisia tabaci* Gen.). *Indian J. Agric. Sci.* 22:75–91

124. Varma, P. M. 1955. Persistence of yellow vein mosaic virus of *Abelmoschus esculentus* (L.) Moench in its vector (*Bemisia tabaci* Gen.). *Indian J. Agric. Sci.* 25:293–302

125. Varma, P. M. 1963. Transmission of plant viruses by whiteflies. *Natl. Inst. Sci. India Bull.* 24:11–33
126. Wilson, K. I. 1972. Chlorotic ring spot of jasmine. *Indian Phytopathol.* 25: 157–58
127. Wolf, F. A., Whitcomb, W. H., Mooney, W. C. 1949. Leaf-curl of tobacco in Venezuela. *J. Elisha Mitchell Sci. Soc.* 65:38–47
128. Yassin, A. M., Abu Salih, H. S. 1972. Leaf curl of tomato. *Tech. Bull. Agric. Res. Corp.,* Gezira and Udeiba Res. Stns. Sudan. 33 pp.

129. Yassin, A. M., El Nur, E. 1970. Transmission of cotton leaf curl virus by single insects of *Bemisia tabaci. Plant Dis. Reptr.* 54:528–31
130. Zaumeyer, W. J., Smith, F. F. 1966. Fourth report of the bean disease and insect survey in El Salvador. *AID Technical Assistance Agreement.* 13 pp. (mimeogr.)
131. Zaumeyer, W. J., Thomas, H. R. 1957. A monographic study of bean diseases and methods for their control. *Tech. Bull. US Dep. Agric. No. 868.* 255 pp.

Copyright © 1976 by Annual Reviews Inc.
All rights reserved

PROSPECTS FOR CONTROL OF ♦3650
PHYTOPATHOGENIC BACTERIA BY
BACTERIOPHAGES AND
BACTERIOCINS

Anne K. Vidaver[1]

Department of Plant Pathology, University of Nebraska, Lincoln, Nebraska 68583

INTRODUCTION

The last review on bacteriophages and bacteriocins of phytopathogenic bacteria was that of Okabe & Goto (98). All but one (21) of the subsequent reviews (11, 12, 60, 94, 104) have dealt primarily with viruses and bacteriocins of bacteria other than phytopathogens. The most recent microbiological compendia that include bacteriophages and bacteriocins are of limited value for phytobacteriologists because the authors generally overlook or ignore such agents (29, 47, 77, 119). However, Adam's classic book (4), the review by Bradley (11), and the bibliography of Raettig (102) are still of value to phytobacteriologists as are reviews dealing principally with methods (7, 39, 65, 66, 83, 99).

In this review, the current and future prospects for control of bacterial plant pathogens with phages and bacteriocins are evaluated. Some general comments on the phages and bacteriocins of phytopathogenic bacteria are presented, but biological, chemical, and physical characteristics are not discussed in detail. Readers interested in these areas may refer to the supplementary bibliography.[2]

TERMINOLOGY AND TAXONOMY

Nomenclature of phytopathogenic bacteria is unsettled, particularly for *Pseudomonas* and *Xanthomonas* species. In this review, I employ nomenclature useful to

---

[1] Published with the approval of the Director as paper no. 5049, Journal Series, Nebraska Agricultural Experiment Station. The work was conducted under Nebraska Agricultural Experiment Station Project No. 21–21.

[2] For supplementary bibliographic material (113 refs., 14 pgs.) order NAPS document 02863 from ASIS, c/o Microfiche Publications, 440 Park Ave. S., NY, NY 10017. Remit in advance for each NAPS accession number $1.50 for microfiche or $5.00 for each photocopy. Make checks payable to Microfiche Publications.

451

plant pathologists. Thus, the nomenspecies of Dye et al (33) are used for all *Pseudomonas* species designated as *P. syringae* by Doudoroff & Palleroni in the eighth edition of Bergey's manual (30). Likewise, I cite the nomenspecies listed in Addendum I of Dye & Lelliot (34) for all the *Xanthomonas* grouped into *X. campestris*. Retention of many of these *Xanthomonas* nomenspecies is supported by DNA homology data (89). Other genera and species are presented as in the eighth edition of Bergey's manual.

Most investigators agree on what constitutes a bacteriophage, although it is not yet clear where noninfectious particles resembling phages or phage components fit into current phage classification schemes (3). In any case, since relatively few bacteriophages of phytopathogenic bacteria have been characterized sufficiently for inclusion into current distinguishing categories (3), it is simplest here to discuss them in relation to their hosts.

What constitutes a bacteriocin is still arguable (11, 104), but in this review I use Nomura's (94) definition of bacteriocins as nonreplicating, bactericidal protein-containing substances which are produced by certain strains of bacteria and are active against some other strains of the same or closely related species. This definition includes heterogeneous substances ranging from low-molecular-weight compounds to high-molecular-weight particles resembling bacteriophage protein components. It includes complete phage-like particles and small-headed "killer particles" containing both protein and nucleic acid (2, 126) as well as low molecular weight (2×10^5) substances sensitive to proteases and nucleases (112). Ackermann & Brochu (2) summarize the grounds for considering the large phage-like particles simply as defective phages but refer to them as particulate bacteriocins. Bradley (11) classified bacteriocins into two broad types: a group of low molecular weight trypsin-sensitive, thermostable bacteriocins and a group of high molecular weight, trypsin-resistant, thermolabile bacteriocins. Although other groups clearly exist (e.g. 129) and a more satisfactory classification is desirable (94, 104), it is still premature to expand Bradley's classification (11).

Bacteriocin nomenclature is in an unfortunate state because some names are derived from the host genus, others from the host species, e.g. agrocins of *Agrobacterium* species, but syringacins of *Pseudomonas syringae*. Furthermore, names appropriate for bacteriocins have recently been used for completely unrelated compounds, e.g. corynecins for chloramphenicol analogs produced by a *Corynebacterium* species (120). Names derived from the species can be awkward, but nevertheless at least identify the producer. In many cases, it may be expedient simply to use trivial names to refer to bacteriocins produced by particular strains, such as bacteriocin 84 produced by *A. radiobacter* strain 84 (68). Classification and nomenclature will be improved when more bacteriocins are purified and characterized, and a genetic basis for their production established.

NEW AND UNUSUAL PHAGES

Since the reviews of Stolp (116) and Okabe & Goto (98), a number of virulent and temperate phages of phytopathogenic bacteria have been studied (see supplementary bibliography) with respect to biological, chemical, and physical characteristics; disease forecasting; phage typing; and epidemiology. Primary phage hosts not mentioned in the previous reviews include *C. insidiosum* (27), *C. michiganense* (37, 132), *C. nebraskense* (128), *Erwinia uredovora* (55), *E. nigrifluens* (138), *P. caryophylli* (93), *P. tomato, P. marginalis, P. viridiflava, P. savastanoi, P. pastinaceae* (8), *X. carotae, X. vesicatoria* (72), and *X. phaseoli* var. *sojensis* (95). Phages have also been reported for *E. herbicola* (58), a nonpathogen associated with the fireblight organism, *E. amylovora.*

A number of phages of plant pathogens have been reported which are of interest because of unusual morphology or novel biochemical properties. The filamentous phage Xf for *X. oryzae* (74) has a protein coat unusually high in hydrophobic amino acids (79) which may account for its lability to organic solvents and to low concentrations of sodium lauryl sulfate. In phage XP-12 of *X. oryzae,* 5-methylcytosine completely replaces cytosine (75); this replacement imparts unusual physical properties to the DNA (38). The double-stranded RNA, enveloped virus $\phi6$ of *P. phaseolicola* (113, 127), is thus far in a class by itself (3) and is the most extensively characterized phage of phytopathogenic bacteria (26, 42, 70, 71, 87a, 107–109, 113, 115, 121a–124, 127). A number of both temperate and virulent phages of *A. tumefaciens* have been studied (see supplementary bibliography), principally because of the possibility that they carry virulence determinants; this possibility is now considered highly unlikely (32, 110). Electron micrographs show typical bacteriophage particles associated with an unusual host, *Spiroplasma citri,* a helical organism without a true cell wall (25). The virus-like particles found associated with mycoplasma-like organisms of SM Stolbur agent (50), aster-yellows (5), and clover phyllody (52) may belong to a new virus class because they are morphologically unlike any other phages. Viruses of these agents are likely to be even more challenging to work with because of the difficulty of culturing their hosts.

BACTERIOCINS

Few bacteriocins of phytopathogenic bacteria have been discovered, much less characterized. Bacteriocins, or antagonisms between strains that might be due to bacteriocins, have been reported for the following: *A. radiobacter* (68, 126), *A. tumefaciens* (43, 117, 126), *C. michiganense* (35), *C. insidiosum* (90), *E. carotovora* var. *atroseptica, E. carotovora* var. *carotovora, E. chrysanthemi, Erwinia* sp. from sugar beet (W.-L. Hsiang, M. N. Schroth and S. V. Thomson, personal communication), *E. carotovora* (syn. *E. aroideae*), *E. chrysanthemi* (syn. *E. carotovora* var. *zeae*), *E. herbicola* (syn. *E. lathyri*), *E. salicis* (56), *E. uredovora* (55), *E. quercina* (59), *P. aptata, P. lachrymans* (57), *P. glycinea, P. phaseolicola* (129), *P. syringae* (82, 129), *P. morsprunorum* (49, 78, 82), and *P. solanacearum* (1, 28, 96). Specific antagonism between *P. solanacearum* strains was reported by Okabe (96), but it is not clear whether bacteriocins were involved since the inhibitory effects were considered bacteriostatic. Unpublished results are cited by Hamon et al (57) for bacteriocin production by *X. albilineans, X. betiicola* (sic), *X. juglandis, X. phaseoli,* and *X. vesicatoria.*

A number of bacteriocins of *Agrobacterium* species have been studied, but even here the full activity spectrum has not been determined (43, 68, 125, 126). Agrobacteriocin 1 (117) of *A. tumefaciens* and bacteriocin 84 of *A. radiobacter* (W. P. Roberts, personal communication) are both of low molecular weight, perhaps only oligopeptides. If so, these antibiotics may fall outside the current definition of bacteriocins. Bacteriocin 84 may act by interfering with DNA synthesis, since incorporation of thymidine, but not uridine, stopped soon after bacteriocin addition to a sensitive culture (W. P. Roberts, personal communication). Nothing is known of the mode of action of any other bacteriocins of phytopathogenic bacteria.

The majority of the bacteriocins for phytopathogenic bacteria are heat-sensitive or trypsin-resistant or both, suggesting they may be high-molecular weight (11). Such bacteriocins are produced by strains of *C. michiganense* (35), *E. carotovora, E. chrysanthemi, E. herbicola, E. salicis* (56), *P. aptata, P. lachrymans* (57), *P. glycinea, P. phaseolicola, P. syringae* (129), and *P. solanacearum* (28). This group includes both inducible (28, 129) and noninducible bacteriocins (35). Other bacteriocins produced by strains of *C. insidiosum* (90), *C. michiganense* (35), *P. glycinea, P. phaseolicola,* and *P. syringae* (129) apparently belong to the second major group of bacteriocins, those of low molecular weight (11). Additional groups, sensitive to both heat and proteolytic enzymes, resistant to both heat and enzymes, or showing differential sensitivity to trypsin and pronase have also been reported for *P. glycinea, P. phaseolicola,* or *P. syringae* (129).

Particulate contractile bacteriocins have been observed for *P. syringae* 4-A (53) and *P. syringae* W-1 (J. R. Imler and A. K. Vidaver, unpublished results), while killer particles have been described for strains 396 and 0362 of *A. tumefaciens* and strain 8149 of *A. radiobacter* (126). Other high molecular weight bacteriocins remain to be characterized.

Phytopathogenic bacteria thus produce a diverse array of bacteriocins. A wealth of material awaits further investigation.

DESIRABLE ATTRIBUTES OF SELECTIVE CONTROL AGENTS

There are risks in using broad-spectrum chemicals and antibiotics to control plant pathogens. Several commonly used fungicides are mutagenic in both prokaryotic and eukaryotic cells (see 13); all mutagens also have some carcinogenic activity. Streptomycin-resistant mutants of *E. amylovora* have arisen in nature as the result of streptomycin application (see 111). Antibiotic-resistant mutants of plant pathogens are undesirable enough, but agricultural use of antibiotics with medical applications might result in the selection or induction of antibiotic-resistant animal or human pathogens. There are already examples of fertility (F) factor and resistance (R) factor transfers between potential human pathogens and plant pathogens. The R factors are the prototypical drug-resistance vectors and F' factors can transfer a variety of genes between bacterial hosts. Thus F' *lac* has been transferred from *Escherichia coli* to *Erwinia* spp. (18) and different R factors can be transferred from *E. coli* and *Shigella flexneri* to *Erwinia* spp. (19) and from *P. aeruginosa* to a number of plant pathogenic *Pseudomonas* spp. (76, 100). The destruction of beneficial species of bacteria must also be considered.

Clearly there is a need for more selective, less persistent, and more environmentally acceptable agents for control of plant pathogens than very broad-spectrum chemicals and antibiotics. Useful bacteriophages and bacteriocins ought to have most of the attributes considered desirable for microbial control of insects (16), namely reasonable persistence, safety, aesthetic acceptability, ability to control to subeconomic concentrations, predictable control, ease of production, low cost, ease of storage, and ease of application. Bacteriophages and bacteriocins ought to have a reasonably broad killing spectrum; specificity against only one or a few strains of a species is useless for control. In our laboratory we generally limit extensive investigation to bacteriocins which kill a minimum of 40% of related test strains. Ideally, bacteriocins either should be inducible or be produced by derepressed mutants to obtain reasonable yields. A highly undesirable property is the capacity to act as a genetic vector; bacteriophages, which can spread rapidly, could have this property.

BACTERIOPHAGES AS CONTROL AGENTS

Numerous attempts have been made to control human and plant diseases with bacteriophages (98, 103, 116). At present, medical uses are rare, but phages in tablet form are claimed to be effective in the therapy and prophylaxsis of bacillary dysentery (84). There has been more current interest in plant disease control. However, neither recent (10, 20, 22, 24, 73, 118) nor previous investigators (see 98, 116) used purified phage preparations. While it is likely that phages were, indeed, the active agents in the crude lysates, the observed effects might have been due to medium components or metabolic products. When bacteriophages were applied first, all investigators to date have found marked protection from disease symptoms. The phages were inoculated into stems (118), applied as a spray (20, 22, 24) or in a dip (10, 73). In all cases the phage:bacterium ratios were 10:1 or higher. However, application of phage simultaneously with the pathogen or after inoculation of the plant with the pathogen reduced or eliminated the protective effect, for unknown reasons. Bacteriocins, as shown below, can also protect against bacterial infection when added first, but their effectiveness decreases when added simultaneously with, or after the bacteria.

Despite some potential for success as prophylactic control agents, bacteriophages are not recommended as control agents for the following reasons [other considerations have been previously mentioned (98)]. 1. Bacteriophages can mutate from virulent to temperate (69); lysogeny of susceptible cells would make them immune to further attack by the virulent parent phage or related phage. The host can also mutate to resistance. 2. Bacteriophages may transduce various characters from one host to another, possibly including virulence for the plant host. In this regard Okabe & Goto's report (97) of transduction of virulence in *P. solanacearum* needs to be substantiated. The conjugational transfer of virulence to avirulent strains in *Erwinia amylovora* (101) suggests that phages might transduce this character. 3. Bacteriophages can carry genes for toxins resulting in lysogenic conversion of nonpathogens or change pathogen virulence (e.g. 40, 41,

61). Probable lysogenic conversion accompanying loss of virulence has been reported for *X. citri* (136). These mechanisms, (2 or 3) may account in part for the emergence of new pathogens. 4. Bacteriophage action is strongly influenced by environmental conditions such as temperature, which may result in an altered host range (130) or curtail infection (22, 23). 5. Transducing phages can introduce active prokaryotic genes into plant and animal cells (for reviews see 9, 17, 87). While this scenario might seem unlikely because of dominance and the multigenic control of eukaryotic phenotypic characteristics, Zhdanov & Tikchonenko's review (139) and Erskine's results (44) suggest that such a possibility for plant pathogens is not farfetched. Erskine (44) found a complete correlation between toxicity of a phage to rabbits and the virulence of the bacterial host, *E. amylovora*, in which the phage was propagated. The same phage propagated in virulent or weakly virulent strains of *E. amylovora*, or a saprophyte was respectively toxic, less toxic, or nontoxic. Killed whole cells or four types of artificially lysed cells of noninfected bacteria showed little or no toxicity. Confirmation and extension of this study will be of great interest.

BACTERIOCINS AS CONTROL AGENTS

Control of Pathogens by Bacteriocin Producer Strains

Dramatic decreases in disease (up to 99% control) have been obtained by inoculating seeds or roots with nonpathogenic bacteriocin-producing *Agrobacterium radiobacter*. This procedure controlled crown gall of peaches in Australia (62, 67, 92). Comparable results were obtained with inoculation of cherries in the northwestern USA (88). Preliminary results indicate that bacteriocin-producing, nonpathogenic *C. michiganense* can control tomato canker (36). Control in these cases was achieved by prior inoculation with the nonpathogen. If plants were inoculated with pathogens and the control agent was used simultaneously, most disease symptoms were prevented. Control decreased if the pathogen was inoculated first; for example, if strain 84 (a producer of bacteriocin 84) was added to wounds 2 hr after inoculation with pathogenic *A. tumefaciens* 27, gall formation occurred in tomato seedlings. Galls increased in size as application of strain 84 was delayed. Detectable galls were absent if strain 84 was added first (62).

In control of crown gall by *A. radiobacter* strain 84, there is a complete correlation between bacteriocin sensitivity of pathogenic strains of *A. tumefaciens* to bacteriocin 84 in vitro and the biological control in plants, except for an *A. tumefaciens* strain which produces a bacteriocin active against *A. radiobacter* strain 84 (68). Sensitivity to bacteriocin 84 is coded by a plasmid gene in the sensitive, pathogenic strains (43, 133). These sensitive strains harbor a large plasmid (137). However, not all pathogenic plasmid-containing *A. tumefaciens* strains are sensitive to bacteriocin 84 (43).

Nevertheless, it is not certain that control is explained by bacteriocin production. Attempts to detect production of bacteriocin 84 in inoculated plants have so far been unsuccessful (A. Kerr, personal communication). Even if bacteriocin production occurs in the plant, it may not be in amounts that could account for control (but see below). For example, only small amounts of bacteriocin were produced by uninduced *P. syringae* strains in vitro and also in bean leaves inoculated with two different producers (A. K. Vidaver and M. Thomas, unpublished results). Since bacteriocin 84 has not been purified and characterized, it is not known whether the bacteriocin alone would be effective in control of *A. tumefaciens*. A single application of concentrated crude culture fluid sterilized with chloroform is partially effective in control but not as effective as an application of the living strain 84 (W. P. Roberts, personal communication).

The mechanism of control of *A. tumefaciens* strains may be more complex than by simple killing by bacteriocin; pathogenic bacteria that become resistant to bacteriocin 84 simultaneously may lose pathogenicity (68) and the large plasmid or a part of the plasmid that carries the genes for pathogenicity (110). This linkage between pathogenicity and sensitivity to bacteriocin might explain the effectiveness and duration of biological control.

The possibility that production of bacteriocin may not always be sufficient for control of crown gall is suggested by the discovery that *A. radiobacter* strain S1005, which produces bacteriocin S1005, does not control pathogenic *A. tumefaciens* B6 in the plant, although B6 is sensitive to bacteriocin S1005 in vitro (A. Kerr, personal communication). Unlike the case with strain 84, S1005 bacteriocin-resistant colonies are as virulent as the parent strain and also retain the large plasmid (43). Unlike the producer strain 84 (J. Schell, personal communication) the producer strain S1005 does not harbor any detectable plasmid (137).

Obvious questions are whether strain S1005 or other bacteriocin-producing strains that are ineffective in biological control produce bacteriocins in the plant and if so, are these bacteriocins inhibited by plant materials? In vitro studies show that the medium is critical for bacteriocin activity or production (A. Kerr, personal communication). Another possible explanation for ineffective biological control strains is that a close genetic linkage between sensitivity and virulence may not exist for these bacteriocin sensitive strains (125), A. Kerr (personal communication). For other strains, the pathogen may escape control by mutating to resistance.

The suppression of pathogenic *Erwinia amylovora* by nonpathogenic *E. herbicola* (45, 51, 85, 105, 135) and of pathogenic *Pseudomonas syringae* by an otherwise indistinguishable saprophyte (31), and interference between plant mycoplasma-like organisms (e.g. 48, 81) suggest analogies with the crown-gall phenomenon. Further studies should prove of great interest.

Control of Pathogens by Bacteriocin Preparations

Presently only syringacins 4-A (53) and W-1 (J. R. Imler and A. K. Vidaver, unpublished results) have been purified and tested in preliminary fashion against *P. phaseolicola* and *P. glycinea*. Both are inducible, particulate bacteriocins.

When purified syringacin 4-A was sprayed on bean leaves prior to spray inoculation with *P. phaseolicola* strain BE under pressure (water-soaking), as little as 3 ng per leaf reduced the lesion count from 250 (average) to zero. The lesion count was at least 50 if bacteriocin was applied 5 min after pathogen inoculation

(A. K. Vidaver and M. Thomas, unpublished results). Similar prophylactic effectiveness has been reported with crude bacteriocin preparations from *C. michiganense* (36).

Seed treatment is a promising use for bacteriocins because of pathogen dissemination via seed (see 6). Bacteriocin 4-A can protect soybean seed against challenge with *P. glycinea* (10^7 CFU/ml); without bacteriocin protection, germination was reduced about 20%. Cost of materials for such treatment with syringacin 4-A, based on 200,000 seeds per acre is about 0.49 US dollars per hectare ($0.20 per acre). This figure would be reduced about tenfold if the mitomycin C induction step of bacteriocin preparation could be eliminated. We also observed an unexplained increase in germination of poor quality seed (from 68 to 77%, $P = 0.01$) after a ten-minute dip into purified syringacin 4-A. Germination also increased in seed drenched with fungicide (Captan®); the bacteriocin effects were then masked (A. K. Vidaver and R. Carlson, unpublished results).

Little research has been done on the use of bacteriocins in mammalian systems. However, bacteriocin treatment for antibiotic-resistant *P. aeruginosa* [an occasional plant pathogen, (30)] has been effective in mice (54, 86, 114).

CAVEATS

Since I advocate more research on bacteriocins, it is perhaps appropriate to offer some cautionary advice.

Artifacts

Detection of bacteriocins is still an art and an imperfect one. Specific inhibition is not always due to a bacteriocin; for example, the specific inhibition of *Neisseria gonorrhoeae* by strains of the same species was due to production of inhibitory free fatty acids and lysophosphatidylethanolamine (134). Ammonia production by some strains of *P. pseudomallei* (106) and peroxide production by strains of group A streptococci (80) also caused specific inhibition which at first was interpreted as bacteriocin production.

Bacteriocin Inhibitors

A number of substances have been reported to inhibit bacteriocin production or destroy active bacteriocin (14, 46, 63, 90, 129, 131). The inhibitors may be active against bacteriocin produced by the same strain, as in *C. insidiosum* (90), or a different strain, as in *P. syringae* (129). Such inhibitors may be suppressed by altering the medium (14, 131) or destroyed, for example, by heat treatment which can differentially inactivate the inhibitor but not the bacteriocin (90). Purification of the bacteriocin can also remove inhibitors (15).

Avoid Use of Crude Bacteriocin Preparations

Conventional purification procedures generally will assure that the observed effects arise from bacteriocins and not from contaminants. Purification can also remove inhibitors and prevent introduction (into seeds or plants) of products potentially toxic to beneficial microorganisms or plants. In addition, purified bacteriocins will not provide an unwanted growth medium for the target or other microorganism. Purification should also allow detection of nucleic acid, if any. The objections to use of phage also apply to bacteriocins or killer particles containing nucleic acid.

PERSPECTIVES AND RECOMMENDATIONS

Much more needs to be learned about bacteriophages and bacteriocins of phytopathogenic bacteria. These agents are not only of intrinsic interest, but also potentially useful for genetic studies of pathogenicity. While phages are not recommended for use as control agents, both phages and bacteriocins are of value in disease forecasting, identification of strains (typing), and taxonomic studies (see supplementary bibliography).

Research should be expanded on bacteriocins because they possess most of the desirable attributes of control agents, particularly specificity of action against the target organism.

The greatest potential use of bacteriocins at present seems to be prophylactic treatment for seed- or tuber-borne diseases, prevention of secondary spread from infected plants, and protection of high value crops, for example, apple and pear blossoms against *E. amylovora*. In the latter case, a thorough understanding of epidemiology (see 111) should enable careful and timely application. The bacteriocins which have been tested so far show little promise for therapeutic use or prevention of secondary spread of systemic diseases or use against organisms which survive well in soil, for example, *P. solanacearum*.

If bacteriocin producers are used for biological control of pathogens and such control is attributed to bacteriocin production, it should be shown that bacteriocins are the active agent. Both the producer and target organisms should be tested for genetic transfer: in vitro between producer and target organisms; and in plants, from producer to the plant and from target bacterium to producer. This recommendation is made to avoid the introduction of deleterious genes into either the bacteriocin producer strain or plants. Nonvirulent forms of pathogens should be tested for reversion rates to virulence.

The emergence of bacteriocin-resistant mutants should not mitigate against use of bacteriocins. Such mutants should be tested for pathogenicity and altered bacteriocin sensitivity because resistance to bacteriocin may be linked to the loss of pathogenicity (68), or the resistant mutants may gain susceptibility to different bacteriocins, or both. For example, 57 out of 59 syringacin 4-A-resistant mutants arising from three different susceptible species of fluorescent pseudomonads developed sensitivity to the complementary bacteriocin W-1 (A. K. Vidaver and S. Buckner, unpublished results). The potential usefulness of a bacteriocin is retained or increased if either or both of these mechanisms operate. In addition, it would be interesting to know whether bacteriocin production or sensitivity was due to a plasmid or a chromosomal determinant.

Of prime importance is the recommendation that at least two and preferably three serologically unrelated bacteriocins be used simultaneously. No matter how effective a single bacteriocin may be, it should not be used alone because the target organism is likely to develop resistance to it. The use of even two bacteriocins with overlapping activity spectra would decrease the probability of mutation to resistance to a frequency of 10^{-16} per generation, assuming spontaneous mutation frequencies of 10^{-8} per cell generation.

The use of protectants against ultraviolet light or other inactivating agents and microencapsulation (91, 121) should increase the effective lifetime of bacteriocins. Such treatment would help overcome the major disadvantage of bacteriocins, their susceptibility to destruction by many physical and chemical agents in the environment. By analogy with a number of antibiotics, the protein nature of bacteriocins also places some restriction on their manufacture, storage, and application.

Prospective bacteriocins must not only be effective; they must be economically competitive with less selective bactericides. Bacteriocins are likely to be economical to produce and use because current production techniques for traditional antibiotics can be employed, including genetic manipulation for product improvement. In the one instance mentioned above, the estimated material cost was about 0.49 US dollars per hectare ($0.20 per acre).

Finally, the role of patents must be considered as it affects the development and use of bacteriocins. In this connection, a current review is instructive (64).

ACKNOWLEDGMENTS

I thank my colleagues who provided me with information and unpublished manuscripts. I also thank S. Buckner for bibliographic assistance. Work from this laboratory has been supported by grants from the United States Department of Agriculture and the National Institutes of Health.

Literature Cited

1. Abo-El-Dahab, M. K., El-Goorani, M. A. 1969. Antagonism among strains of *Pseudomonas solanacearum.* *Phytopathology* 59:1005–7
2. Ackermann, H.-W., Brochu, G. 1973. Particulate bacteriocins. In *Handbook of Microbiology,* Vol. I, *Organismic Microbiology,* ed. A. I. Laskin, H. A. Lechevalier, 629–32. Cleveland: CRC. 924 pp.
3. Ackermann, H.-W., Eisenstark, A. 1974. The present state of phage taxonomy. *Intervirology* 3:201–19
4. Adams, M. 1959. *Bacteriophages.* New York: Interscience. 592 pp.
5. Allen, T. C. 1972. Bacilliform particles within asters infected with a western strain of aster yellows. *Virology* 47: 491–93
6. Baker, K. F. 1972. Seed pathology. In *Seed Biology,* ed. T. T. Kozlowski, II:318–416. New York: Academic. 447 pp.
7. Billing, E. 1969. Isolation, growth and preservation of bacteriophages. In *Methods in Microbiology,* ed. J. N. Norris, D. W. Ribbons, 3B:315–29. New York: Academic. 369 pp.
8. Billing, E. 1970. Further studies on the phage sensitivity and the determina-

tion of phytopathogenic *Pseudomonas* spp. *J. Appl. Bacteriol.* 33:478–91

9. Bottino, P. J. 1975. The potential of genetic manipulation in plant cell cultures for plant breeding. *Radiat. Bot.* 15:1–16

10. Boyd, R. J., Hildebrandt, A. C., Allen, O. N. 1971. Retardation of crown gall enlargement after bacteriophage treatment. *Plant Dis. Reptr.* 55:145–48 ˙

11. Bradley, D. E. 1967. Ultrastructure of bacteriophages and bacteriocins. *Bacteriol. Rev.* 31:230–314

12. Brandis, H., Smarda, J. 1971. *Bacteriocine und Bacteriocinaehnliche Substanzen* (*Infektionskrankheiten und ihre Erreger*, Bd. 11). Jena, E. Germany: Fischer. 407 pp.

13. Bridges, B. A. 1975. The mutagenicity of captans and related fungicides. *Mutat. Res.* 32:3–34

14. Brubaker, R. R., Surgalla, M. J. 1961. Pesticins. I. Pesticin bacterium interrelationships, and environmental factors influencing activity. *J. Bacteriol.* 82:940–49

15. Brubaker, R. R., Surgalla, M. J. 1962. Pesticins. II. Production of pesticin I and II. *J. Bacteriol.* 84:539–45

16. Burges, H. D., Hussey, N. W. 1971. *Microbial Control of Insects and Mites.* New York: Academic. 861 pp.

17. Chaleff, R. S., Carlson, P. S. 1974. Somatic cell genetics of higher plants. *Ann. Rev. Genet.* 8:267–78

18. Chatterjee, A. K., Starr, M. P. 1972. Genetic transfer of episomic elements among *Erwinia* species and other enterobacteria: F'lac⁺. *J. Bacteriol.* 111:169–76

19. Chatterjee, A. K., Starr, M. P. 1972. Transfer among *Erwinia* spp. and other enterobacteria of antibiotic resistance carried on R factors. *J. Bacteriol.* 112:576–84

20. Civerolo, E. L. 1970. Comparative relationships between two *Xanthomonas pruni* bacteriophages and their bacterial host. *Phytopathology* 60:1385–88

21. Civerolo, E. L. 1972. Interaction between bacteria and bacteriophages on plant surfaces and in plant tissues. *Proc. Third Int. Conf. Plant Pathog. Bact.*, Wageningen, April 14–21, 1971, 25–37. Wageningen: PUDOC

22. Civerolo, E. L. 1973. Relationships of *Xanthomonas pruni* bacteriophages to bacterial spot disease in *Prunus*. *Phytopathology* 63:1279–84

23. Civerolo, E. L. 1974. Temperature effects on the relationships between *Xanthomonas pruni* and its virulent phages. *Phytopathology* 64:1248–55

24. Civerolo, E. L., Keil, H. L. 1969. Inhibition of bacterial spot of peach foliage by *Xanthomonas pruni* bacteriophage. *Phytopathology* 59:1966–67

25. Cole, R. M., Tully, J. G., Popkin, T. J., Bove, J. M. 1973. Morphology, ultrastructure, and bacteriophage infection of the helical mycoplasma-like organism (*Spiroplasma citri* gen. nov., sp. nov.) cultured from "stubborn" disease of citrus. *J. Bacteriol.* 115:367–86

26. Coplin, D. L., Van Etten, J. L., Koski, R. K., Vidaver, A. K. 1975. Intermediates in the biosynthesis of double-stranded ribonucleic acids of bacteriophage φ6. *Proc. Natl. Acad. Sci. USA* 72:849–53

27. Cook, F. D., Katznelson, H. 1960. Isolation of bacteriophages for the detection of *Corynebacterium insidiosum*, agent of bacterial wilt of alfalfa. *Can. J. Microbiol.* 6:121–25

28. Cuppels, D., Hanson, R., Kelman, A. 1975. Production of bacteriocin-like compounds by *Pseudomonas solanacearum*. *Abstr. Ann. Meet. Am. Phytopathol. Soc.*, Aug. 10–14. No. 155

29. Daltog, A. J., Haguenau, F. *Ultrastructure of Animal Viruses and Bacteriophages: An Atlas.* 1973. New York: Academic. 413 pp.

30. Doudoroff, M., Palleroni, N. J. 1974. Genus *Pseudomonas*. In *Bergey's Manual of Determinative Bacteriology*, ed. R. E. Buchanan, N. E. Gibbons, 217–243. Baltimore: Williams & Wilkins. 1268 pp. 8th ed.

31. Dowler, W. M. 1972. Inhibition of *Pseudomonas syringae* by saprophytic bacterial isolates in culture and in infected plant tissues. See Ref. 21, pp. 307–11

32. Drlica, K. A., Kado, C. I. 1975. Crown gall tumors: are bacterial nucleic acids involved? *Bacteriol. Rev.* 39:186–96

33. Dye, D. W. et al 1975. Proposals for a reappraisal of the status of the names of plant pathogenic *Pseudomonas* species. *Int. J. Syst. Bacteriol.* 25:252–57

34. Dye, D. W., Lelliott, R. A. 1974. Genus *Xanthomonas*. See Ref. 30, pp. 243–49

35. Echandi, E. 1976. Bacteriocin production by *Corynebacterium michiganese*. *Phytopathology* 66:430–32

36. Echandi, E. 1975. Biological control

of bacterial canker of tomato with bacteriocins from *Corynebacterium michiganense.* See Ref. 28, No. 154

37. Echandi, E., Sun, M. 1973. Isolation and characterization of a bacteriophage for the identification of *Corynebacterium michiganese. Phytopathology* 63:1398–1401

38. Ehrlich, M., Ehrlich, K., Mayo, J. A. 1975. Unusual properties of the DNA from *Xanthomonas* phage XP-12 in which 5-methyl-cytosine completely replaces cytosine. *Biochem. Biophys. Acta* 395:109–19

39. Eisenstark, A. 1967. Bacteriophage techniques. In *Methods in Virology,* ed. K. Maramorosch, H. Koprowski, 1:450–524. New York: Academic. 640 pp.

40. Eklund, M. W., Poysky, F. T. 1974. Interconversion of type C and D strains of *Clostridium botulinum* by specific bacteriophages. *Appl. Microbiol.* 27:251–58

41. Eklund, M. W., Poysky, F. T., Meyers, J. A., Pelroy, G. A. 1974. Interspecies conversion of *Clostridium botulinum* type C to *Clostridium novyi* type A by bacteriophage. *Science* 186:456–58

42. Ellis, L. F., Schlegel, R. A. 1974. Electron microscopy of *Pseudomonas* φ6 bacteriophage. *J. Virol.* 14:1547–51

43. Engler, G. et al 1975. Agrocin-84 sensitivity: A plasmid determined property in *Agrobacterium tumefaciens. Mol. Gen. Genet.* 138:345–50

44. Erskine, J. M. 1973. Association of virulence characteristics of *Erwinia amylovora* with toxigenicity of its phage lysates to rabbit. *Can. J. Microbiol.* 19:875–77

45. Erskine, J. M., Lopatecki, L. E. 1975. In vitro and in vivo interactions between *Erwinia amylovora* and related saprophytic bacteria. *Can. J. Microbiol.* 21:35–41

46. Foulds, J. D., Shemin, D. 1969. Concomitant synthesis of bacteriocin and bacteriocin inactivator from *Serratia marcescens. J. Bacteriol.* 99:661–66

47. Fraenkel-Conrat, H. 1974. Descriptive catalogue of viruses. *Comprehensive Virology,* ed. H. Fraenkel-Conrat, R. R. Wagner, Vol. 1. New York: Plenum. 191 pp.

48. Freitag, J. H. 1964. Interaction and mutual suppression among three strains of aster yellows virus. *Virology* 24:401–13

49. Garrett, C. M. E., Panagopoulos, C. G., Crosse, J. E. 1966. Comparison of plant pathogenic pseudomonads from fruit trees. *J. Appl. Bacteriol.* 29:342–56

50. Giannotti, J., Devauchelle, G., Vago, C. Marchoux, G. 1973. Rod-shaped virus-like particles associated with degenerating mycoplasma in plant and insect vector. *Ann. Phytopathol.* 5:461–66

51. Goodman, R. N. 1965. In vitro and in vivo interactions between components of mixed bacterial cultures isolated from apple buds. *Phytopathology* 55:217–21

52. Gourett, J. P., Maillet, P. L., Gouranton, J. 1973. Virus-like particles associated with mycoplasmas of clover phyllody in the plant and in the insect vector. *J. Gen. Microbiol.* 74:241–49

53. Haag, W. L., Vidaver, A. K. 1974. Purification and characterization of syringacin 4-A, a bacteriocin from *Pseudomonas syringae* 4-A. *Antimicrob. Agents Chemother.* 6:76–83

54. Haas, H., Sacks, T., Saltz, N. 1974. Protective effect of pyocin against lethal *Pseudomonas aeruginosa* infections in mice. *J. Infect. Dis.* 129: 470–72

55. Hamon, Y., Peron, Y. 1962. Les bacteriocines, elements taxonomiques eventuels pour certains bacteries. *C. R. Acad. Sci.* 254:2868–70

56. Hamon, Y., Peron, Y. 1961. Les proprietes antagonistes reciproques parmi les *Erwinia.* Discussion de la position taxonomique de ce genre. *C. R. Acad. Sci.* 253:913–15

57. Hamon, Y., Veron, M., Peron Y. 1961. Contribution a l'etude des proprietes lysogenes et bacteriocinogenes dans le genre *Pseudomonas. Ann. Inst. Pasteur Paris* 101:738–53

58. Harrison, A., Gibbons, L. N. 1975. The isolation and characterization of a temperate phage Y46/(E2), from *Erwinia herbicola* Y46. *Can. J. Microbiol.* 21:937–44

59. Hildebrand, D. C., Schroth, M. N. 1967. A new species of *Erwinia* causing the drippy nut disease of live oaks. *Phytopathology* 57:250–53

60. Holloway, B. W., Krishnapillai, V. 1975. Bacteriophages and bacteriocins. In *Genetics and Biochemistry of Pseudomonas.* ed. P. H. Clarke, M. H. Richmond, 99–132. New York: Wiley. 366 pp.

61. Holmes, R. K., Barksdale, L. 1969. Genetic analysis of tox$^+$ and tox$^-$

bacteriophages of *Corynebacterium diphtheriae. J. Virol.* 3:586–98

62. Htay, K., Kerr, A. 1974. Biological control of crown gall: seed and root inoculation. *J. Appl. Bacteriol.* 37:525–30

63. Ikeda, M. 1970. Pyocin inactivation by cell free supernatant of *Pseudomonas aeruginosa* culture. Its quantitative assay and approaches to the established method. *Nagasaki Igakkai Zasshi* 45:621–31

64. Irons, E. S., Sears, M. H. 1975. Patents in relation to microbiology. *Ann. Rev. Microbiol.* 29:319–32

65. Kay, D. 1972. Methods for studying the infectious properties and multiplication of bacteriophage. See Ref. 7, 7A:191–262

66. Kay, D. 1972. Methods for the determination of the chemical and physical structure of bacteriophages. See Ref. 7, 7A:263–313

67. Kerr, A. 1972. Biological control of crown gall: seed inoculation. *J. Appl. Bacteriol.* 35:493–97

68. Kerr, A., Htay, K. 1974. Biological control of crown gall through bacteriocin production. *Physiol. Plant Pathol.* 4:37–44

69. Kleczkowska, J. 1969. The mutation of some virulent rhizobiophages into the temperate form. *Can. J. Microbiol.* 15:1055–59

70. Kleinschmidt, W. J., Boeck, L. D., Van Frank, R. M., Murphy, E. B. 1973. Interferon production by φ6 *Pseudomonas phaseolicola* phage and its double-stranded RNA. *Proc. Soc. Exp. Biol. Med.* 144:304–7

71. Kleinschmidt, W. J., Van Etten, J. L., Vidaver, A. K. 1974. Influence of molecular weights of bacteriophage φ6 double-stranded ribonucleic acids on interferon induction. *Infect. Immun.* 10:284–85

72. Klement, Z. 1959. Some new specific bacteriophages for plant pathogenic *Xanthomonas* spp. *Nature* 184:1248–49

73. Kuo, T.-T., Chang, L.-C. Yang, C.-M., Yang, S.-E. 1971. Bacterial leaf blight of rice plant. IV. Effect of bacteriophage on the infectivity of *Xanthomonas oryzae. Bot. Bull. Acad. Sinica* 12:1–9

74. Kuo, T.-T., Huang, T.-C., Chow, T.-Y. 1969. A filamentous bacteriophage from *Xanthomonas oryzae. Virology* 39:548–55

75. Kuo, T.-T., Huang, T.-C., Teng, M. H. 1968. 5-Methylcytosine replacing cytosine in deoxyribonucleic acid of a phage of *X. oryzae. J. Mol. Biol.* 34:373–75

76. Lacy, G. H., Leary, J. V. 1975. Transfer of antibiotic resistance plasmid RP1 into *Pseudomonas glycinea* and *Pseudomonas phaseolicola in vitro* and *in planta. J. Gen. Microbiol.* 88:49–57

77. Laskin, A. I., Lechevalier, H. A. 1973. See Ref. 2, pp. 573–650

78. Lazar, I., Crosse, J. E. 1969. Lysogeny, bacteriocinogeny, and phage types in plum isolates of *Pseudomonas morsprunorum* Wormald. *Rev. Roum. Biol. Ser. Bot.* 14:325–33

79. Lin, J.-Y., Wu, C.-C., Kue, T.-T. 1971. Amino acid analysis of the coat protein of the filamentous bacterial virus Xf from *Xanthomonas oryzae. Virology* 45:38–41

80. Malke, H., Starke, R., Jacob, H. E., Kohler, W. 1974. Bacteriocine-like activity of group-A streptococci due to the production of peroxide. *J. Med. Microbiol.* 7:367–74

81. Marchoux, G., Giannotti, J. 1971. Interferences entre deux mycoplasmoses vegetales. *Physiol. Veg.* 9:595–610

82. Matthews, P. 1965. Bacteriocin activity in *Pseudomonas morsprunorum* and *P. syringae. John Innes Inst. Ann. Rept.* 1964:35–36

83. Mayr-Harting, A., Hedges, A. J., Berkeley, R. C. W. 1972. Methods for studying bacteriocins. See Ref. 7, 7A:315–422

84. Mazacek, M., Petera, A., Mach, J. 1969. Die Bakteriophagie in der Therapie und Prophylaxe der Infektionskrankheiten. *Zentralbl. Bakteriol. Parasitenkd. Infektionskr. Hyg. Abt. 1* 211:385–94

85. McIntyre, J. L., Kuć, J., Williams, E. B. 1973. Protection of pear against fire blight by bacteria and bacterial sonicates. *Phytopathology* 63:872–77

86. Merrikin, D. J., Terry, C. S. 1972. Use of pyocin 78-C2 in the treatment of *Pseudomonas aeruginosa* infection in mice. *Appl. Microbiol.* 23:164–65

87. Merril, C. R., Stanbro, H. 1974. Intercellular gene transfer Z. *Pflanzenphysiol.* 72:371–88

87a. Mindich, L., Cohen, J., Weisburd, M. 1976. Isolation of nonsense suppressor mutants in *Pseudomonas. J. Bacteriol.* 126:177–82

88. Moore, L. W. 1975. Biological control of crown gall with an antagonistic

Agrobacterium species. See Ref. 28, No. 157

89. Murata, N., Starr, M. P. 1973. A concept of the genus *Xanthomonas* and its species in the light of segmental homology of deoxyribonucleic acids. *Phytopathol. Z.* 77:285–323

90. Nelson, G. A., Semeniuk, G. 1964. An antagonistic variant of *Corynebacterium insidiosum* and some properties of the inhibitor. *Phytopathology* 54:330–35

91. Neogi, A. N., Allan, G. G. 1974. Controlled release pesticides: concepts and realization. *Adv. Exp. Med. Biol.* 47:195–224

92. New, P. B., Kerr, A. 1972. Biological control of crown gall: field measurements and glasshouse experiments. *J. Appl. Bacteriol.* 35:279–87

93. Nishimura, J., Wakimoto, S. 1971. Ecological studies on bacterial wilt disease of carnation (*Dianthus caryophyllus*). 1. Some characteristics and mode of multiplication of *Pseudomonas caryophylli* phage. *Ann. Phytopathol. Soc. Jpn.* 37:301–6

94. Nomura, M. 1967. Colicins and related bacteriocins. *Ann. Rev. Microbiol.* 21:257–84

95. Numic, R. 1970. Bakteriofagi fitopatogenih bakterija Soje. *Sarajevo Univ. Poljopriuredno-sum Fak. Rad.* 19:187–224

96. Okabe, N. 1954. Studies on *Pseud. solanacearum.* V. Antagonism among the strains of *P. solanacearum. Rept. Fac. Agric. Shizuoka Univ.* 4:37–40

97. Okabe, N., Goto, M. 1955. Studies on *Pseudomonas solanacearum.* X. Genetic change of the bacterial strains induced by the temperate phage T-c200. *Rept. Fac. Agric. Shizuoka Univ.* 5:57–62

98. Okabe, N., Goto, M. 1963. Bacteriophages of plant pathogens. *Ann. Rev. Phytopathol.* 1:397–418

99. Ozeki, H. 1968. Methods for the study of colicins and colicinogeny. See Ref. 39, 4:565–92

100. Panopoulos, N. J., Guimaraes, W. V., Cho, J. J., Schroth, M. N. 1975. Conjugative transfer of *Pseudomonas aeruginosa* R factors to plant pathogenic *Pseudomonas* spp. *Phytopathology* 65:380–88

101. Pugashetti, B. K., Starr, M. P. 1975. Conjugational transfer of genes determining plant virulence in *Erwinia amylovora. J. Bacteriol.* 122:485–91

102. Raettig, H.-J. 1958. *Bakteriophagie.* Stuttgart: Fischer. 2 vols. 215 pp.,

344 pp.

103. Rautenshtein, Y. I. 1967. The role of bacteriophage for agricultural microbiology *Skh. Biol.* 2:499–507

104. Reeves, P. 1972. *The Bacteriocins.* New York: Springer. 142 pp.

105. Riggle, J. H., Klos, E. J. 1972. Relationship of *Erwinia herbicola* to *Erwinia amylovora. Can. J. Bot.* 50:1077–83

106. Rogul, M., Carr, S. R. 1972. Variable ammonia production among smooth and rough strains of *Pseudomonas pseudomallei:* Resemblance to bacteriocin production. *J. Bacteriol.* 112:372–80

107. Sands, J. A. 1973. The phospholipid composition of bacteriophage ϕ6. *Biochem. Biophys. Res. Commun.* 55:111–16

108. Sands, J. A., Cupp, J., Keith, A., Snipes, W. 1974. Temperature sensitivity of the assembly process of the enveloped bacteriophage ϕ6. *Biochem. Biophys. Acta* 373:277–85

108a. Sands, J. A., Lowlicht, R. A. 1976. Temporal origin of viral phospholipids of the enveloped bacteriophage ϕ6. *Can. J. Microbiol.* 22:154–58

109. Sands, J. A., Lowlicht, R. A., Cadden, S. C., Haneman, J. 1975. Assembly of the enveloped bacteriophage ϕ6 in environments which perturb the host cell membranes. *Can. J. Microbiol.* 21:1287–90

110. Schell, J. 1975. The role of plasmids in crown-gall formation by *A. tumefaciens.* In *Genetic Manipulations with Plant Materials,* ed. L. Ledoux, 163–81. New York: Plenum. 601 pp.

111. Schroth, M. N., Thomson, S. V., Hildebrand, D. C., Moller, W. J. 1974. Epidemiology and control of fire blight. *Ann. Rev. Phytopathol.* 12:389–412

112. Schwinghamer, E. A. 1975. Properties of some bacteriocins produced by *Rhizobium trifolii. J. Gen. Microbiol.* 91:403–13

113. Semancik, J. S., Vidaver, A. K., Van Etten, J. L. 1973. Characterization of a segmented double-helical RNA from bacteriophage ϕ6. *J. Mol. Biol.* 78:617–26.

114. Sezen, I. Y., Blobel, H., Scharmann, W. 1974. Possible use of pyocins against *Pseudomonas aeruginosa. Zentralbl. Bakteriol. Parasitenkd. Infektionskr. Hyg. Abt. 1 Orig. Reihe A* 229:205–8

115. Sinclair, J. F., Tzagoloff, A., Levine, D., Mindich, L. 1975. Proteins of

bacteriophage φ6. *J. Virol.* 16:685–95

116. Stolp, H. 1956. Bacteriophagenforschung und Phytopathologie (ein Sammelreferat). *Phytopathol. Z.* 26: 171–218

117. Stonier, T. 1960. *Agrobacterium tumefaciens* Conn. II. Production of an antibiotic substance. *J. Bacteriol.* 79: 889–98

118. Stonier, T., McSharry, J., Speitel, T. 1967. *Agrobacterium tumefaciens* Conn. IV. Bacteriophage PB2₁ and its inhibitory effect on tumor induction. *J. Virol.* 1:268–73

119. Tikhonenko, A. S. 1970. *Ultrastructure of Bacterial Viruses.* New York: Plenum. 294 pp.

120. Tomita, F., Nakano, H., Honda, H., Suzuki, T. 1974. Production of corynecins by chloramphenicol resistant mutants of *Corynebacterium hydrocarboclastus. Agriç. Biol. Chem.* 38: 2183–90

121. Vandegaer, J. E. 1974. *Microencapsulation: Processes and Applications.* New York: Plenum. 180 pp.

121a. Van Etten, J., Lane, L., Gonzalez, C., Partridge, J., Vidaver, A. 1976. Comparative properties of bacteriophage φ6 and φ6 nucleocapsid. *J. Virol.* 18:652–58

122. Van Etten, J. L., Vidaver, A. K., Koski, R. K., Burnett, J. P. 1974. Base composition and hybridization studies of the three double-stranded RNA segments of bacteriophage φ6. *J. Virol.* 13:1254–62

123. Van Etten, J. L., Vidaver, A. K., Koski, R. K., Semancik, J. S. 1973. RNA polymerase activity associated with bacteriophage φ6. *J. Virol.* 12: 464–71

124. Van Frank, R. M., Kleinschmidt, W. J. 1973. Concentration and purification of two small RNA viruses: mycophage PS-1 and bacteriophage φ6. *An. Biochem.* 55:601–8

125. Van Larebeke, N. et al 1975. Acquisition of tumour-inducing ability by non-oncogenic agrobacteria as a result of plasmid transfer. *Nature* 255: 742–43

126. Vervliet, G., Holsters, M., Teuchy, H., Van Montagu, M., Schell, J. 1975. Characterization of different plaqueforming and defective temperate phages in *Agrobacterium* strains. *J. Gen. Virol.* 26:33–48

127. Vidaver, A. K., Koski, R. K., Van Etten, J. L. 1973. Bacteriophage φ6: a lipid-containing virus of *Pseudomonas phaseolicola. J. Virol.* 11:799–805

128. Vidaver, A. K., Mandel, M. 1974. *Corynebacterium nebraskense,* a new orange-pigmented phytopathogenic species. *Int. J. Syst. Bacteriol.* 24: 482–85

129. Vidaver, A. K., Mathys, M. L., Thomas, M. E., Schuster, M. L. 1972. Bacteriocins of the phytopathogens *Pseudomonas syringae, P. glycinea,* and *P. phaseolicola. Can. J. Microbiol.* 18:705–13

130. Vidaver, A. K., Schuster, M. L. 1969. Characterization of *Xanthomonas phaseoli* bacteriophages. *J. Virol.* 4:300–8

131. Wahba, A. H. 1963. The production and inactivation of pyocines. *J. Hyg.* 61:431–41

132. Wakimoto, S., Vematsu, T., Mizukami, T. 1969. Bacterial canker disease of tomato in Japan. 2. Properties of bacteriophages specific for *Corynebacterium michiganense* (Smith) Jensen. *Ann. Phytopathol. Soc. Jpn.* 35:168–73

133. Watson, B., Currier, T. C., Gordon, M. P., Chilton, M-D., Nester, E. W. 1975. Plasmid required for virulence of *Agrobacterium tumefaciens. J. Bacteriol.* 123:255–64

134. Walstad, D. L., Reitz, R. C., Sparling, P. F. 1974. Growth inhibition among strains of *Neisseria gonorrhoeae* due to production of inhibitory free fatty acids and lysophosphatidylethanolamine: absence of bacteriocins. *Infect. Immun.* 10:481–88

135. Wrather, J. A., Kuc, J., Williams, E. B. 1973. Protection of apple and pear fruit tissue against fireblight with nonpathogenic bacteria. *Phytopathology* 63:1075–76

136. Wu, W. C. 1972. Phage-induced alterations of cell disposition, phage adsorption and sensitivity, and virulence in *Xanthomonas citri. Ann. Phytopathol. Soc. Jpn.* 38:333–41

137. Zaenen, I., Van Larebeke, N., Teuchy, H., Van Montagu, M., Schell, J. 1974. Supercoiled circular DNA in crowngall inducing *Agrobacterium* strains. *J. Mol. Biol.* 86:109–27

138. Zeitoun, F. M., Wilson, E. E. 1969. The relation of bacteriophage to the walnut-tree pathogens, *Erwinia nigrifluens* and *Erwinia rubrifaciens. Phytopathology* 59:756–61

139. Zhdanov, V. M., Tikhonenko, T. I. 1974. Viruses as a factor of evolution: exchange of genetic information in the biosphere. *Adv. Virus Res.* 19: 361–94

AUTHOR INDEX

Greelen, J. L. M. C., 5
Greeley, L. W., 191
Green, G., 183-85
Green, G. J., 34, 98, 104,
105, 109, 180, 183
Green, R. J. Jr., 249, 250
Greenaway, W., 408
Greene, S. W., 356
Greenhalgh, F. C., 135
Greenham, C. G., 314-16
Greenland, D. J., 64, 65
Greenwood, A. D., 146, 151,
153, 155, 156
Greenwood, D. J., 59, 66
Greenwood, R. M., 134
Gregory, K. F., 266, 292
Gregory, P. H., 33, 46
Greig-Smith, P., 121, 122
Greppin, H., 284-86
Gresshoff, P. M., 273, 284
Grierson, D., 269, 273, 284
Griffee, P. J., 421
Griffin, D. M., 59, 126, 127,
131, 136
Griffin, G. D., 328, 329
Griffin, G. J., 258
Griffiths, D. A., 145, 149,
151, 153, 156
Griffiths, E., 123
Griliches, Z., 389
Grill, D., 315
Grillo, H. V. S., 21
Grindle, M., 101, 103, 107,
417
Grisham, L. M., 416
Grobbelaar, N., 68
Grogan, R. G., 80, 250
Gross, A. E., 369
Grossenbacher, K., 124
GROTH, J. V., 177-88; 177,
183, 186, 259
Grove, P. L., 134
Grylls, N. E., 200, 202
Guckert, A., 124
Guderian, R. H., 295
Guillé, E., 287, 288
Guimaraes, W. V., 454
Gull, K., 416
Gunn, K. B., 135
Guthrie, W. D., 184
Guttenberg, H., 146, 149,
151, 153, 157
Gutter, Y., 421
Guyot, H., 18
Guzmann, J., 98, 108
Gynn, R., 367

H

Haag, W. L., 454, 457
Haas, D. L., 151
Haas, H., 458
Habgood, R. M., 184
Hadley, G., 145, 155
Hadlow, A. J., 367
Hagborg, W. A. F., 191
Hagedorn, D. J., 249, 258

Hagen, K. S., 382
Haggag, M. E. A., 184
Häggman, J., 293
Haguenau, F., 451
Hahn, F. E., 274
Haigh, J. C., 108
Haldane, J. B. S., 32
Hall, C. B., 78
Hall, D. C., 382-85
Hall, D. H., 362
Hall, I. V., 361
Hall, M. A., 67
Hall, R., 5
Hallbauer, D. K., 238
Halliwell, R. S., 146, 147,
151
Hallman, R. G., 370
Halpin, J. E., 258
Halter, A. N., 386
Ham, T. H., 165
Hamblin, C. O., 363
Hamdi, Y. A., 126
Hamilton, R. D., 62
Hamilton, R. H., 273, 275,
280
Hamilton, R. I., 191
Hamilton-Miller, J. M. T.,
417
Hammerschlag, F., 258
Hammerschlag, R. S., 408
Hamon, Y., 453, 454
Hams, A. F., 421
Hanchley, P., 146, 149, 150,
155, 311, 319
Hancock, J. G., 67, 125,
127, 133, 250, 251, 255
Haneman, J., 453
Hanounik, S. B., 343
Hansen, E., 53, 58
Hanson, E. W., 149, 150,
258
Hanson, H. C., 356
Hanson, R., 453, 454
Harder, D. E., 147, 149,
150, 155, 203
Harding, P. R., 421
HARDISON, J. R., 355-79;
364-66, 368
Hardwick, N. V., 151, 156
Hardy, R. W. F., 66
Hare, W. W., 86
Hargreaves, A. J., 421
HARLAN, J. R., 31-51; 40-
43, 45, 48, 224
Harpaz, I., 429, 431-33,
437
Harper, D. M., 248
Harper, S. H. T., 57-59
Harris, J. O., 58
Harris, J. R., 134
Harris, K. F., 82
Harris, S. A., 398
Harrison, A., 453
Harrison, B. D., 82, 86,
197
Harrison, D. J., 77, 88
Hart, K., 99, 112

Hartel, V. O., 315
Hartmann, H. D., 335
Hartzler, E., 214
Harvey, G. M., 359
Hashizume, T., 286
Hassan, M. S., 442
Hastie, A. C., 411, 412, 414
Hattori, T., 59
Hauman, L., 19
Hawkes, J. G., 216, 217
Hawkins, J. H., 90
Hawksworth, F. G., 358
Hayden, R. E., 314
Hayden, R. I., 311, 312,
314
Hayes, H. K., 223
Hayes, J. D., 184
Hayman, D. S., 135
Headley, J. C., 382, 386,
390, 391, 397
Heagle, A. S., 98, 108-11
Heald, C. M., 320
Heald, F. D., 260
Heath, I. B., 149, 150
Heath, M. C., 145, 149,
150, 153, 155, 156
Heberlein, G. T., 266, 271,
273, 274, 286, 292
Hebert, T. T., 192, 205
Hecq, L., 13
Hedén, A., 363
Hedges, A. J., 451
Hedges, R. W., 282
Hegeman, G. D., 280
Hegewald, E., 156, 157
Heidelberger, M., 279
Heijbroek, W., 334
Heimann, M., 335
Heinselman, M. L., 356,
360
Held, A. A., 150, 151, 156,
157
Hellinga, J. J. H., 327
Hely, F. W., 134
Henderson, V. E., 327
Hendrix, F. F. Jr., 251,
254
Hendrix, J. W., 215
Henis, Y., 127, 133, 410,
412, 413, 416, 419
Hepting, G. H., 357, 359
Hermans, A. K., 275, 278,
280
Hernalsteens, J. P., 275,
278, 280
Herold, F., 189, 203
Heskett, M. G., 268, 269,
272, 274
Hesler, T. L., 409, 415
Hesling, J. J., 333, 334
Heslop-Harrison, J., 154
Hess, D., 2
Heta, H., 197
Heuberger, J. W., 77
Heyn, R. F., 269
Hickey, K. D., 384
Higinbotham, N., 136, 311,

SUBJECT INDEX

A

Abacarus hystrix
  virus transmission, 195
Abbott, H., 24
Abies
  see Balsam fir
Abrego, 23
Absidia glauca, 240
Abutilon
  thompsonii, 430
  variegation, 436
Abutilon mosaic, 431
  epidemiology, 439
  mechanical transmission, 434
  physical properties, 435
Aceria
  tosichella
    virus transmission, 194-95
  tulipae, 194-96
Acetylene
  nitrogen fixation assay, 66
Acquired resistance to fungicides
  conclusions, 423
  definition, 405
  fungicide use
    application means, 422
    new fungicide development, 420-22
  genetic resistance origin, 410-11
    aromatic hydrocarbons, 411
    benzimidazole and thiophanates, 412-13
    cycloheximide, 412
    dodine, 412
    oxathiins and related compounds, 412
    selective or mutagenic action, 413-14
  in vitro resistance emergence
    chitin synthesis, 415
    energy production, 416
    fungicide comparisons, 417-18
    fungicide type, 414
    mitosis, 416-17
    nucleic acid synthesis, 415
    pathogen type, 418
    permeability, 417
    protein synthesis, 415
    sterol biosynthesis, 415
  resistance in field, 418
    disease type, 419
    resistant mutants fitness, 419-20
  resistance mechanism
    binding, 408

circumvention, 409-10
compensation, 410
conversion lack, 407-8
decreased permeability, 406-7
inactive compound formation, 407
other effects, 410
reactive site affinity, 408-9
Acridine orange, 122, 274
Actinomycetes
  ethylene sensitivity, 57
  oxygen-ethylene cycle, 60
Aegilops
  comosa
    disease resistance transfer, 226
  speltoides
    interspecific hybrids, 226
  squarrosa
    disease resistance transfer, 224
  umbellulata
    disease resistance transfer, 225
Aeneolamia varia saccharina, 362
Agaricites, 241
Agaricus bisporus
  ethylene stimulation, 56
Aggressiveness, 106-7
  definition, 104
Agrobacteriocin 1, 453
Agrobacterium
  radiobacter, 452
    bacteriocin production, 456-57
    cell wall attachment, 277
    crown gall control, 456-57
    thymidine transfer, 280
    virulence transfer, 280
  tumefaciens, 134, 453
    arginine derivitives formation in host, 295
    bacteriocin control, 456-57
    cell site attachment, 277-79
    cell wall differences, 278
    cytochromes, 296
    DNA uptake by plant cells, 282-84
    fibril release, 280
    genetic expression in plant, 284
    glycine attenuation, 275
    host-bacterium homologous DNA regions, 292
    intercellular location, 279

L-forms, 277
mRNA, 284
phages, 273-74
plant DNA hybridization, 288-90
plasmids, 274-76
plasmid transfer, 275, 281
polysaccharide, 279
RNA, 285
RNA in host, 291
virulence transfer, 280
see also Crown gall induction
Agrobacterium spp.
  relation to Rhizobium, 266
Agrocins, 452
Agropyron
  elongatum
    disease resistance transfer, 225, 227
  intermedium
    disease resistance transfer, 225
  repens
    fungal growth along root, 131
  see also Western wheatgrass
Agropyron mosaic virus
  description, 195
Agrostis
  see Bentgrass
Agrostis stolonifera
  fungal growth along roots, 131
Alandia, 20
Aleurothrixus manihotis, 436
Alfalfa (Medicago)
  disease control by burning, 364
  dodder control by burning, 368
  electrical conductivity
    Corynebacterium toxin, 316
  Pythium infecting power, 249
Alfalfa mosaic virus, 316
Allium
  see Onion
Allomyces
  Rozella penetration and papillae, 156
Alternaria
  alternata
    host resistance, 221
  brassicae, 15
  kikuchiana

mechanical, 434
spread, 439
rust sporulation and pustule
number, 110
Bean dwarf mosaic
symptoms, 430
Bean yellow mosaic virus
root exudation increase, 133
Beet curly top virus
bean symptoms, 430
Beet pseudo-yellows, 435
symptoms, 431
Beet western yellows, 369
Beet yellows, 369
Bell, 20
Belonolaimus longicaudatus
corn recovery, 346
Bemisia tabaci, 430-31, 438,
441
antifactors, 444
breeding hosts, 440
control, 442-43
disease agent persistence,
432-33
disease transmission, 433
Euphorbia disease, 435
temperature effect, 441
tomato yellow leaf curl transmission, 432
transmission effectiveness,
439
traps, 443
vector importance, 431
weed reservoirs of diseases,
439
Benecke, 20
Benomyl [Methyl-1-(butylcar-
bamoyl)-2-benzimidazole-
carbamate], 422
genetic effect, 414
resistance, 408
Sigatoka control, 18
thiabendazole cross-resistance, 413
tolerance, 410, 416
Bentgrass (Agrostis)
disease control by burning,
367
heat sensitivity, 366-67
Benzimidazole
fungus resistance, 412-14
Bermudagrass (Cynodon)
burning
disease control, 364
effect, 363-64
Bertoni, 24
Beta
procumbens
nematode resistance, 216
vulgaris, 216
see also Sugar beet
Betula
see Birch
Bews, 9
Bhendi yellow vein mosaic
crop loss, 440
source, 439

Bianchini, 22
Biopotentials
in intact plants, 318-19
measurements, 313
in plant cells, 319-20
Bipolaris sorokiniana
perfect stage importance,
253
Birch (Betula)
vigor measurement, 318
Birsiomyces, 242
Bitancourt, A. A., 20-21
Black cherry (Prunus)
electrical resistance
crown gall, 316
Blackman, V. H., 8
Blasticidin-S
fungus resistance, 419
Piricularia resistance, 406
Blitecast system, 173
Blue grama (Bouteloua)
chromosome races, 45
Bordeaux mixture
Sigatoka control, 18
Borlaug, N., 23
Botrytis
cinerea, 6-7
aromatic hydrocarbon
tolerance, 411
PCNB conversion, 407
penetration requirement,
250
Boutelous
see Blue grama
Bovista plumbea, 241
Brachycaudus helichrysi
virus transmission, 197
Brachysporium, 244
Bradley grass (Cynodon)
nematode control by burning,
370
Brassica
campestris
disease resistance trans-
fer, 221
napus, 221
see also Cabbage; Rape
Bremia lactucae
spore collection, 98
Broad bean (Vicia)
fungal migration along root,
131
Brome mosaic virus
description, 192
host range, 193-94
Bromus
disease control by burning,
366-67
Burgoa, 241
Burning
see Control by burning

C

Cabbage (Brassica)
papillae, 152-53
formation, 156

plasmalemma region, 154-
55
Plasmodiophora stimula-
tion, 150
nature, 151
Caglevic, 21
Callose
identification, 152
papillae component, 151
solute passage restriction,
154
Calluna
see Heath
Camacho, 21
Capsicum
see Pepper
Carbon dioxide, 136
importance in soil
anaerobic microsites, 66-
67
Carboxin (5,6-Dihydro-2-
methyl-1,4 oxathiin-3-
carboxanilide)
energy production inhibition,
416
resistance, 410
cross resistance, 422
genes, 412
mutants fitness, 419
Cardamom mosaic streak
virus
description, 197
occurrence, 198
Cardenas, M., 20
Carrot (Daucus)
nematode density and crop
growth, 338
tumor induction
inhibition, 286
nucleic acid, 272
Casein
TMV inhibition, 86
Cassava mosaic
disease agent nature, 435-
36
epidemiology, 439
quarantine, 441
resistance, 441
Cassytha filiformis, 12
Castanea
see Chestnut
Castro, D., 22
Cecidophyopsis ribis
host resistance, 216
Cellulose
fibrils
Agrobacterium, 280
papillae component, 153
Cephaleuros virescens, 243
Cephalosporium gramineum,
373
selection pressure, 257
Ceratocystis
fagacearum
endemic balance, 32
spread rate estimate, 398
ulmi

CUMULATIVE INDEXES

CONTRIBUTING AUTHORS VOLUMES 10-14

CHAPTER TITLES VOLUMES 1-14